MATERIALS SCIENCE RESEARCH
Volume 4

KINETICS OF REACTIONS IN IONIC SYSTEMS

MATERIALS SCIENCE RESEARCH

MATERIALS SCIENCE RESEARCH • Volume 4

KINETICS OF REACTIONS IN IONIC SYSTEMS

*Proceedings of an International Symposium on Special
Topics in Ceramics, held June 18-23, 1967,
at Alfred University, Alfred, New York*

Edited by

T. J. Gray

Director, Atlantic Industrial Research Institute
Nova Scotia Technical College
Halifax, Nova Scotia

and

V. D. Fréchette

SUNY College of Ceramics
Alfred University, Alfred, New York

Springer Science+Business Media, LLC 1969

Library of Congress Catalog Card Number 63-17645

© 1969 Springer Science+Business Media New York
Originally published by Plenum Press in 1969.
Softcover reprint of the hardcover 1st edition 1969

ISBN 978-1-4899-6224-9   ISBN 978-1-4899-6461-8 (eBook)
DOI 10.1007/978-1-4899-6461-8

Foreword

The kinetics of reactions in ionic solids is of profound importance not only in the understanding of the fundamental principles involved but also in many aspects of industrial research and development. These relationships, which are necessarily complex, cover a very broad spectrum, from the initial interaction through all stages of reaction to nucleation and growth of each individual phase. Their academic and applied implications touch on every varied aspect of solid-state reactions and merit even more attention than they have yet been accorded.

This Conference was held at Alfred University in June 1967 under the sponsorship of the U. S. Office of Naval Research and the U. S. Army Research Office, Durham, and was attended by more than 120 scientists. Ten foreign scientists attended and presented papers, including representatives of Sweden, Germany, France, the United Kingdom, Australia, and Canada. It was the fourth of a series of conferences on ceramic science. The previous conferences were titled "The Role of Grain Boundaries and Surfaces in Ceramics," at North Carolina State University at Raleigh, 1964, "Sintering," at the University of Notre Dame, 1965, and "Ceramic Microstructures," at the University of California at Berkeley, 1966.

The subdivisions of the text were determined on a very broad basis. The introductory lecture by J. G. Cohn reviews the many implications of reaction kinetics and is followed by a historical paper by the one scientist preeminently qualified to review what in many respects is autobiographical, Professor J. A. Hedvall of the Silikatforskningsinstitut, Göteborg, Sweden. A group of analytical papers follows, covering the kinetics of generalized solid-state reactions, including diffusion, reaction, nucleation, and crystal-growth kinetics. The kinetics of sintering is the subject of a second group of papers, while a third group deals with specific solid-state reactions, including the oxidation process.

We wish to express our sincere appreciation to the authors and other conference participants who made this volume possible. While preprinting for the conference was most arduous, our efforts were immeasurably assisted by C. H. Bloomquist of the State University of New York, College of Ceramics, who was responsible for the preparation of the photo-offset preprints, which materially aided the lively discussions and were subsequently of considerable assistance to the Editors in producing the final manuscript.

The cooperation of the many faculty and staff members from both Alfred University and the State University of New York, Agricultural and Technical College, who contributed unstintingly of their time and efforts to make the conference a success, is most gratefully acknowledged. The support of the Air Preheater Corporation of Wellsville, N. Y. and the Corning Glass Center of Corning, N. Y., and of Mr. W. Taylor, Jr., of the Pleasant Valley Wineries of Hammondsport, N. Y., with respect to our social activities, was appreciated by all. A special acknowledgment is due Dr. H. M. Davis of AROD, Dr. W. G. Rauch and Dr. A. M. Diness of ONR, and Dr. Cyrus Klingsberg of the National Academy of Sciences, both as representatives of sponsoring agencies and for their personal interest and assistance.

T. J. Grey
V. D. Fréchette

Alfred, N. Y.
October, 1968

Contents

Chapter 1

General Introduction: Kinetics of Reactions in Ionic Systems

J. G. Cohn

Engelhard Industries
A Division of Engelhard Minerals and Chemical Corp.
Newark, New Jersey

A review is given of studies of reactions in ionic solid systems and of the implications of these studies for industrial applications. Work on the kinetics of solid-state reaction systems is discussed, as are studies of reaction mechanisms and of the effects of process variables on product characteristics. As examples of the significance of these studies for industry the formation of ferrites and of other spinels by reaction in the solid state, the use of catalytic processes employing such solid catalysts as zeolites, and the development of batteries and fuel cells using solid-state electrolytes are described.

Due to their theoretical and practical importance numerous investigations of reactions involving the motion of the building units of ionic solids have been carried out. It might therefore be of interest to briefly highlight some of the historical aspects of such investigations in relation to current research as well as to illustrate their industrial significance by a few examples.

In the early period dating from 1912 to about 1930 a variety of solid-state reaction systems were studied ([1]). The pioneering work in this area has been carried out by J. A. Hedvall. Formation of solid solutions has been reported for such systems exhibiting complete miscibility as CoO–ZnO, CoO–MgO, CoO–MnO, CoO–NiO, NiO–MnO, NiO–MgO, CaO–CdO, BaO–SrO, and Fe_2O_3–Cr_2O_3 as well as for such systems of partial solubility as Al_2O_3–Cr_2O_3, Al_2O_3–Fe_2O_3, Fe_2O_3–Mn_2O_3, CdO–MnO, and MgO–MnO.

Many instances of actual compound formation have been described, as illustrated by the following cases. Acidic oxides react additively with basic oxides to form the corresponding salts. Table I shows examples of various interacting components, whereas Table II refers specifically to the formation of spinels by additive reactions. This type of reaction is of considerable practical interest, and, accordingly, has been extensively explored. The mechanism of some additive reactions (Fe_2O_3 + MgO, NiO, ZnO; Fe_2O_3 + Cd_2O_3) is discussed elsewhere in this volume ([2,3]). Similarly reactions

TABLE I

Compound Formation
from Acidic and Basic Oxides

Acid	Base
SiO_2	BaO
TiO_2	BeO
ZrO_2	CaO
WO_3	CdO
MoO_3	CuO
V_2O_5	FeO
Sb_2O_3	MgO
As_2O_3	NiO
	PbO
	SrO
	ZnO

TABLE II

Additive Spinel Formation

Reaction	Temperature of first noticeable reaction (°C)
$MgO + Al_2O_3 \longrightarrow MgAl_2O_4$	800
$MgO + Cr_2O_3 \longrightarrow MgCr_2O_4$	600
$MgO + Fe_2O_3 \longrightarrow MgFe_2O_4$	500
$CaO + Al_2O_3 \longrightarrow CaAl_2O_4$	800
$CaO + Fe_2O_3 \longrightarrow CaFe_2O_4$	550
$ZnO + Al_2O_3 \longrightarrow ZnAl_2O_4$	700
$ZnO + Cr_2O_3 \longrightarrow ZnCr_2O_4$	400
$ZnO + Fe_2O_3 \longrightarrow ZnFe_2O_4$	500
$CdO + Fe_2O_3 \longrightarrow CdFe_2O_4$	800
$NiO + Al_2O_3 \longrightarrow NiAl_2O_4$	1000
$NiO + Fe_2O_3 \longrightarrow NiFe_2O_4$	500
$CoO + Al_2O_3 \longrightarrow CoAl_2O_4$	550
$PbO + Fe_2O_3 \longrightarrow PbFe_2O_4$	500
$2MgO + SnO_2 \longrightarrow SnMg_2O_4$	1400
$2CaO + SnO_2 \longrightarrow SnCa_2O_4$	900

occur between oxides and salts decomposable into a solid and a gas. Again, such reactions are of considerable practical utility. Table III shows the formation of spinels by this route. This variety of additive reactions is represented in this volume by chapters on the reaction between TiO_2 and $SrCO_3$ and between SiO_2 and $CaCO_3$[4,5]. In these modern studies experimental techniques have been employed which were not available to the earlier workers.

It is of interest to note that the initial kinetic studies investigated reactions

TABLE III

Additive Spinel Formation with Reactant Decomposition

Reaction	Temperature of first noticeable reaction (°C)
$CaCO_3 + Al_2O_3 \longrightarrow CaAl_2O_4 + CO_2$	600
$CaCO_3 + Fe_2O_3 \longrightarrow CaFe_2O_4 + CO_2$	600
$SrCO_3 + Al_2O_3 \longrightarrow SrAl_2O_4 + CO_2$	900
$BaCO_3 + Fe_2O_3 \longrightarrow BaFe_2O_4 + CO_2$	650
$BaSO_4 + Al_2O_3 \longrightarrow BaAl_2O_4 + SO_2 + \frac{1}{2}O_2$	1200
$CoCO_3 + 2Al(OH)_3 \longrightarrow CoAl_2O_4 + CO_2 + 3H_2O$	840

of this kind—for example, the reaction between $BaCO_3$ and SiO_2. For reaction in powder mixtures with diffusion-controlled rates the following rate expression was derived [6]:

$$(1 - \sqrt[3]{1 - \alpha})^2 = (C/R^2)t$$

where α is the fraction of completion and R the grain radius of the minority component being surrounded by the excess component. A number of refinements of the kinetic expressions have been developed, many of which have been reviewed by Hulbert and Popowich [4]. Subsequently expressions for the kinetics controlled by nucleation or phase boundary reactions have also been derived.

In the important case of reactions between powders conditions are unavoidably nonisothermal, due to the exothermic nature of solid-state reactions and due to the low heat conductivity of the components involved. Hence these conditions affect the kinetics, as already recognized by Jander [6] in his second equation:

$$X^2 = 2 k_i t \exp(-C'X)$$

where X is the thickness of the product layer and k_i the rate constant at initiation temperature.

Another class of reaction which had been thoroughly explored in the early period is represented by base exchange of the type $Me'O + MeXO_n$. Table IV lists base exchange reactions of BaO, SrO, and CaO with various salts, and Table V shows that this type of reaction may also lead to the formation of spinels. The reactions with the alkaline earth oxides (Table IV) exhibited certain regularities. Carbonates, sulfates, phosphates, and silicates react with BaO around 350–370°C, with SrO about 100°C higher and with CaO between about 520–540°C. The method used for defining initiation temperatures was thermal analysis, which could also be utilized to determine the

TABLE IV
Reaction Temperatures of Exchange Reactions between Alkaline Earth Oxides and Salts of Oxygen-Containing Acids

Salt component	$T(°C)$ with BaO*	Reaction products	$T(°C)$ with SrO*	Reaction products	$T(°C)$ with CaO*	Reaction products
Carbonates:						
$SrCO_3$	395	$BaCO_3 + SrO$	465	$SrCO_3 + CaO$		
$CaCO_3$	345	$BaCO_3 + CaO$	455	$SrCO_3 + MgO$		
$MgCO_3$	345	$BaCO_3 + MgO$			525	$CaCO_3 + MgO$
Sulfates:						
$SrSO_4$	370	$BaSO_4 + SrO$	450	$SrSO_4 + CaO$		
$CaSO_4$	370	$BaSO_4 + CaO$	440	$SrSO_4 + MgO$		
$MgSO_4$	370	$BaSO_4 + MgO$	425	$SrSO_4 + ZnO$	540	$CaSO_4 + MgO$
$ZnSO_4$	340	$BaSO_4 + ZnO$	420	$SrSO_4 + CuO$	520	$CaSO_4 + ZnO$
$CuSO_4$	345	$BaSO_4 + CuO$			515	$CaSO_4 + CuO$
Phosphates:						
$Sr_3(PO_4)_2$	350	$Ba_3(PO_4)_2 + SrO$	450	$Sr_3(PO_4)_2 + CaO$		
$Ca_3(PO_4)_2$	340	$Ba_3(PO_4)_2 + CaO$	455	$Sr_3(PO_4)_2 + PbO$		
$Pb_3(PO_4)_2$	335	$Ba_3(PO_4)_2 + PbO$	465	$Sr_3(PO_4)_2 + CoO$	525	$Ca_3(PO_4)_2 + PbO$
$Co_3(PO_4)_2$	355	$Ba_3(PO_4)_2 + CoO$	465	$Sr_3(PO_4)_2 + Cr_2O_3$	520	$Ca_3(PO_4)_2 + CoO$
$CrPO_4$	340	$Ba_3(PO_4)_2 + Cr_2O_3$	450	$Sr_3(PO_4)_2 + Ag_2O†$	515	$Ca_3(PO_4)_2 + Cr_2O_3$
$Ag_4P_2O_7$	330	$Ba_3(PO_4)_2 + Ag_2O†$			510	$Ca_3(PO_4)_2 + Ag_2O†$
Silicates:						
$CaSiO_3$ (Wollastonite)	355	Barium silicate $+ CaO$	455	Strontium silicate $+ CaO$		
$MgSiO_3$ (Enstatite)	355	Barium silicate $+ MgO$	455	Strontium silicate $+ MgO$	560	Calcium silicate $+ MgO$
$MnSiO_3$ (Rhodonite)	355	Barium silicate $+ MnO$	465	Strontium silicate $+ MnO$	565	Calcium silicate $+ MnO$
Al_2SiO_3 (Sillimanite)	355	Barium silicate $+ Al_2O_3$	430	Strontium silicate $+ Al_2O_3$	530	Calcium silicate $+ Al_2O_3$

*T = reaction temperature.
†Dissociates subsequently into $Ag + O_2$.

TABLE V

Spinel Formation by Exchange Reaction

Reaction	"Takeoff" temperatures (°C)
$CaO + CuAl_2O_4 \longrightarrow CaAl_2O_4 + CuO$	760
$CaO + Fe_3O_4 \longrightarrow CaFe_2O_4 + FeO$	525
$SrO + CuAl_2O_4 \longrightarrow SrAl_2O_4 + CuO$	420
$SrO + ZnAl_2O_4 \longrightarrow SrAl_2O_4 + ZnO$	427
$SrO + FeCr_2O_4 \longrightarrow SrCr_2O_4 + FeO$	403
$SrO + CoCr_2O_4 \longrightarrow SrCr_2O_4 + CoO$	403
$SrO + CoAl_2O_4 \longrightarrow SrAl_2O_4 + CoO$	435
$BaO + ZnAl_2O_4 \longrightarrow BaAl_2O_4 + ZnO$	345
$BaO + FeCr_2O_4 \longrightarrow BaCr_2O_4 + FeO$	347
$BaO + CoCr_2O_4 \dashrightarrow BaCr_2O_4 + CoO$	331
$BaO + CoAl_2O_4 \longrightarrow BaAl_2O_4 + CoO$	350

degree of completion based on the extent of heat liberation and as a rough measure of reaction rate. This principle of analysis as applied to solid-state reactions has been expanded to furnish more exact data in the chapter by Campbell in this volume [7].

Although it was found nearly 45 years ago that the initiation temperatures depended essentially on the nature of the reacting basic oxide, there is still today no clearcut interpretation of the reaction mechanism. Some more recent investigations indicate that the motion of larger neutral groups, e.g., P_2O_5, along grain boundaries may be involved, but more research of these interesting systems is required. It had already been observed by Hedvall that these reaction temperature regularities did not apply when the salt component was undergoing a crystallographic transition at a temperature below the normal reaction temperature. An example is given in Table VI. Silver nitrate,

TABLE VI

Induction of Reaction by Crystallographic Transition

Basic oxide	Reactant			
	$AgNO_3$		Ag_2SO_4	
	Transition temperature (°C)	Reaction temperature (°C)	Transition temperature (°C)	Reaction temperature (°C)
BaO	160	170	411	342
SrO	160	172	411	422
CaO	160	164	411	422

which has a transition at 160°C, reacts with the alkaline earth oxides BaO, SrO, and CaO at approximately that temperature, whereas silver sulfate, with a transition temperature of 411°C, reacts with barium oxide at the regular reaction temperature of 340°C. With strontium oxide and calcium oxide reaction occurs near the transition temperature. The reaction does not only occur at the transition temperature, but reaction rates are exceptionally high and result in a yield maximum, a feature which has been shown for a number of reactions. (See also the maximum of the self-diffusion coefficient in connection with crystallographic transition cited in the chapter in this volume by Hedvall.) This phenomenon, that a lattice becomes more reactive during the occurrence of a transformation, has been designated Hedvall's rule, and also applies to other changes of the lattice such as thermal decomposition or any other mode of forming a new lattice. The reactivity of freshly formed phases is of considerable practical interest, for instance, for the promotion of sintering or of reaction.

In the middle 1920's Hedvall observed the first evidence that reactions may be carried out by the motion of lattice ions. This was concluded from the fact that reaction commenced to take place at about the same temperatures at which ionic conductivity becomes noticeable. These conditions are shown in Table VII for the reaction between barium oxide and copper halides.

Among the important developments of this period was the well-known interpretation by Wagner of the formal rate expressions in terms of the gradient of chemical potential, mobility of ionic particles in solids, and of the lattice disorder models as postulated primarily by Frenkel and by Schottky. This early work provided the foundation of our present knowledge.

In this volume discussions are presented on a variety of aspects of ionic solid systems: kinetics of reaction, of diffusion and sintering, of crystallization, of nucleation and crystal growth, of precipitation, and of the destruction of crystals by evaporation or by thermal decomposition of a solution. Unquestionably such studies will provide valuable aid in furthering the practical utilization of reactions in ionic systems. However, as pointed out in the chapter by Stringer et al.[8], theoretical models are still, in general, inadequate for

TABLE VII
$BaO + 2CuX \longrightarrow BaX_2 + Cu_2O$

Halide	Starting temperatures (°C)	
	Reaction	Conductivity
CuCl	270	260
CuBr	310	290
CuI	340	350

real cases, and more data and observations are needed as a broader basis for kinetic analysis. Real systems are usually in a state of considerable imperfection. In fact, enhanced and desirable surface activity and lattice reactivity are usually obtained by deliberately producing a solid in the form of imperfect crystals. This can be achieved by using specific preparation techniques, for example, by quenching of melts, by preparation at relatively low temperatures, by the incorporation of controlled impurities, by mechanical treatment, etc. In addition to the presence of dislocations and stacking faults in a lattice, imperfections may also occur as unstable modifications, particles of colloidal dimensions, stretched or otherwise deformed lattices, or amorphous states.

Obviously, all these structure-sensitive factors influence the reaction kinetics in practical applications to such systems as refractories, ceramics, cement, glass, luminescent materials, semiconductors, catalysts, and pigments. The reactivity of the raw mixes of the constituents, the effects of calcination or burning conditions, and the various types of diffusing species in the course of solid reactions influence the quality of the final product. From an industrial viewpoint the reactions in ionic solid systems have a universal importance. This is so vast a field that for an illustration of this significance an arbitrary selection is necessary. It is hoped, however, that the examples discussed will convey the impact of the reactivity of ionic solids on modern industrial processes and products.

FERRITES

The ferrite industry involves many aspects of the factors discussed. Ferrites of the general composition $MeOFe_2O_3$ combine the high electrical resistance of ceramic materials with magnetic permeabilities comparable with those of metallic systems. Even before the application of ferrites to solid-state electronics (computers, microwave applications, lasers), the formation of ferrites and of other spinels by reaction in the solid state had been studied extensively (Fricke, Hüttig, etc.) to elucidate phenomenologically the various stages of interaction between two solids leading ultimately to the formation of a new crystalline material. Since ferrites have become practical materials of commerce many more studies have been carried out. In this volume there are chapters on the kinetics of the formation of gadolinium garnet [3] and on the ferrites of nickel, zinc, and magnesium [2].

The composition and structural properties of ferrites vary with the end use. For square-loop ferrites the $MnO–Fe_2O_3$ combination is fundamental, together with CuO, MgO, or other oxides as well as with oxides added in "doping" concentrations. In soft ferrites (high initial permeability, low demagnetizing field, narrow hysteresis loop) manganous oxide and zinc oxide are the principal constituents; hard ferrites (high energy product) are composed of barium ferrite, hexagonal $BaFe_{12}O_{19}$. Other ferrites, for example, magnetostrictive or ferromagnetic ferrites, contain other oxides; garnets are employed for microwave applications such as yttrium or gadolinium garnet.

Figure 1 shows a typical flow sheet[9] for the manufacture of polycrystalline ferrites, the usual form used, which is analogous to the manufacture of many other ceramic materials. Regardless of individual formulation, the starting materials are mostly used in the form of salts, such as hydroxides, nitrates, carbonates, oxalates, etc., which upon thermal decomposition yield highly active oxides in order to intensify the subsequent ferrite formation.

The scientific foundation for these processes, that is, the general kinetic laws of the formation and growth of nuclei and of possible modes of aggregation, are considered in several chapters of this volume [10-12].

In the practice of ferrite production proper particle sizes are obtained by proper conditions of precipitation, drying, and calcination. Ferric oxide, for instance, prepared in less than $1\ \mu$ size and of acicular shape is more reactive than larger sized oxide of oblong shape[13]. Sizes below $1\ \mu$ are generally preferred for the other components as well. Precipitation of barium carbonate onto ferric oxide, for instance, has also been suggested to optimize the contact for the ensuing reaction in the solid state.

As an example of the effect of process variables Fig. 2 shows the influence of presintering temperature on the spinel content in the case of Mn–Zn ferrite, and also the greater rate of densification when a more reactive type of ferric oxide is being used, the density of the fired ferrite being about 4.9 [14]. The effect of presintering temperature on saturation induction is shown by

Fig. 1. Flow sheet for ferrite manufacture [Brown (9)].

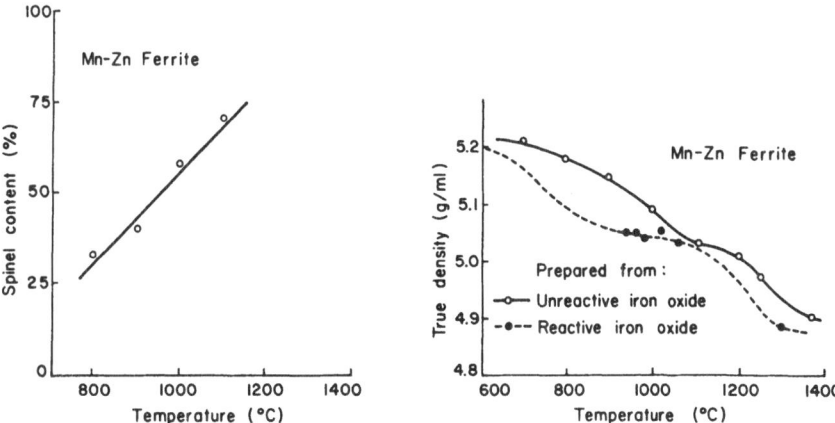

Fig. 2. Effect of presintering temperature on spinel content and density of Mn–Zn ferrite [Swallow and Jordan ([14])].

Fig. 3. Saturation induction, however, is also dependent on the atmosphere present during the ferrite production, mainly the oxygen concentration levels at various stages.

It is of interest that the Curie temperature of Mn–Zn ferrites is strictly a function of composition in accordance with the relation ([15]):

$$T_c \quad (°C) = 12.8(X - \tfrac{2}{3}Z) - 354$$

where Fe_2O_3 $(X) = 50$–55.5 mol.%, MnO $(Y) = 16.5$–35.5 mol.%, and

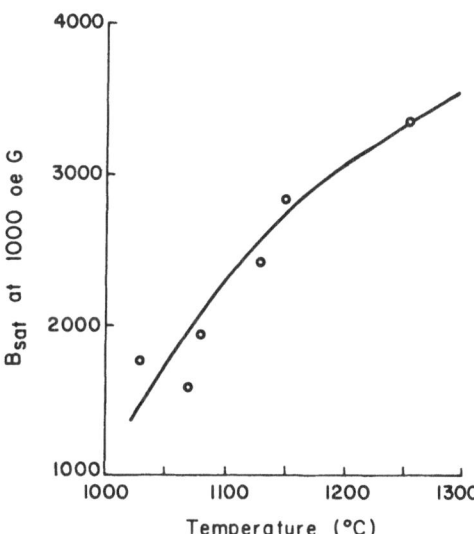

Fig. 3. Effect of presintering temperature on saturation induction of Mn–Zn ferrite [Swallow and Jordan ([14])].

TABLE VIII

Correspondence of Activation Energies*

Oxide	Activation energy (Kcal/mole)		
	Sintering†	Creep	Self (cation) diffusion
BeO	—	114	115
Al_2O_3 (Sapphire)	165–180	180	—
Fe_2O_3	123	—	120

*From Chang [21].

†The sintering data refer to bulk diffusion of cations. In the case of sintering of alumina the mechanism and the nature of the diffusing species are not clearly established and appear to depend on factors such as the content of specific impurities. Robertson and Ekstrom[18] interpret their data to indicate sintering due to surface diffusion, whereas according to the work of Johnson[19] grain-boundary diffusion seems to be predominant.

ZnO (Z) 14–28 mol.%. On the other hand the initial permeability can be varied by about one order of magnitude between 1000 and 10,000 Oe by variation of the preparation conditions.

Sintering to the final density occurs during the last stages of the ferrite formation. A substantial part of this volume is devoted to the problem of sintering and crystal growth[8,16-20] with respect to measuring techniques, to bulk diffusion, and to surface and grain-boundary diffusion. In interpreting the findings on the densification of refractory oxides Morgan[20] concluded that the rate is not controlled by bulk diffusion but by transport via dislocations. However, that at least in the initial phases of sintering ionic self-diffusion may occur is indicated by data given in Table VIII showing identical activation energies for sintering, creep, and self-diffusion of the cations in some oxides.

At any rate, in practice all the means available to influence sintering rates—for example, doping, generating defect structures, deviation from stoichiometry, adjustment of the surrounding atmosphere as, for instance, shown by Gray[22]—can be utilized in developing the desired ferrite properties.

CATALYSIS

Catalytic processing is becoming increasingly important. The National Research Council of the National Academy of Science estimated that in 1966 18% of all manufactured goods were derived through catalysis. Most new developments in petroleum, chemicals, and petrochemicals involve catalytic steps.

The majority of industrial catalytic processes are heterogeneous and employ solid catalysts. All factors affecting the reactivity of solids are important: for the preparation of highly active catalysts, for the prevention of sintering (retention of high surface areas), for the preparation of specific pore structures, for the adjustment of the surface to a suitably low value in order to control selectivity and to prevent interfering reactions which would cause deactivation of the catalyst, or to provide hardness to the catalyst in order to avoid attrition. Although preparative techniques are usually proprietary, much useful information can be found in the patent literature. There is a vast quantity of information on the use of thermal decomposition to prepare active catalysts, on the use of reactions between solids or the formation of solid solutions to produce catalysts, on the incorporation of promoters or structural stabilizers, on the generation of lattice defects by doping, and on the introduction of structural changes in a catalyst. In certain cases, for instance, catalytic oxidation or hydrogenation, the reaction mechanism may involve a change of the valence state at local surface sites. These considerations apply to promotion as well as to the inhibition which may be desirable in order to enforce selectivity of the reaction path.

An example of catalysts comprising ionic solids are zeolite-containing catalysts, which in recent years have been introduced for such important industrial processes as cracking, isomerization and polymerization, hydrocracking, Fischer–Tropsch reaction, and others. Zeolites, "molecular sieves," are crystalline aluminosilicates possessing very open, often rigid anionic frameworks composed of $SiAlO_4$ anions. The net electronegative charge of these frameworks is neutralized by the electrochemically equivalent number of cations located within cavities of the framework. The intracrystalline space may furthermore be occupied by water molecules. The variation of the silica–alumina ratio with the corresponding change in the cations required for electroneutrality permits flexibility in designing a proper catalyst or catalyst support. The group KAl, for instance, would be equivalent to Si, the group CaAl to NaSi. For catalytic action to be possible the internal channels should be large enough for all reacting molecules to enter or leave the catalyst structure, that is, in general, the diameters should be about 10 Å or larger, in contrast to the sizes required for the use of zeolites as molecular sieves.

Faujasites are of particular importance for industrial catalysis. As do all zeolites, they exhibit ionic conductivity, by motion of the cations in the internal channel structure. The cationic mobility is considerable, as shown by Table IX. The conductivity of zeolites is greater than that of most other crystals. At a temperature of 25°C the diffusion coefficient of the sodium ion in hydrated zeolite is only three orders of magnitude smaller than that of simple ions in aqueous solution. The conductivity increases with temperature; at 335°C the conductivity of the anhydrous zeolite is greater than that of the

TABLE IX
Diffusion Coefficients at 25°C*

Ion	D
Ions in aqueous solution	$\sim 10^{-5}$
Na^+ in Faujasite (Mol. Sieve X), hydrated	$\sim 10^{-8}$
Na^+ in Mol. Sieve X, anhydrous	$\sim 10^{-12}$
Na^+ in Na-Analcite	$\sim 10^{-13}$
Cs^+ in Cs-Analcite	$\sim 10^{-24}$
Ca^{2+} in α-$CaSiO_3$	$\sim 10^{-42}$
Ca^{2+} in α-Ca_2SiO_4	$\sim 10^{-77}$

*From Barrer [23].

hydrated zeolite, the latter being 10^4 times greater at room temperature. The activation energies for conduction are dependent on ionic radius and charge, the divalent anions having larger activation energies than the monovalent ions [24] (see Fig. 4).

The possibility of incorporating cations of varying sizes and charges into zeolites allows the generation of specific catalyst activity. In the hydrated state cations are solvated by water molecules. In the anhydrous state, obtained upon heating to about 335°C, zeolites show a remarkable temperature resis-

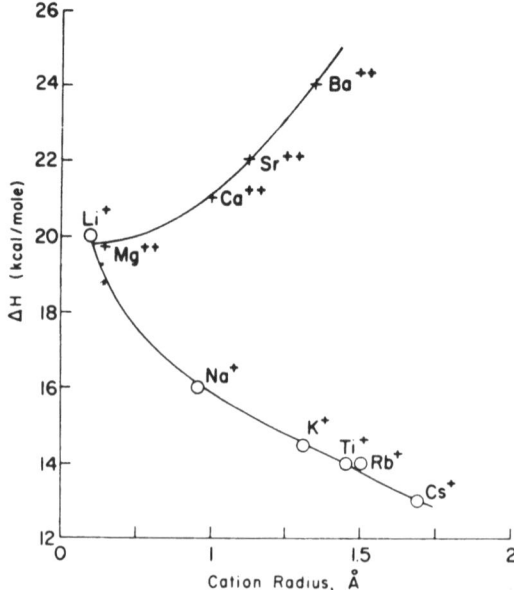

Fig. 4. Activation energies for cation conduction in zeolite [Breck [24]].

TABLE X

Catalytic Cracking: Conventional Silica–Alumina Catalysts versus Metal-Acid Faujasites*

		Catalyst		
	Silica–alumina	$Al_2O_3 \cdot 2.5\ SiO_2$		
		0.7 CaO	0.67 MnO	0.28 Rare earth
Conditions:				
LHSV†	1	10	16	16
Catalyst/Oil	6	0.6	0.38	0.38
% Conversion	50	54.7	53.1	65.4
C_{5+} Gasoline (vol. %)	35.5	48	48	52.3
C_{4+} (vol. %)	12.5	9.9	8.4	14.7
Advantage over silica-alumina:				
C_{5+} Gasoline (vol. %)	—	+8.1	+9.3	+7.1
C_{4+} (vol. %)	—	−4.4	+5.3	−3.5
Dry gas (wt. %)	—	−3.1	−2.6	−2.1
Coke (wt. %)	—	−1.0	−1.9	−2.1

*From the data of Plank et al. [25].
†LHSV = Liquid Hourly Space Velocity.

tance of the crystallinity. Temperatures of 700°C are required to destroy the crystal structure. These properties make them useful as catalysts.

The performance of zeolite catalysts is illustrated for the case of catalytic cracking of naphtha in Table X. The operation refers to the cracking of a Mid-Continent gas oil having a boiling range (ASTM) of 497–750°F. The cracking obtained with conventional silica–alumina cracking catalyst is compared with the cracking over Ca, Mn, and rare-earth zeolites. The activity of the zeolites is much greater than that of conventional catalyst as expressed by the higher space velocities and lower catalyst to oil ratios. Indeed, for the Mn^{2+} or rare-earth activated zeolite the activity ratio is about 100. Besides having greater activity, the yield on C_5 gasoline, the most valuable product, is substantially greater than with the conventional catalyst, as shown by the "advantage" section. Less C_4 and dry gas is produced, which is no particular penalty since the value of these products is less than the value of motor gasoline. The zeolites produce considerably more coke than the conventional silica–alumina catalyst; however, the zeolites remain operative at coke levels which would completely deactivate conventional cracking catalysts. This is possibly due to a much higher concentration of acidic sites in the zeolites than in silica–alumina. The greater activity and selectivity make the zeolite catalysts far superior to conventional catalysts. Accordingly, the majority of catalytic cracking is carried out nowadays with zeolite-based catalysts.

ELECTROCHEMISTRY

A deeper insight into electrochemical reaction mechanisms is possible by electrochemical studies employing solid electrolyte instead of liquid electrolyte [26]. With a solid electrolyte having preponderantly only one mobile ionic species electrode polarization can be studied under thermodynamically well-defined conditions without superimposed side effects by solvents and without the complications created by the presence of hydrated films or hydrolytic layers. Such measurements can be used, for instance, for the study of electrodeposition, formation of monolayers or of dendrites due to nucleation, for the study of polarization phenomena in ionic solids, solid-state reaction kinetics, transport phenomena, thermodynamics or constitutional diagrams, and for the development of practical devices.

A variety of practical applications are conceivable. Two developments of recent date may serve to illustrate this.

Batteries. A device based on an ionic solid electrolyte is the recently announced sodium–sulfur secondary battery, which is shown schematically in Fig. 5. This battery employs a ceramic electrolyte which is described as being a modification of $Na_2O \cdot 11Al_2O_3$ having a layer structure in which sodium ions occupy relatively open planes together with bridging oxygen ions, resulting in a structure with nearly optimum mobility for the sodium ions. This electrolyte has a specific resistivity of about 6 ohm-cm at a temperature of 300°C. High power densities are possible with an 0.8-mm thickness of the electrolyte structure. Figure 6 shows the electrical performance of this cell. If this battery could be developed into a practical form it would have much higher power density than that obtainable with conventional batteries;

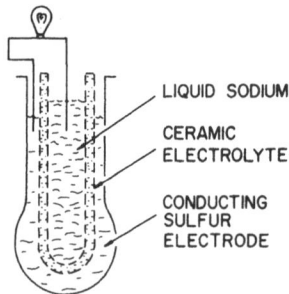

LIQUID SODIUM

CERAMIC ELECTROLYTE

CONDUCTING SULFUR ELECTRODE

Fig. 5. Secondary sodium–sulfur galvanic cell with solid-state electrolyte [Kummer and Weber [27]].

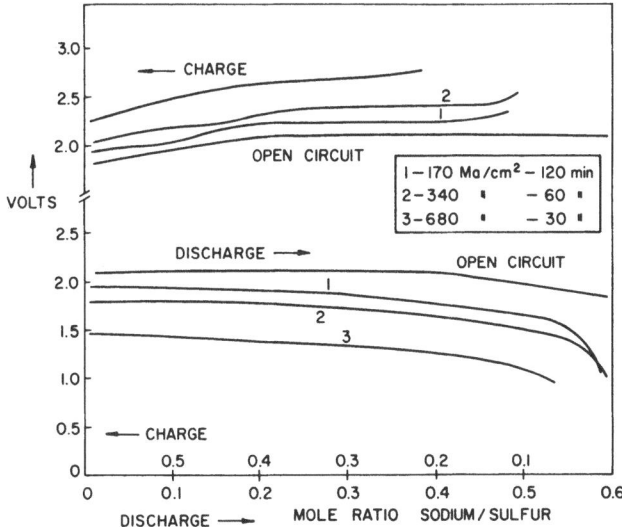

Fig. 6. Performance characteristics of sodium–sulfur solid electrolyte secondary cell [Kummer and Weber [27]].

for instance, the sodium–sulfur battery for a 5-hr discharge rate would have 15 times the power density of a conventional lead-acid battery.

Fuel Cells. A considerable effort is presently being expended by several research groups to develop fuel cells employing a solid-state electrolyte[28-31]. The electrolyte so far discovered for this purpose is ZrO_2 stabilized in its cubic high-temperature modification by incorporation of minor amounts of calcium oxide, magnesium oxide, or rare-earth oxides such as the oxides of Y, Yb, Ce, and La. In this condition ZrO_2 becomes a fair conductor in the temperature range 750–1000°C with a transport number of unity for the oxygen ions. The resistivity at 1000°C is of the order of 50–100 ohm-cm. Since practical fuel cells must employ oxygen (air) as oxidant, ion conduction is required for fuel cell operation.

A basic configuration of a solid-state fuel cell is shown in Fig. 7. The thickness of the electrochemically active section of the electrolyte is 0.04 cm. Experimental fuel cells have been produced with this particular structure by stacking individual cells into series of rows operating at 1000°C on H_2–CO mixtures as fuel. A process design for a 100 Kw coal-burning fuel cell power system has recently been described[33] which is projected, based on present technology, to operate with an overall efficiency approaching 60%. With further refinements higher efficiencies could be reached.

Many technical difficulties have still to be overcome before solid-state electrolyte fuel cells become practical power plants. In particular, the task

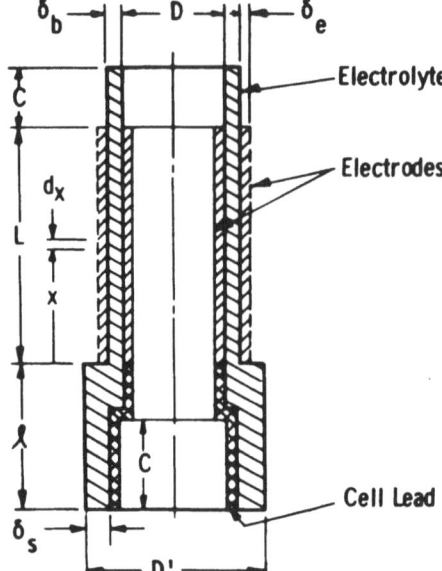

Fig. 7. Solid-state bell-and-spigot type fuel cell unit [Archer et al.[32]].

of developing other oxide electrolytes having better anionic conductivity than the zirconia electrolytes presently used represents a challenging problem for the research workers concerned with the properties of solid ionic systems.

However, a solid-state fuel cell device operated as high-speed response indicator for oxygen in gases has been announced to be commercially available suitable for monitoring the O_2 content in flue gases [34].

These remarks may suffice to indicate the extent to which industrial uses of ionic solids depend on the outcome of scientific research. In conclusion a statement by Frechette[35] may be paraphrased that "progress made in general theory by experimental studies in model systems, supplemented by experimental studies with practical systems offers the greatest promise for advancement in the fabrication of improved or of new industrial products."

REFERENCES

1. J. A. Hedvall, *Reaktionsfähigkeit fester Stoffe*, Leipzig, 1938; *Einführung in die Festkörperchemie*, Braunschweig, 1943; G. Cohn, *Chem. Rev.* **42**: 527 (1948).
2. P. A. Venkatu and G. C. Kuczynski, this volume, Chapter 17.
3. E. A. Giess and R. M. Potemski, this volume, Chapter 29.
4. S. F. Hulbert and M. J. Popowich, this volume, Chapter 25.
5. M. R. Montierth, R. S. Gordon, and I. B. Cutler, this volume, Chapter 32.
6. W. Jander, Z. Anorg. Chem. **163**: 1 (1927) and **166**: 31 (1927).
7. W. B. Campbell, this volume, Chapter 6.
8. R. K. Stringer, C. E. Warble, and L. S. Williams, this volume, Chapter 4.

9. C. S. Brown, *Proc. Brit. Ceram. Soc. No. 2*, p. 55 (1964).
10. P. W. M. Jacobs, this volume, Chapter 3.
11. W. J. Dunning, this volume, Chapter 7.
12. R. H. Campbell and M. O'Keeffe, this volume, Chapter 24.
13. P. Erzberger, *Proc. Brit. Ceram. Soc. No. 2*, p. 19 (1964).
14. D. Swallow and A. K. Jordan, *Proc. Brit. Ceram. Soc. No. 2*, p. 1 (1964).
15. E. Röss and E. Moser, *Z. Angew. Phys.* **13**: 247 (1961).
16. H. J. Oel, this volume, Chapter 13.
17. R. H. Condit, this volume, Chapters 15 and 20.
18. W. M. Robertson and F. E. Ekström, this volume, Chapter 14.
19. D. L. Johnson, this volume, Chapter 18.
20. C. S. Morgan, this volume, Chapter 19.
21. R. Chang, *Fifth Nuclear Eng. Sci. Conf. Preprint V*, Cleveland, 1959, p. 109.
22. T. J. Gray, *J. Am. Cer. Soc.* **37**: 378 (1954).
23. R. M. Barrer, *Proc. Brit. Ceram. Soc. No. 1*, p. 145 (1964).
24. D. W. Breck, *J. Chem. Educ.* **41**: 678 (1964).
25. C. J. Plank, F. J. Rosinski, and W. P. Hawthorne, *Ind. Eng. Chem., Product Res. and Dev.* **3**: 165 (1964).
26. D. O. Raleigh, *J. Phys. Chem.* **71**: 1785 (1967).
27. J. T. Kummer and N. Weber, *Soc. Automotive Engineers Preprint No. 670179* (1967).
28. D. H. Archer, L. Elikan, and R. L. Zahradnik, in: *Hydrocarbon Fuel Cell Technology* (B. S. Baker, ed.) Academic Press, New York, 1965.
29. H. Tannenberger and H. Siegert, *Am. Chem. Soc. Div. Fuel Chem. 154th National Meeting, Preprints* **11** (3): 197 (1967).
30. D. T. Bray, L. D. La Grange, U. Merten, and C. D. Park, U.S. Patent No. 3,300,344.
31. T. Takahashi, *Denki Kagaku (Japanese Electrochemical News and Patents)* **1** (6): 5 (1967).
32. D. H. Archer, R. L. Zahradnik, E. F. Sverdrup, W. A. English, L. Elikan, and J. J. Alles, *Proc. 18th Annual Power Sources Conf.* 36 (1964).
33. D. H. Archer and R. L. Zahradnik, *Am. Chem. Soc. Div. Fuel Chem. 154th National Meeting, Preprints* **11** (3): 212 (1967).
34. W. M. Hickam and J. F. Zamaria, *Instruments and Control Systems* **40** (Aug): 87 (1967).
35. V. D. Fréchette, *Trabajos Reunión Intern. Reactividad Solidos* 2(3) Madrid, 1956, p. 189.

Chapter 2

A Historical Review of Solid-State Chemistry, Mechanisms, and Corrosion Problems

J. A. Hedvall

Silikatforskningsinstitut
Göteborg, Sweden

A discussion is presented of problems associated with the corrosion of nonmetallic materials and of aspects of solid-state reactions related to corrosion processes. The effects of the various characteristics of surfaces and of attacking agents are considered, and the kinetics and some possible mechanisms of solid-state reactions are briefly reviewed. The effects of transition states, through their influence on reactivity, and of extreme environmental stresses are noted, as are epitaxial effects and effects produced by adsorbed gases. Emphasis is placed on the need for further research on problems of the corrosion of such materials as glasses, ceramics, plastics, and natural and synthetic stones, as well as on the need for interdisciplinary cooperation to help combat these problems.

Every kind of interaction between a solid phase and other solid, liquid, or gaseous phases naturally begins at surfaces (external or internal). The mechanism of all type of physicochemical processes at such boundaries will be influenced by the properties of the more or less unsaturated surface particles, their dipole moments, their migration capacity, and the conditions of electronic exchange between the phases in question. This also holds true for the rather badly defined group of processes which are broadly described as corrosion (Figs. 1–3).

In practice corrosion investigations have generally dealt with such pro-

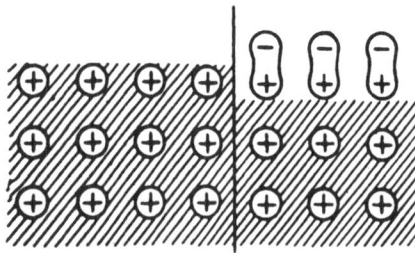

Fig. 1. Surface and surface molecules (From J. H. de Boer, *Elektronentheorie und Adsorptionserscheinungen*, Barth, Leipzig, 1937.)

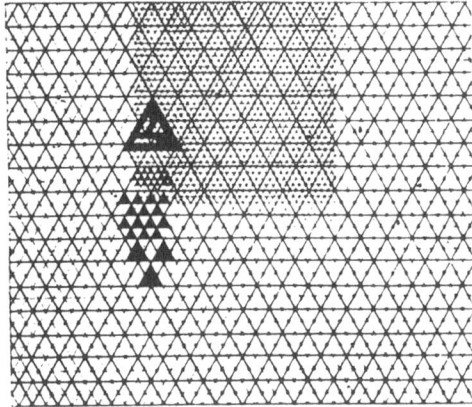

Fig. 2. Surface mosaic structure (Goetz). (From J. A. Hedvall, *Einführung in die Festkörperchemie*, Vieweg, Brunschweig, 1952, p. 36.)

cesses as the corrosion of copper, iron, and alloys. It is natural that most of these experiments have been carried out under conditions simulating those found in actual service. However, the extremely important problems of corrosion of nonmetallic materials have not so far attracted investigation to the same extent (Fig. 4). Thus mineralography is as necessary as her older sister metallography.

Considering the great variety of materials and production methods used, the multiplicity of applications, and the great differences in climate and local conditions of service it is impossible in a brief survey to present more than the principles of research achieved or projected. Limiting this chapter

Fig. 3. Corrosion (black) at phase boundaries.

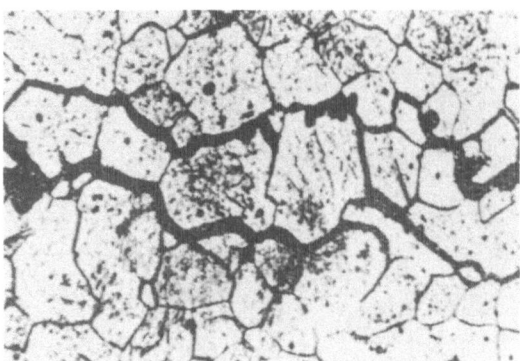

Fig. 4. Corrosion at phase contacts of concrete.
[From Hedvall([2]).]

to that extent, some experiences will be mentioned from the author's work in solid-state chemistry ([1-4]) and some research will be cited from other laboratories. Only limited mention will be made of the work carried out in the U.S.A., since this work and its applications are generally well known.

First of all it must be remembered that the different processes covered by the term corrosion are chemical reaction starting at surfaces in the form of adsorption or chemisorption, a concept which implies the formation of something like molecules which may be called surface molecules or moleculoids (Fig. 1). In studying such processes it is therefore necessary to take into account the many peculiarities of surfaces, e.g., individual properties depending on the chemical composition, deviations from strict stoichiometric formulas, lattice defects, the still very imperfect studies of the state of crystallographic transition in the surface, the thermal conditions, etc. In the second place the nature of the attacking agents must be considered. Problems of this

Fe $Fe O$ $FeO + Fe_3O_4$ Fe_3O_4 Fe_2O_3 O_2

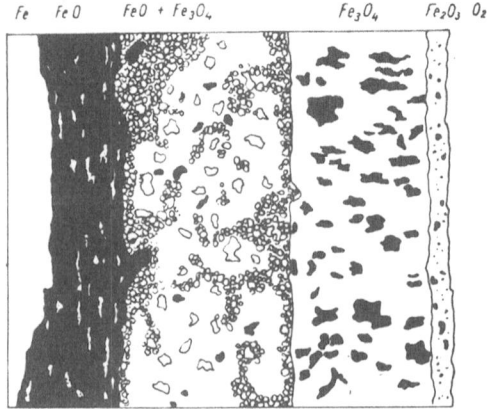

Fig. 5. Oxidation of iron [Jost ([6])].

type have been studied by a number of scientists. Wagner was one of the first to show that when metals are attacked by ambient gases, e.g., oxygen or sulfur, a combined migration of metal ions and electrons toward the gas takes place [5]. This is a very complex process, especially if more than one product is formed, as for example in the oxidation of iron [6] (Fig. 5).

At a symposium in Darmstadt in 1928 the mechanisms of reactions in the solid state were discussed in detail for the first time. In studies of the reactions in mixtures of oxides and copper or silver halides at the Chalmers Institute it was qualitatively demonstrated that there was a connection between the ionic conductivity of a salt and its reactivity[7]. Some years later Wagner calculated quantitatively the yield of the reactions in such systems[8].

In the early twenties a new reaction type was observed, at that time called exchange reactions; for example,

$$BaO + CaSiO_3 = BaSiO_3 + CaO$$

where, in some mixtures, rapidly occurring surface processes start at such low temperatures that there is little possibility of either electronic or ionic conductivity. [[9], also [4], p. 8].

At the same symposium it became clear that it was necessary to calculate reaction kinetics considering at least two types of mechanism, surface migration and lattice diffusion, $dx/dt = C$ and $dx/dt = C/x$, respectively. It was also emphasized that each of these formulas must be regarded as the sum of

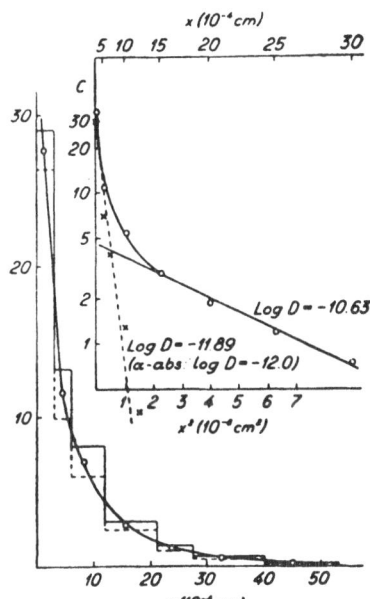

Fig. 6. Diffusion process in $PbSiO_3$ measured by a tracer method; lower part (linear) total amount, upper part the measurements split up in surface and lattice diffusion (logarithmic). [From Hedvall[2].]

a number of partial processes, changing locally and structurally. By means of tracer elements it was possible to distinguish between surface transport and lattice diffusion ([10]) (Fig. 6).

Some years later the complex nature was further illustrated by the investigations by Jagitsch and his collaborators at the Chalmers Institute. In studying a great number of place exchange reactions it was shown that sometimes uncharged particles in the form of "moleculoids" can play an important part in the transport of matter in solid state [([11]), also ([2]), p. 110].

A number of other results from these investigations of the reactivity of solids may be mentioned in this short survey. Even if they are not primarily concerned with corrosion processes the results are important because there is a natural connection between real reactions starting at surfaces and processes involving adsorption, catalysis, passivation, and overall mechanism. This fact is worth mentioning because even scientists sometimes overlook the. close connection between apparently separate phenomena. It is urgent that such an approach be adopted because of the appearance of a rapidly growing group of new materials, new applications, and new techniques. Once more it is strongly emphasized that the problems of the corrosion of substances

Fig. 7. Damage to black roof tiles from differences in chemical composition between tiles and roof material, damage being facilitated by the heat-absorbing black color. [From Hedvall([2]).]

Fig. 8. Damages in the Etruscian Tomba dei Rilievi (Cerveteri) caused by the attack of percolating solutions forming thin $CaCO_3$ films and then thick layers mellowing the tomb material. (From J. A. Hedvall, *Chemie im Dienst der Archaologie.*)

Fig. 9. Oak logs mouldering in the air from the lakeside fortified village of Biskupin.

other than metals have not received sufficient attention. The damage to buildings, archaeological specimens, and ancient monuments, for instance, have a claim to modern research. There is often little if any cooperation between the experts on these topics and chemists. The personnel at excavation sites and the staffs of museums usually have little or no knowledge about the specific nature of their materials, the mechanism of weathering or biological attacks as, for example, on marble and textiles, or about the many forms of corrosion to which objects of metals, carbonates, and siliceous products are more or less susceptible. Often the situation is the same in many countries in relationship to architects and building contractors (Figs. 7–9).

Synthetic stone materials such as concrete products, plastics (with regard to which special mention should be made of the great many silicone resins that recently have been developed for use at high temperatures), and many types of glass are increasing in use and demand detailed consideration. The silicone resins are already widely used in electrical insulation. It often seems to be forgotten that there are innumerable kinds of glass, ceramics, and plastics with individual properties influencing their interaction with contacting materials (Fig. 10). In this connection it is necessary to suggest that not only the synthetic materials but also, and to no lesser extent, natural stones such as granite, marble, limestone, sandstone, etc. exist with greatly different qualities. The differences are due to variable chemical composition, structure, grain size, amount and kind of accessory minerals, etc. All of these substances are more or less heterogeneous, with a great number of phase boundaries and surfaces sensitive to attacks by ambient media. Consequently in many cases there is no equilibrium in the interior. At elevated temperatures, often in "cooperation" with vibrations, there is a trend to establish equilibria, causing processes facilitating the attacks by surrounding agents. As an instance of this, mention may be made of damage to building elements of lightweight

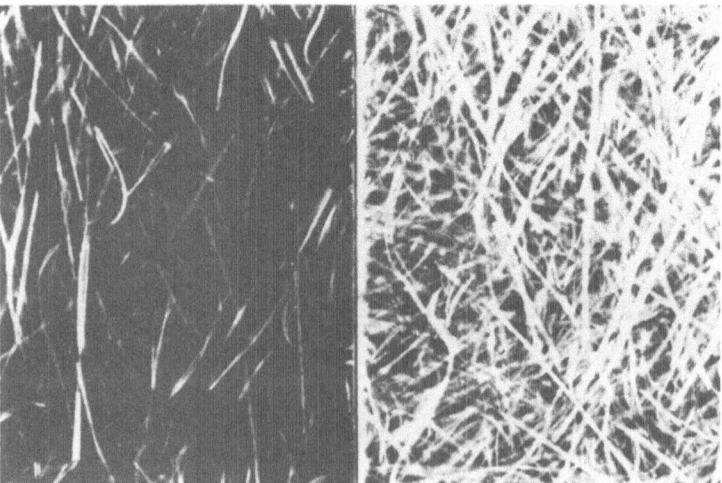

Fig. 10. Destruction of glass-fiber-reinforced plastic (left) starting at the contact between glass and fiber. Undamaged material is shown at right. (From J. A. Hedvall, *Chemie im Dienst der Archaologie.*)

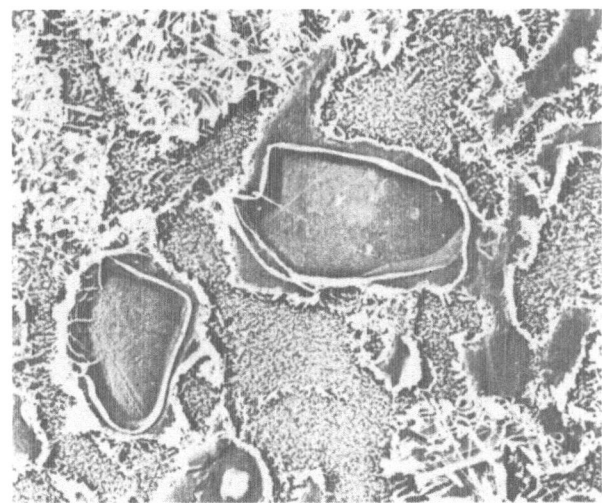

Fig. 11. Damage caused by vibrations may take place and can be observed in many silica products.

concrete and other silica products caused by vibrations (Fig. 11). Another example is the so-called glass disease, where crystallization and chemical

Fig. 12. The lower part of the vase is damaged by interior processes ("weathering" and crystallization). The upper part is restored by sucking out these products in vacuum and impregnating with a plastic preparation having the same optical properties as the glass.

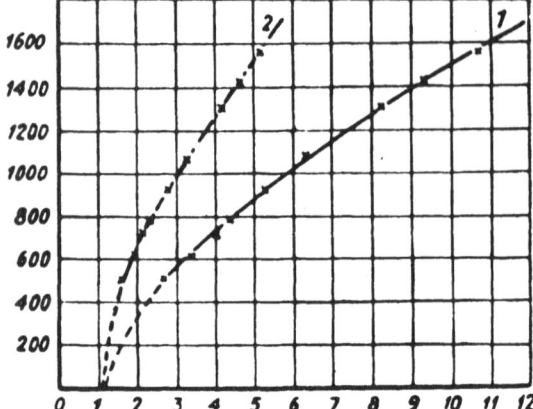

Fig. 13. Effect of ultrasonic treatment (Curve 2) on the tarnishing of Cu. Time, 0–12 min, thickness of tarnishing film in microns.

reactions are caused by the cooperative effect of weathering factors and vibration (Fig. 12).

In this connection it may be pointed out that the coating of metals with adherent insulating films can be facilitated by subjecting the metal to ultrasonic vibration. This technique is used for insulating special types of electrical transformers (Fig. 13).

Some experiments which are at present in progress in the author's laboratory will now be discussed. They concern the possibility that metal wires carrying electric currents or metals producing a field, as in transformers, may exhibit an activity toward surrounding media different from that of

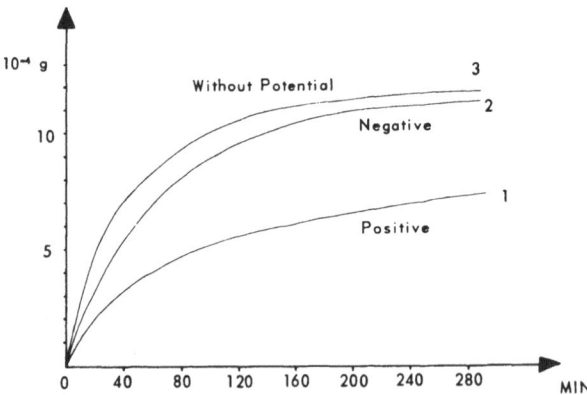

Fig. 14. Influence of an electric field on rate of corrosion of silver.

metals in an uncharged electric state, a possibility that appears plausible because reactions or interaction processes are generally connected with some type of electron exchange. These experiments derive from observations on high-voltage power lines. It has been established that the expected influence exists and that it takes place also at low potentials. These experiments are proceeding. Silver and copper rods were placed in a small tube through which a current of air with a low concentration of H_2S (0.001 atm, 25°C) was flowing. A potential of 100 V was applied to the rod with (1) a positive charge of 100 V, (2) a negative charge of 100 V, and (3) no charge (Fig. 14). Because of the dipole moment the results were as expected.

It is now a well-known general rule that all kinds of transition states in solids involve an increase of their physicochemical activity and this implies that a state of equilibrium cannot exist during the formation of new phases, especially when the rate of the transformation or decomposition process is so high that the lattices of the phases formed show imperfections of structure. The effect of transition states appears also in measuring the adsorption capacity on heating, e.g., pholerite (Fig. 15), and also during the transformation interval of glass in measuring the capillary constant([12]). In this connection it is interesting to mention that Lindner([10]), carrying out diffusion measure-

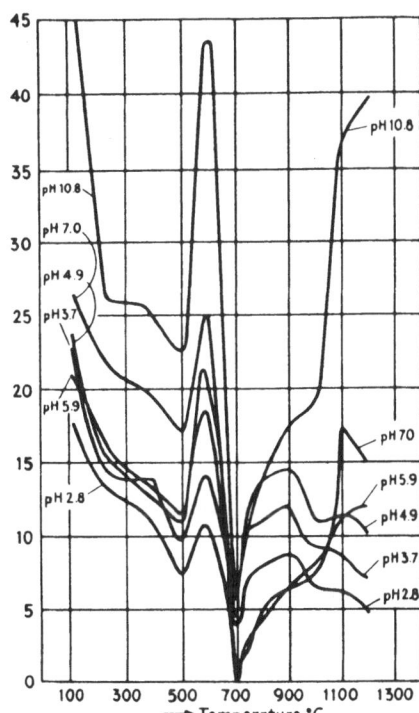

Fig. 15. Adsorption of methylene blue at different pH values as dependent on the preheating temperature of pholerite.

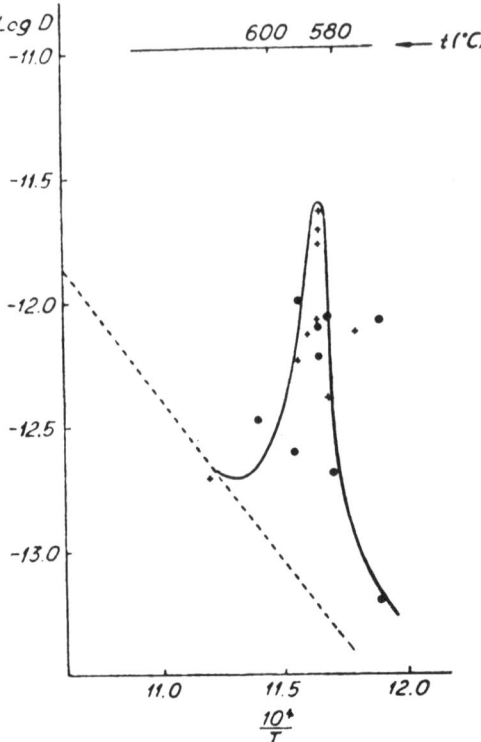

Fig. 16. Influence of crystallographic change on diffusion in
$PbSiO_3$ [after Lindner[10]]. [From
Hedvall[4].]

ments, found a change of crystallographic structure of $PbSiO_3$ (Fig. 16).

Naeser, of the Mannesmann Research Institute, has recently published a summary of the results obtained in experiments with powders which had been ground in a special type of grinding machine to an extremely small grain size. The surface to mass ratio was high, and the surface reaction $(dx/dt = C)$ consequently played a more important role than in ordinary powder mixtures. The results represent some rather surprising phenomena originating, at least in part, from the influence of this factor[13] (Fig. 17).

In a survey of this kind one must also refer to the numerous problems that arise in the construction of equipment for rocket propulsion units and for atomic reactors, where new metallic materials are subjected to conditions of extremely high stress. These questions were given special attention at the international congress of powder metallurgy in Reutte (1961) and some very interesting results were then disclosed from various investigations in progress, of which only two instances will be given here. Goetzel and Landler[14] stated that the most practical method of protecting tungsten against oxidation at 1815°C was to place it in direct contact with a siliciding agent in order to produce a silicide coating. Protective layers of zirconium oxide with intermediate layers of tungsten silicide were of no particular advantage. In another paper Muller[15] elucidated the migration of tungsten particles on the surface of tungsten crystals. The transport of matter at a surface is also of basic importance in corrosion processes. The heavy corrosion on containers used for storing radioactive waste from reactors also presents entirely new problems. Here also it must be remembered that the corrosion attack is of quite another type.

Turning to another area, mention will be made of the use of tin cans for various preserved foods. The "tin" may not always be made of tin plate; there are other metals and alloys currently in use. In a recent publication from the Applied Research Laboratory of the U. S. Steel Corporation, R. P. Frankenthal[16] has described some most interesting results regarding corrosion of these materials. It must be remembered that cold-working exerts a considerable influence on the susceptibility of a metal to corrosion; such effects have been known for quite a time. A great number of results from investigations into the role of deformation and surface treatment have been published in late years.

Coming back to the effects of transition processes, it may be mentioned that the ductile modification of tin is transformed into the brittle modification at temperatures below about 15°C. Although normally a very slow process, the rate of transformation is accelerated by vibrations, temperature fluctuations, and the presence of some liquids. This process also includes an increased rate of oxidation and is popularly known as "tin pest" and "bronze pest" (Figs. 18 and 19).

Every kind of change of structure when new phases are formed affects

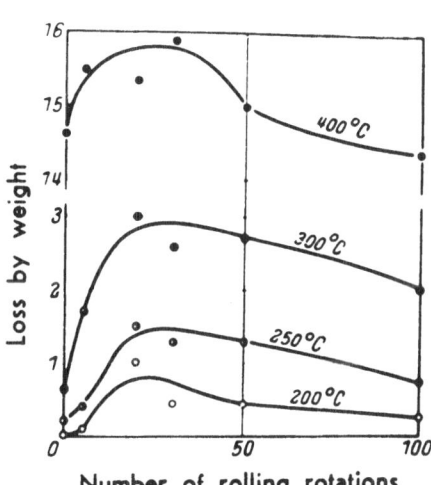

Fig. 17. The influence of increasing the surface/mass ratio on the reduction of Fe_2O_3 at different temperatures [Naeser[13]].

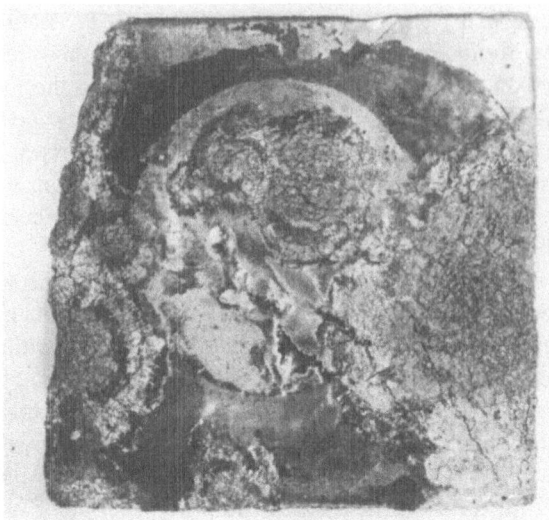

Fig. 18. Piece of tin partly attacked by "tin pest."
(From J. A. Hedvall, *Chemie im Dienst der Archaologie.*)

the reactivity. Some forty years ago the Dutch chemist Cohen observed that changes in the surface structure occur in copper, bismuth, and antimony at very low temperatures. No other metals were investigated. The transition of copper at 70°C and bismuth at 75°C do not correspond to any change of the lattice structure. In order to determine whether such transitions involving the surface structure also exert an influence on the physicochemical activity experiments were conducted with thin sheets of these metals. Figure 20 illustrates the fact that the role of increased activity is valid in these cases. These results have been put to use in some powder metallurgical processes at relatively low temperatures. Doubtless there are other metals showing simi-

Fig. 19. "Bronze pest" corrosion on piece of antique bronze caused by transition processes and phase boundary attacks. (Belonging to the collection of the King of Sweden.)

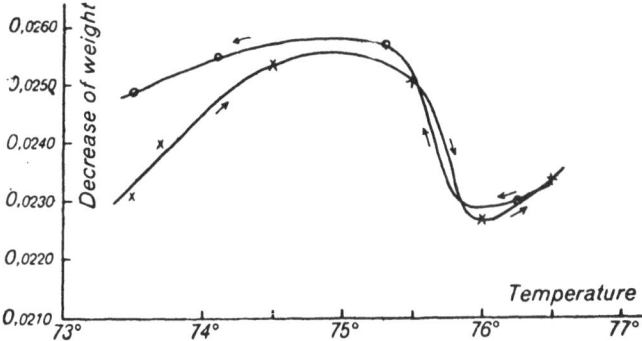

Fig. 20. Reaction of bismuth with a solution of HNO_3. Time of reaction $\frac{1}{2}$ hr.

lar effects. It has also been established that there is a marked susceptibility to corrosion at the temperatures where such changes of structure take place [(4), p. 52]. It should be mentioned that very preliminary studies on the role of binding substances has been undertaken, e.g., Co in cutting tools. It is

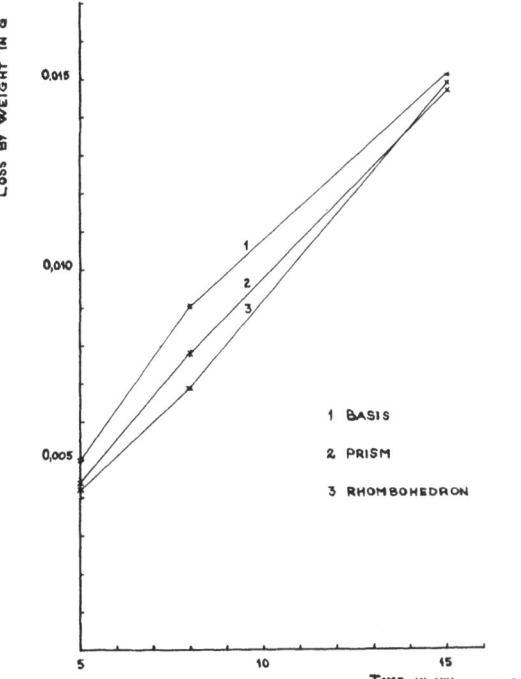

Fig. 21. The influence of different crystal surfaces on the velocity of thermal decomposition of $CaCO_3$ at varying duration of the experiments and at 600°C.

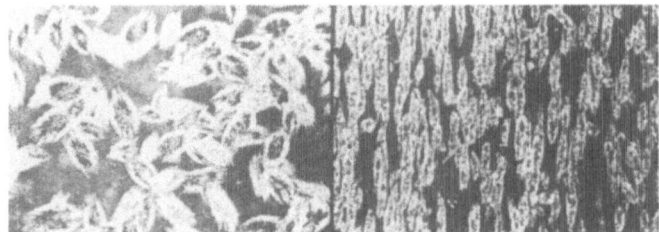

Fig. 22. Structural orientation of absorbed substances on different surfaces of the support. (From H. Seifert, *Structure and Properties of Solid Surfaces*, University of Chicago Press, 1953, p. 319 ff.)

likely that it would be possible to achieve informative results by using the electron microprobe analyzer for diffusion observations.

Interest is being shown in the problems of topochemistry and topotaxis. An early example was the measurement of the different rates of thermal decomposition on different crystallographic surfaces of calcium carbonate. Such processes do not depend solely on geometric differences as was once believed [(4), p. 40]. They are related, as are other phase boundary processes, to the energy state of crystallographically different surfaces and to differences in the electron exchange propensity. It has been mentioned that the effects of different surfaces of crystallites may be observed, especially in metallic systems, even though they are chemically but not structurally homogeneous (Fig. 21).

The concepts of supports and epitaxis must naturally be considered

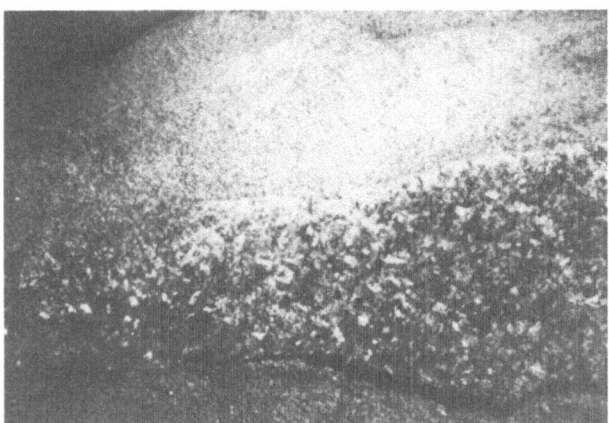

Fig. 23. Contact between two different granites. The lower part of the picture represents the species with larger grain size and higher concentration of accessory minerals. Because of these factors it weathers more rapidly through the attack of plants.

when discussing surface activity problems. Epitaxis is an interaction between different phases, solid–solid, solid–liquid, or solid–gaseous, and was observed more than a century ago. The phenomenon could also be roughly described as structural resonance between different phases and it implies that the adsorption or crystallization of one phase on the other depends on an orientation specific to the surface. In addition to normal lattice structures defects of one type or another may play an important role. Recent studies are to be found in the transactions of the symposia on catalysis in Philadelphia, 1956, in Paris, 1960, and in Amsterdam, 1964, and further reports are given in papers by Seifert and by Schwab and Seifert ([17]) (Fig. 22).

In connection with epitaxial effects attention should be drawn to the comparatively new field of biocrystallography. This field is of special importance in regard to the many cases of corrosion caused by attacks by different species of plants and microbes. It is very often overlooked that the first steps in damage to buildings and archaeological objects depend on such pro-

Fig. 24. One of the Mexican pyramids extensively damaged by biological attacks.

Fig. 25. Rock paintings in Sahara from different epochs ("palimpsests") exposed to attacks by wind erosion and plants. (From L'Hote, *A La Découvert des Fresques di Tassili, p.134.*)

cesses, for instance the attacks by the acids produced by lichens (Figs. 23–25).

Other phenomena concern the influence of ambient gases which are not corrosive in the usual sense of the term, an influence which has also proved to be of considerable interest in relation to corrosion and especially to the problems of passivation. A long series of experiments was performed in which it was established that the reactivity and the adsorption capacity of a great number of substances, perhaps of all substances, is considerably influenced by chemically inactive gases. Small quantities of gas may be dissolved, as, for instance in silica and alumina, first in the outermost layers and then gradually penetrate to deeper levels.

The problems of passivation should now be considered. The formation of a protective oxide film is assumed, for instance, in the case of iron and aluminum. This explanation does not appear satisfactory, as such "surface oxides" are generally of more or less faulty constitution and consequently not very resistant to chemical attack. The phenomena related to passivation are complicated and present different aspects according to the properties of the substances and the methods employed in inducing passivation. It seems plausible to assume that in many cases gases such as oxygen or nitrogen are dissolved in the metal, forming thin protective barriers. A general principle of passivation is to render difficult the transport of surface particles.

Such gas effects have to be considered in the electric lamp industry and in the case of semiconductors which are now used in numerous devices susceptible to corrosion. Sparney and Ruler ([18]) of the Philips laboratories in Eindhoven recently published a paper on this subject, describing a new method of studying the surface electrical properties of semiconductors. The method depends on conducting resistivity measurements for pressures of oxygen ranging from 10^{-9} to 10^{-2} mm Hg adsorbed on thin germanium crystals. By this method the roles played by adsorbed gases can easily be distinguished. These authors establish a close relationship between the adsosption of gases and the electrical properties of the surface. They emphasize that the influence of adsorbed or dissolved gases is of great interest and far too little investigated, particularly in relation to corrosion and passivation. At the Chalmers Institute Sandford [19] has found some interesting results which confirm the great influence of dissolved gases. Upon the heating of Fe_2O_3 to about 1000°C in O_2 at 100 atm presure the oxide still retained its beautiful orange color but it did not react with CaO. Normally the bright color is lost at 850°C and such preparations react at about 500°C with CaO.

A number of other factors influencing surface activity could be mentioned, for example, guest particles, generally loosening up the lattice, and also irradiation, which is very active in cases of biological attack.

It is hoped that this short survey of an area of chemistry of extreme theoretical and practical importance has given some glimpses of what we know and do not know. It is the current task in present investigations to establish a much better contact between, on one the hand physical chemistry, crystallography, and biology, and on the other hand building techniques, archaeology, and "museology." The costs of neglecting such cooperation are enormous and to a large extent both unnecessary and out of fashion.

REFERENCES

1. J. A. Hedvall, *Reaktionsfähigkeit fester Stoffe*, Ambros. Barth, Leipzig, 1938 and Edward Brothers, Ann Arbor, 1943.
2. J. A. Hedvall, *Einführung in die Festkörperchemie*, Vieweg, Braunschweig, 1952.
3. J. A. Hedvall, *La Chimie des Solides*, Univ. Libre Press, Bruxelles, 1954.
4. J. A. Hedvall, *Solid State Chemistry, Whence, Where, Whither*, Elsevier Publ. Co., Amsterdam 1966.
5. C. Wagner, *J. Corrosion* (Sept.–Oct.) p. 9 (1948); *J. Metals* (Feb.) p. 214 (1952); *Z. Elektrochem.* **66**: 502 (1962).
6. W. Jost, *Diffusion und Chemische Reaktionen in fester Stoffe*, Steinkopf, Dresden, 1937.
7. J. A. Hedvall and E. Gustafsson, *Z. Anorg. Allgem. Chem.* **170**: 71 (1928).
8. C. Wagner, *Z. Phys. Chem.* **B36**: 321 (1936).
9. J. A. Hedvall and J. Heuberger, *Z. Anorg. Allgem. Chem.* **122**: 181 (1922).
10. R. Lindner, *Acta Chem. Scand.* **5** (2): 735 (1951).
11. R. Jagitsch and O. Perlstrom, *Arkiv Kemi* vol. **22A**, No. 5, (1946), (Ver. Akad. Stockholm).
12. J. A. Hedvall, *Z. Glass u. Hochvacuumtechnik*, Vol. 3/4 (1952).
13. G. Naeser, *Z. Kolloid* **156**: 1 (1958).
14. C. G. Goetzel, and P. Landler, 4th. Intern. Plansee Seminar 1961 and 1964.
15. W. Muller, 4th Intern. Plansee Seminar 1961.
16. R. P. Frankenthal, P. R. Carter, and A. N. Laubscher, *J. Agr. Food Chem.* **7**: 441–2 (1959).
17. H. Seifert *et al.*, *Structure and Properties of Solid Phases*, Univ. Press, Chicago 1953,

p. 318; *Chem. Eng. Techn.* **33**: 210 (1961); G. M. Schwab *et al.*, *Angew. Chem.* **71**: 101 (1959).

18. M. J. Sparnay, *Koninkl. Ned. Acad. Wetenschap. Proc. Ser. B.* **66**: 64, 70 (1963).

19. F. Sandford and B. Liljegren, *Trans. Chalmers Univ.* No. 282 (1963).

Chapter 3

Kinetics of the Thermal Decomposition of Solids

P. W. M. Jacobs

Department of Chemistry
University of Western Ontario
London, Ontario, Canada

The kinetics of the thermal decomposition of solids are reviewed, with emphasis on topological considerations. The general model of nucleation in the bulk of the reactant is explored in detail and the kinetic equations appropriate to this model are derived. It is pointed out that a multistage nucleation process leads to a power law whenever the characteristic time for nucleus formation is long compared with the observation time, and that the assumption of equal rate constants for successive steps is unnecessarily restrictive. The problem of the induction period is examined and two possible reasons for the critical time t_0, namely the use of an incorrect model, and time-dependent growth rates (including, as a special case, aggregation without chemical decomposition) are advanced. Finally, the consequences of nucleation only on the surface of the reactant are mentioned briefly.

INTRODUCTION

The kinetics of the thermal decomposition of solids are dominated by topochemical considerations. Thus the traditional concepts of order and molecularity, which play an important role in the kinetics of gas phase and liquid phase reactions, have little application in considerations of the kinetics of the thermal decomposition of solids. Experimental observations of the isothermal reaction are conveniently represented in the form of a plot of the fractional decomposition α against time t. The objective of a physicochemical study of the thermal decomposition of a solid is then the devising of a mechanism for the chemical reaction which can be formulated in mathematical terms leading to a theoretical representation of the $\alpha(t)$ curve which is in complete agreement with the experimental observations. Such an idealized procedure is seldom realized completely in practice and one often has to be content instead with an empirical analysis of the kinetics.

Although the general form of the $\alpha(t)$ curves can appear to vary widely from one reaction to the next [1] all these forms are special cases of the generalized decomposition curve [2] shown in Fig. 1. This shows the following

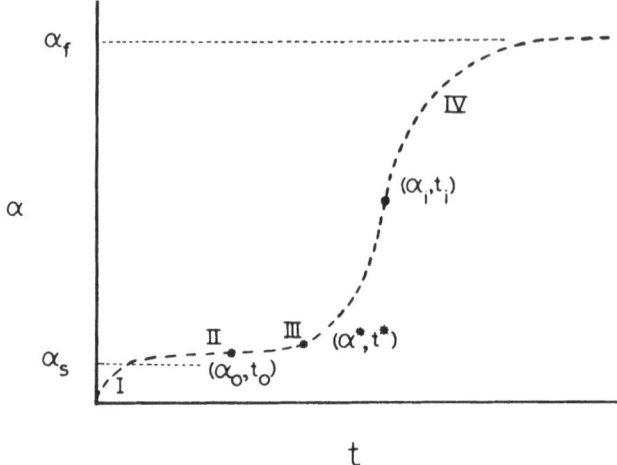

Fig. 1. General form of the $\alpha(t)$ decomposition curve. Here α_s represents the extent of the initial process (I), t_0 marks the end of the induction period (II), and (α_i, t_i) are the coordinates of the point of inflection separating the acceleratory period (III) from the decay period (IV). The significance of (α^*, t^*) is discussed in the text.

features. (I) represents an *initial process* which involves typically $0.01 < \alpha < 0.05$. The gas evolution during (I) may be due to desorption of physically adsorbed gases or to a limited amount of thermal decomposition, e.g., a reaction which involves only a few atomic layers near the surface of the reactant. The extent of reaction during the *induction period* (II) is very limited, but after some critical time t_0 the reaction rate increases rapidly during the *acceleratory period* (III), finally reaching a maximum value at the point of inflection, α_i, t_i. For $t > t_i$ the reaction rate decreases steadily during the *deceleratory period* (IV), finally falling to zero as the reaction ceases. The fractional decomposition α_f at this stage may or may not correspond to almost complete decomposition of starting material. In at least one remarkable material, ammonium perchlorate (AP), thermal decomposition (at low enough temperatures) leaves a residue which is itself AP and which constitutes approximately 70% of the initial reactant.

The above phenomenological aspects of the kinetics of the thermal decomposition of solids can receive a formal explanation in terms of the concepts of the *formation* and *growth* of *nuclei* [3,4]. The essential basis of this theory is that chemical reaction is confined to the interface between the solid reaction product and the undecomposed reactant. If the rate of this interface reaction is either constant or a unique function of the area of the interface, then the kinetics of the reaction will be governed by the geometrical area of the reaction interface.* It is presumed, and experimental observation in favorable cases confirms, that decomposition commences at certain selective points called nucleus forming sites. The reaction then propagates by growth of these nuclei. The surface area of the interface between the nuclei and the reactant matrix, and hence the reaction rate, is thus controlled by the laws governing the formation and growth of the nuclei. Below we distinguish certain straightforward cases of special importance. Since nucleus growth is essentially a three-dimensional phenomenon it may be isotropic or nonisotropic. If the length that a nucleus grows in a particular direction in time Δt is Δl_j, then

$$\lim_{\Delta t \to 0} (\Delta l_j / \Delta t) = G_j \tag{1}$$

is the rate of nucleus growth in the jth direction. Growing nuclei are, in general, characterized by three different growth functions G_1, G_2, and G_3, corresponding to growth along three principal (not necessarily perpendicular) crystallographic directions. At least for large (microscopically observable) nuclei experimental evidence where available suggests that G_1, G_2, and G_3 are constant, that is, independent of the size of the nuclei. The extrapolation of this law to nuclei of submicroscopic size may not, however, always be justified.

NUCLEATION

If there are N_0 potential nucleus-forming sites and the number of these formed in time t is n_1, then if each of the N_0 sites has the same a priori probability of being converted into a nucleus,

$$dn_1/dt = k_0(N_0 - n_1)$$

or

$$n_1 = N_0[1 - \exp(-k_0 t)] \tag{2}$$

which is the law of random nucleation at a finite number of sites. If the characteristic time for nucleus formation $\tau \, (= 1/k_0)$ is very short compared with the time of observation, then the nuclei are formed effectively instantaneously and so $N = N_0$. Conversely, if τ is much larger than the time of the whole experiment then

---

*In the special case of the low-temperature decomposition of AP, which forms only gaseous products, the residue apparently plays the role of a solid product.

$$dn_1/dt \approx k_0 N_0 \tag{3}$$

which represents a linear law of nucleus formation.

In the above simple analysis it has been implicitly assumed that a single molecular decomposition results in a growth nucleus. This assumption may or may not be justified. In phase transitions, which are subject to much the same laws of nucleus formation and growth as solid-state decompositions[5], small ("germ") nuclei will be unstable[6] and may either reform the reactant phase or form stable growth nuclei by acquiring new molecules of product. In solid-state decompositions the free energy change associated with the chemical transformation is generally too large for the reaction to be reversed and so nucleus formation is irreversible. It does not necessarily follow, however, that a single molecular decomposition will produce a growth nucleus. The most obvious example is perhaps that in which the solid product is a metal. The energy levels in a single atom, and in a group of atoms large enough to be considered a metal, will be quite different, and thus the catalytic properties of nuclei containing very few product atoms and those consisting of a large number of atoms may also be different. (It is as well to remark that the mechanism of the interface reaction has not really been elucidated in every detail for any known reaction.)

The kinetics of nucleus formation in which an active growth nucleus requires the accumulation of at least r product atoms has been considered by Bagdassarian[7,2]. A nucleus with less than this number of product atoms is a germ nucleus, which may become a growth nucleus by adding atoms of product in successive reaction steps. If n_i denotes the number of nuclei containing i atoms then

$$dn_i/dt = k_{i-1}n_{i-1} - k_i n_i \tag{4}$$

with $n_i = 0$ for $i < 0$, $n_i = 0$ for $i > 0$ at $t = 0$, $n_i = N_0$ for $i = 0$ at $t = 0$, and $n_i = 0$ for $i = 0$ at $t = \infty$. The solutions to the set of coupled differential equations (4) are for $r = 0, 1, 2$

$$n_0 = N_0 e^{-k_0 t} \tag{5}$$

$$n_1 = \frac{k_0 N_0}{k_1 - k_0}[e^{-k_0 t} - e^{-k_1 t}] \tag{6}$$

$$n_2 = \frac{k_1 k_0 N_0}{(k_1 - k_0)(k_2 - k_0)(k_2 - k_1)}$$
$$\times [k_2 - k_1)e^{-k_0 t} - (k_2 - k_0)e^{-k_1 t} + (k_1 - k_0)e^{-k_2 t}] \tag{7}$$

Solutions for higher values of r may be obtained by straightforward mathematical techniques, but the expressions for n become somewhat lengthy. When $k_i t \ll 1$ for all i we may expand the exponentials in (6) and (7) to obtain

$$n_1 = k_0 N_0 t \tag{8}$$

and

$$
\begin{aligned}
n_2 &= \frac{k_1 k_0 N_0}{(k_1 - k_0)(k_2 - k_0)(k_2 - k_1)} \\
&\quad \times \left[(k_2 - k_1)\frac{k_0^2 t^2}{2!} - (k_2 - k_0)\frac{k_1^2 t^2}{2!} + (k_1 - k_0)\frac{k_2^2 t^2}{2!} \right] \\
&= K_2 N_0 t^2
\end{aligned}
\tag{9}
$$

with $K_2 = k_1 k_0/2!$.

Thus the number of growth nuclei (by definition a growth nucleus cannot contain zero product atoms) vary linearly with time for $r = 1$ and as the square of the time for $r = 2$. These formulas may be considerably simplified by the approximation $k_0 = k_1 = k_2 = \ldots k_r$, in which case the set of equations (4) yields the solutions

$$n_0 = N_0 e^{-k_0 t} \tag{5}$$

$$n_1 = N_0 (k_0 t) e^{-k_0 t} \tag{10}$$

$$n_2 = N_0 \frac{(k_0 t)^2}{2!} e^{-k_0 t} \tag{11}$$

and in general

$$n_r = N_0 \frac{(k_0 t)^r}{r!} e^{-k_0 t} \tag{12}$$

or

$$n_r = N_0 \frac{(k_0 t)^r}{r!}, \qquad \text{for} \quad k_0 t \ll 1 \tag{13}$$

Bagdassarian's model for the formation of growth nuclei by product accumulation in successive steps is thus capable of explaining the general power law of nucleation

$$n_r = K_r N_0 t^r \tag{14}$$

It should be noted, however, that the assumption that all the k's are equal

TABLE I

Expressions for $n_r(t)$, the Number of Growth Nuclei which Contain r Product Atoms

I. Single step nucleation	
a. Random nucleation at N_0 sites	$n_1 = N_0[1 - e^{-k_0 t}]$
b. "Instantaneous" nucleation	$n_1 = N_0$
c. "Slow" nucleation ($k_0 t \ll 1$)	$n_1 = k_0 N_0 t$
II. *a.* $k_r t \ll 1$	$n_r = K_r N_0 t^r$
b. $k_r t$ not $\ll 1$	Complex formulas, e.g. Eq. (7)
c. $k_r t$ not $\ll 1$, but	$n_r = N_0 \dfrac{(k_0 t)^r}{r!} e^{-k_0 t}$
$k_i = k_0$ for $0 < i \le r$	

until a growth nucleus is formed is hardly consonant with the idea of multiple step nucleation, and it is, in fact, dictated more by the desire for compact mathematical formulas than by physical reasonableness. Fortunately, the limiting case $k_r t \ll 1$ for all r is not only physically reasonable but also often realized in practice and this leads to a power law [e.g. Eq. (9)] without the need for unsatisfactory assumptions.

The laws of nucleation are summarized in Table I.

Reactions in which the number of growth nuclei formed in a given time have been observed experimentally are rather few in number, and for many decompositions the law of nucleus formation has been inferred from the kinetics rather than determined by direct observation. Notable exceptions are barium azide[8]: $n_r \propto t^3$; nickel sulphate heptahydrate[9]: $n_r \propto t^2$; and copper sulphate pentahydrate [10] and chrome alum [11]: $n_r \propto t$.

GENERAL KINETIC EQUATIONS

If the experimental observations of a constant growth rate for visible nuclei can be extrapolated right back to the stage when germ nuclei ($i < r$) first become growth nuclei ($i = r$), then the development of a general kinetic equation from the known or assumed laws of nucleation and growth is a straightforward matter provided nucleation is random and the growth rate constant. It is apparent that certain of the N_0 potential nucleus-forming sites will never become active growth nuclei because they have already been consumed by the growth of other nearby nuclei. Similarly growing nuclei will eventually impinge on one another and the material common to two or more nuclei can, of course, decompose only once! Such impingement is the cause of the decay period in the fractional decomposition curve.

The ingestion of phantom nuclei by growth nuclei and the overlap of growing nuclei can be allowed for in a general way by utilizing the concept of the extended fractional decomposition, α_{ex}. Here α_{ex} is simply the fractional decomposition which would have occurred in time t had ingestion and overlap not occurred. Because both nucleation and growth are random in the sense that topochemically equivalent elements of reactant all have the same probability of decomposition, in a given time increment dt

$$d\alpha = d\alpha_{ex}(1 - \alpha) \tag{15}$$

This equation clearly satisfies the appropriate boundary conditions: $d\alpha/dt = d\alpha_{ex}/dt$ at $\alpha = 0$, $d\alpha/dt = 0$ but $d\alpha_{ex}/dt$ finite at $\alpha = 1$. A trivial integration ($\alpha_{ex} = 0$, $\alpha = 0$ at $t = 0$) yields

$$-\ln(1 - \alpha) = \alpha_{ex} \tag{16}$$

The extended fractional decomposition is given by the expression

$$\alpha_{ex} = \frac{V(t)}{V(\infty)} = \frac{\sigma}{V(\infty)} \int_0^t \int_u^t G_1(x')\, dx' \int_u^t G_2(y')\, dy' \int_u^t G_3(z')\, dz' \left[\frac{dn}{dt}\right]_{t=u} du \tag{17}$$

where $V(t)$ is the volume of all growth nuclei at time t, σ is a shape factor (e.g., $4\pi/3$ for spherical nuclei), $G_1(t)$ is the growth function for the x direction [that is, the rate of growth in the x direction at time $t = x'$ is $G_1(x')$], and dn/dt is the rate of formation of growth nuclei.

For constant growth rates (17) reduces to

$$\alpha_{ex} = \frac{\sigma}{V(\infty)} \int_0^t G_1 G_2 G_3 (t - u)^3 \left[\frac{dn_r}{dt}\right]_{t=u} du \tag{18}$$

For multiple-step nucleation with $k_r t \ll 1$ Eq. (14) holds, $n_r = K_r N_0 t^r$ and Eq. (18) becomes

$$\alpha_{ex} = \frac{\sigma}{V(\infty)} \int_0^t G_1 G_2 G_3 (t - u)^3 r K_r N_0 u^{r-1}\, du$$

$$= \frac{6\sigma G_1 G_2 G_3 K_r N_0 t^{r+3}}{(r + 1)(r + 2)(r + 3)V(\infty)} \tag{19}$$

Combining (19) and (15) gives the Erofeev equation[12,13]

$$-\ln(1 - \alpha) = (kt)^n \tag{20}$$

with k, the rate constant, given by

$$k^n = (6\sigma G_1 G_2 G_3 K_r N_0)/[(r + 1)(r + 2)(r + 3)V(\infty)] \tag{21}$$

and the exponent n equal to $r + 3$. For small values of α we have $\alpha \approx \alpha_{ex}$, so that the early stages of the acceleratory period ought to be described by a simple power law. If the growth rate for one direction is either very fast or much slower than those for the other two directions, then this growth rate does not control the reaction kinetically. For example, if $G_3 \ll G_1, G_2$, then the nuclei would grow two-dimensionally while remaining effectively of constant thickness d_3; in this situation Eq. (20) would still hold but with $n = r + 2$ and

$$k^n = (2\sigma G_1 G_2 d_3 K_r N_0)/[(r + 1)(r + 2)V(\infty)] \tag{22}$$

Similarly if $G_3, G_2 \ll G_1$, then growth in one dimension predominates and

$$k^n = (\sigma G_1 d_2 d_3 K_r N_0)/[(r + 1)V(\infty)] \tag{23}$$

with $n = r + 1$.

It is therefore clear that in the absence of information derived from direct observation of the nuclei ambiguities exist concerning the interpretation of the exponent n which is the sum of two terms, r the number of molecular decompositions required to form a viable growth nucleus and λ the number of dimensions of active nucleus growth. There are interpretative difficulties about assuming that nuclei grow much more slowly in one or two directions in that a very much larger number of nuclei are required to ensure consumption of the reactant than would be so for three-dimensional growth. If $G_3 \gg G_1, G_2$, however, the nuclei would grow very rapidly in one direction and thereafter steadily in the other two directions. Thus $\lambda = 2$, but the difficulties regarding consumption of the reactant do not obtain. Finally we note that Ic and Ib in Table I are formally special cases of the general problem which we have considered (IIa): this means that the same formulas hold with $r = 1$ ($K_1 \equiv k_0$) and r equal to zero ($K_0 \equiv 1$), respectively.

The other important limiting case is that of random nucleation with constant growth rates, where $k_r t$ is not $\ll 1$. We consider only single-step nucleation, case Ia, because of criticisms already leveled againt case IIc, multiple-step nucleation with equal rate constants for each step. The general case IIb is also soluble but leads to a complex formula of the same general type as that now to be derived for case Ia. The general expression (18) for α_{ex} becomes

$$\alpha_{ex} = \frac{\sigma}{V(\infty)} \int_0^t G_1 G_2 G_3 (t - u)^3 k_0 N_0 e^{-k_0 u} \, du \tag{24}$$

On integrating and using (16) we have

$$-\ln(1 - \alpha) = \frac{6\sigma G_1 G_2 G_3 N_0}{V(\infty)k_0^3} \left\{ e^{-k_0 t} - 1 + k_0 t - \frac{(k_0 t)^2}{2!} + \frac{(k_0 t)^3}{3!} \right\} \tag{25}$$

which we shall call the Avrami equation. This analysis suggests that over the early part of the acceleratory period ($\alpha \ll 1$) if $k_0 t \ll 1$

$$-\ln(1 - \alpha) = (kt)^4 \approx \alpha \qquad (26)$$

a result already contained in (20) and (21), since $r = 1$ ($K_r \equiv k_0$). In the decay period if $k_0 t \gg 1$ then (25) reduces to

$$-\ln(1 - \alpha) = (kt)^3 \qquad (27)$$

which is again the Erofeev equation corresponding to a fixed number of nuclei ($r = 0$, $K_r \equiv 1$). Exponents < 4 or 3, respectively, would result from nucleus growth being kinetically controlling in < 3 dimensions, as already explained.

In summary, the above analysis suggests that the kinetic equation of most practical use ought to be the Erofeev equation (20), $-\ln(1 - \alpha) = (kt)^n$, where n might have any integral value from 1 to $r + 3$. Strictly, when $k_0 t$ is comparable with 1, the Avrami equation (25) ought to be used instead, but it is doubtful if experimental data of sufficient precision exist to make this a worthwhile exercise at present. In addition we note that the assumptions that the shape and the growth rates are independent of nucleus size (σ, G_1, G_2, G_3 not functions of t) may not always be justified.

THE INDUCTION PERIOD

Equation (20) is tested by plotting $F(\alpha) \equiv [-\ln(1 - \alpha)]^{1/n}$ against t for various integral powers of n and selecting that value for n which gives the most linear plot. When this is done it is frequently found that the Erofeev equation (20) does not hold down to $t = 0$, that is,

$$F(\alpha) = [-\ln(1 - \alpha)]^{1/n} = k(t - t_0) \qquad (28)$$

where t_0 provides a quantitative measure of the induction period. If $\alpha_0 = \alpha(t_0)$ is not negligibly small then α in Eq. (28) should be replaced by $\alpha' = (\alpha - \alpha_0)/(1 - \alpha_0)$.

It is not generally realized that this may be due, at least in part, to the approximation made in replacing

$$e^{-k_0 t} - 1 + k_0 t - \frac{(k_0 t)^2}{2!} + \frac{(k_0 t)^3}{3!}$$

of the Avrami equation by $(kt)^n$. When $k_0 t$ is of order unity this is a poor approximation, and yet the resulting deviations would be hard to detect using data of the precision normally obtained. In Fig. 2 plots of

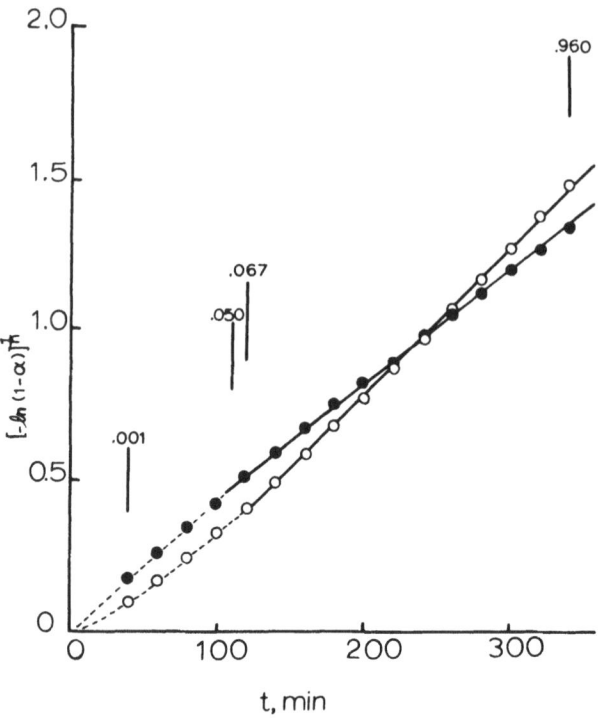

Fig. 2. Plots of $[-\ln(1 - \alpha)]^{1/n}$ against t for $\alpha(t)$ data which satisfy the extended Avrami equation (25). The data are approximately fitted by straight lines over a substantial range of α, as shown by the vertical lines on the figure.

$[-\ln(1 - \alpha)]^{1/n}$ are shown for $n = 3$ and $n = 4$ using $\alpha(t)$ data generated from Eq. (25) with

$$k_0 = 1 \times 10^{-2} \quad \text{min}^{-1} \quad \text{and} \quad C = \frac{6\sigma G_1 G_2 G_3 N_0}{V(\infty)k_0^3} = 1$$

It is observed that the data are well fitted by straight lines over the range $0.06 < \alpha < 0.96$. Experimental data will, of course, show some scatter, making it even more difficult to decide whether or not Eq. (28) is a good representation of the data. For the data used in Fig. 2 for $n = 3$ we have $t_0 = -8$ min and for $n = 4$ we have $t_0 = +39$ min. It would therefore be an extremely difficult problem to verify the kinetic equation appropriate to the physical situation k,t not $\ll 1$.

Unfortunately one cannot simply use the existence of a finite value for t_0 as a criterion for distinguishing between Eqs. (20) and (25), since this may be due to other causes. It has been shown by Young [15] that the coalescence of a cluster of small nuclei can account for a negative value of t_0 in Eq. (28) for the special case of the decomposition of neutron-irradiated silver oxalate. Positive values of t_0 have traditionally been explained in terms of the concept of slow growth [16,17]. The idea that small nuclei might grow less rapidly than ones which are lärger than a certain critical size is an attractive one. Very small nuclei will form a coherent interface with the reactant matrix, the misfit being accommodated initially by elastic strain. The strain will clearly increase with the size of the nucleus and if the free energy of activation depends significantly on the strain energy a variation of $\Delta G\ddagger$ with nucleus size is to be expected. Thus the growth rates G_1, G_2, and G_3 will be functions of $(t - u)$, where u is the time at which a nucleus becomes an active growth nucleus. Thomas and Tompkins [16] applied the idea of a variable growth rate in its simplest form by using the approximation of two constant growth rates

$$G_1(t - u) = G_1 \qquad (u < t < u + t^*)$$
$$G_1(t - u) = G_1' \qquad (t > u + t^*) \tag{29}$$

where $G_3 = G_2 = G_1$ (spherical symmetry) was further assumed. Jacobs and Kureishy [17] considered a somewhat more sophisticated, but still empirical, model for the decomposition of nickel oxalate in which G_1 was assumed to be proportional to either the perimeter or the surface area of a nucleus, provided it was below a critical size, with G_1 constant thereafter. The kinetic data fitted Eq. (28) with $n = 2$, and so two-dimensional growth from a fixed number of nuclei formed early in the reaction was assumed. Since the nuclei are all formed together at $t = 0$, $u = 0$, and they also all reach critical size approximately) together at $t = t^*$, the fractional decomposition at this stage being α^* (cf. Fig. 1), the two versions of this model led to an exponential law

$$\ln \alpha - \ln \alpha_0 = k(t - t_0) \qquad (\alpha < \alpha^*) \tag{30}$$

and the equation

$$\alpha_0^{-1/2} - \alpha^{-1/2} = k(t - t_0) \qquad (\alpha < \alpha^*) \tag{31}$$

In both these equations α_0 is the fractional decomposition which accompanies the initial formation of the nuclei. [The analysis of the kinetics of the thermal decomposition of nickel oxalate is complicated by an initial deceleratory process like that depicted in Fig. 1. Here α_0 is the decomposition corresponding to the formation of growth nuclei in the bulk in addition to that occurring in the initial process, α_s (Fig. 1).] Comparison with experiment showed that Eq. (31) was to be preferred.

The combination of Eqs. (31) and (28) certainly provided an excellent fit to the nickel oxalate data, but the model used is by no means unique. Dominey *et al.*([18]) showed that the $\alpha(t)$ curves during the acceleratory period could also be fitted by a cube law ($n = 3$) succeeded by a square law ($n = 2$). This analysis accounts beautifully for the fractional decomposition curves but leaves us with the problem of t_0. Dominey *et al.* propose that t_0 represents the time taken for the nickel atoms formed during the initial surface reaction to diffuse over the surface to face edges and there form growth nuclei. This model is thus the ultimate in slow growth since it involves the formation of growth nuclei by a purely physical process which is not accompanied by chemical decomposition. Thus $u = 0$, $t^* = t_0$, and $\alpha^* = \alpha_0 = \alpha_s$ in this model, whereas in the model based on a finite slow-growth rate $\alpha^* > \alpha_0 > \alpha_s$. The most appealing of these three models physically is that of Dominey *et al.* and it would be interesting to see if this model, which interprets the induction period as the time taken for the formation of growth nuclei by aggregation, could be used for calcium([19]) and strontium([20]) azides, which decompose according to a t^3 law with very little gas evolution during the induction period.

VARIABLE SHAPE FACTORS

It has been shown that the induction period can be interpreted as a period of nucleus growth in which the nuclei attain a critical size either by a process of physical aggregation (diffusion) or by growth involving chemical decomposition at a slower rate than that characteristic of nuclei which have attained the critical size. Since the total amount of decomposition occurring in many hours of photolysis at low temperatures may represent only α of the order of 10^{-5}, photochemical decomposition offers a method of studying the early stages of decomposition without the complicating features of impingement and overlap. While the photolysis of many solid materials has been investigated, its relation to the thermal decomposition of ionic materials has been insufficiently stressed. In one significant study([21]) the equivalence of photochemical and thermal decomposition was established for barium azide by showing that the photolysis rates attained after some significant decomposition had occurred were independent of whether this had been carried out thermally or photochemically. Further studies of this sort are needed.†

For sodium azide Jacobs and Kureishy([22]) showed that the kinetics of photolysis were explicable in terms of a model involving random nucleation at a fixed number of sites followed by linear growth of these nuclei. It was further tentatively suggested that the direction of growth might be along dis-

---

†Note in this connection the valuable kinetic work being done on neutron-and γ-irradiated materials, e.g. ([18-20]).

location lines. If the thermal decomposition initiates in the same fashion, then rapid nucleation followed by linear growth would lead to filamentary nuclei which would subsequently grow two-dimensionally $(G_1 = G_2 \ll G_3)$, leading to kinetics described by the Erofeev equation with $n = 2$[23]. Thus the induction period t_0 corresponds, at least in part, to the initial period of linear nucleus growth, although other contributory factors may be a finite nucleation rate (Fig. 2) and physical aggregation. While the latter is always an attractive possibility where the nucleus is metallic in nature, aggregation by surface diffusion is unlikely in sodium azide in view of the volatility of sodium.

Another example of changing nucleus shape arises when nucleation occurs initially in three dimensions but, owing to the initial distribution of the nucleus-forming sites (along dislocation lines?), coalescence results in cylindrical nuclei which again grow two-dimensionally. Thus the kinetics would change from a cube law, $n = 3$, to a square law, $n = 2$, as the reaction progresses, a situation reported for the decomposition of nickel oxalate[18].

EXPONENTIAL ACCELERATORY PERIODS

There are a number of substances, exemplified by silver oxalate and mercury fulminate, which decompose according to an exponential law

$$\alpha = Ce^{kt} \tag{32}$$

when freshly prepared. Originally this was explained by a model involving linear, branching nuclei[24]. Clearly any model which contains an autocatalytic feature will predict

$$d\alpha/dt = k\alpha \tag{33}$$

and thus lead to an exponential kinetic law. There is therefore nothing unique about the various models which have been used. The modern version of Garner's branching chains envisages compact microclusters of nuclei all of which become growth nuclei when infected by another growing nucleus[2]. This can occur either by chemical reaction spreading down the boundary between two neighboring crystallites in an essentially linear manner or by the propagation of strain in the crystal leading to the production of dislocations or even cracks which in turn facilitate nucleus formation, or by the actual diffusion of product ahead of the reaction interface. The production of cracks due to product-reactant misfit was first emphasized by Prout and Tompkins[25], who also allowed for interference between branching nuclei. Clearly as the extent of reaction becomes appreciable some of the existing nuclei will attempt to infect potential nucleus-forming sites which have already reacted and so will terminate. Thus $d\alpha$ in Eq. (33) ought to be replaced by $d\alpha_{ex} = d\alpha/(1 - \alpha)$, giving

$$d\alpha/dt = k\alpha(1 - \alpha) \tag{34}$$

A more detailed analysis[1] shows that the special conditions which lead to this equation are (1) a branching coefficient independent of α (which is reasonable if this simply represents the number of individual nuclei in each microcluster, but surely less plausible in terms of a model in which branching is due to strain or cracking); (2) a termination coefficient which is equal to the branching coefficient times α/α_i (this should result in the same rate constant for acceleratory and decay periods, whereas they usually differ and sometimes even have different temperature coefficients); and (3) a symmetrical decomposition curve with $\alpha_i = 0.5$ (potassium permanganate is the best example of this). Equation (34) presupposes the existence of some initial decomposition (nucleus formation) during the induction period; integration between limits (α_0, t_0) and (α, t) yields the Prout–Tompkins equation[25]

$$\ln\frac{a}{1 - a} - \ln\frac{a_0}{1 - a_0} = k(t - t_0) \tag{35}$$

This equation certainly provides an excellent fit to the decomposition data for a number of salts, notably the permanganates, but should really be regarded as an approximate equation unless the three criteria referred to above are satisfied.

Aging of materials (for example, silver oxalate) which obey an exponential law when freshly prepared can lead to kinetics described by a power law (Erofeev equation) with $n = 3$, due to slow chemical decomposition at potential nucleus-forming sites. These then undergo three-dimensional growth from the beginning of the reaction and no infection (chain branching) occurs. Light doses of reactor radiation produce the equivalent effect.

SURFACE NUCLEATION

When nuclei are formed only on the surface nucleus growth and overlap lead to the formation of a coherent interface between the undecomposed reactant in the interior of the particles and the product layer outside this. If this interface propagates inward at a constant rate, as it will if every molecule of reactant in the interface has the same probability of decomposing, then it follows that

$$1 - (1 - \alpha)^{1/n} = kt \tag{36}$$

with $n = 3$. For reactions spreading in two dimensions $n = 2$.

Equation (36) often provides an excellent fit to the decay period and should obviously be applied when direct observation indicates the presence of a coherent interface propagating inward at a constant rate. It may, of course, be applied only to the decay period, and is particularly effective when this constitutes virtually the whole of the reaction, i.e., when nucleation and surface growth are both much faster than inward growth. Various modifications to this equation can be developed by retaining the concept of inward propagation from surface nuclei and combining this with either (or both) random nucleation and a finite rate of surface growth. This work has been done mainly by Bradley et al.[27] and by Mampel[28] and is reviewed in greater detail elsewhere [1,2].

REFERENCES

1. P. W. M. Jacobs and F. C. Tompkins, in: *Chemistry of the Solid State* (W. E. Garner, ed.) Butterworths, London, 1955, Chapter 7.
2. D. A. Young, *Decomposition of Solids*, Pergamon, Oxford, 1966.
3. I. Langmuir, *J. Am. Chem. Soc.* **38**: 2263 (1917).
4. W. Fraenkel and W. Goez, *Z. Anorg. Allg. Chemie* **144**: 45 (1925).
5. W. G. Burgers and L. J. Groen, *Discussions Faraday Soc.* **23**: 183 (1957).
6. J. H. Hollomoñ and D. Turnbull, *Progress in Metal Physics* **4**: 333 (1953).
7. Kh. S. Bagdassarian, *Acta Physicochim. URSS* **20**: 441 (1945).
8. A. Wischin, *Proc. Roy. Soc. (London) Ser. A* **172**: 314 (1939).
9. W. E. Garner and W. R. Southon, *J. Chem. Soc.* p. 1705 (1935).
10. N. F. H. Bright and W. E. Garner, *J. Chem. Soc.* p. 1872. (1934).
11. J. A. Cooper and W. E. Garner, *Trans. Faraday Soc.* **32**: 1739 (1936).
12. M. Avrami, *J. Chem. Phys.* **7**: 1103 (1939); **8**: 212 (1940); **9**: 177 (1941).
13. B. V. Erofeev, *Compt. Rend. Acad. Sci. URSS* **52**: 511 (1946).
14. P. W. M. Jacobs and A. R. T. Kureishy, in: *Proceedings of the Fourth International Symposium on the Reactivity of Solids*, Elsevier, Amsterdam, 1961, p. 352.
15. D. A. Young, *Nature* **204**: 281 (1964).
16. J. G. N. Thomas and F. C. Tompkins, *Proc. Roy. Soc. (London) Ser. A* **210**: 111 (1951).
17. P. W. M. Jacobs and A. R. T. Kureishy, *Trans. Faraday Soc.* **58**: 551 (1962).
18. D. A. Dominey, H. Morley, and D. A. Young, *Trans. Faraday Soc.* **61**: 1246 (1965).
19. E. G. Prout and M. E. Brown, *Nature* **205**: 1314 (1965).
20. E. G. Prout and D. J. Moore, *Nature* **205**: 1209 (1965).
21. P. W. M. Jacobs, F. C. Tompkins, and V. R. Pai Verneker, *J. Phys. Chem.* **66**: 1113 (1962).
22. P. W. M. Jacobs and A. R. T. Kureishy, *Can. J. Chem.* **44**: 703 (1966).
23. P. W. M. Jacobs and A. R. T. Kureishy, *J. Chem. Soc.* p. 4718 (1964).
24. W. E. Garner and H. R. Hailes, *Proc. Roy. Soc. (London) Ser. A* **139**: 576 (1933).
25. E. G. Prout and F. C. Tomkins, *Trans. Faraday Soc.* **40**: 488 (1944).
26. R. M. Haynes and D. A. Young, *Disc. Faraday Soc.* **31**: 229 (1961).
27. R. S. Bradley, J. Colvin, and J. Hume, *Proc. Roy. Soc. (London) Ser. A* **137**: 531 (1932).
28. K. L. Mampel, *Z. Physik. Chem.* **A187**: 43, 235 (1940).

DISCUSSION

R. S. Gordon, (*University of Utah*): I am curious about these models by Avrami. Do you have to make the assumption that this is a volume nucleation or can it be a random nucleation on a surface? If it is a volume nucleation, you undoubtedly have problems getting the gases away, and then how applicable are these models for a thermal decomposition reaction?

Answer: It is an implicit assumption that the gas can escape and that this is not kinetically controlling. There has been a little work done on the question of what happens to the kinetics when the diffusion of gaseous products out of the solid phase is kinetically significant. You need not have nucleation throughout the bulk in order to have three-dimensional growth. A hemispherical nucleus will result from formation on the surface and then grow inward in a three-dimensional manner. I am quite sure that in many reactions nucleation does occur throughout the bulk. Perhaps one of the most obvious examples is sodium azide. Sodium is very volatile at the temperatures at which the kinetics of decomposition of sodium azide can be followed. Therefore it seems to me that nucleation must occur in the bulk and that the nitrogen gas must be able to escape, possibly by diffusing along grain boundaries or dislocations or along some other sort of easy diffusion path. Maybe it is only after a certain amount of decomposition takes place that this easy escape of gas becomes possible. That may be one of the reasons why reaction is slow in the early stages. In summary, I do not think you have to restrict reaction to the surface in order to apply these concepts.

P. J. Herley (*Brookhaven National Laboratory*): I would like to ask Professor Jacobs if he would comment on the indefinite decay period where the reverse of nucleation occurs at the other end of the thermal decomposition. I notice that you were talking of say 5–96% coverage. What happens from 96–100%?

Answer: That, of course, referred to the particular data that I was using as an example. Commonly, things do rather fade away toward the end and I am sure you know that one seldom gets the equations to fit right to the end of the reaction. I think that the cause of this is that after so much of the reactant is consumed one is left with isolated pockets or streaks of undecomposed reactant and that this is why the reaction tends to progress more slowly toward the end. I suppose one could think of it formally as a kind of nucleation in reverse and then perhaps try to analyze it that way. One usually tends to ignore the end and concentrate one's efforts on trying to understand the beginning.

Chapter 4

Phenomenological Observations During Solid Reactions

R. K. Stringer, C. E. Warble,* and L. S. Williams

Division of Applied Minerology, Chemical Research Laboratories
Commonwealth Scientific and Industrial Research Organization
Melbourne, Australia

Research is presented on the characteristics of refractory oxide powders used in sintering processes. Oxide powder characterization studies using electron microscopy and selected-area electron diffraction are described. The morphological and crystallographic features of fine powders are examined, as are surface conditions and the effects of impurities. The nature and extent of precursor–oxide orientation relationships and the decomposition and recrystallization stages in conversion to the requisite form of the oxide are considered. Work on decomposition, growth, and sintering features which might be relevant to sintering and hot-pressing processes is also discussed.

INTRODUCTION

There has been increasing recognition in recent years of the need for detailed knowledge of the characteristics of oxide powders used in sintering studies[3,40,49,50,66,67]. It is perhaps not so widely recognized that instrumental characterization studies, notably by optical and electron microscopy and by X-ray and electron diffraction, can yield vital information on basic growth and sintering processes from examination of the products of various calcination conditions.

Objective assessment of the extensive literature of the theory and practice of sintering leads to the conclusion that a firm foundation of direct observation is particularly important at the present juncture. Great strides have been made through the model concepts and kinetic treatments of, *inter alios*, Kuczynski[39], Clark and White[11], Kingery and Berg[37], Coble[12,13], and Johnson and Cutler[36], but the task of reconciling theory and practice is proving formidable indeed, despite the laudable approach of the Rhines school[55,56] in seeking to bring "simplified reality" into basic treatments through their topological sintering model; the most recent reports of this enlightened study available to the present writers show a markedly esoteric trend despite continuing efforts to work from the basis of real sintering situa-

---

*Now with Division of Chemical Physics, CSIRO.

tions. Down-to-earth assessments of the present capabilities and limitations of sintering models and kinetic treatments([8,14,53,66]) amply illustrate the need for direct observation of basic phenomena.

Reeve makes two particularly pertinent observations: (1) for the most part densification and grain growth have been studied separately, and there is as yet no adequate theory that allows the interrelation between the kinetics of these processes to be explained quantitatively; and (2) a sounder theoretical basis for controlling microstructure must await fuller knowledge of the nature and condition of fine powders and either development of sintering and grain growth models which completely describe these processes in a real powder or development of powders which approach the ideal starting material of sintering models.

Prior to 1950 "characterization" of a powder (if the term was used at all) was commonly limited to information gained by particle size and surface area measurements, optical microscopy, X-ray diffraction, and traditional chemical analysis. With increasing emphasis on fine particles (usually in the context of high activity) and high purity, more refined and subtle methods of characterization were applied, such as X-ray line-broadening for particle size assessment, spectrographic analytical techniques, and various absorption and exchange measurements which could be related to sintering activity; the accurate measurement of heat of solution warrants special mention as one of the few means available for the assessment of the excess energy of active powders ([3]).

The mushrooming awareness during the past few years of the need to know a great deal more about the morphology, crystallographic detail, and surface condition of fine oxide (and other) powders stems from increased understanding of the influence of these factors on all aspects of fabrication and the attainment of optimum properties. In short the science of fabrication has evolved, largely replacing art, and providing essential support for technology. No longer is it sufficient to deduce mean crystallite sizes from X-ray line-broadening, and hence calculate specific surface areas on the assumption, often inadequate, that the crystallites are essentially equiaxed. It is necessary to *see* the ultimate particles, and to distinguish their detail at the highest attainable magnification and resolution. The high-resolution electron microscope is meeting this need, and when conjoined with selected-area electron diffraction providing identification and structural data, constitutes the primary characterizing tool and undoubtedly offers more comprehensive and valuable information than any other single instrument.

The general belief in the essentially equiaxed form of fine oxide crystallites has not died easily. Murray([51]) stated that electron microscopy observations on fine powders of magnesia, beryllia, and thoria (limited, of course, by the resolution then attainable) suggested that the crystallite shape in no case deviated markedly from an equiaxed form. J. Williams[67] reported as recently as 1965 that information on crystallite shape obtained by X-ray diffraction techniques for the same three oxides again indicated an equiaxed form. However, both Murray and Williams emphasized the pseudomorphic and orientation relationships existing between derived oxides and their precursors, and so provided the stimulus for the present study, whereby description of refractory oxide powders more definitively and in finer detail than hitherto reported was sought; a particular objective of the study would be to trace the derivation of oxide powders from their parent compounds through identification of distinctive shape, orientation, and crystallographic features at various stages.

Thus the continuing investigation for which progress is now reported was initiated some two years ago, with electron microscopy as the primary tool, closely supplemented by electron diffraction. Up to that time the combined potential of these two techniques for oxide powder characterization studies had apparently not been generally recognized. Goodman[27] is one worker to have relied entirely on electron microscopy and diffraction in studying the decomposition of magnesium hydroxide in the electron beam, and Anderson and Livey[3] recognized these techniques as complementary to X-ray diffraction for oxide powder studies. Electron microscopy itself, alone or with X-ray diffraction, was of course already well established, with magnesia frequently the subject material[18,24,53,59].

While the work reported has been primarily concerned with detailed characterizing observations as outlined above, a watching brief has been maintained for decomposition, growth, and sintering features that might impinge on selected decomposition-sintering and pressing experiments intended to serve as pointers toward determining optimum fabrication conditions. Such phenomenological observations can help provide a sounder basis for kinetic analysis.

SUBJECT MATERIALS

It is intended to work progressively through a representative range of refractory oxides of several structural types by studying appropriate precursor–oxide pairs. The three combinations considered in reasonable depth to date are magnesium hydroxide–magnesia, basic magnesium carbonate–magnesia and basic aluminum sulfate–alumina. It should be stated at this point that only the variable of temperature of decomposition has been utilized to any considerable extent in the work reported; it is of course recognized that rate of heating to decomposition range, time of heating in this range, depth of bed, ventilation, and control of atmosphere (with special emphasis on water vapor and vacuum) are influential factors which must be considered in due course, but it has become equally evident that detailed characterizing observations cannot be hurried, and the limited effort available must continue to be devoted to a deeply searching approach to each phase of the work.

The decomposition of magnesium hydroxide, in the form of both brucite and chemically prepared hydroxide, has been studied extensively elsewhere. The recent papers by Gordon and Kingery[29,30] present and reference concisely the well-established structual relationships between the hydroxide and the oxide, as well as report detailed observations by electron and optical microscopy, electron diffraction data, and decomposition kinetics. The present work is believed to provide a valuable supplement to the Gordon and Kingery investigation, particularly through high-resolution electron microscopy (see next section).

Magnesium hydroxide has been the parent compound for magnesia in many sintering, hot-pressing, and decomposition-pressing studies[31,44,48] and can be calcined in vacuum in the vicinity of 280°C to yield magnesia which is very active in the sintering sense, with specific surface areas around 300 m^2/g[3]; the ultimate crystallites could not be resolved but were assumed to be cubic in shape, and thus a mean crystallite size of approximately 50 Å was deduced. It has also been demonstrated that basic magnesium carbonate can be calcined to yield highly sinterable magnesia powder; thus Hyde and Duckworth[35] reported that very active magnesia was produced by calcining in air at 1100°F (593°C), and Koester et al.[38] found by adsorption and exchange measurements that maximum surface activity and maximum total activity were exhibited by magnesia obtained by calcining in flowing nitrogen at or a little above 500°C. In the present instance decomposition-pressing and decomposition-sintering experiments indicated greater sinterability with basic magnesium carbonate as the precursor than with magnesium hydroxide, but interrelationships were not known to have been studied in detail for the basic carbonate–magnesia pair, not had the oxide product been well characterized. Accordingly, a large share of the available effort has been devoted to this phase of the investigation.

Selection of an alumina precursor for detailed study was guided by a combination of fundamental value and local interest. Basic aluminum sulfate is an intermediate product in the "acid alumina" process[60] developed in the Mineral Chemistry Division of these Laboratories, and the thermal decomposition of this material has been studied by Davey et al.[20]. While morphological and crystallographic aspects of the conversion to alumina were not a primary objective of their study, valuable pointers are given; in addition an explanation of the difficulty of eliminating traces of sulfate is made, based on structural considerations, and the likely effect of sulfate residues in delaying the alumina transformations is noted. The final transformation to α-Al_2O_3 is from very small η-Al_2O_3 crystallites. It is interesting

TABLE I

Analytical Data for Precursor Materials*

Impurity	A.R. grade basic magnesium carbonate†			Technical grade basic magnesium carbonate**		Magnesium hydroxide‡	Magnesium oxide§	Basic aluminum sulfate§§	
	SA	CCA	AAS	SA	AAS	SA	SA	TDA	CCA
Cl	0.002	—	—	0.04	—		0.01	—	—
SO_3	—	—	—	—	—		—	40.5	—
NO_3	0.003	—	—	—	—		—	—	—
SO_4	0.005	—	—	0.6	—		0.001	—	—
NO_4	0.002	—	—	0.04	—		0.005	—	0.06
Ca	0.02	—	—	—	—		0.02	—	—
Pb	0.001	—	—	—	—		0.005	—	0.03
N	—	—	—	—	—	Supplier's label states "extra pure"	0.002	—	—
Zn	—	—	—	—	—		0.005	—	—
Ba	—	—	—	—	—		0.005	—	—
Na	—	—	—	—	—		0.2	—	—
K	—	—	—	—	—		0.005	—	—
Hg	—	—	—	—	—		—	—	0.09
H_2O	—	—	—	—	—		—	20.0	—
Al_2O_3	—	—	—	—	—		—	39.5	—
CaO	—	0.008	0.015	—	0.20		—	—	—
Na_2O	—	0.006	0.07	—	0.21		—	—	—
K_2O	—	0.0005	0.003	—	0.003		—	—	—
Li_2O	—	—	<0.001	—	<0.001		—	—	—
MgO	—	0.051	—	—			—	—	—
CO_3	—	present	—	—	—		—	—	—
Soluble salts	0.4	0.145	—	—	—		—	—	—
Loss on ignition	—	—	57.2	—	57.4		3	—	—
Residue on ignition	—	—	—	42–45	—		—	—	—
Substances soluble in H_2O	—	—	—	1	—		0.5	—	—
Substances insoluble in HCl	—	—	—	—	—		0.005	—	—
Substances insoluble in HCl + NH_4OH	0.02	—	—	—	—		—	—	—

*All figures are maximum percent of as-received material. SA = suppliers analysis. CCA = conventional chemical analysis. TDA = thermal decomposition analysis. AAS = atomic absorption spectroscopy. A.R. = Analytical Reagent.

†Mallinckrodt; label formula "approximately" $4MgCO_3 \cdot Mg(OH)_2 \cdot 4H_2O$.

**Hopkin and Williams; label formula "approximately" $3MgCO_3 \cdot Mg(OH)_2 \cdot 3H_2O$.

‡Merck.

§Merck. Additional hydroxide produced from this material by rehydration.

§§Australian Minerals Development Laboratories; empirical formula $3Al_2O_3 \cdot 4SO_3 \cdot 9H_2O$.

to consider prospects for eliminating sulfate residues and better utilizing the sintering activity of the material in appropriate states of calcination.

Available impurity data for the various starting materials are given in Table I, with particular emphasis on basic magnesium carbonate. It is recognized that anion impurities may be no less important than cation species. The basic carbonate used in the present work was the A. R. material unless otherwise stated.

CHARACTERIZING OBSERVATIONS BY ELECTRON MICROSCOPY AND DIFFRACTION

Morphological and Crystallographic Relationships

In a recent note[46] reporting progress in studying the basic magnesium carbonate–magnesia pair, it was mentioned that a parallel study with magnesium hydroxide as the magnesia precursor was in progress. Gordon and Kingery[29,30] have now reported their detailed study of the hydroxide–oxide

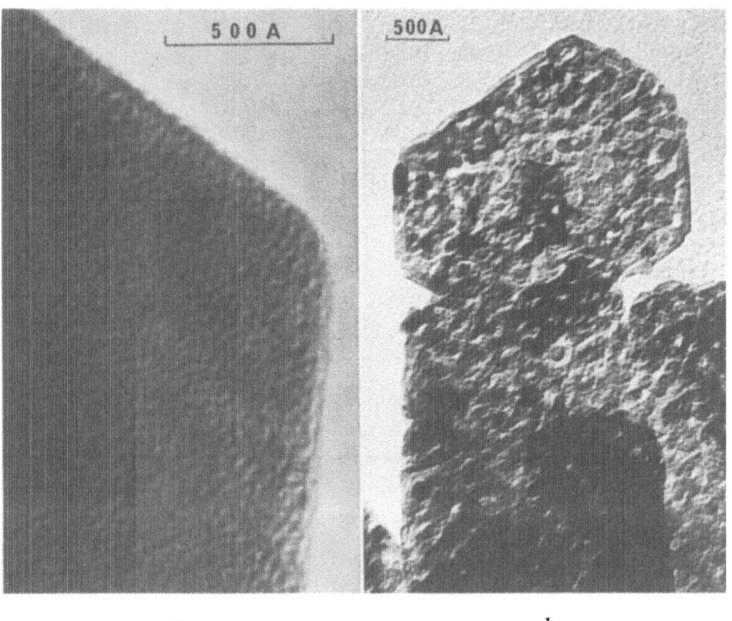

a   b

Fig. 1. (a) Pseudomorphic MgO (lightly calcined from rehydrated oxide in the electron beam) with indications of ultrafine development of hexagonal crystallites. (b) Pseudomorphic MgO (600°C calcine) from rehydrated oxide showing well-developed hexagonal plate-like nature of crystallites.

system, and in any case several of the recrystallization, growth, and sintering features the present writers wish to report are better illustrated in the "looser" powders derived from basic magnesium carbonate; thus only one pair of electron micrographs of magnesia obtained from the hydroxide is reproduced [Fig. 1(a) and (b)]. The starting material in this case was the rehydrated oxide (Table I). As pointed out by other workers [2,29] the physical mode of decomposition of the hydroxide is apparently not much influenced by the size of the parent crystallites, at least in the normal powder range, and this was so in the present instance; the micrographs of Fig. 1(a) and (b) are substantially representative of both rehydrated oxide and precipitated hydroxide source materials. With the exception of a few samples decomposed in the electron beam, the magnesia samples from both hydroxide and basic carbonate were produced by calcination of approximately 5 g of the precursor as a relatively shallow bed (ca. 3 mm) in an electric furnace with a large heated space but barely adequate ventilation. Thus the problem of accumulation of decomposition products was not entirely eliminated. The temperature of calcination was generally approached at a rate of 250°C/hr. Samples were prepared for examination at 75 keV in a Hitachi HU11-A electron microscope by ultrasonic dispersion in acetone and fogging onto carbon films or directly onto 2000 mesh copper grids; the latter procedure favors observation of surface detail[45].

The lightly decomposed material of Fig. 1(a) was obtained by heating in the electron beam and that of Fig. 1(b) by calcination at 600°C for 1 hr. The hexagonal platelet morphology of the CdI_2-structured hydroxide crystallites previously established[2,3] is clearly illustrated here. In the present instance the hydroxide platelets formed by rehydration of oxide powder were relatively thick, of the order of 1000 Å, and several layers of magnesia crystallites of lateral dimension around 30 Å [Fig. 1(a)] and 120 Å [Fig. 1(b)] are apparently stacked in the pseudomorphed material. This stacking is also evident in thinner relics from precipitated hydroxide. Known structural features of the oriented conversion to oxide are summarized and some controversial points resolved by Gordon and Kingery[29] who have also studied the kinetics of decomposition in vacuum[30]. The decomposition was interpreted as a normal nucleation and growth process in which magnesium oxide nuclei form coherently with the hydroxide matrix, introducing large strains and causing extensive fissuring. It is well known [e.g., Brindley[66]] that the close-packed (111) planes of oxygen ions in the NaCl-cubic magnesia structure form in parallel orientation with the close-packed (0001) planes of hydroxyl ions in the hydroxide during lateral fissuring and collapse of the layer-structured parent platelets, and the [001] and [100] directions of the hydroxide become the [111] and [110] directions of magnesia. The (0001) planes of magnesium ions in the hydroxide simply contract without rearrangement to become (111) planes in magnesia. However, even in the latest work cited above the actual morphology of the magnesia crystallites is obscure. The present work has shown these to be essentially plate-like and in [111] orientation with major surfaces also (111),* arrested in the course of growth into cubes balanced on a body diagonal. This is already indicated by the tendency to develop an ultrafine hexagonal pattern exhibited by the sample lightly calcined in the beam [Fig. 1(a)]. Even though 600°C is a relatively high calcination temperature for magnesium hydroxide, appropriate tests have shown that the oriented plate-like nature of the magnesia crystallites is evident to a marked degree in the 600°C product [Fig. 1(b)].

The several observations and techniques employed in identifying the crystallographic orientation and plate surfaces of the crystallites which constitute the pseudomorphs followed the pattern described in a progress note of the basic magnesium carbonate study[46] and are mentioned below in recounting the carbonate work. Suffice to say at this point that selected-area electron diffraction played a vital part. The matter of surface detail is also treated more fully later, but it should be noted here that the plate surfaces of the magnesia crystallites are properly described as "mean" (111) surfaces minutely stepped with (100) facets (see p. 68).

Referring again to Fig. 1(b), it is further observed that cracking within the pseudomorphs tends to follow a grid pattern related to [100] cleavage directions. The crystallites are extensively interjoined, to such a degree that Balmbra et al.[4] have suggested that brucite relics may be considered as highly imperfect single crystals. This characteristic of magnesia derived from the hydroxide tends to make the material less active in sintering than crystallite size and surface area would indicate. Decomposition-sintering and decomposition-pressing results presented later illustrate this point.

A close orientational relationship is again apparent in the conversion of basic magnesium carbonate to magnesia. This precursor–oxide pair does not appear to have been studied to any great extent from the standpoint of morphological and crystallographic relationships, apart from its inclusion in the investigation of several magnesia precursors by Royen and Trömel[59] using X-ray diffraction and electron microscopy. Certainly no report utilizing electron microscopy and selected-area diffraction has been encountered. The progress note already cited on the present study established the crystallographic orientation of the derived magnesia crystallites as [111] and the relevant mean surfaces of these plate-like particles as (111) (cf. the hydroxide case) by a combination of morphological observations in the electron microscope, definitive electron diffraction data including "fine structure" analysis[17,63] gold shadowing experiments, and moiré patterns in electron micrographs[34].

---

*But see comment on (100) growth facets, p. 68.

A search of the normal literature sources of structural data has not disclosed the complete crystal structure of basic magnesium carbonate, much less any statement on crystallographic and morphological aspects of decomposition comparable with the well-documented hydroxide–oxide case. At least the structural type and lattice parameters are known, orthorhombic with $a_0 = 8.98$ Å, $b_0 = 9.32$ Å, and $c_0 = 8.42$ Å ([22]). The a_0 and c_0 values were confirmed by electron diffraction in the present study, but a tilting stage was not available at the time, and thus the value of b_0 could not be checked. The fact that water of crystallization has to be driven off and both hydroxyl and carbonate stages of decomposition have to be negotiated complicates the issue, and there is also some variability in the compositional formula of the basic carbonate, or more correctly hydro-magnesite since it is the native compounds that are discussed in the reference cited ([19]).

However, two reports are available of decomposition studies of the A. R. material used in the present investigation [label formula "approximately" $4MgCO_3 \cdot Mg(OH)_2 \cdot 4H_2O$]. Koester et al.([38]) discussed thermogravimetric data for the decomposition in terms of the stages mentioned above, and lend weight to Dell and Weller's([21]) finding by differential thermal analysis and supporting compositional evidence that water in some form is retained well into the carbonate decomposition range. With a heating rate of 540°C/hr Dell and Weller obtained a first endothermic peak around 330°C, which was identified as loss of water and occurred with little increase in surface area. The product was amorphous to X rays and retained approximately 0.5 mole of water per mole of magnesia. A sharp exotherm at 510°C was identified as recrystallization of magnesium carbonate, which decomposed to magnesia endothermically above 520°C. Loss of CO_2 occurred from about 400°C onward.

In discussion of these findings Dell and Weller emphasized the remarkable stability of the amorphous intermediate product and noted its close-knit pseudomorphic character. They cited these factors as indicative of the difficulty of movement of the large carbonate ions, and interpreted the thermally activated process involved in sudden recrystallization of magnesium carbonate above 500°C as simply the unhindered movement of these ions.

Figure 2 shows a typical pseudomorph produced by calcining A. R. basic magnesium carbonate at 550°C for 1 hr, consisting of a highly oriented assemblage of magnesia platelets within the outline of the original flake. It was established by observing the decomposition process in the electron beam that the parent flakes contract overall as well as "crazing" in the course of multistage decomposition, and that they "collapse" in a manner reminiscent of the layer-structured hydroxide. The original flakes are remarkably uniform in thickness (~ 200 Å), but vary widely in lateral dimensions from a few hundred angstroms to several microns. Although the detailed structure of the basic carbonate has not been reported, the plate-like habit suggests a layer structure when taken in conjunction with the ease of cleavage along the (010) basal plane and the fact that decomposition products can be observed to move out laterally from flakes aligned normal to the electron beam. Recalling that Dell and Weller established that the "amorphous carbonate" product of the first stage of decomposition contained 0.5 mole of water for each mole of magnesia, it would seem likely that carbonate and hydroxyl groups could be associated in some type of composite layer in the basic carbonate structure. This is suggested by analogy with Cowley's[16] electron diffraction study of the basic lead carbonate structure, and some support is provided by Wells[65], who envisages that the basic magnesium carbonate may actually contain hydroxy-carbonate components. The proposal is also consistent with the evidence that the product of primary dehydration is surprisingly stable and the residual "water" is apparently tightly held, so that recrystallization of the $NaNO_3$ (distorted NaCl)-structured carbonate is delayed until the necessary movement of the large carbonate ions can be thermally activated. The occurrence of recrystallization is probably structurally influenced also, as carbon dioxide and water are progressively evolved.

As noted earlier the preferred crystallographic orientation of the magnesia platelets within the pseudomorphs has been established as [111] and the mean surfaces of the platelets, which are substantially in the plane of an original flake, are (111). The predominance of 60° and 120° angles showing a tendency to assume hexagonal and trigonal outlines is consistent with preferred [111] orientation, and among the several ways in which crystallogra-

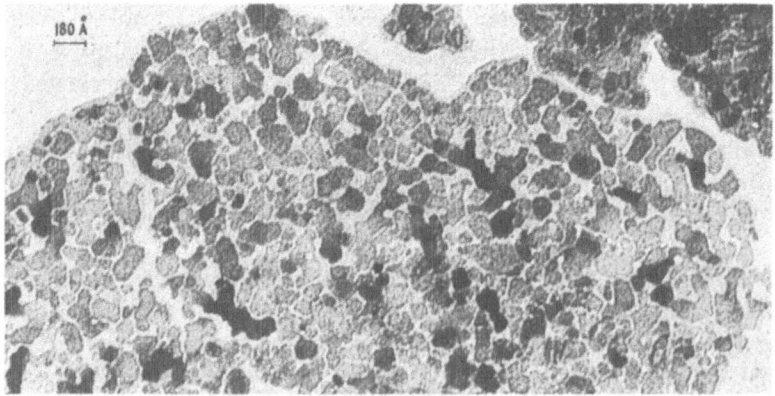

Fig. 2. Pseudomorphic MgO (550°C calcine) from A.R. basic magnesium carbonate showing lightly interjoined plate-like crystallites.

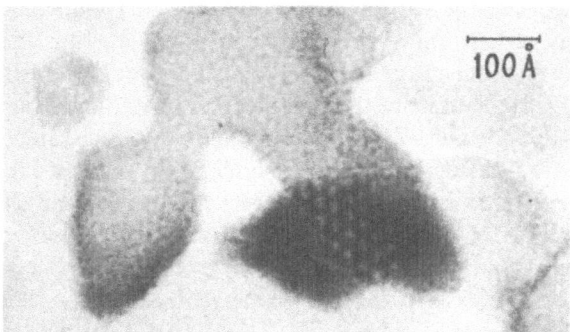

Fig. 3. MgO from A.R. basic magnesium carbonate (700°C calcine) showing two-dimensional moiré pattern from over-lapped crystallites.

phic and surface orientation were established and checked the observation of two-dimensional moiré patterns provided the most unequivocal evidence. Overlapping crystallites in a 700°C calcine having flat surfaces in near perfect contact but rotated a few degrees out of crystallographic alignment produced the hexagonal pattern shown in Fig. 3; this effect is extremely sensitive to alignment normal to the electron beam, and the fact that the grid angles are so close to 60° shows that the crystallites are fortuitously almost precisely normal.

The faithful retention of the outline and shape of the original basic carbonate flakes together with the high degree of orientation and strong tendency to assume hexagonal and trigonal shapes observed for the magnesia platelets within the pseudomorph suggest that there is a simple orientation relationship between the precursor and the oxide, despite the fact that the three major hurdles of primary decomposition, carbonate recrystallization, and conversion to oxide have to be negotiated in the course of decomposition. Structural data for the basic carbonate and the intermediate product are required in support of the following suggestion, but as a working hypothesis consistent with the foregoing observations it is proposed that the magnesium ions retain the same two-dimensional arrangement throughout the decom-position and recrystallization stages, perhaps with some transient "rumpling" within the layers, while the requisite expulsion and rearrangement of anion species occur. A layer structure closely related to that of the basic carbonate is readily envisaged for the intermediate composition, and a simple relation-ship exists between the terminal carbonate and oxide structures such that the morphology of the carbonate should be transferable to the oxide. The car-bonate has the rhombohedral $NaNO_3$ structure which may be regarded as distorted NaCl-type, or alternatively the structure of the oxide may be viewed as having a pseudotrigonal unit cell related to the rhombohedral $NaNO_3$-type, which is perhaps more appropriate in this case. Thus the critical step would appear to be the retention of plate-like morphology during recrystallization of the carbonate from the intermediate phase, and structural studies of this phase and the parent basic carbonate are scheduled in order to establish the feasibility of retaining a direct orientation relationship throughout, based on preservation of magnesium layers free from major disturbance.

The magnesia crystallites in the basic carbonate pseudomorphs appear to be much less subject to mutual constraint through close packing and intergrowth than in the case of the hydroxide relics, and this "looser" texture favors a higher degree of sinterability.

Continuing the theme of precursor–oxide relationships, further up the scale of difficulty is the basic aluminum sulfate-alumina case. In addition to the relative complexity of the decomposition of the basic sulfate[20] to the primary alumina product which is amorphous to X rays, there are the further stages of recrystallization to η-alumina and transformation to α-alumina to be considered. A limited study by electron microscopy and diffraction has been made of the conversion of the basic sulfate to α-alumina, and the few observations so far available combined with the structural essence of the work cited above provide a useful sketch of the subject system.

The basic sulfate under investigation has the nominal formula $3Al_2O_3 \cdot 4SO_3 \cdot 9H_2O$, and is a member of the rhombohedral alunite family. The material was obtained in laboratory preparations in the form of plate-like particles varying in lateral dimension from around 1000 Å to a few microns; the thickness has not been positively determined, but is estimated to be below 200 Å and essentially uniform. Thermal decomposition starts with loss of interstitial water below 300°C, followed by loss of hydroxyl water. Most of the combined water is lost by 600°C without perceptible decomposition of sulfate groups; a lower-hydrate intermediate phase is formed toward the end of this stage. The sulfate content is largely eliminated in the temperature range 600–900°C depending on several factors, including time of heating and nature of the atmosphere. However, anhydrous aluminum sulfate is formed in readily detectable amounts and persists to higher temperatures. Above 800°C alumina amorphous to X rays is the primary product of desulfation, transforming in the range 900–1000°C to η-Al_2O_3. There are indications that this transformation would take place at lower temperatures but for the presence of residual SO_3, which also has the effect of stabilizing η-Al_2O_3 at higher temperatures where α-Al_2O_3 would otherwise be formed. The final phase change from η- to α-Al_2O_3 only takes place rapidly above 1000°C, with the slow elimination of SO_3, which may persist to at least 1200°C.

Based on Hendrick's[32] projection of the alunite structure on (0001) the structure of basic aluminum sulfate consists of layers parallel to (0001)

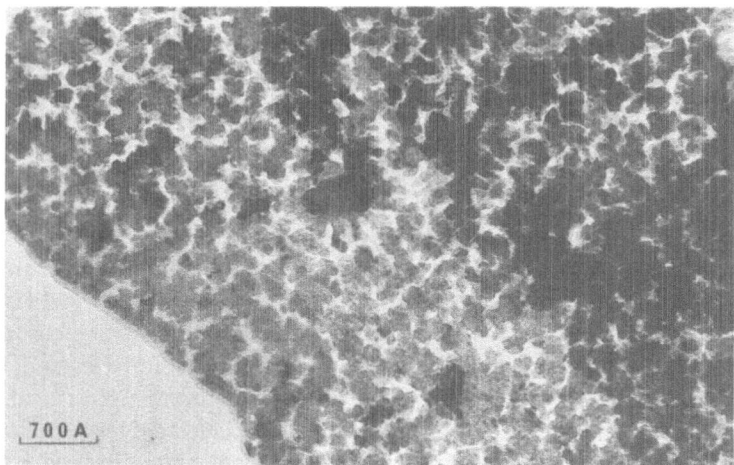

Fig. 4. Pseudomorphic η-Al$_2$O$_3$ (800°C calcine from basic aluminum sulfate); crazing within the pseudomorph shows marked hexagonal influence.

comprising triangular groupings of aluminum, sulfate, and hydroxyl. After interstitial water has been lost elimination of water from the hydroxyl bridges is believed to bring about collapse of the layer structure with increasing disorder. The strongly bonded sulfate groups then progressively break down, following the pattern described above.

An appropriate calcination schedule to arrive at this stage is several hours at 800°C, and in view of the postulated extensive disorder during the preceding 300°C or more, the neatly pseudomorphed appearance in the electron microscope of a sample calcined for 3 hr at 800°C (Fig. 4) was rather unexpected. The original flake outlines were well preserved and the crazing withing the relics showed a marked hexagonal influence; the fragments exhibited hexagonal tracery with subunits very uniform in size around 60 Å. The next structural event reported by Davey et al.[20] is reconstructive transformation to η-Al$_2$O$_3$, and areas of the same micrograph clearly reveal this occurrence, the resulting crystallites being larger (~ 200 Å) than the units of the hexagonal networks and apparently approaching an equiaxed form.

The structure of η-Al$_2$O$_3$ has been described as distorted spinel[15]. In the spinel structure one-quarter of the octahedral and one-quarter of the tetrahedral interstitial sites in the close-packed oxygen lattice are occupied in an ordered way by the cations. In terms of the spinel structure of overall formula $M_{24}O_{32}$, η-Al$_2$O$_3$ would be represented as $Al_{21^1/_3}O_{32}$, i.e., with one-ninth of the allowed cation sites unoccupied. The coordination of oxygen atoms around sulfur in the sulfate ion is very similar to the arrangement around the tetrahedrally coordinated cations in the spinel array, so that although the

SO_4 groups or SO_3 molecules are too large to fit interstitially in the lattice, the oxygens could substitute for some spinel oxygens and bind in the sulfate residues as part of the framework structure. In this situation it would be very difficult to remove the sulfate residues by the normal mechanism of diffusion because of the highly directional and covalent sulfur–oxygen bonds. Furthermore, as the sulfur atoms would now occupy normal interstitial lattice sites, the rate-determining cation diffusion into octahedral sites required for transformation from the disordered η-Al_2O_3 structure to the highly ordered α-Al_2O_3 array and the attendant rearrangement of the anion layer sequences from cubic to hexagonal close-packed would be inhibited. In normal circumstances this thermally activated recrystallization to the denser corundum structure is accompanied by a large increase in crystallite size, and when a higher activation energy is required because of the postulated critical location of precursor residues it is to be expected that accelerated recrystallization and growth at higher temperatures will yield an even coarser and more interjoined aggregate of corundum crystallites.

Figure 5 depicts alumina obtained by calcining basic aluminum sulfate for 1 hr at 1100°C. The focus was preferentially on the larger (α-Al_2O_3) particles but pseudomorphed η-Al_2O_3 aggregates are still evident, although, as noted earlier, close orientation relationship between precursor and product within the pseudomorph boundaries apparently did not survive recrystallization to the η modification. However, the platy nature of the primary product of recrystallization in the corundum form is consistent with retention of the external morphology of the parent flakes during decomposition of the basic

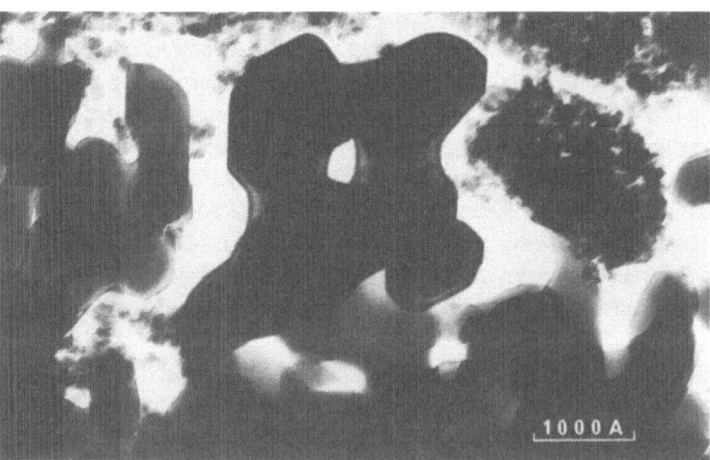

Fig. 5. Al_2O_3 from basic aluminum sulfate (1100°C calcine) showing η-Al_2O_3 pseudomorphic aggregates (small particles) and platy and blocky α-Al_2O_3 (large crystallites).

sulfate and the subsequent recrystallization. Particles of η-Al_2O_3 are seen to be "feeding" corundum wafers. Further recrystallization and growth produces corundum of a more blocky nature, exemplified by the intergrown cluster of crystallites in the center of the micrograph, which exhibits thickness fringes characteristic of the development of three-dimensional shape from primary wafers.

Growth and Sintering Observations

The preceding observations have been primarily concerned with the nature and extent of precursor–product orientation relationships and with the decomposition and recrystallization stages in conversion to the requisite form of the oxide. Although the present work has not had the advantage of a hot-stage facility for the electron microscope, information relating to growth and sintering processes has been obtained. The subject material has been chiefly magnesia derived from the basic carbonate, but some comments germane to the present heading can be made on the alumina micrograph just described (Fig. 5) before proceeding to discuss the extensive observations for magnesia. Considering only the corundum portion for the present purpose, it is not too difficult to envisage crystallographic influences at work in the development of the wafer material and its further growth into the blocky form. Part-hexagonal outlines are apparent, and these outlines in conjunction with thickness fringes are consistent with the development of crystallographic faces by growth on (0001) basals. Close study reveals serrated edges and face detail suggestive of stepwise growth in the manner predicted by Frank[25]. Evidence of the development of neck junctions and filling of reentrants as envisaged for both growth and sintering processes can be seen, and in the central group of better developed and extensively intergrown crystallites a classical "genus 1" situation[55] has evolved, enclosing a hole which in the loose powder could well be eliminated by further intergrowth, but could develop into a closed pore if contacted with other particles in a higher genus situation in a sintering compact.

Turning now to the more detailed study of magnesia, observations have been made primarily on samples derived from the basic carbonate by heating in air for 1 hr at 550°, 700°, 900°, and 1100°C. It may assist the presentation of selected observations if an outline of important findings is first given without detailed reference to the micrographs.

As already noted, when small samples are calcined at low temperatures in a large but rather poorly ventilated electrically-heated chamber,* the A. R.

---

*It must be emphasized that all the observations reported relate to these calcination conditions, and that the phenomena observed and processes deduced may be importantly influenced by the presence of moderate accumulations of decomposition products.

basic carbonate forms pseudomorphs consisting of arrays of magnesia crystallites well oriented parallel to the [111] direction. The individual particles are single crystals substantially perfect within themselves, but bounded by stepped nonequilibrium faces which are evidently one important source of high activity in lightly calcined powder. Superficially the particles appear irregular in shape, but crystallographic influence is indicated by a marked tendency to assume hexagonal and trigonal outlines. The particles are initially interconnected fairly extensively by necks which are rarely more than a few unit cells in lateral dimension (i.e., 20 Å or less). With increasing temperature of calcination the nonequilibrium faces tend to grow out; that is, the crystallites develop toward their equilibrium configuration of cubes balanced on body diagonals by differential growth on higher index mean planes. Thus the (111) mean planes are eliminated more rapidly than the (110) mean planes. "Mean" is used here to indicate that only (100) growth facets in fact exist; micrographs showing steps down to the level of one unit cell ($\sim$4.2 Å) establish that the essential growth mechanism can be understood at least qualitatively in terms of Frank's kinematic theory of crystal growth. For simplicity Frank's treatment does not consider the influence of dislocations once a growth wave has been initiated.

At 700°C a high proportion of mean (111) face still exists, as shown, for example, by two-dimensional moiré patterns (Fig. 3). The predominant morphology up to 900°C is that of crystallites bounded by (111) and other nonequilibrium faces of identifiable type, and much less frequently (100) plates and embryo cubes are observed, often bounded by (110) wedges. These morphologies may intergrow.

As growth proceeds, other developing particles are contacted. Smaller particles then form "bunches," in Frank's sense, on larger ones, and these entities either feed the larger crystallites or initiate new cube growth regimes, thereby ensuring further contacts. The morphology of the complex associations established has been elucidated by means of electron diffraction "fine structure," thickness fringes, and growth step patterns.

Extremely fine detail is now observed directly in the micrographs; for instance, "kinks" of the type suggested by Frenkel[26] and further discussed by Dunning[23] may be observed along the growth steps. Dunning comments that these should contribute significantly to the surface energy and constitute preferred sites for adsorption. The identification of adsorbed species is clearly of vital concern, and while their direct observation at the molecular level cannot yet be inferred from present findings, positive observation and identification are believed to be in prospect.

The finest surface detail now observed is tentatively interpreted in terms of "neutral units"[45] on the basis that the standard model for a unit cell, which is derived on the assumption of an infinite lattice, breaks down for surface observations on the present fine scale and is probably replaced by accumulations of uncharged entities, or "neutral units." One such neutral unit could, for instance, consist of four oxygen and four magnesium ions in adjacent (100) "layers" of a normal unit cell. Experimental evidence([28,71]) indicates the O^{--} ion to be stable only in a cage of positive charge, and this makes the model of a neutral unit consisting of only two oxygen and two magnesium ions improbable.

Growth waves of the type envisaged by Frank have been observed to develop from the original points of contact between particles. The necks formed at contact points may consist of higher-index mean planes filling out the reentrants but conforming to the same general growth pattern, so that, for instance, the mean bounding faces of the neck may become largely (110) or even (100). The valley formed between the crystallites then fills under the action of growth waves. Pores generated by both genus 1 and genus 3 contacts in Rhines' sense have been observed to fill by the same mechanism. Sintered junctions not infrequently exhibit contrast characteristic of strain fields surrounding dislocations, and such contrast is found to be consistent with known configurations in magnesia.

The accumulated evidence indicates that surface diffusion is the predominant transport mechanism at temperatures up to 900°C, to which the above observations apply. The very nature of the overall growth process ensures a high probability of neighboring crystallites assuming like orientations and virtually eliminates the possibility of observing well-developed separate cubes since sintering intergrowth starts before decomposition is complete. Because of the stepwise growth process producing tiny (100) facets on the nonequilibrium mean faces the geometric surface area is greatly increased and is much higher, for instance, than that estimated from X-ray line-broadening measurements of mean particle size.

Between 900°C and 1100°C the overall morphology undergoes a marked change. Whereas crystallites produced at 900°C or below have nonequilibrium mean faces based largely on (111) and (110), those produced at 1100°C exhibit the influence of some more rapid and less precise transport process in crystallite shapes more closely approaching cubes but bounded by coarsely stepped faces; vapor phase transport is likely.

Representative micrographs illustrating some important aspects of the foregoing outline are now presented. More detailed commentaries on specific morphological, growth, and sintering features are in preparation. The present micrographs are restricted to 900°C and 1100°C calcines (1 hr in the conditions noted earlier) since the features now described are generally better developed in 900°C calcines than in samples prepared at lower temperatures, and the 1100°C calcines illustrate a striking change in morphology.

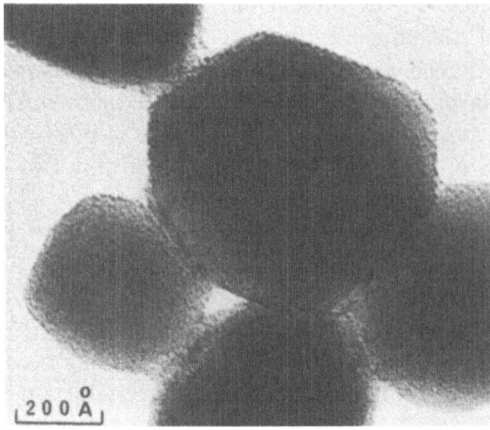

Fig. 6. MgO from A.R. basic magnesium carbonate
(900°C calcine) on conventional carbon support film.

Figures 6–10 are micrographs of samples calcined at 900°C. In Fig.
6 the conventional use of supporting carbon film blots out all but the most
prominent surface detail([45]). Growth steps approaching unit cell size can be
seen in profile on particle edges. One triple junction is essentially closed apart
from a probable narrow crevice terminating at the junction, while another
closure appears to be developing in an adjacent area. Running into this area

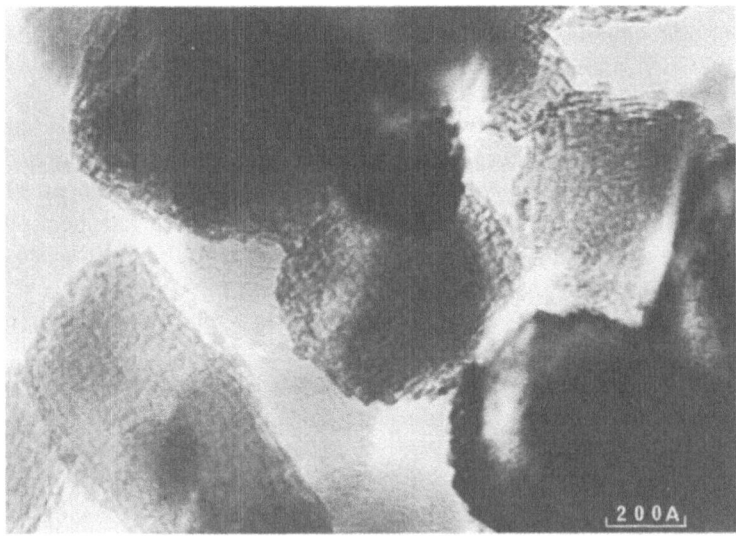

Fig. 7. MgO from A.R. basic magnesium carbonate (900°C calcine) without
carbon support film.

is a low-angle boundary along which a crack approximately two unit cells wide is evident; this is probably due to the misfit of growth steps which are sweeping out from the original contact point. Directly above this feature is a neck or bridge developing between two crystallites; there are indications of stepwise growth in the filling of the reentrants, which superficially appear smoothly curved in this and other appropriate areas of the micrograph.

Figure 7 illustrates the advantage of dispensing with the carbon support film (provided contamination and astigmatism are of a sufficiently low order), thus enabling the stepped detail of the surfaces to be clearly seen. "Terraced" regions are apparent, fitting the pattern of Frank's growth treatment, and the stepped nature of reentrants can now be plainly observed. A portion of a crystallite clearly showing both profile and face detail is depicted in Fig. 8.

The main subject of Fig. 9 is an irregularly shaped particle coalesced from several smaller crystallites. The major surface was identified as (100). Thickness fringes confined to the perimeter are due to wedges formed by (110) faces and indicate the conglomerate to be essentially plate-like rather than blocky. It is unusual to observe platy morphology with (100) as the major surface in this material. Dislocation contrast observed along some junctions is consistent with known configurations in magnesia; a dislocation dipole can be seen near the center of the micrograph. Some of the joins between particles have been achieved without obvious evidence of strain. A crystallite at an advanced stage of growth toward the cube form is in [100] orientation with thickness fringes at the corners due to wedges formed by (110) facets, and appears to have joined near perfectly to a (110) wedge face of the large plate-like particle. It is unlikely that a purely electrostatic attachment in this precise orientation would have survived ultrasonic dispersion of the sample

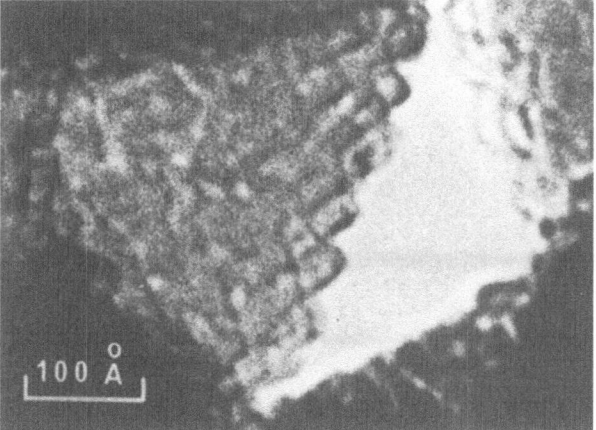

Fig. 8. MgO from A.R. basic magnesium carbonate (900°C cal-cine) showing both profile and face detail.

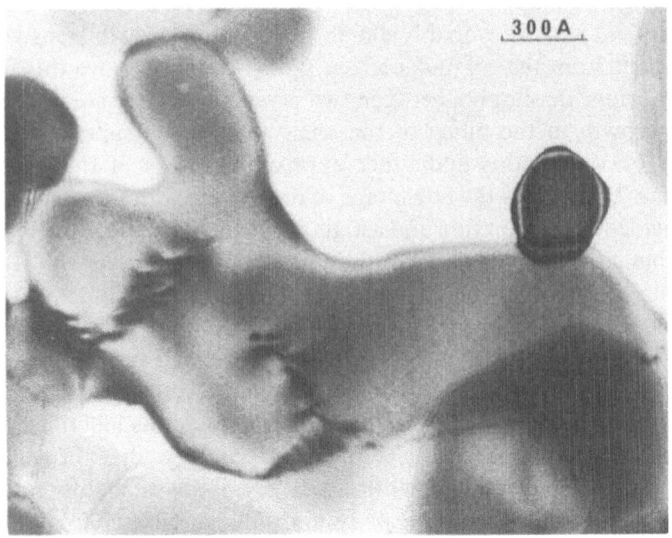

Fig. 9. MgO from A.R. basic magnesium carbonate (900°C calcine) showing irregularly shaped particle coalesced from several smaller particles.

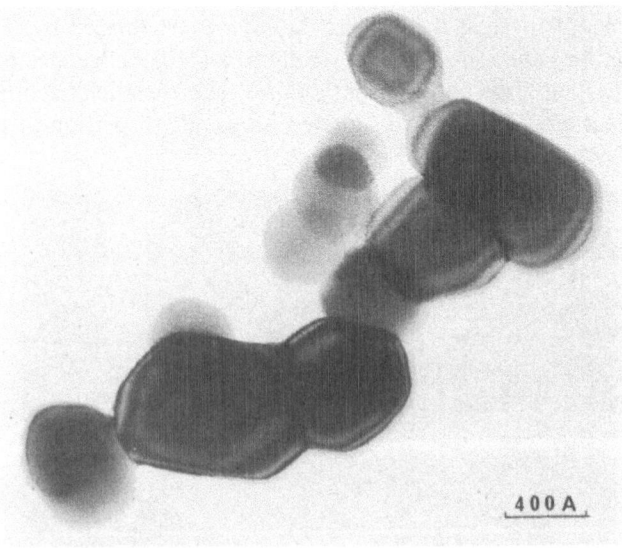

Fig. 10. MgO from A.R. basic magnesium carbonate (900°C calcine)
showing variations in development and degree of perfection of crystallite junctions.

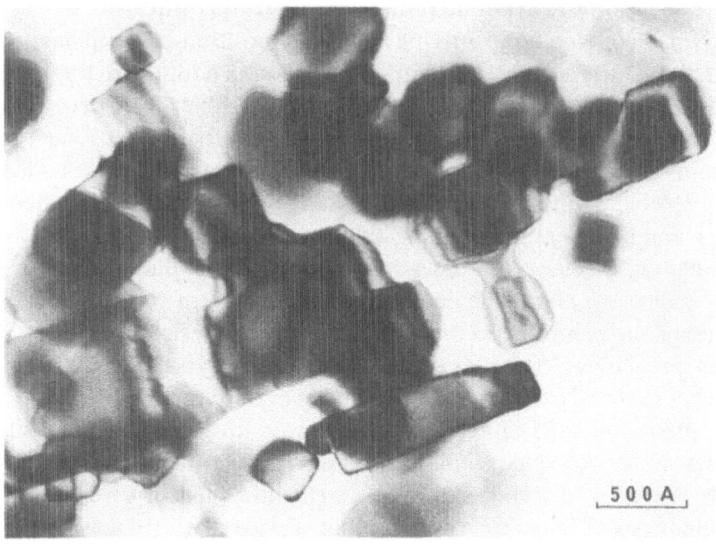

Fig. 11. MgO from A.R. basic magnesium carbonate (1100°C calcine) showing coarse growth steps and predominance of near-cube particles.

in acetone. Figure 10 further illustrates variations in the development and degree of perfection of junctions between several crystallites which are tending toward a blocky character, as indicated by thickness fringes. An interesting feature of this micrograph is the buildup of contamination on the "windward" front of the group, which obscures step-growth profiles. The observation of such profiles on the sheltered perimeter of the group was one of the first direct indications of stepwise growth obtained in this study.

The clusters of attached and intergrown particles in Fig. 11 (1100°C calcine) exhibit marked development of blocky morphology as the crystallites approach crude forms, seemingly under the influence of a less orderly process than the stepwise surface diffusion mechanism indicated to be the main transport system up to 900°C. At temperatures in the vicinity of 1100°C vapor phase transport is considered likely in the calcination conditions employed. The pattern of thickness fringes in several crystallites is characteristic of development toward the cubic form.

Influence of Impurities

Attention has been drawn in the preceding section to the nature of crystallite surfaces, and before presenting further micrographs and comments relating to the presence and influence of impurities it may be appropriate to review briefly some relevant literature. These remarks will also serve as background for decomposition-sintering and decomposition-pressing observations presented in a later section. It is emphasized that, in addition to foreign cation and anion species, it is proper to regard anion residues from decomposition of the precursor compounds themselves as impurities in the product oxides.

Dunning[23] draws attention to the fact that present knowledge of surface structure and properties is based almost entirely on theoretical analysis, and he makes the obvious but very pertinent comment that a better understanding of many processes involving crystal surfaces would result if more were known about the detailed structure of such surfaces. Further, in considering surface-influenced phenomena, he emphasizes that many of the surfaces are "living"; for example, a growing crystal face, a rupturing interface, or a catalyst surface may not be in even quasi-equilibrium. Several of the preceding micrographs go some way toward meeting Dunning's requirement for knowledge of the detailed structure of crystal surfaces, and his reference to living surfaces is well illustrated in several instances. The main thesis of Dunning's paper is that surface structure may be modified in a significant way by the nature of the phase with which it is in contact, and the particular aspect his analysis considers, namely the effect which an adsorbed gas may have on the equilibrium surface, is of special concern in the context of separate and combined compaction and sintering of oxide powders. Observations given later on the next group of micrographs are relevant to this point in a preliminary way, but formidable problems involved in more precise detection and positive identification of adsorbed species have to be overcome before more meaningful observations can be made.

In presenting selected aspects of the physical chemistry of surfaces having special significance in ceramic research, Anderson[1] points out that in powders having specific surface areas greater than $100 \ m^2/g$, around 10% of the atoms reside in the surface, and thus consideration of such materials in terms of the usual "bulk" properties must prove inadequate. He makes the particularly cogent observation that a monolayer of adsorbed gas represents an impurity content in the order of 5 to 10%. Kuhn[40] emphasizes that very fine or "active" powders may readily be deactivated on exposure to gases and vapors that become adsorbed or chemically produce films on the particles, and J. Williams[65] notes that in the case of $200 \ m^2/g$ magnesia prepared from hydroxide some water vapor can still be present to the extent of approximately 3 wt. $\%$ as an adsorbed monolayer; this could equally be true for gases present during the thermal decomposition of other salts, such as carbonates, sulfates and oxalates. Nielsen and Leipold[52] found that dense hot-pressed compacts of magnesia of 99.95% initial purity[42] retained hydroxyl in amounts from 100 to 1000 ppm (atomic) after refiring as high as $2200°C$, even though the samples had undergone high-vacuum baking and ion bombardment in the mass spectrograph in order to reduce surface effects.

In the light of the foregoing comments the claim of Brown ([7]) to have prepared and sintered active magnesia of 99.999 % purity would seem difficult to sustain literally, particularly when the limit of detection of several common magnesia impurities in spectrographic analysis was at least 0.001 %. The matter of anion impurities was not mentioned. However, the important impurities calcium and sodium were, respectively, just on the limit of spectrographic detection (0.0001 %), and below this level as checked by flame photometry. Evidently the material was of exceptionally high purity by ceramic standards, and thus the finding of no measurable grain growth at sintering temperatures up to 1500°C and times up to 100 hr is of patricular interest. By contrast Leipold and Nielsen([43]), claiming a more modest but still creditable 99.95 % as the purity of laboratory-produced magnesia powder, reported significant grain growth in essentially dense compacts hot-pressed in alumina dies at 900°C and higher, and gross coarsening of the grain structure on reheating to 1700°C or more in oxygen. The fluorine levels disclosed by analysis of the starting material and hot-pressed products could be significant in this respect, and the much earlier and more complete densification achieved by hot pressing would provide better opportunity for rapid grain growth than in the relatively porous sintered material. However, the primary purpose in seeking to compare the findings of these two investigations is to indicate that it is no simple matter to assess the influence of impurities, or in this instance the relative lack of them, on anything approaching an absolute basis.

To conclude this short review it is worthy of note that microprobe and optical analysis of the hot-pressed 99.95 % magnesia by Leipold([42]) revealed appreciable segregation of the impurities silicon, aluminum, and calcium at grain boundaries after reheating in air at 2200°C. The degree of segregation was such as to appear inconsistent with bulk analysis and reported solubility data. Iron was not observed to segregate, and the calcium and silicon on the one hand and aluminum on the other formed distinct and apparently mutually incompatible phases with magnesia. Thus grain boundaries, like surfaces, are indicated to have a special affinity for impurities; in the case of grain boundaries a degree of selectivity is noted, influenced by factors such as relative ease of accommodation in the bulk lattice and compound-forming opportunities.

Reference to Table I shows that while the A. R. basic magnesium carbonate has some reasonable claim to purity, the technical grade material is distinctly inferior on the basis of the limited data available, particularly in respect to the levels of lime, soda, and foreign anion species. Thus the A. R. basic carbonate has been used almost exclusively in the present investigation, but some earlier observations on the lightly calcined technical grade material are relevant to the issue of the influence of impurities. Figure 12 depicts a pseudomorph obtained by calcination at 550°C for 1 hr reminiscent of Fig. 2, which illustrated the pseudomorphed condition of the A. R. material when

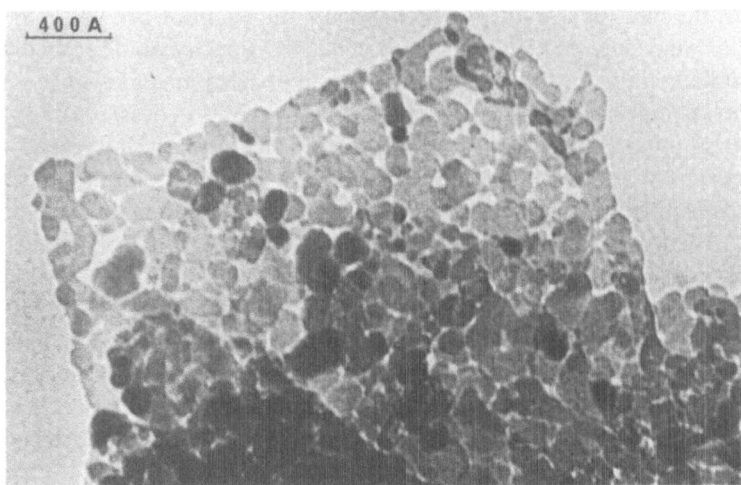

Fig. 12. Pseudomorphed MgO (550°C calcine) from technical grade basic magnesium carbonate showing extensively interjoined plate-like crystallites.

similarly calcined. The crystallite sizes and shapes are generally not too dissimilar. However, the technical grade basic carbonate forms relic structures which are typically less "open," with the joins between individual crystallites tending to be substantial bridges rather than the fine necks observed in the A. R. product. This difference is thought to be impurity-influenced, and the more substantial nature of the interparticle junctions in the technical grade pseudomorphs can be demonstrated by flexing behavior when the intensity

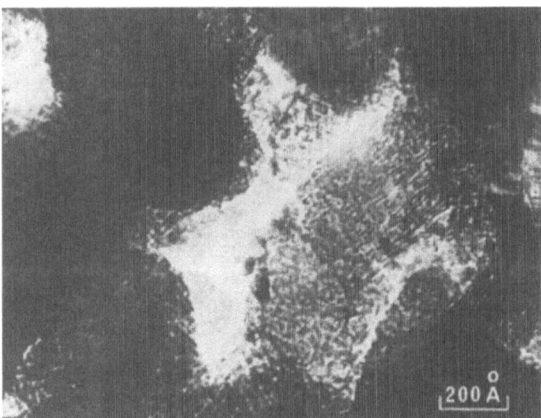

Fig. 13. MgO from A.R. basic magnesium carbonate (900°C calcine) showing unidentified adsorbed species on centrally located crystallites.

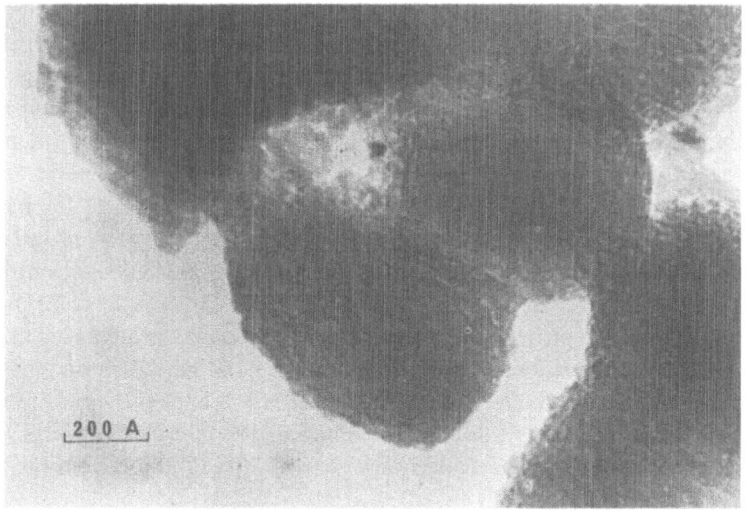

Fig. 14. MgO from A.R. basic magnesium carbonate (900°C calcine) showing impurities concentrated at grain junctions.

of the electron beam is varied; the tighter pseudomorphs exemplified in Fig. 12 tend to flex as single entities, whereas the looser relics shown in Fig. 2 "twinkle" in the beam as the crystallites respond individually to the varying intensity. Thus the A. R. product can be expected to exhibit a closer approach to individual crystallite behavior during fabrication and sintering.

The dark spots shown on crystallite surfaces in Fig. 13 (A. R. basic carbonate, 900°C calcine) are unidentified adsorbed species, whereas in the similarly prepared sample of Fig. 14 the impurities could either have concentrated at the grain junctions from the bulk material or been swept into the junctions by surface diffusional flow. Finally, referring back to Fig. 6, one of three surfaces defining the triangular gap at the developing triple junction appears inactive by comparison with the two "flowing" wedges adjoining it, and alternative interpretations are that this surface has been "poisoned" by some adsorbed species or is simply well developed crystallographically and therefore close to equilibrium.

SINTERING AND HOT-PRESSING STUDY INVOLVING DECOMPOSITION

Purpose of Study

This phase of the investigation was intended to encompass only selected experiments developed from the foregoing characterizing observations, with continuing emphasis on precursor–oxide relationships and unequilibrated "active" products of decomposition reactions. Experimental conditions have been chosen to illustrate and exaggerate problems in some instances, as well as to indicate progress in effectively employing representative precursor compounds and active oxide products of their decomposition. Although the pattern of these experiments has not remained quite as selective and definitive as was intended, the "basic phenomenology" philosophy of the overall investigation has been preserved, and certainly there has been no attempt to collect data for kinetic analysis nor to make sophisticated deductions from the experimental findings. However, it is felt that the type of information obtained can help to provide the firmer basis required for the advanced kinetic and mechanistic treatments that have dominated the literature on sintering in recent years, and which in some instances are seemingly in advance of adequate experimentation and detailed observation. A logical starting point is to present findings of direct decomposition-sintering and decomposition-pressing experiments.

Decomposition-Sintering and Decomposition-Pressing without Additives

The use of active oxide powders, both in pre-calcined form and obtained by decomposition *in situ* of all or part of a hot-pressing charge, has been a continuing interest of our group for several years (e.g., L. S. Williams[69], also earlier unpublished work with refractory oxides). The potential of the approach of combining decomposition and hot pressing in the same fabrication cycle has been well recognized in more recent investigations by two separate schools [Chaklader[9], Chaklader and McKenzie[10]; Morgan and Scala[47,48], Morgan and Schaeffer[49]]. Spriggs and Atteraas[61] discuss these developments along with other novel pressure-sintering techniques.

Having observed the oriented conversion of basic magnesium carbonate to magnesia in the present investigation, it seemed reasonable that a compact of basic magnesium carbonate could be converted to magnesia by decomposition-sintering without cracking or undue distortion and without the benefit of pressure during a properly designed heating cycle. Hepworth and Rutherford[33] had prepared high-density calcium oxide shapes directly from slip-cast calcium hydroxide, but had been less successful with magnesium hydroxide. In the present instance A. R. basic magnesium carbonate casting dispersions were prepared in non-aqueous vehicles after the fashion of earlier work with calcium oxide slips[70]. The cast shapes, mainly in the form of crucibles and rods, were decomposition-sintered to densities approximately 95 % of theoretical* without cracking and with negligible distortion despite

---

*Theoretical density for MgO taken as 3.58 g/cm^3.

linear shrinkage of the order of 60%. A sintering cycle of 24 hr was employed, with arrests or very slow heating rates through critical decomposition and recrystallization stages, and a soak of several hours at the final temperature of 1500°C. This temperature was adopted as standard for various magnesia sintering and refiring treatments, being sufficient to densify "active" material without excessive grain growth and adequate to eliminate most residues and/or show up reheat-bloating problems.

The next point of interest was to apply the successful decomposition-sintering procedure to relatively chunky samples, typically end-pressed cyclindrical compacts $\frac{5}{8}''$ to $1''$ diameter and of height around one-third to one-half of the diameter. The basic carbonate formed strong compacts at moderate pressure in the range 13,000 to 20,000 psi. The compacts achieved relative densities of only 32 to 35%, but with due care they survived decomposition and sintering strains without apparent damage, with sintered densities again at or near 95% of theoretical. Visual observation during the heating cycle indicated that sintering shrinkage was not appreciable below 900°C. Optical examination revealed a fairly uniform microstructure with a median grain size of 25 to 30 μm and well distributed porosity largely at the grain boundaries. The characteristic of slot-like porosity lying along the grain boundaries distinguished the decomposition-sintered material from the common run of moderately well-sintered magnesia exhibiting porosity concentration at and near grain junctions.

Precipitated hydroxide similarly compacted and decomposition-sintered attained a density of only 64% of theoretical, despite the initial advantages of much higher relative density as-compacted (approaching 60%) and a reduction of only 35% in molal volume from the hydroxide to the oxide. This result is in keeping with assessment of the relative inherent sinterability of the basic carbonate and the hydroxide from examination of the pseudomorphed structures of the calcined materials.

Decomposition-pressing experiments were in two categories: (1) preliminary treatments, mostly in graphite pressing dies at 3300 psi, to attain relative densities in the vicinity of 90%, followed by sintering at 1500°C for 40 min to achieve further densification; and (2) pressure-densification at higher temperatures (in graphite dies at 3300 psi) and higher pressure (19,000 psi in a titanium carbide die) to high relative densities followed by reheating at 1500°C for 40 min in order to observe bloating and grain-growth characteristics. Times and temperatures at the decomposition stage were generally minimal, favoring sinterability in the product oxide but tending to introduce problems of incomplete decomposition and retention of decomposition residues.

The electron replica micrographs of Fig. 15(a) and (b) depict relatively dense areas of a compact pressed at 900°C in vacuum in a graphite die after

Fig. 15. Basic magnesium carbonate decomposition-pressed at 900°C in vacuum showing relic structures, well developed crystallites, and porosity.

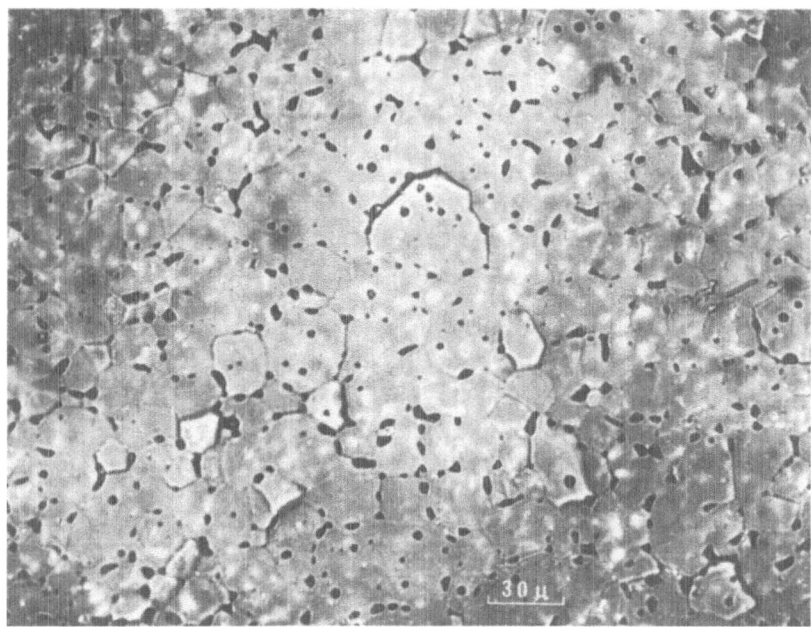

Fig. 16. Basic magnesium carbonate decomposition-pressed at 1000°C and refired to 1500°C showing pore morphology typical of reheated material.

decomposition *in situ* at 600°C for 1 hr. The overall density of the compact was 88% of theoretical. Various basic carbonate relics can be seen, notably "asterisk" clusters in Figure 15(*a*) and several stacks and plates in 15(*b*). Well developed crystallites around 1 μm in size are embedded in a matrix which appears relatively structureless apart from pseudomorph traces and fine pores. The small crystallites exhibit definite crystallographic forms, and some are evidently oriented on body diagonals.

The compact of Fig. 16 was decomposition-pressed in graphite to a finishing temperature of 1000°C but without the benefit of vacuum, and was then refired at 1500°C, improving its relative density from 92% to 96%. The grain boundary perforations and rounded pores within grains are typical of pore morphology in reheated material. The possibility that pores were enlarged and rounded during polishing was investigated, but such effects were shown to be slight. Figure 17 provides an interesting comparison; in this case the basic carbonate was predecomposed and then conventionally hot-presssed at 1000°C before refiring at 1500°C. The density was 2% lower at the pressing stage but attained the same final figure of 96%. The average grain size is markedly lower, around 10 to 15 μm. The coarsening effect of confinement in the die during decomposition and of application of pressure during the early stages of recrystallization and growth is thus indicated in the former case.

Increasing the finishing temperature of vacuum decomposition-pressing in graphite to 1200°C brought about an improvement in relative density to 99%, but the problem of decomposition residues is illustrated by Fig. 18, showing extensive grain boundary erosion in the fine-grained product during polishing in an aqueous medium. This problem was not encountered when a similarly prepared compact was refired at 1500°C before polishing.

The effect of a large increase in pressure (to 19,000 psi) was assessed by decomposition-pressing in a laboratory-produced bonded titanium carbide die. With finishing temperatures of 900° and 1000°C relative densities were variable in the range 94 to 97% and were seemingly very sensitive to minor variations in the decomposition stage of the cycle. An extreme comparison is provided by the electron replica micrographs of Fig. 19(a) and (b). The specimens were both pressed in vacuum at 1000°C, but decomposed for 1 hr at 600° and 300°C, respectively. Well developed crystallites of the order of 1 μm are apparent in Fig. 19(a), with some evidence of fine porosity, but the material of Figure 19(b) appears to be poorly recrystallized and there is abundant evidence of fine porosity, most probably caused by trapped decom-

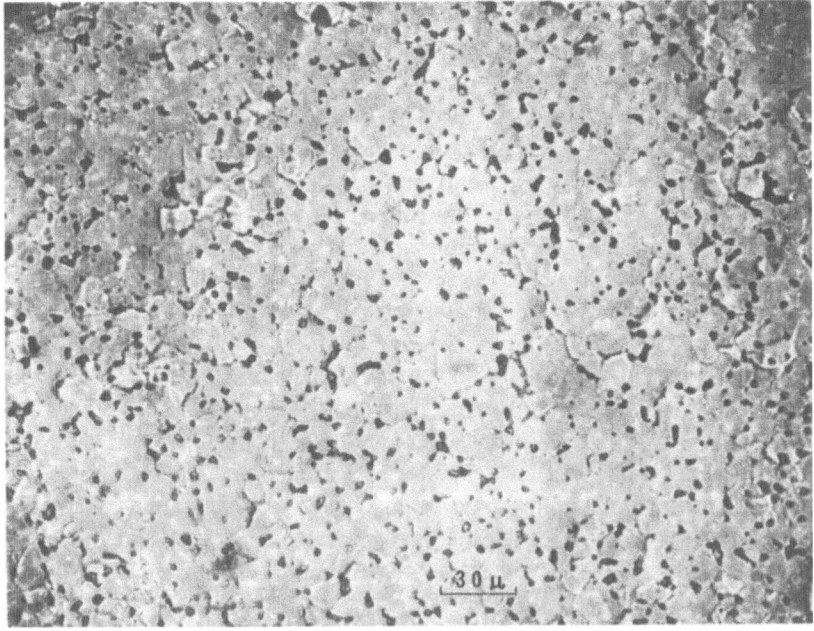

Fig. 17. MgO derived from basic magnesium carbonate conventionally hot-pressed at 1000°C and refired to 1500°C showing finer grain structure than decomposition-pressed material.

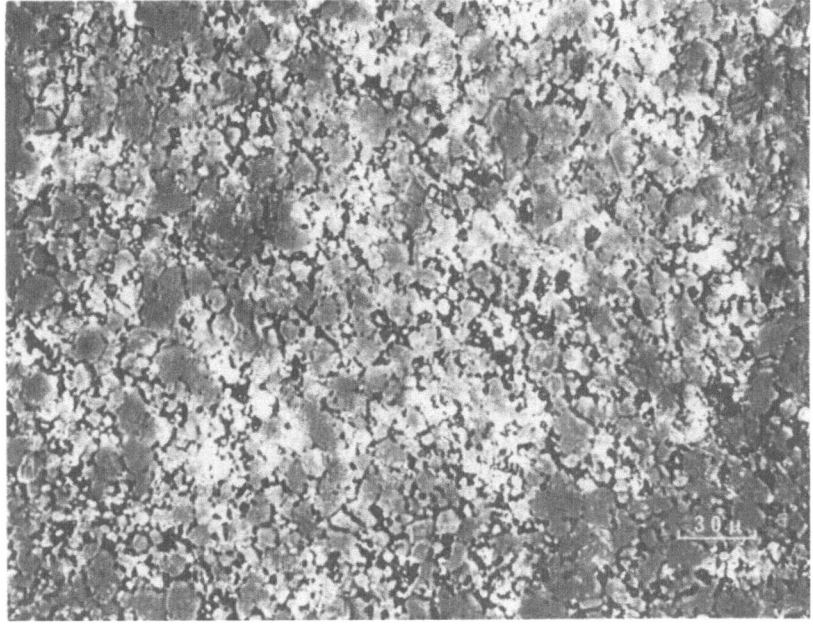

Fig. 18. Basic magnesium carbonate decomposition-sintered at 1200°C in vacuum showing grain boundary erosion from polishing.

position products. Such material invariably bloats on reheating to 1500°C, as evidenced by loss of density and occasional blistering, whereas material which has been more adequately decomposed before the hot-pressing stage of the cycle commonly densifies to 98% of theoretical on reheating to 1500°C.

The superior sinterability of magnesia derived from basic magnesium carbonate is again indicated by the fact that neither the precipitated hydroxide nor the rehydrated oxide could be decomposition-pressed to relative densities appreciably higher than 50% at 900° and 1000°C in graphite dies, but increasing the pressure to 19,000 psi in vacuum in the titanium carbide die achieved a dramatic increase in density to better than 98% of theoretical.* However, these samples suffered a small decrease in density on reheating to 1500°C, as indicated by the pore phase in Fig. 20. The average grain size around 15 to 20 μm is slightly smaller than that generally observed in similarly treated compacts obtained from basic magnesium carbonate.

Direct decomposition-sintering and decomposition-pressing of alumina

---

*Morgan and Schaeffer[49] observed broadly comparable pressure dependence in the course of a more sophisticated decomposition-pressing ("pressure calcintering") study of the $Mg(OH)_2$–MgO pair, but also noted the sensitivity of the process to impurity levels and various aspects of precursor preparation and pretreatment, and obtained excellent results with due attention to these factors.

a

b

Fig. 19. Basic magnesium carbonate decomposition-pressed at 1000°C in vacuum showing effect of holding temperature at decomposition stage. Holding temperature: (a) 600°C and (b) 300°C.

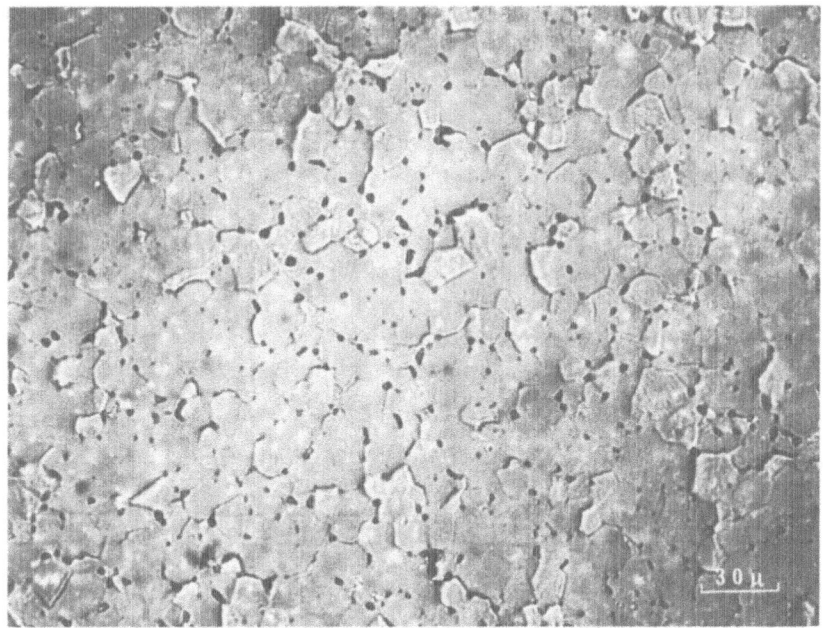

Fig. 20. Rehydrated MgO decomposition-pressed at 1000°C in vacuum and refired to 1500°C showing pore morphology typical of reheated material.

from basic aluminum sulfate appeared unpromising in terms of densification, and retention of sulfate residues was particularly troublesome in the case of decomposition-pressing. Decomposition-sintered compacts were crack-free and strong, but densities were little above 70% of theoretical with a final treatment of 2 hr at 1550°C.

Effect of Additives and Additional Treatments

Following Rice[57,58] lithium fluoride additions were employed in seeking to improve densification and achieve transparency in the magnesia products. Rice noted the lubricating effect of LiF in facilitating particle movement and rearrangement, and hence compaction at moderate temperatures; he also considered the contribution of LiF to continued sintering during the post-heating stage of his process, listing formation of defects, particularly in the grain boundary region, as one likely factor. Benecke et al.[5] worked from the standpoint of pressure-dependent processes in seeking to explain how LiF affects densification after initial compaction has occurred in the hot-pressing stage of the Rice process.

In considering the experimental findings of Beneck et al. two points of particular interest were noted: (1) the rapid acceleration of densification approaching 600°C and (2) the lack of any evident discontinuity in the lowest-pressure densification curve around the melting point of LiF (given as 846°C) although the density was still 5% short of theoretical at this stage. Accordingly the *modus operandi* of LiF additions is being considered further. Lack of complete thermodynamic and phase-equilibrium data invalidates any attempt at definitive comment at the time of writing, but the existence of a eutectic in the LiF–Li$_2$CO$_3$ system near 600°C is considered significant. Feasible reactions between LiF, MgO, residual carbonate, and other cation and anion impurities open up a range of possibilities for the formation of a complex low-melting phase, and may also have an important bearing on the matter of elimination of lithium and fluorine during the post-heating stage of the Rice process. (More generally, association of carbonate and other anion species with impurity cations such as calcium and sodium could well be an important factor in densification and reheat bloating problems in magnesia fabrication; the literature reveals an unfortunate tendency to restrict consideration to the precursor anion and ignore impurity anions.)

It was anticipated that rather more than the 2 wt.% of LiF commonly used with conventional MgO powders in the Rice process could be required to promote near-theoretical densification during decomposition-sintering

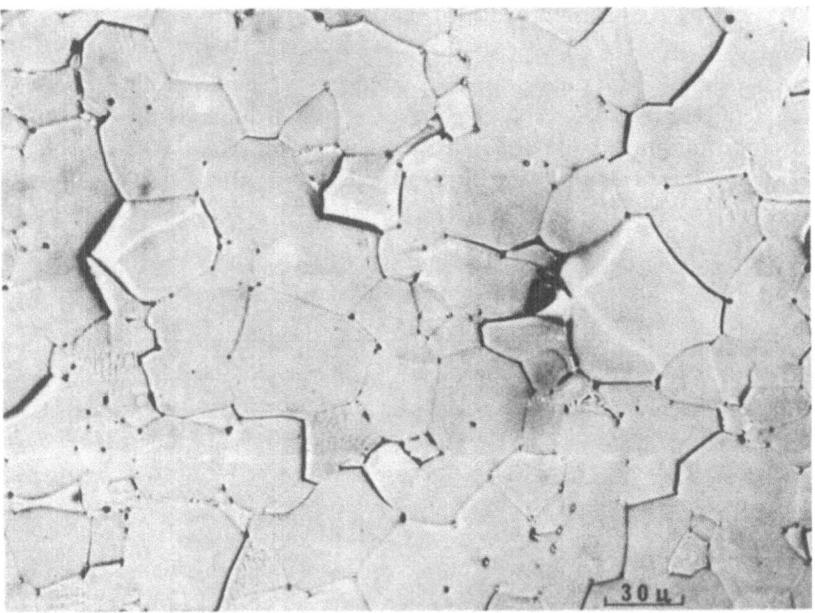

Fig. 21. Basic magnesium carbonate containing 6 wt.% LiF decomposition-sintered to 1200°C.

Fig. 22. Basic magnesium carbonate containing 6 wt.% LiF decomposition-sintered to 1500°C.

from the basic carbonate. Accordingly two batches of basic magnesium carbonate compacts containing 0.5, 2.0, and 6.0 wt.% of LiF introduced by mechanical blending in isopropanol were decomposition-sintered to final conditions of 4 hr at 1200°C and 40 min at 1500°C. The relative densities of the 1200°C compacts were 46.1%, 97.1%, and 98.1%, and those of the 1500°C compacts 62.9%, 97.4%, and 98.1%, respectively. Thus an optimum level of addition above 2 wt.% is indicated for the materials and processing conditions used. Highly translucent areas were observed in the higher density products. It was particularly interesting to observe that the rate of densification of the compacts containing 6 wt.% LiF, and to a lesser degree those with 2 wt.%, increased markedly around 600°C by comparison with the earlier observation of a similar increase around 900°C for plain basic carbonate compacts. Figure 21 depicts a 1200°C product which originally contained 6 wt.% LiF, and in Fig. 22 the same starting composition has been taken to 1500°C. The marked coarsening of the grains in these two cases indicates that the lithium fluoride phase, probably considerably adulterated as noted earlier, constitutes a very effective medium for grain growth.

Compacts of precipitated hydroxide densified quite well during decomposition-pressing with the aid of 2 wt.% of LiF, but elimination of the fluoride-

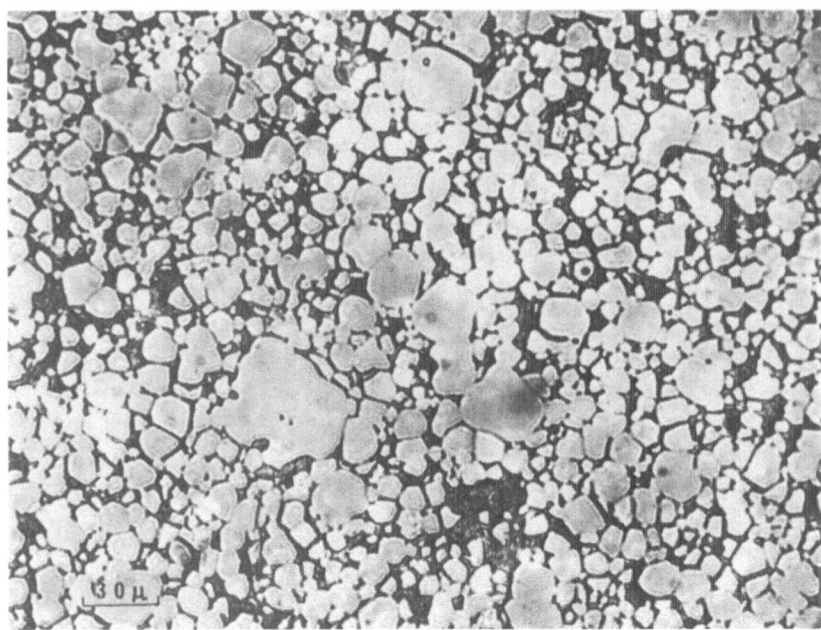

Fig. 23. Precipitated magnesium hydroxide containing 2 wt.% LiF decomposition-sintered to 1150°C showing rounding of grains and erosion of grain boundary phase.

containing phase proved to be more of a problem than in the case of the basic carbonate, which again indicates that carbonate may play an important part in the fluoride-densification process. Figure 23 illustrates an extreme manifestation of the problem in the case of a compact decomposition-sintered from precipitated hydroxide containing 2 wt.% LiF, the final heating stage being 4 hr at 1150°C. Note the marked rounding of the grains and extensive erosion of the grain boundary phase. The as-sintered density of the compact was 96.0% of theoretical.

Decomposition-pressing of basic carbonate containing 6 wt.% of LiF followed by reheating in air produced essentially dense transparent material, typified by the micrograph of Fig. 24. The compact had been reheated for 4 hr at 1200°C after decomposition-pressing at 900°C in a graphite die, and attained a relative density above 99.8%. With 2 wt.% of LiF both basic carbonate and precipitated hydroxide could be decomposition-pressed at 900°C and reheated at 1200°C to yield somewhat blotchy compacts with transparent peripheral zones; final densities were 99.4% and 99.2% of theoretical, respectively. Figures 25 and 26 are representative micrographs.

The densities reported throughout are overall figures, determined to within ±0.1% by mercury volume-meter ([62]) and/or flotation in Clerici solu-

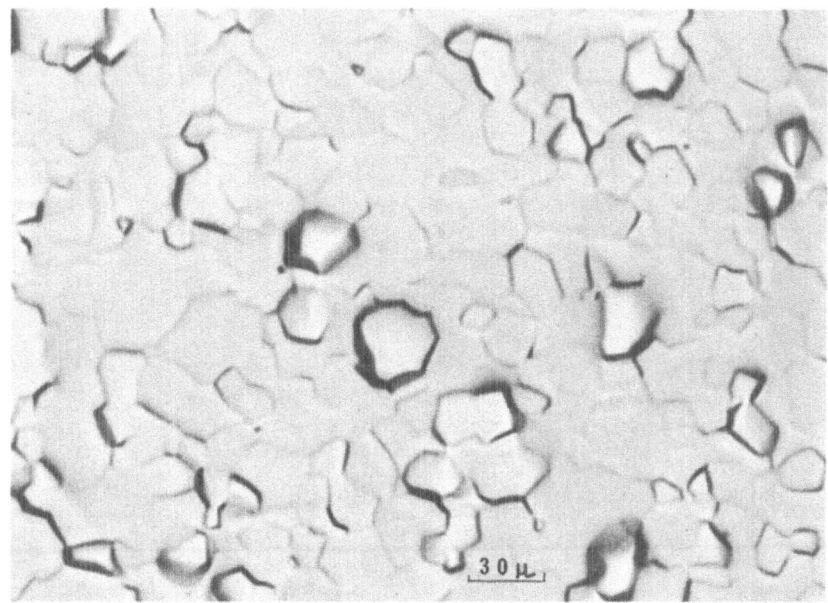

Fig. 24. Basic magnesium carbonate containing 6 wt.% LiF decomposition-pressed at 900°C and reheated to 1200°C. Compact was transparent.

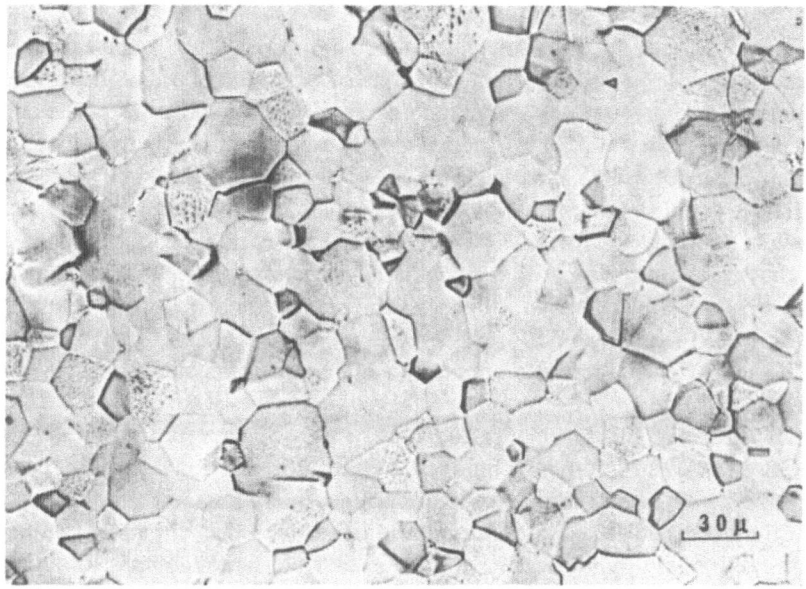

Fig. 25. Basic magnesium carbonate containing 2 wt.% LiF decomposition-pressed at 900°C and reheated to 1200°C. Compact was partially transparent.

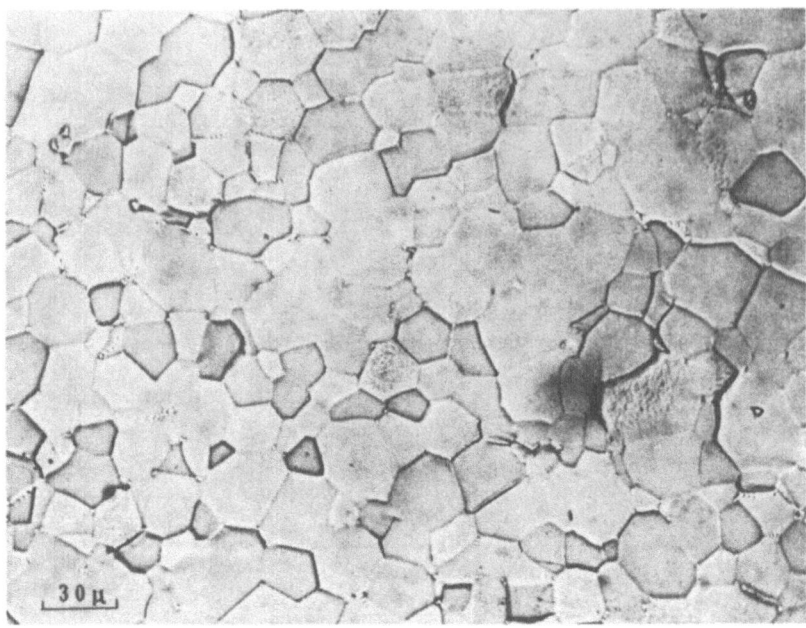

Fig. 26. Precipitated magnesium hydroxide containing 2 wt.% LiF decomposition-pressed at 900°C and reheated to 1200°C. Compact was partially transparent.

tion([64]); the latter technique is particularly appropriate for small specimens sintered to the closed-pore stage. In several instances where the overall density was appreciably less than theoretical quite large portions of the compacts, particularly with LiF present, were transparent or markedly translucent, indicating that more detailed study of materials factors and process variables should yield essentially dense products in these cases. For example, there is ample scope for profitable variation in the method of adding LiF; in the work reported relatively coarse as-received powder was introduced by mechanical blending in isopropanol. Improvement can very likely be made both in homogeneity and in reducing the quantity of LiF required.

The matters of increased orientation and compaction of basic magnesium carbonate flakes in end-pressed specimens were investigated briefly. Differential shrinkage measurements and observation of preferred orientation in fracture surfaces by X-ray techniques showed that additional pressing cycles (up to 50) increased the degree of preferred orientation, but that methanol additions reduced the extent of orientation, although they enhanced compaction, presumably by lubricating complete stacks of flakes and assisting them to pack better by sliding and rotation. However, the extremes of exposure to a damp atmosphere (90% R.H.) and vacuum drying both markedly increased preferred orientation as well as improving compaction; the damp atmosphere treatment was particularly effective in the latter respect. Thus it is inferred that either substantial removal of water molecules from the contacting surfaces of the flakes or introduction of sufficient water to lubricate the faces increased the tendency of the flakes to act as individual entities during compaction instead of as relatively equiaxed stacks. This phase of the work impinges on the objective of introducing an appreciable degree of preferred crystallographic orientation into sintered or hot-pressed compacts.

Attempts to obtain dense, fine-grained alumina compacts by direct decomposition-sintering or decomposition-pressing from basic aluminum sulfate with the aid of various fluoride additives have not been successful. However, the more complex pattern of the precursor–oxide relationships in this case affords the opportunity of achieving the desired result by combining the effects of simple preliminary treatments and the occurrence of the η- to α-Al_2O_3 transformation during hot pressing. Ultrafine η-Al_2O_3 was prepared by thermal decomposition of the basic sulfate and was water-washed for removal of residual sulfate until the remaining liquor gave no reaction to barium chloride. The dried product, in some cases recalcined to eliminate hydroxide, was hot pressed at 3300 psi for periods of $\frac{1}{2}$ to 2 hr at temperatures in the range 1350–1500°C. Relative densities varied from around 98% to better than 99%, the higher densities being obtained when the washed material was not recalcined, and thus the reactive effect of decomposing a small amount of rehydrated material *in situ* was added to the hot-pressing treatment. The predominant grain size varied with material preparation and hot-pressing conditions from around 2 to 10 μm, but occasional individual or grouped large columnar grains suggested that sulfate traces had not been completely eliminated; this would not be unexpected in view of the comments given earlier, in particular on pp. 65–66, on stereochemical aspects of retention of sulfate residues.

Utilization of the η-Al_2O_3 product of calcination of the basic sulfate warrants fuller investigation, and the particular problem of eliminating the final sulfate traces may be expected to yield to appropriate treatment with hot gases, notably hydrogen.

CONCLUSION

It is believed that a greater investment of effort in "detailed phenomenology" of fine oxide powders and precursor–product relationships will return a handsome dividend in terms of both fundamental knowledge and improved understanding and control of fabrication variables. The combination of high-resolution electron microscopy and selected-area diffraction constitutes a powerful characterizing tool. In the particular context of this volume phenomenological observations of the type presented should help to provide a sounder basis for kinetic appraisal of fundamental processes in oxides.

ACKNOWLEDGMENTS

The writers are particularly appreciative of the guidance of Mr. A. F. Moodie of Division of Chemical Physics, Chemical Research Laboratories, C.S.I.R.O., whose skill in interpreting electron micrographs and electron diffraction data has contributed vitally to the progress of this investigation. Two of us (C.E.W. and L.S.W.) wish to express gratitude to the Army Research Office, Durham, for travel grants to attend this Conference.

REFERENCES

1. P. J. Anderson, *J. Brit. Ceram. Soc.* **3**(3): 423 (1966).
2. P. J. Anderson and R. F. Horlock, *Trans. Faraday Soc.* **58**(478): 1993 (1962).
3. P. J. Anderson and D. T. Livey, *Powder Met.* **7**: 189 (1961).
4. R. R. Balmbra, J. S. Clunie, and J. F. Goodman, *Nature* **209**(5028): 1083 (1966).
5. M. W. Benecke, N. E. Olson, and J. A. Pask, *J. Am. Ceram. Soc.*, **50**(7): 365 (1967).
6. G. W. Brindley, *Progress in Ceramic Science*, Vol. 3 (J. E. Burke, ed.), Pergamon Press, Oxford, 1963, p. 1.
7. R. A. Brown, *Bull. Am. Ceram. Soc.* **44**(6): 483 (1965).
8. J. E. Burke, *J. Brit. Ceram. Soc.* **3**(1): 5 (1966).
9. A. C. D. Chaklader, *Nature* **206**(4982): 392 (1965).
10. A. C. D. Chaklader and L. G. McKenzie, *J. Am. Ceram. Soc.* **49**(9): 477 (1966).
11. P. W. Clark and J. White, *Trans. Brit. Ceram. Soc.* **49**: 305 (1950).
12. R. L. Coble, *J. Am. Ceram. Soc.* **41**(2): 55 (1958).
13. R. L. Coble, *J. Appl. Phys.* **32**(5): 787, 793 (1961).
14. R. L. Coble, and J. E. Burke, *Progress in Ceramic Science*, Vol. 3 (J. E. Burke, ed.), Pergamon Press, Oxford, 1963, p. 197.
15. J. M. Cowley, *Acta Cryst.* **6**: 846 (1953).
16. J. M. Cowley, *Acta Cryst.* **9**: 391 (1956).
17. J. M. Cowley, P. Goodman, and A. L. G. Rees, *Acta Cryst.* **10**(1): 19 (1957).
18. I. A. Cutter, J. H. Hensler, and G. V. Cullen, "Sintering of Magnesium Oxide," presented at Eighth Commonwealth Mining and Metallurgical Congress, Australia, preprint No. 62, 1965.
19. J. D. Dana, *System of Mineralogy*, Vol. II (C. Palache, H. Berman, and C. Frondel, eds.), J. Wiley and Sons, New York and London, 1963, p. 271.
20. P. T. Davey, G. M. Lukaszewski, and T. R. Scott, *Australian J. Appl. Sci.* **14**(2): 137 (1963).
21. R. M. Dell and S. W. Weller, *Trans. Faraday Soc.* **55**: 2203 (1959).
22. J. Donnay, *Crystal Data: Determinative Tables*, 2nd Ed., American Crystallographic Assoc., New York, 1963.
23. W. J. Dunning, *Proc. Brit. Ceram. Soc.* **5**(Dec.): 59 (1965).
24. W. R. Eubank, *J. Am. Ceram. Soc.* **37**(8): 225 (1951).
25. F. C. Frank, *Growth and Perfection of Crystals* (R. H. Doremus, B. W. Roberts, and D. Turnbull, eds.), J. Wiley and Sons, New York, 1958, p. 411.

26. J. Frenkel, *J. Phys. USSR* **9**: 392 (1945).
27. J. F. Goodman, *Proc. Roy. Soc. (London) Ser. A* **247**(1250): 346 (1958).
28. P. Goodman and G. Lehmpfuhl, *Acta Cryst.* **22**(1): 14 (1967).
29. R. S. Gordon and W. D. Kingery, *J. Am. Ceram. Soc.* **49**(12): 654 (1966).
30. R. S. Gordon and W. D. Kingery, *J. Am. Ceram. Soc.* **50**(1): 8 (1967).
31. K. Hamano, *J. Ceram. Assoc. Japan* **74**(5): 144 (1966).
32. S. B. Hendricks, *Am. Mineral.* **22**: 773 (1937).
33. M. A. Hepworth and J. Rutherford, *Bull. Am. Ceram. Soc.* **43**(1): 18 (1964).
34. P. B. Hirsch, A. Howie, R. B. Nicholson, D. W. Pashley, and M. J. Whelan, *Electron Microscopy of Thin Crystals*, Butterworth and Co., London, 1965.
35. C. Hyde and W. H. Duckworth, "Investigation of Sinterable Oxide Powders and Ceramics Made from Them," WADD Technical Report 61–262 (1961).
36. D. L. Johnson and I. B. Cutler, *J. Am. Ceram. Soc.* **46**(11): 541, 545 (1963).
37. W. D. Kingery and M. Berg, *J. Appl. Phys.* **26**(10): 1205 (1955).
38. D. W. Koester, E. B. Cornelius, and J. J. Donovan, "Investigation of Physical Parameters of Non-Electrically Conducting Fine Particulates," Technical Documentary Report No. ASD–TDR–62–843 (1963).
39. G. C. Kuczynski, *J. Metals*, **1**: 169 (1949).
40. G. C. Kuczynski, in: *Powder Metallurgy* (W. Leszynski, ed.), Interscience, New York, 1961, p. 11.
41. W. E. Kuhn, in: *Ultrafine Particles* (W. E. Kuhn, ed.), John Wiley and Sons New York, 1963, p. 104.
42. M. H. Leipold, *J. Am. Ceram. Soc.* **49**(9): 498 (1966).
43. M. H. Leipold, and T. H. Nielson, *Bull. Am. Ceram. Soc.* **45**(3): 281 (1966).
44. D. T. Livey, B. M. Wanklyn, M. Hewitt, and P. Murray, *Trans. Brit. Ceram. Soc.* **56**: 217 (1957).
45. A. F. Moodie and C. E. Warble, *Phil. Mag.*, **16**(143): 891 (1967).
46. A. F. Moodie, C. E. Warble, and L. S. Williams, *J. Am. Ceram. Soc.* **49**(12): 676 (1966).
47. P. E. D. Morgan and E. Scala, "High Density Oxides by Decomposition Pressure Sintering of Hydroxides," presented at 67th Annual Meeting, The American Ceramic Society, Philadelphia, May 3, 1965.
48. P. E. D. Morgan and E. Scala, "Production of Refractories by Pressure Calcintering," presented at 68th Annual Meeting of American Ceramic Society, Washington, D.C., May 11, 1966.
49. P. E. D. Morgan and N. C. Schaeffer, "Chemically Activated Pressure Sintering of Oxides," AFML–TR–66–356 (1966); Contract No. AF 33 (615)–3065.
50. P. Murray, in: *Agglomeration* (W. A. Knepper, ed.), Interscience, New York, 1962, p. 93.
51. P. Murray, *J. Brit. Ceram. Soc.*, No. 1 (October) p. 113 (1963).
52. T. H. Nielson and M. H. Leipold, *J. Am. Ceram. Soc.* **49**(11): 626 (1966).
53. H. J. Oel, in: *Agglomeration* (W. A. Knepper, ed.), Interscience, New York, 1962, p. 271.
54. K. O. Reeve, *J. Australian Ceram. Soc.* **2**(2): 38 (1966).
55. F. N. Rhines, in: *Plansee Proceedings* (F. Benesovsky, ed.), Pergamon Press, 1958 (Metallwerk Plansee AG, Reutte/Tyrol, 1959), p. 38.
56. F. N. Rhines, Annual Progress Report, Metallurgical Research Laboratory of the Engineering and Industrial Experiment Station, University of Florida, Gainesville, Florida, 1966.
57. R. W. Rice, "Production of Transparent MgO at Moderate Temperatures and Pressures," presented at 64th Annual Meeting, The American Ceramic Society, New York, New York, April 30, 1962.

58. R. W. Rice, "The Role of Grain Boundaries and Surfaces in Ceramics," in: *Materials Science Research*, Vol. 3 (W. W. Kriegel and H. Palmour III eds., Plenum Press, New York, 1966, p. 387.

59. P. Royen and M. Trömel, *Berichte der Bunsengesellschaft für Physikalische Chemie* 67(9/10): 908 (1963).

60. T. R. Scott, *J. Metals* 14: 121 (1962).

61. R. M. Spriggs and L. Atteraas "Densification of Single Phase Systems Under Pressure," presented at 3rd International Materials Symposium on Ceramic Microstructures— Their Analysis, Significance and Production, held at Univ. of Calif., Berkeley, Calif., June 13–16, 1966.

62. R. K. Stringer, *J. Australian Ceram. Soc.* 2(2, Nov.): 47 (1966).

63. L. Sturkey and L. K. Frevel, *Phys. Rev.* 68: 56 (1945).

64. H. E. Vassar, *Am. Mineral.* 10: 123 (1925).

65. A. F. Wells, *Structural Inorganic Chemistry*, 3rd Ed. Oxford University Press, London, 1962, p. 561.

66. J. White, *Proc. Brit. Ceram. Soc.* No. 3, Sept., p. 155 (1965).

67. J. Williams *Science of Ceramics*, Vol. 2 (G. H. Stewart, ed.), Academic Press, London and New York, 1965, p. 3.

68. J. Williams, *Proc. Brit. Ceram. Soc.* No. 3, Sept., p. 1 (1965).

69. L. S. Williams, *Mechanical Properties of Engineering Ceramics*, (W. W. Kriegel and H. Palmour III eds.), Interscience, New York and London, 1961, pp. 346, 379.

70. L. S. Williams, *Bull. Am. Ceram. Soc.* 42(6): 340 (1963).

71. J. Yamashita and M. Kojima, *J. Phys. Soc. Japan* 7: 261 (1952).

DISCUSSION

C. S. Morgan (*Oak Ridge National Laboratory*): Does the fact that you do not see very many dislocations in the MgO specimens examined by electron transmission microscopy mean that they are not formed during crystal growth or could it be that they form but move out?

Answer: It is difficult to give a satisfactory answer to this question solely on the basis of observations made during this investigation. Our evidence shows that the majority of dislocations remaining at the time of examination in the microscope are concentrated at crystallite boundaries, but whether they were generated there or moved out to the boundaries is uncertain. These observations of dislocations are relatively infrequent, and it should be noted that rotation into precise orientation for strain-free growth is favored by the "looseness" and fineness of the powder formed by low-temperature calcination of basic magnesium carbonate. Where this is prevented by the constraint of several contact points with other crystallites the usual observation is that of strain-contrast indicative of considerable misorientation. The situation is complicated by the precise requirements to be met for observation by transmission electron microscopy of dislocations at low-angle boundaries, so that our impression that the situations of near-perfect alignment or relatively gross misorientation (high-angle boundaries) are preferred is difficult to substantiate quantitatively. The rare observation by moiré pattern of a dislocation "frozen" some 20 Å from the edge of an MgO crystallite ([46]) could be interpreted in terms of either occurrence of dislocations during recrystallization or generation during growth by interaction with an adjacent crystallite.

J. H. Hensler, (*University of Melbourne*): In the course of our work on the sintering of magnesium oxide, electron microscopy of material derived from calcination of the carbonate at 1000°C showed crystallites exhibiting thickness fringes similar to those observed in your work. I think this tends to support the conclusions drawn in your paper from such fringes.

Answer: We agree that it is a well-established fact in electron microscopy that wedge-shaped crystals having a high degree of internal perfection will exhibit thickness fringes.

J. B. Wachtman, Jr. (National Bureau of Standards): Where you able to distinguish the degree of crystallinity of the reaction products as a function of heat treatment from the character of the electron diffraction patterns?

Answer: The MgO crystallites developed during calcination at 550°, 700°, and 900°C were, with the exception of occasional dislocations at boundaries, perfect in themselves. Sharp diffraction patterns were taken of the starting materials and, after decomposition, of the resulting oxide. Internal crystal perfection was shown by the presence of fine structure in the diffraction patterns of the oxide. Thickness fringes, which are critically dependent upon internal crystal perfection, were further proof of this.

It was not a primary purpose of the work at the time to follow the decomposition process with electron diffraction, but this could, however, be done with due time and care. We did observe the diffraction pattern during decomposition in the electron beam, but did not take photos. The diffraction patterns of both magnesium hydroxide and the basic carbonate start as single crystal. The hydroxide goes to an MgO pseudo-single-crystal pattern, i.e., it retains the outline and shape of the parent crystal but is composed of many slightly disoriented individual MgO crystallites. The basic carbonate pattern goes to halos, then to halos with faintly superimposed MgO rings, then to strong MgO ring patterns.

Chapter 5

Factors Affecting the Kinetics of Grain Growth and Densification in Ceramic Bodies Containing a Liquid Phase

J. White

University of Sheffield
Sheffield, England

Work is discussed on processes occurring during densification and grain growth in ceramic bodies containing a liquid phase at the firing temperature. Discussion is limited primarily to grains of approximately spherical shape. Grain growth is considered first in dispersed systems of particles and then in two-phase (solid plus liquid) systems in which the solid grains are in contact with each other. Evidence is presented that geometrical similarity is maintained during the growth process. Effects of such factors as grain boundary migration and the presence of a second solid phase on growth and densification are described, and the effect of the dihedral angle at which the surfaces of adjacent grains intersect is examined.

INTRODUCTION

This chapter describes conclusions based on observations that have been made during investigations into the factors determining the microstructure of ceramic bodies containing a liquid phase at the firing temperature ([1-4]). In all the bodies the liquid contents were such that as grain growth and densification proceeded the solid grains were in contact with each other, i.e., the bodies did not consist of a dispersion of solid particles in a liquid matrix but rather of an aggregate of solid grains, with the liquid phase and pores in the interstices between them. Throughout this chapter discussion is further limited primarily to bodies containing rounded grains of approximately spherical shape.

The processes occurring during the densification of bodies of this kind (both metallic and ceramic) have been discussed by Price *et al.* ([5]), Gurland and Norton ([6]), Cannon and Lenel ([7]), and Kingery ([8,9]). For complete densification Cannon and Lenel considered that appreciable solubility of the solid in the liquid phase and complete wetting of the solid by the liquid (in the sense that the liquid phase should penetrate completely around the solid grains) were necessary.

Three steps in the densification process were also distinguished: (1) formation of the liquid phase and flow of the latter into the interstices between the solid grains while they are drawn together by the capillary forces exerted by the pores, (2) a slower stage involving a solution-precipitation process and transport of material through the liquid phase, resulting in closer packing of the solid grains and further densification, and (3) the formation of a solid skeleton involving direct grain-to-grain contact with a further slowing up or complete cessation of densification. It was considered that after the initial rapid stage the last stage would be the predominant one if the solid grains were not completely wetted by the liquid.

Kingery considered the second stage, which is the one that depends on solubility in the liquid phase, to be critical for the achievement of complete densification in such systems. According to Price et al. densification during this stage results from solution of the smaller grains and growth of the larger. This process, it was considered, would occur because, as predicted by the Freundlich–Thomson relation between surface curvature and solubility (see next section), the solubility of the smaller grains would be greater than that of the larger grains.

Kingery on the other hand considered that the difference in solubility over the range of grain sizes usually present was too small to account for the rates of material transport required and also pointed out that if the growing grains remained spherical in shape the degree of close-packing necessary to achieve complete densification would not be achieved. As an alternative he suggested that the capillary pressure exerted by the pores would cause a pressure to be exerted at the points of contact between the solid grains which would increase their solubility at these points. Consequently material would dissolve in the liquid at points of contact and be deposited elsewhere on the grains, permitting the distances between the centers of adjacent grains to decrease. In support of this mechanism he pointed out that in the samples sintered by Price et al. the grains tended to become more angular as densification proceeded and that the angularity increased as the liquid content decreased. He deduced that shrinkage by this mechanism would be proportional to the 1/3 power of the time and inversely proportional to the 4/3 power of the initial particle size, whereas during the first stage, when viscous flow of the liquid is the rate-determining process, shrinkage would be proportional to a power of time slightly greater than unity. Since densification during the second stage was assumed to result from increased solubility in the liquid phase at contacts between the grains, he found it necessary to assume that a thin film of liquid capable of transmitting a compressive load existed between them (i.e., wetting was complete).

GRAIN GROWTH IN DISPERSED SYSTEMS OF
PARTICLES (OSTWALD RIPENING)

Closely similar expressions for particle (grain) growth in dispersed systems of widely-separated particles by solution of the smaller particles in the matrix and precipitation on the larger particles have been derived by Greenwood [10], Wagner [11], Lifshitz and Slezov [12], and Li and Oriani [13]. All these authors assumed the particles to be spherical, so that the variation of solubility with particle size was given by the Thomson–Freundlich equation,

$$\ln \frac{S_1}{S_2} = \frac{2M\gamma}{\rho RT}\left(\frac{1}{r_1} - \frac{1}{r_2}\right) \tag{1}$$

where M is the molecular weight of the dispersed phase, and ρ is its density. Here γ is the interfacial tension between the phases, and S_1 and S_2 are, respectively, the solubilities of particles of radius r_1 and r_2 in the matrix (the activities if solution is nonideal).

When S_1/S_2 does not differ much from unity

$$\ln (S_1/S_2) \approx (S_1/S_2) - 1 \tag{2}$$

so that the logarithmic term can be eliminated from Eq. (1). The difference between the solubility of a particle of radius r ($r_1 = r$) and the saturation concentration over a flat surface ($r_2 = \infty$) is then

$$S_r = S[1 + (2M\gamma/r\rho RT)] \tag{3}$$

where S is the solubility over the flat surface.

From these relationships, assuming that diffusion to and from the particles occurred radially with spherical symmetry and that the volume of the solid remained constant, Greenwood derived an equation for the change in size of an individual particle:

$$\frac{dr}{dt} = \frac{2DSM\gamma}{\rho^2 RT}\frac{1}{r^2}\left(\frac{r}{\bar{r}} - 1\right) \tag{4}$$

where $\bar{r}$ is the mean particle radius, D the diffusion coefficient in the matrix, and S is the solubility in g/cm³. It will be seen that at any instant particles having radii greater than $\bar{r}$ will be growing, particles with radii less than $\bar{r}$ will be dissolving, while particles of radius $\bar{r}$ will neither be growing nor dissolving. Over a period of time, however, $\bar{r}$, which is the mean radius of the population, will increase, so that some particles will grow initially only to dissolve later.

It follows from (4) that at any time particles having twice the mean radius will have the maximum growth rate, assuming that the population contains particles of this size. By substituting $r = 2\bar{r}$ in (4) and integrating Greenwood obtained the equation

$$r^3 - r_0^3 = (6\,DSM\gamma/\rho^2 RT)t \tag{5}$$

for the growth of particles of twice the mean size (r_0 being the value of r at zero time) and suggested that since the cube of r occurred it should predict growth rates for the population which are too high but of the correct order of magnitude. Although it had been derived for widely dispersed particles, he found that it did, in fact, predict particle growth rates of the correct order in settled slurries of UPb_3 in liquid lead. He suggested that agreement under such conditions might be due to the fact that while close contact between the particles would increase the concentration gradients locally, a nonuniform distribution of sizes, particularly where particles of similar size occurred clustered together, would cause them to decrease.

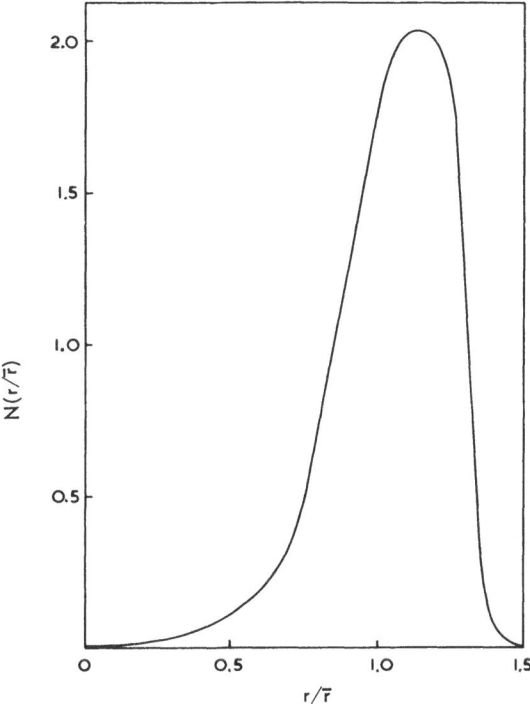

Fig. 1. Time-independent normalized size-distribution function given by Lifshitz and Slezov[12]. Here r is the particle radius, $\bar{r}$ is the mean particle radius.

Lifshitz and Slezov on the other hand showed that in a dispersion in which grain growth was occurring according to Eq. (4) a time-independent normalized size distribution function would be approached in which the radius of the largest particles would be only 1.5 $\bar{r}$ (See Fig. 1.) From this steady-state distribution function, however, they derived an equation that predicts growth rates of the same order as Greenwood's equation, viz.,

$$\bar{r}^3 - \bar{r}_0^3 = \frac{8}{9} \frac{DSM\gamma}{\rho^2 RT} t \tag{6}$$

where r_0 is the mean particle radius at zero time, which is defined as the time at which steady-state behavior begins.

GRAIN GROWTH IN TWO-PHASE (SOLID PLUS LIQUID) SYSTEMS IN WHICH THE SOLID GRAINS ARE IN CONTACT WITH EACH OTHER

The principles governing the equilibrium distribution of the liquid phase in two-phase mixtures of this kind were first established by the work of Smith ([14]) on metallic alloys. In particular the condition that the liquid phase should penetrate completely between the solid grains is that the surface tension of the solid–solid grain boundaries should be equal to or greater than twice that of the solid–liquid interface, i.e., $\gamma_{aa} \geq 2\gamma_{al}$. When $\gamma_{aa} < 2\gamma_{al}$ complete penetration will not occur. Instead, a balance of forces will be reached when

$$\gamma_{aa} = 2\gamma_{al} \cos(\phi/2) \tag{7}$$

where ϕ is the dihedral angle at which the surfaces of adjacent grains (solid–liquid interfaces) intersect each other. This equation is only strictly valid when the forces are tangential to the surfaces and independent of crystallographic orientation. Smith found that these assumptions were justified for the metallic systems he studied and are probably admissible when the solid grains are rounded in shape.

He also showed that as ϕ increases from zero the ability of the liquid phase to penetrate between the solid grains decreases, although up to $\phi = 60°$ it should still be capable of penetrating along three-grain edges. When $\phi > 60°$ it should occur as discrete inclusions at four-grain junctions.

In practice the sizes of the dihedral angles observed in a polished section vary with the orientation of the line of intersection of the surfaces defining the angle relative to the plane of the section. If, however, a large number of angles is measured and a histogram constructed the most frequently occurring angle should be the true angle ([15]). Alternatively the median of the distribution can be used as a measure of the true angle ([16,17]).

Other parameters obtainable from measurements on polished sections ([2]) are the mean grain diameter, $\bar{d} = 1.225\bar{l}$, where $\bar{l}$ is the mean diameter of the grains observed in the section, and the ratio of the contact area (grain boundary area) between grains to their total surface area (grain boundary area plus solid–liquid interface area). The latter is obtained by traversing the section and counting the number of intersections with grain boundaries (N_{aa})

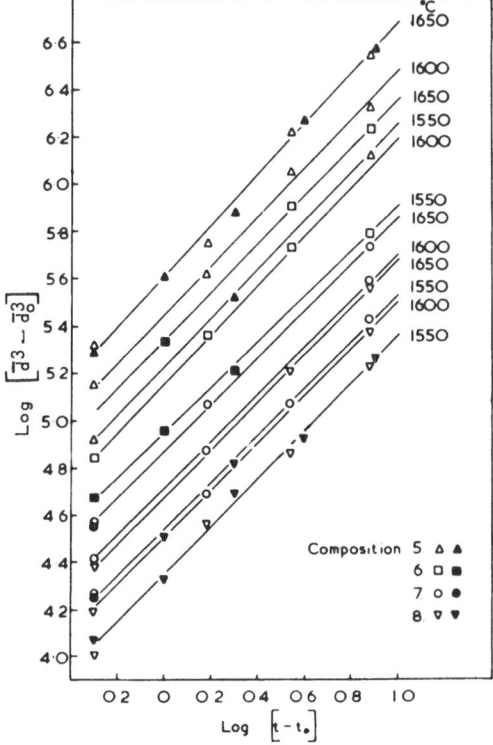

Fig. 2. Plots of $\log(\bar{d}_0^3 - \bar{d}^3)$ against $\log(t - t_0)$ for periclase-liquid mixtures fired at various temperatures in air. Here $\bar{d}$ and $\bar{d}_0$ are the mean grain diameters (in microns) at times t and t_0 (in hours). Open symbols are for $t_0 = \frac{1}{2}$ hr; darkened symbols for $t_0 = 0$, $\bar{d}_0$ being neglected. Composition 5: 85 MgO; 15 $Ca_2Fe_2O_5$ ($\phi < 10°$). Composition 6: 80 MgO; 15 $CaMgSiO_4$; 5 Fe_2O_3 ($\phi = 20°$). Composition 7: 85 MgO; 15 $CaMgSiO_4$; ($\phi = 25°$). Composition 8: 80 MgO; 15 $CaMgSiO_4$; 5 Cr_2O_3 ($\phi = 40°$). (All in wt. %.) [Buist et al.([2]).]

and phase boundaries (N_{al}). Then

$$\frac{\text{Grain boundary area}}{\text{Total boundary area}} = \frac{N_{aa}}{N_{aa} + N_{al}} = \frac{N_{aa}}{N} \tag{8}$$

In the course of the present investigations 16 two-phase systems containing periclase, lime, and corundum grains in a variety of liquid phases have been studied [2], the mean grain size being determined as a function of time at the firing temperature, and it has been found that in all cases the cube of the mean grain diameter increased linearly with time as predicted by Eqs. (5) and (7) for dispersed particles. This behavior is illustrated in Fig. 2, in which $\log(\bar{d}^3 - \bar{d}_0^3)$ is plotted against $\log(t - t_0)$ for four bodies consisting of solid periclase grains with about 15% liquid by volume. All the plots are straight lines having slopes approaching unity (between 0.95 and 1.05).

This finding suggests that even at these relatively low liquid contents grain growth is still controlled by solution of the smaller grains and deposition on the larger and transport through the liquid phase. To test the feasibility of this conclusion Greenwood's equation was used by Buist et al. [2] to obtain values of D in those bodies for which the solubilities of the solid in the liquid phase (in g/g) could be obtained from published phase diagrams. For this purpose $\bar{d}$ was assumed equal to r in Eq. 5, r_0 being neglected, and plausible values of the density of the liquid phase (needed to convert the solubilities

TABLE I

Estimated Values of Diffusion Coefficient in the Liquid Phase from Observed Rates of Grain Growth

Mixture (wt. %)	Temperature (°C)	Solubility (g/g)	Assumed density (g/cm³) Solid	Liquid	D (cm²/sec)	Activation energy (kcal/mole)
85 MgO	1550	0.235	3.58	3.2	9.04×10^{-5}	38
15 CaMgSiO$_4$	1650	0.265			1.56×10^{-4}	
	1725	0.290			2.25×10^{-4}	
95.8 CaO	1550	0.615	3.32	3.0	1.33×10^{-4}	53
3.3 Al$_2$O$_3$	1600	0.625			1.71×10^{-4}	
0.9 SiO$_2$	1650	0.635			2.85×10^{-4}	
96.2 CaO	1550	0.610	3.32	3.0	8.40×10^{-5}	48
2.8 Al$_2$O$_3$	1600	0.625			1.19×10^{-4}	
0.9 SiO$_2$	1650	0.635			1.81×10^{-4}	
85 Al$_2$O$_3$	1550	0.427	3.96	2.8	1.55×10^{-5}	59
15 CaAl$_2$Si$_2$O$_8$	1700	0.495			3.98×10^{-5}	
	1800	0.607			7.29×10^{-5}	

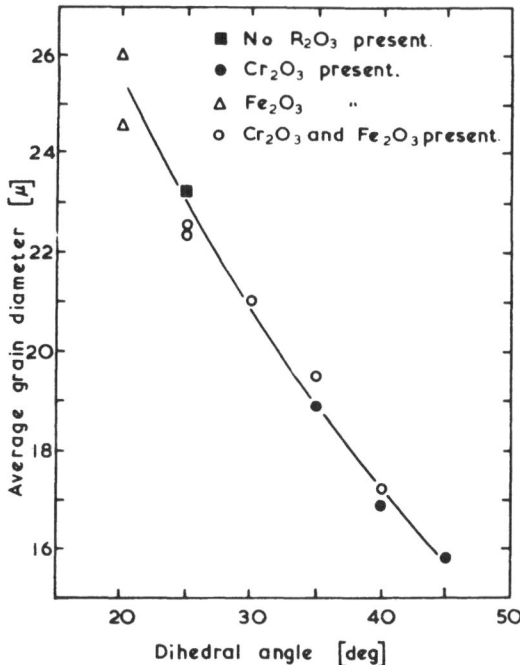

Fig. 3. Variation of mean grain diameter with equilibrium dihedral angle in periclase-liquid mixtures after firing 2 hr in air at 1550°C [Jackson et al.([1])].

to g/cm³) and of the solid phase were assumed; γ was taken as 200 dyn/cm.

The calculated values of D are shown in Table I and appear to be of the right order for diffusion in the liquid phase. Other indications that transport in the liquid phase controlled grain growth were that large increases in the liquid content tended to cause a decrease in the rate of grain growth and that in Fig. 2 the slowest growth rates are for mixtures in which the liquid phase contained silica. In the latter the solution and growth of the lime grains would involve displacement of silica in the liquid phase.

Another significant observation has been that the rate of grain growth in such mixtures is quite markedly influenced by the size of the equilibrium dihedral angle, decreasing as the dihedral angle increases. The effect is illustrated in Fig. 3, which shows the average diameter of the magnesia grains in a series of MgO-liquid mixtures after firing for 2 hr at 1550°C plotted against the equilibrium dihedral angle. The series was obtained by adding Cr_2O_3, which increased the angle, and Fe_2O_3, which decreased the angle, to a body consisting initially of 85% MgO and 15% CaMgSiO₄ by weight, the additions being made as replacements for equal weights of MgO. The marked influence

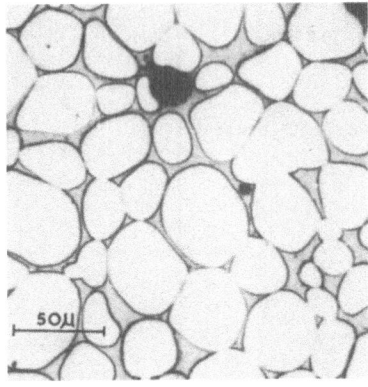

Fig. 4., Micrograph of mixture containing 80%
MgO, 15% CaMgSiO₄, 5% Fe₂O₃ after firing
8 hr at 1550°C in air. $\phi = 20°$.

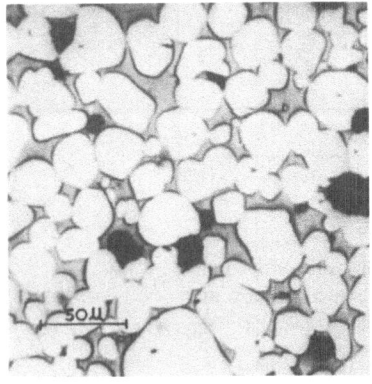

Fig. 5. Micrograph of mixture containing 80%
MgO, 15% CaMgSiO₄, 5% Cr₂O₃ after firing
8 hr at 1550°C in air. $\phi = 40°$.

on grain growth of a 5% addition of Cr_2O_3 to the composition 85% MgO:
15% CaMgSiO₄ can be seen by comparing Fig. 4 and 5.

To account for this effect Buist *et al.* [2] pointed out that in an assemblage
of spherical grains with finite contact areas the surface curvature would tend
to decrease as ϕ increased, and it was assumed that the diameters of the
necks (contact areas) between the grains remained constant; they showed
that if a suitable geometry was assumed the observed effects on grain growth
could be accounted for.

EVIDENCE FOR RETENTION OF GEOMETRICAL SIMILARITY DURING GRAIN GROWTH

Lifshitz and Slezov [12] deduced that in a dispersion of spherical particles
in which grain growth is occurring by diffusion-controlled solution precipitation a time-independent size-distribution function is approached asymptot-

ically. The following indications that solid–liquid systems in which the solid grains are in contact with each other exhibit similar behavior have been obtained in the course of the present work.

1. Buist *et al.* ([2]) found that in bodies containing periclase, lime, and corundum the ratio of the grain boundary area to the total boundary area as measured by N_{aa}/N usually tended to remain constant after 1–2-hr firing, suggesting that geometrical similarity was being maintained as grain growth proceeded.

2. Jackson *et al.* ([1]), who used an approximate method to estimate mean neck diameters in bodies consisting of periclase and a liquid phase, found that values of ϕ calculated from the ratio of the mean neck diameter to the mean grain diameter on the assumption that the grains could be regarded as equal interpenetrating spheres generally agreed within 5° with the values obtained by construction of a histogram. Subsequently Buist *et al.* ([2]), who used an expression due to Fullman ([18]) for determining the mean diameter of randomly oriented circular disks to estimate the mean neck diameter, found that in periclase-liquid mixtures fired for periods between 2 to 8 hr the ratio of the mean neck diameter to the mean grain diameter tended to remain constant as grain growth proceeded, suggesting that geometrical similarity was being maintained. The values of ϕ estimated from the sections, the mean grain diameters, and the mean neck diameters, and the ratio of the latter are shown in Table II. It will also be seen that the value of the ratio decreases as ϕ decreases.

These findings suggested that the geometry achieved and maintained during grain growth might be a minimum energy one. This possibility has been examined ([4,19]) by considering the change in total surface energy that

TABLE II

Comparison of Experimental and Predicted Values of the Ratio of Mean Neck Diameter to Mean Grain Diameter

Mix	Firing treatment	ϕ (deg)	Mean grain diam. (μ)	Mean neck diam. (μ)	Ratio mean grain to mean neck diameter	$A = \gamma_{aa}/\gamma_{al}$	y/r
5	1550°C/8 hr	20	39.0	9.0	0.23	1.97	0.17
	1725°C/2 hr	20	35.2	9.0	0.26		
	1725°C/8 hr	20	55.2	13.7	0.25		
4	1550°C/4 hr	25	28.8	7.2	0.24	1.95	0.22
	1550°C/8 hr	25	33.8	8.8	0.26		
6	1550°C/4 hr	40	23.0	7.9	0.34	1.88	0.33
	1550°C/8 hr	40	28.0	8.7	0.31		

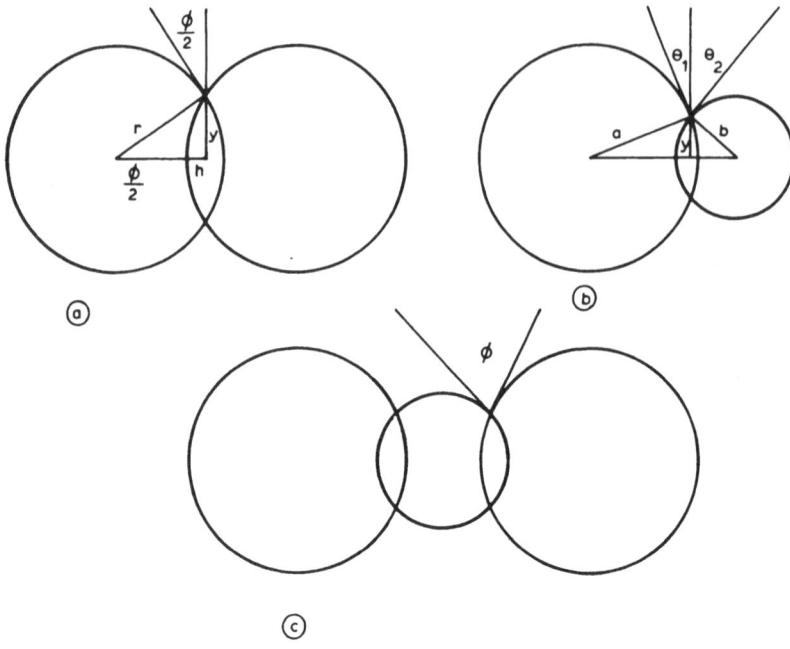

Fig. 6. (a) Section through two equal spherical grains in contact with a flat circular grains boundary of radius y in the neck. (b) Similar section through two unequal spherical grains of radii a and b. (c) Small spherical grain situated between two larger grains with its surfaces with the latter at the equilibrium dihedral angle.

would occur when a regular assemblage of uniform spheres which were initially in point contact with each other and having a liquid phase in the interstices were brought closer together causing flat contact areas (grain boundaries) to develop between neighboring grains (see Fig. 6a). It was further assumed that the volume of each grain remained constant during this process and that its surface, except for the flat contact areas, remained spherical.

Then if each sphere makes n contacts with its neighbors and the radius of the curved surface is r, the volume of each grain will be that of a sphere of radius r with n spherical caps removed, i.e., with a volume

$$(n/3)\pi h^2(3r - h) \qquad (9)$$

removed, where h is the height of a cap.

Then from the volume constraint, if the initial radius is r_0

$$\frac{r_0^3}{r^3} = 1 - \frac{3np^2}{4} + \frac{np^3}{4} \qquad (10)$$

where $p = h/r$.

Again the area of the curved surface of a spherical cap is

$$2\pi rh = 2\pi r^2 p \tag{11}$$

and the area of the boundary in the neck is

$$\pi y^2 = \pi h(2r - h) = \pi r^2(2p - p^2) \tag{12}$$

where y is the radius of the neck.

Then, since the surface energy of each grain boundary is shared equally between the grains in contact, E_γ the total surface energy per sphere is

$$
\begin{aligned}
E_\gamma &= 4\pi r^2 \gamma_{al} - 2n\pi r^2 p\gamma_{al} + \frac{n}{2}\pi r^2(2p - p^2)\gamma_{aa} \\
&= 4\pi r_0^2 \gamma_{al}\left\{\frac{1 - (n/2)p + (n/8)(2p - p^2)A}{[1 - (n/4)(3p^2 - p^3)]^{3/2}}\right\} \\
&= 4\pi r_0^2 \gamma_{al} f(p) \tag{13}
\end{aligned}
$$

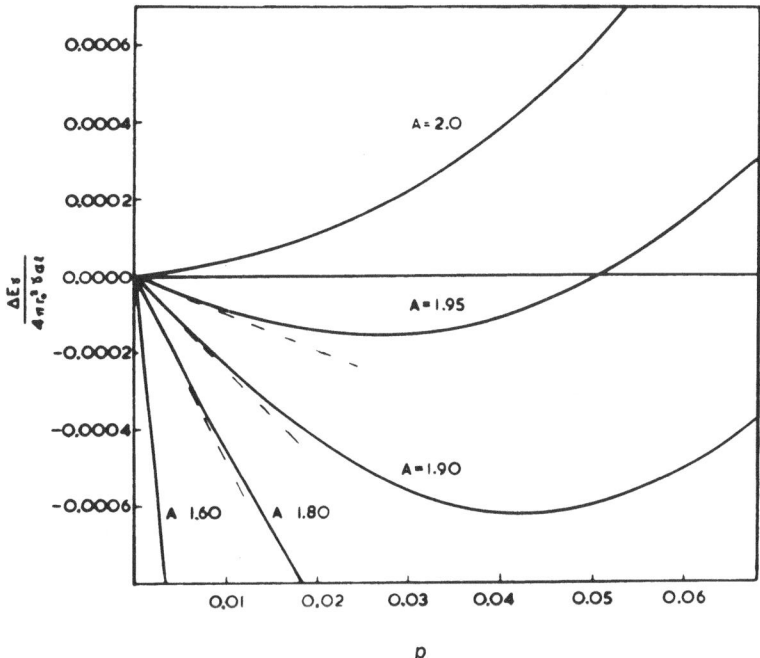

Fig. 7. Change in total surface energy per sphere with change in p at various values of A for the case of two spherical grains in contact with a flat grain boundary in the neck (see Section IV and Fig. 6a).

where γ_{aa} is the surface energy of the grain boundary per cm² and γ_{al} that of the solid-liquid interfaces, and $A = \gamma_{aa}/\gamma_{al}$.

Figure 7 shows how $f(p)$, which is equal to E_γ divided by the initial energy per sphere, varies with p for various values of A at $n = 1$. (Changing the value of n changes the depth of the minima but not the values of p at which they occur.) Since

$$p = h/r \approx h/r_0$$

p is equal to the fractional decrease in the distance between the centers of the spheres, so that the plots show how the energy varies as the latter distance decreases. It will be seen that when $A = 2$ the energy increases continually as p increases, which would be expected, since, when $\gamma_{aa} = 2\gamma_{al}$ the liquid phase will penetrate completely between the spheres. For all lower values of A the energy decreases to a minimum and then rises again as p increases.

Again, from Eq. (12)

$$y/r = (2p - p^2)^{1/2}$$

where y/r is the neck/grain diameter ratio (see Fig. 6a), while

$$A = \gamma_{aa}/\gamma_{al} = 2\cos(\phi/2)$$

The values of y/r at minimum energy at the values of A corresponding to the three experimental values of ϕ are shown in the final column of Table II. Agreement between the measured and predicted ratios is surprisingly good (except in the case of the mixture having the lowest value of ϕ, which is discussed below) in spite of the fact that in the actual bodies a range of grain sizes would exist. Hence the conclusion that a minimum-energy geometry is approached in such structures seems justified.

The predicted values of y/r at minimum energy also agree well within the limits of accuracy of the method used to obtain them with those that would have been predicted at the experimental values of ϕ from the relation

$$y/r = \sin(\phi/2)$$

which follows from the geometry of Fig. 6a. Hence the assumption that the surfaces of the grains are spherical satisfies the condition that the surface forces must be in equilibrium when the energy is a minimum.

Figure 6b shows the situation that exists when grains of different radii (a and b) are in contact. From the geometry $\sin\theta_1 = y/a$ and $\sin\theta_2 = y/b$, and for any assumed ratio of a to b we can evaluate y from the condition that $\theta_1 + \theta_2 = \phi$. The effect of varying b/a on the ratio of the neck diameter to the *mean* of the two grain diameters when $\phi = 25°$ ($A = 1.95$) is shown by the following:

$$(b/a): \quad 1.0 \qquad 0.5 \qquad 0.3 \qquad 0.2$$

$$[2y/(a + b)]: \quad 0.22 \qquad 0.19 \qquad 0.16 \qquad 0.12$$

Hence the ratio of the diameters decreases as b/a decreases fairly slowly down to $b/a = 0.5$ and then more rapidly. The same changes in the ratio will also occur as b/a changes in an assemblage consisting of two sizes of spheres in equal numbers so long as each sphere touches only spheres of the other size, while if a certain number of contacts between spheres of the same size occur the ratio will still decrease but to a smaller extent. On the other hand an increase in the diameter ratio will only occur when the number of small spheres is appreciably greater than that of the large spheres and when on the average the number of small spheres touching each large sphere is greater than the number of large spheres touching each small sphere.

In most of the sections examined in the course of the present investigation, with firing times of the order of 2 hr or more, the number of grains much smaller than the mean size has been relatively small, so it seems unlikely that the latter condition would be fulfilled (see Figs. 4 and 5). Further, the number of contacts between grains of widely different size has also been small, so that no marked lowering of the mean neck to mean grain diameter ratio would be expected.

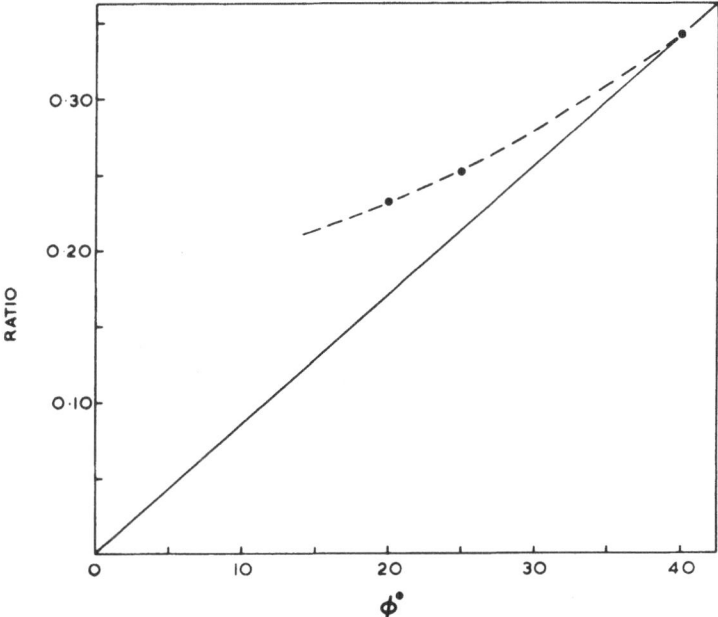

Fig. 8. Plot of experimental values of mean neck/grain diameter ratios shown in Table II against ϕ. Continuous line shows predicted relationship for equal spheres.

Actually, as shown in Table II, the experimental values of the latter ratio tend to be higher than the predicted values for equal spheres and become increasingly so as ϕ decreases. This trend is shown more clearly in Fig. 8, which shows that the experimental values when plotted against ϕ lie on a curve which would apparently extrapolate to a finite value of the ratio when $\phi = 0$, while the predicted curve passes through the origin. This behavior is what would be expected if shrinkage continued after the equilibrium neck diameters had been reached, necessitating a closer approach of the grain centers and further neck growth. (This could be considered as shrinkage during Cannon and Lenel's stage III.)

Such shrinkage becomes possible if when the neck diameters have reached their equilibrium values porosity still remains and if the decrease in surface energy associated with shrinkage of the pores is greater than the increase associated with growth of the necks beyond their equilibrium values.

The relative magnitudes of the energies involved have been estimated [4] for the case of an assemblage of equal spheres initially in point contact in simple cubic packing and having a spherical pore in each of the interstices, while the remaining space is occupied by the liquid phase. There will then be one pore per sphere. It is further assumed that the initial pore radius a_0 when point contacts exist between the particles is the maximum that can be accommodated in the interstices, i.e., $0.732\, r_0$, where r_0 is the radius of the spheres.

As each pore shrinks the volume of the cube defined by the centers of the eight spheres surrounding it will decrease by the same volume. Hence when the pore radius has shrunk to a the edge of the cube, which was originally $2r_0$, will have shrunk by an amount

$$2r_0 - \{(2r_0)^3 - (4/3)\pi(a_0^3 - a^3)\}^{1/3} \approx 2h \approx 2r_0 p$$

where h and p are as defined earlier.

Curve (a) in Fig. 9 is the plot of the surface energy per pore, $4\pi a^2 \gamma_l$, against $r_0 p$ for an assemblage of spheres of initial radius $r_0 = 10^{-4}$ cm, with γ_l, the surface energy of the liquid phase, assumed to be 300 ergs/cm². Curves (b), (c), (d), and (e) are plots of E_γ, the sum of the interfacial and grain boundary energies per sphere [see Eq. (13)] against $r_0 p$ for $\phi = 0°$, 25°, 36°, and 52°, respectively, γ_{sl} being taken as 200 ergs/cm² and n as 6, the value for cubic packing.

It will at once be obvious that the energy minima associated with neck growth between the particles are too shallow to create a minimum in the total energy of the system. Consequently shrinkage to zero porosity should always be possible. With the cubic packing and pore size assumed, however, the minima on curves (c) and (d) (i.e., with $\phi = 25°$ and 36°) are reached before densification is complete, so that further densification involving neck

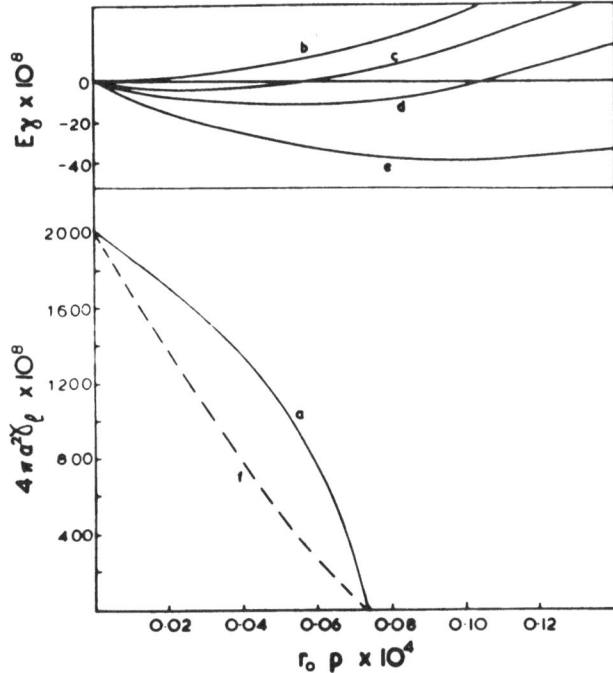

Fig. 9. Curve (*a*) shows change in surface energy per pore with linear shrinkage for an assemblage of uniform spherical grains in cubic packing having liquid in the interstices and one spherical pore per interstice (see Section IV). Curves (*a*), (*b*), (*c*), and (*d*) show changes in sum of grain boundary energy plus solid–liquid interfacial energy per grain with linear shrinkage of the assemblage when $\phi = 0°$, $25°$, $36°$, and $52°$, respectively. Upper energy scale magnified 10 times to show positions of minima. Curve (*f*) shows possible form of curve (*a*) in practice.

growth to give neck diameter/grain diameter ratios greater than those corresponding to the minimum values of E_γ will be possible. When $\phi = 52°$ on the other hand densification is complete and the interstices are completely filled with liquid before the minimum value of E_γ is reached.

The model therefore provides a possible explanation (although the mode of packing and initial pore size were arbitrary) of the divergence between the experimental and predicted values of the neck diameter/grain diameter ratio at low values of ϕ. It can also, it will be evident, provide an explanation for Kingery's observation that the angularity of the grains in bodies containing a liquid phase at the firing temperature increased as the liquid content decreased.

There are, however, a number of limitations to the use of the model as an accurate description of the behavior of the actual bodies. Thus the effect of gas in the pores has been ignored. In addition in the absence of such gases the system of isolated pores in a continuous liquid phase envisaged is a highly unstable one since $2\gamma/a$, the capillary pressure tending to cause closure of the pores, increases as the pore radius decreases. Hence any deviation from a strictly uniform pore size would result in collapse of the smaller pores and expansion of the larger, with viscous flow of the liquid. Such a redistribution of the pore volume would lower the surface area and surface energy of the pores without a corresponding decrease in porosity, which would change the shape of curve (a), possibly to one more resembling curve (f). It would also lower the densifying pressure exerted by the pores (and might even cause bloating), and it is this fall in pressure rather than the achievement of complete densification that is likely to terminate densification and the increase in the neck diameter/grain diameter ratio associated with it in mixtures of low dihedral angle. Residual porosity in the form of a relatively small number of large pores has generally been observed in the micrographs even after long periods of firing.

No attempt has been made to identify the mechanism by which densification and neck growth occur under the influence of pressure exerted by the pores, but it will be evident that Kingery's suggestion that transport occurs in a liquid film between the grains becomes untenable when the dihedral angle is greater than zero.

The most probable mechanism would appear to be that proposed by Herring [20] for neck growth between single-crystal particles having a grain boundary between them and having their surfaces intersecting at the equilibrium dihedral angle. He showed that under these conditions the tendency of the system to minimize its surface energy causes a pressure to be exerted at the grain boundary which causes a decrease in the concentration of lattice vacancies in its vicinity and sets up a diffusion flux that results in neck growth and a contraction in the center-to-center distance.

In the presence of a liquid phase the surface tension operating at the grain surfaces will be replaced by the solid–liquid interfacial tension, but the lower pressure exerted by the latter across the grain boundary will be supplemented by the capillary pressure exerted by the pores.

INFLUENCE OF DIHEDRAL ANGLE ON RATE OF GRAIN GROWTH AND DENSIFICATION

As was mentioned previously Buist *et al.* concluded that the effect of ϕ on the rate of grain growth was in close-packed aggregates of grains most probably due to its effect on the curvature of the grain surfaces. They pointed out that if a central smaller grain formed finite contact areas with a number of grains arranged symmetrically around it the surface curvature of the central grain would decrease as ϕ increased if it were assumed that the diameters of the necks did not change as ϕ changed. By assuming a suitable geometry they found that an effect of the correct order of magnitude was predicted.

It will, however, be evident that the assumption of a constant neck diameter is inconsistent with the evidence discussed above that the neck/grain diameter ratio varies with ϕ.

A modified explanation that is not inconsistent with this finding is indicated in Fig. 6c, which shows a section through a small spherical grain situated between two larger grains with which it forms necks of the equilibrium size. (The large grains are shown situated diametrically opposite each other for convenience.) Then for a given size and separation of the large grains there will be only one size of the small grain that will enable its surface to remain spherical while intersecting the surfaces of the larger spheres at the equilibrium angle, and the larger ϕ is the larger will be the size of the smaller sphere that fulfils this condition. For any value of ϕ if the small sphere is larger than this critical size the necessity to achieve the equilibrium angle will cause its surface curvature, and hence its solubility, to increase, while, if it is smaller its curvature and solubility will be decreased. Hence a large value of ϕ will inhibit solution of the smaller grain below a critical size which will increase as ϕ increases.

This mechanism implies that there is a constraint on the movement of the larger grains that prevents the distance between them from changing as the smaller grain dissolves. This constraint will be imposed by the neigh-

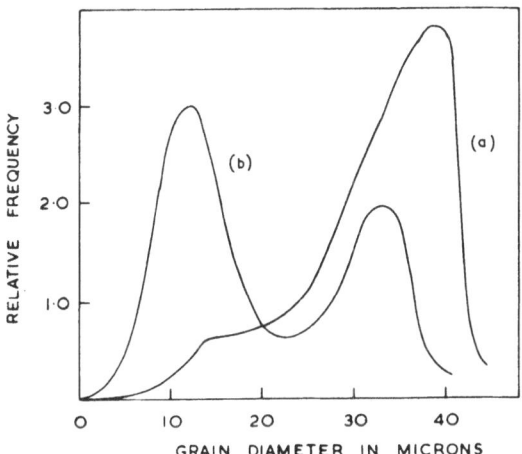

Fig. 10. Estimated grain-size distributions in periclase-liquid bodies whose micrographs are shown in (a) Fig. 4 and (b) Fig. 5. ϕ was 20° in (a) and 40° in (b). Both bodies fired 8 hr at 1550°C.

boring grains with which they are in contact and by the fact that adjustments in the center-to-center distances between grains in contact is a slow process relative to the solution-deposition process.

An indication that the proportion of smaller grains in bodies containing a liquid phase at the firing temperature does in face increase as ϕ increases is provided by the fact that the frequency with which small grain sections occur in Fig. 4 ($\phi = 20°$) is less than that with which they appear in Fig. 5 ($\phi = 40°$). To test this conclusion an attempt has been made to estimate the grain size distributions in the two bodies from the micrographs, Scheil's method ([21]) being used to obtain the true distributions from the distributions observed in the sections. The distribution curves derived in this way are shown in Fig. 10 and although they are not claimed to be precise because of the limitations of this method they appear to provide a clear indication that the tendency to retain small grains in the structure increased as ϕ increased.

Evidence that the degree of densification achieved during the firing of

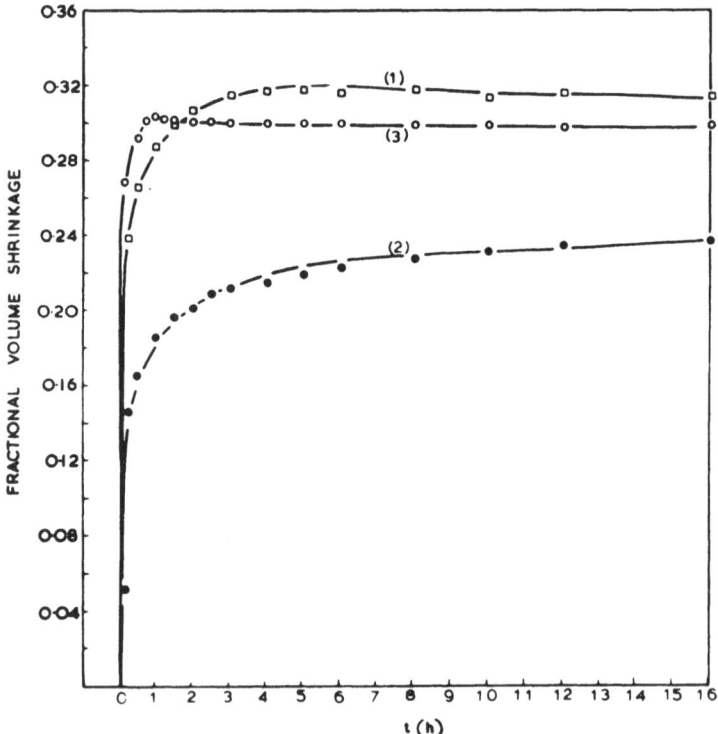

Fig. 11. Shrinkage/time curves at 1550°C of mixtures containing (1) 85 MgO, 15 CaMgSiO$_4$, (2) 80 MgO, 15 CaMgSiO$_4$, 5 Cr$_2$O$_3$; (3) 80 MgO, 15 CaMgSiO$_4$, 5 Fe$_2$O$_3$. (Wt. %.) Values of ϕ were 25°, 40°, and 20° respectively.

solid–liquid mixtures is affected by the size of the dihedral angle, decreasing as ϕ increases, has also been obtained in the course of the investigations ([1,4]), recalling the suggestions of Cannon and Lenel ([7]) and Kingery ([8]) that complete wetting of the solid by the liquid is necessary for complete densification. (A large value of ϕ indicates a low tendency for the liquid to penetrate between the grains, i.e., low wetting.)

The nature of this effect is illustrated in Fig. 11, which shows shrinkage/time curves at 1550°C for three mixtures in which ϕ was, respectively, 25°, 40°, and 20°. It will be seen that the inverse relation between ϕ and densification occurs primarily during the initial rapid stage of densification (Cannon and Lenel's stage I) when both the rate and the magnitude of the shrinkage decrease as ϕ increases.

An explanation of this effect based on the energy plots of Fig. 7 has recently been suggested ([4]). It follows from the latter that if shrinkage depended on the growth of necks between grains already in contact, as in the case during the sintering of single-phase powders, it would increase as ϕ increased, i.e., as A decreased. The fact that the reverse is the case in the initial stage of densification, when the liquid phase has just formed and is being drawn into the pores, suggests that the magnitude of the shrinkage is determined by the ability of the solid grains to rearrange themselves. Any tendency of the grains to stick together would then oppose shrinkage.

A measure of the tendency of the capillary forces that are responsible for neck growth to cause cohesion can be obtained from Eq. (13). Thus if two equal spheres of radius r_0 are in point contact the force tending to cause cohesion is

$$F = 2 \, dE_\gamma/dx = -2 \, dE_\gamma/2r_0 \, dp = -4\pi r_0 \, df(p)/dp$$

where x is the distance between the grain centers and $p = h/r_0$ as before. (The factor 2 occurs because E_γ was defined as the surface energy associated with one sphere only.)

The required values of $-df(p)/dp$ are given by the initial slopes of the curves in Fig. 7, which are indicated by the dashed straight lines. The magnitudes of these slopes and the cohesive forces calculated for spherical particles of radius 10^{-4} cm and having $\gamma_{sl} = 200$ ergs/cm^2 are shown in Table III.

Although the forces may seem small, they appear to be of the same order as those responsible for coagulation in colloidal systems and are probably large enough to account for the observed behavior. (They are actually of the order of 1 kg/cm^2 calculated over the projected area of the particles.) In addition the magnitude of the force increases as ϕ increases. Hence the probability that particles which come into contact during the initial period should cohere will increase as ϕ increases.

Once two growing grains cohere growth of the neck between them by

TABLE III

ϕ (deg)	A	$-\dfrac{df(p)}{dp}$	F (dyns)
0	2.0	—	No attraction
25	1.95	0.01	1.3×10^{-3}
36	1.90	0.02	2.8×10^{-3}
52	1.80	0.05	6.0×10^{-3}
74	1.60	0.10	1.3×10^{-2}
120	1.00	0.30	3.9×10^{-2}

transport through the liquid phase will be rapid (unless ϕ is zero) and a continuous solid skeleton will be built up. Further shrinkage, assuming that the interstices are not then completely filled with liquid, will depend on a contraction of the center-to-center distance between adjacent grains by the process discussed above and will be a much slower process.

ROLE OF GRAIN BOUNDARY MIGRATION DURING GRAIN GROWTH AND THE INHIBITING EFFECT OF A SECOND SOLID PHASE ON GRAIN GROWTH

Although the evidence so far presented appears consistent with the conclusion that the rate of grain growth is determined primarily by a solution-deposition process occurring at the solid–liquid interfaces it will be evident from the micrographs shown in Figs. 4 and 5 (and also from a consideration of Fig. 6b) that the growth of large grains in contact with small grains will, if the former are to remain isometric, involve not only growth at these interfaces but also displacement of the grain boundaries that they form with adjacent grains.

To account for their observation that in bodies containing periclase or lime the grains remained isometric during growth Buist et al. [2] pointed out that when grains of different sizes are in contact at a grain boundary deposition on the surface of the larger grain and solution at the surface of the smaller (or different rates of deposition or solution at the two surfaces) would displace the root of the neck between the grains, and hence the periphery of the grain boundary, toward the smaller grain. Consequently the boundary would migrate away from the center of the growing grain.

Support for this proposal has been obtained from studies of systems containing two solid phases and a liquid phase at the firing temperature. In such systems it has been found that each solid phase obstructs the growth of the other. This behavior is illustrated in Figure 12, which shows how the mean sizes of the periclase and lime grains varied with the ratio of CaO

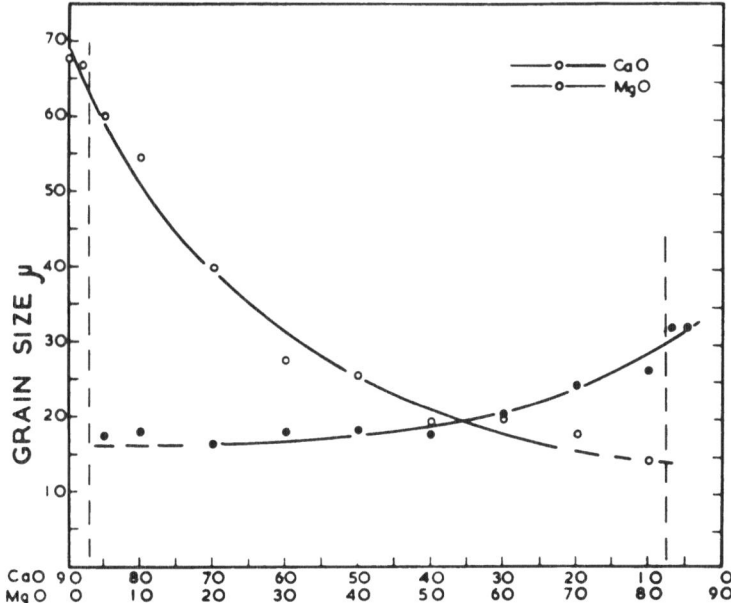

Fig. 12. Curves showing mean grain size of lime and periclase grains in CaO–MgO–Fe$_2$O$_3$ mixtures after firing for 2 hr at 1550°C in air. Fe$_2$O$_3$ content was 10% by weight. Dashed vertical lines show limits of three-phase area at 1550°C.

to MgO in a series of CaO–MgO–Fe$_2$O$_3$ mixtures containing 10% Fe$_2$O$_3$ by weight after firing for 2 hr at 1550°C in air. Over the composition range between the dashed lines solid lime and periclase would be present together at the firing temperature, the periclase content increasing from left to right

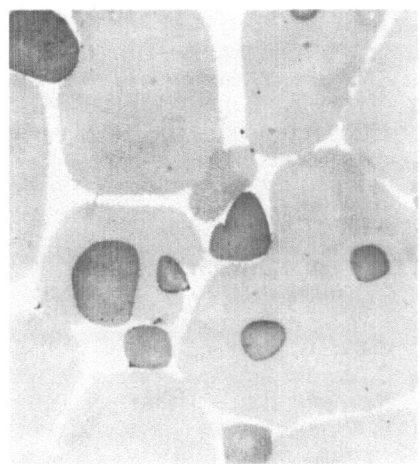

Fig. 13. Microstructure of mixture containing 80% CaO, 10% MgO, 10% Fe$_2$O$_3$ (by weight) after firing for 2 hr at 1550°C, showing lime grains growing around periclase grains.

and the lime content from right to left. In each case the greatest fall in the growth rate occurred at low contents of the second phase, showing that the effect was due to obstruction by the grains of the latter and not to the fact that the separation between like grains was increasing.

Examination of the microstructures has thrown light on the nature of this effect, since it has shown that in mixtures in which the grains of one species are growing much faster than the other the faster growing grains tend to grow around the more slowly growing grains. One effect of this behavior is that the curvature of the former is increased relative to what it would be if the obstructing grain were not present and this should tend to inhibit growth. This phenomenon is illustrated in Fig. 13 for a mixture in which lime was the faster growing species. It seems certain, from the curvature of the interfaces produced, that in the analogous case of two like grains, grain boundary migration would have occurred.

REFERENCES

1. B. Jackson, W. F. Ford, and J. White, *Trans. Brit. Ceram. Soc.* **62**: 577 (1963).
2. D. S. Buist, B. Jackson, I. M. Stephenson, W. F. Ford, and J. White, *Trans. Brit. Ceram. Soc.* **64**: 173 (1965).
3. B. Jackson and W. F. Ford, *Trans. Brit. Ceram. Soc.* **65**: 19 (1966).
4. I. M. Stephenson and J. White, *Trans. Brit. Ceram. Soc.,* **66**: 443 (1967).
5. G. H. S. Price, C. J. Smithells, and S. V. Williams, *J. Inst. Metals*, **6**: 239 (1938).
6. J. Gurland and J. T. Norton, *J. Metals* p. 1051 (1952).
7. H. S. Cannon and F. V. Lenel, in: *Plansee Proceedings*, 1952, (F. Benesovsky, ed.), Metalwork Plansee, Reutte, 1953, p. 106.
8. W. D. Kingery, in: *Ceramic Fabrication Processes* (W. D. Kingery, ed.), Technology Press, M.I.T., 1958, p. 131.
9. W. D. Kingery, in: *Kinetics of High Temperature Processes* (W. D. Kingery, ed.), Technology Press, M.I.T., 1959, p. 187.
10. G. W. Greenwood, *Acta Met.* **4**: 243 (1956).
11. C. Wagner, *Z. Elektrochemie* **65**: 581 (1961).
12. I. M. Lifshitz and V. V. Slezov, *J. Phys. Chem. Solids* **19**: 35 (1961).
13. Che-Yu Li and R. A. Oriani, in: Proceedings of the Bolton Landing Conference on Oxide Dispersion Strengthening, Bolton Landing, N.Y., June 1966.
14. C. S. Smith, *Trans. AIME* **175**: 15 (1948).
15. B. Harker and E. R. Parker, *Trans. Am. Soc. Metals* **34**: 156 (1945).
16. O. K. Riegger and L. H. Van Vlack, *Trans. Met. Soc. AIME* **218**: 933 (1960).
17. C. J. Ball, *Trans. Brit. Ceram. Soc.* **65**: 41 (1966).
18. R. L. Fullman, *Trans. AIME* **197**: 447 (1953).
19. J. White, Paper presented at Third International Materials Symposium, University of California, Berkeley, June 1966.
20. C. Herring, in: *Physics of Powder Metallurgy* (C. E. Kington, ed.), McGraw-Hill, New York, 1951, p. 143.
21. E. Scheil, *Z. Metallkunde* **27**: 199 (1935).

DISCUSSION

J. H. Hensler (University of Melbourne): Could Professor White give the basis of the factor of 1.225 used in section III for converting mean intercept diameter to mean grain diameter?

Work we have been doing indicates a factor of 1.27 (equals $4/\pi$), as used by other workers, and also indicates that this factor applies not only to uniform grain sizes, but to skewed and bimodal grain diameter distributions. It also shows a method for getting the actual grain size distribution from the intercept distribution, as attempted in section V of Professor White's paper.

Answer: The factor of 1.225 is the factor given by R. L. Fullman[22], for the case of uniform spherical particles. It is the factor by which the mean neck diameter in the section should be multiplied to get the true diameter.

For the case of non-uniform spherical particles Fullman gives

$$\bar{r} = \frac{\pi}{4\bar{m}}$$

where $\bar{m}$ is the average of the reciprocals of the diameters in the section and $\bar{r}$ is the mean particle radius. In an earlier paper we compared the results obtained by the two methods with bodies consisting one solid phase and a liquid and found they agreed within a few percent and since then we have generally preferred to use the simpler one to save time. Actually I am doubtful whether it is justifiable to strive for too much refinement in analyzing such structures. Consistency is more important than absolute accuracy.

This also applies to grain size distributions. The method I used in this case was a very old one (Scheil) which requires nothing more sophisticated than a slide rule, and although we have tried out a more recent method that involves the use of a computer, I suspect that its advantages are more apparent than real. Actually we find, with our bodies, that the form of the size distribution can be deduced from the size distribution in the section without converting it to the "true" size distribution.

H. J. Oel (Max Planck Institut für Silikatforschung, Würzburg, Germany): Have distributions of the Lifshitz–Slezov type shown in your Fig. 1 been determined experimentally? Attempts to prove the existence of distributions of this type in precipitates, where they would be expected to occur, have not been too successful, and I think that your Figure may be the first experimental evidence for this type of distribution.

Answer: The distribution shown in Fig. 1 was determined experimentally from the apparent size distributions in the sections, using the method of Scheil. The micrographs we have published in earlier papers, however, show that this type of distribution, in which the maximum frequency occurs at the coarse end of the distribution, is common in bodies consisting of a single solid phase and a liquid that have been fired for periods of an hour or more in that the proportion of grains having cross sections much smaller than the average is very small. Exceptions are bodies that have been fired for shorter times and, as shown in Fig. 10, bodies in which the dihedral angle is large. I was not aware that our observations may have provided the first experimental evidence for the occurrence of the Lifschitz–Slezov type of distribution.

P. E. D. Morgan (Franklin Institute): In effect you say that redistribution of the solid via solution-precipitation allows the movement of boundaries. I wonder if the situation is not reversed, i.e., that the moving boundaries are dragging the liquid with them. It is known that boundaries can drag pores and solid particles with them, so why not liquid phases? This is easy to visualize when the liquid is the discontinuous phase and in effect fills the pores. Grain growth is then controlled by the movement of the liquid-filled pores, redistribution of the material across the pores occurring by solution-precipitation. As with unfilled pores, this gives grain growth according to the relation $D^3 - D_0^3 = kt$. Although it is more difficult to visualize this process when the liquid phase is not discontinuous, nevertheless the same mechanism can apply. In conclusion the movement of boundaries due to curvature is the driving force for the solution-precipitation, not the other way round.

Answer: The conditions under which grain boundaries can drag pores and solid or liquid inclusions with them are very different from those existing in our bodies. In the first instance the liquid content of our bodies is always relatively large and forms a continuous phase. A second important difference is that the curvature of the grain surfaces is always convex and growth of the larger grains occurs "against the curvature", as it were, simply because they are less soluble in the liquid phase than the smaller grains. The situation envisaged by Dr. Morgan occurs only when the volume of the pores or the content of the second phase is small. Each grain boundary is then convex towards the grain on one side of it and concave towards the grain on the other side and migrates under the influence of the pull exerted by its curvature.

ADDITIONAL REFERENCES

22. R. L. Fullman, *Trans. AIME* **175**: 447 (1953).

Chapter 6

Dynamic Thermal Analysis

W. B. Campbell

Department of Ceramic Engineering
The Ohio State University
Columbus, Ohio

Methods for the study of changes of state in systems under dynamic thermal conditions are examined. The effects of heat transfer rate on thermal analysis are discussed, and both exothermic and endothermic reactions are considered. The theoretical basis for the measurement of heating and reaction rates using dynamic differential calorimetry is discussed.

INTRODUCTION

Differential thermal analysis (DTA), thermogravimetric analysis (TGA), and dynamic differential calorimetry (DDC) are major techniques for the identification and investigation of rapid changes of state under dynamic thermal conditions. The difference between DTA and TGA is well delineated in technique and analysis; however, the distinction between DDC and DTA remains poorly established in the experimental literature. It must be noted that DTA and DDC differ in their use of a homogeneous sample block and separated sample cups, respectively. These arrangements produce isolated thermal peaks by the former method and quantitative reaction heat determinations by the latter. In each analytical method, atmospheric effects must be considered in as much detail as are heat transfer conditions, specimen characteristics, and other physical parameters.

Arens ([1]) presents an empirically detailed examination of DTA and suggests a number of factors which may affect thermal analysis. These are:

(a) Rate of furnace heating.
(b) Material and type of sample holder.
(c) Depth and radius of sample cavities.
(d) Measurement sites of all temperatures.
(e) Nature and proportions of thermocouples.
(f) Nature of inert standard.
(g) Packing density of sample and standard.

(h) Effect of sealed or partially sealed sample cavities.
(i) Composition of furnace atmosphere.
(j) Particle size of specimens.
(k) Degree of crystallization.
(l) Presence of admixtures.

Arens concluded that recorded results are determined primarily by heating rate, sample cavity dimensions, temperature measurement sites, packing density, cavity sealing, furnace atmosphere, particle size, and degree of crystallization.

According to Arens, change in sample heating rate produces a systematic difference between weight-loss and weight-stable reactions. Weight-loss reactions show increased peak temperatures, peak heights, and peak areas with increased heating rates even though the peak range, measured as the time of reaction, decreases.

The effect of heating rate also depends on the manner of containing the specimen, i.e., on the type of sample holder. If only a change of state is involved, the properties of the sample holder are singularly important for heat transfer considerations (²). Using rates of 6, 12, 18, and 21°C per minute, Arens found no shift of transition temperature for either α to β quartz or the kaolin to metakaolin decomposition. Weight-loss reactions showed peak shifts to be associated with the time-dependent nature of decomposition product effusion from the specimen.

Increased peak areas at higher heating rates may be due to heat transfer limits and/or sample permeability. For a loosely packed sample, the differential signal is dependent on the amount of material reacting at any given time; in turn, the reaction volume depends on the rate of heat transfer. For a tightly packed sample the permeability may become the rate-limiting step. The limiting step may be identified by the effect of different atmospheres on the experimental curves.

REACTION MODELS

The effect of heat transfer rates on the type of reaction may be considered by way of the following models. Crystallization and reaction are principal exothermic occurrences. Gaseous product formation, phase transitions, and melting represent the major endothermic reactions. The surface of each sample will be considered to be instantaneously heated by a constant heat source. Only the effects during a heating cycle are discussed for the homogeneous specimen.

Below the reaction temperature, the thermal gradient in the specimen will change as required by its thermal diffusivity. Above the reaction tempera-

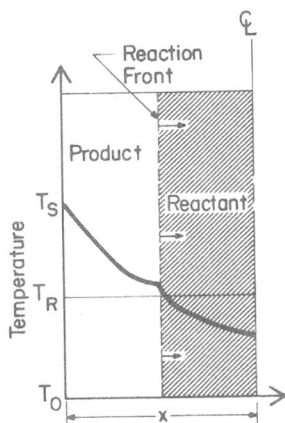

Fig. 1. Exothermic reaction system in terms of surface (T_S), reaction (T_R), and initial (T_0) temperatures, and distance (x).

ture, the thermal gradient will be controlled by the thermal diffusivity of the product or the reactant and the amount of each present at any time. Consequently the differential signal may be controlled more by the thermal diffusivity of the material than by the surface temperature.

For exothermic reactions the diffusivity of the unreacted material will limit the rate of reaction. This is shown schematically in Fig. 1. If the surface temperature T_s is always higher than the reaction temperature T_r and the reaction is exothermic, the reaction front will accelerate until the diffusivity of the reactant becomes limiting. At such time the reaction rate should remain constant and should be unaffected by surface temperature. The effect of increased surface temperatures should be to decrease the time necessary to achieve a constant reaction rate. (Note that it is assumed that an AB reaction occurs between very fine particles and that mass transport processes are not reaction rate-limiting; however, if such is the case the rate is less than the limiting diffusivity value.)

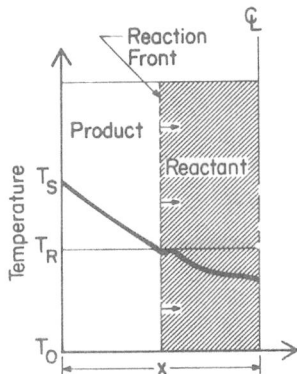

Fig. 2. Endothermic reaction system in terms of surface (T_S), reaction (T_R), and initial (T_0) temperatures, and distance (x).

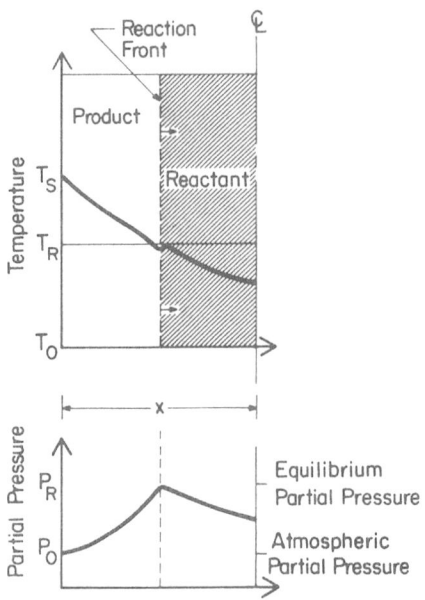

Fig. 3. Endothermic reaction with gaseous products.

An endothermic reaction is represented in Fig. 2. It is obvious that the diffusivity of the product may become rate limiting. If a gaseous product is evolved during the reaction, the partial pressure may become rate limiting, as shown in Fig. 3. For the simple endothermic reaction an increased surface temperature results in an increased reaction rate until the diffusivity of the product becomes rate limiting and the reaction rate remains constant. The complex endothermic reaction evolving gaseous products behaves like the simple reaction until the partial pressure at the reaction front becomes great enough to affect the reaction system. For an $ABC = AB + C\uparrow$ reaction the driving force for decomposition decreases as the partial pressure of C increases. Since gaseous permeation occurs in the direction of lowest pressure, the partial pressure at the reaction front increases as the distance from the free surface increases. When the equilibrium pressure is reached at the reaction front, the gaseous permeation rate toward the free surface becomes the rate-limiting factor. Thus in the complex reaction, product diffusivity is initially controlling but its effect is continually decreased by an increasing partial pressure which makes permeability (or solubility) the final rate limiting factor.

EXPERIMENTAL PRINCIPLES

It is desirable to measure the heat changes in the system with respect to time because the time derivative of the reaction heat is indicative of the reaction rate. Any suppression of the rate will be evidenced by an extended time for heat evolution. Thus a method sensitive to heat changes in the specimen is required.

The magnitude of heat flow between reference and sample accounts for the principle difference between DTA and DDC. In the former, heat is permitted to flow between cavities to return the differential signal to zero as soon as possible and enable the detection of successive reactions. Although this arrangement facilitates the determination of "reaction temperature" the measurement signal and total sensitivity are reduced. For DDC, where the heat transfer between cavities is principally by the radiation mode, the sensitivity and magnitude of the differential signal is greatly enhanced at the expense of temperature specificity.

From the above considerations DDC is preferable for the investigation of the effect of heating rate on the experimental heat of reaction. Since differential measurements will be a function of time, the heat change involved will be a function of the reacting volume and the reaction rate. Thus the peak area,

$$A = \int_{t_1}^{t_2} \Delta T \, dt$$

represents the heat involved in the reaction. The time limits (t_1, t_2) are established by the reaction rate as well as heating rate, heat capacities, thermal diffusivities, and other intrinsic material parameters. For a series of identical specimens, increased heating rates should provide an increased reacting volume and increased reaction rates, at least initially. Partial pressure effects and thermal diffusivity limits will become more pronounced at higher rates of heating. When the limiting heating rate is reached the peak height will maximize; at higher rates peak heights will decrease and the time limits will be extended. A noticeable change in curve shape is also to be expected.

The peak area in DDC is

$$\int_{t_1}^{t_2} \theta \, dt = mq/G$$

where m is the mass of sample, G is the heat transfer coefficient between the cups and the system, q is the heat of reaction per unit volume, and θ is the differential temperature. Neither the volume nor the thermal conductivity affect the peak area; however, the heating rate and sample size may cause peak overlaps in sequential reaction systems. A knowledge of sequential reaction temperatures may be used to circumvent this experimental difficulty.

The determination of q for complex endothermic reactions requires the consideration of the reaction product pressures. Since the peak area is a measure of the reaction heat ΔH, and the reaction product pressures are dependent on the reaction temperature, the reaction should obey the van't Hoff relation,

$$d(\ln P)/dT = \Delta H/RT^2$$

However, in powder compacts and at high heating rates gas permeation causes a pressure differential within the specimen and ΔH is not time constant. Taking the relations

$$(d \Delta H/dT)_p = \Delta C_p$$

and

$$\Delta H = \Delta H_0 \, dT + \int_{T_1}^{T_2} \Delta C_p \, dT$$

for a simple decomposition we obtain

$$\frac{d \ln K_p}{dT} = \frac{d(\ln P)}{dT} = \frac{\Delta H_0}{RT^2} \, dT + \int_{T_1}^{T_2} \frac{\Delta C_p \, dT}{RT^2}$$

Assuming that ΔC_p changes very little with T, i.e., that mass is conserved in the system,

$$\int_{T_1}^{T_2} \frac{d(\ln P)}{dT} = \int_{T_1}^{T_2} \frac{\Delta H_0}{RT^2} \, dT + \int\int_{T_1}^{T_2} \frac{\Delta C_p}{RT^2} \, dT$$

and

$$\ln \frac{P_2}{P_1} = -\frac{\Delta H_0}{R}\left(\frac{1}{T_2} - \frac{1}{T_1}\right) + \frac{\Delta C_p}{R} \ln \frac{T_2}{T_1}$$

Since the ΔC_p term is usually greater than the ΔH_0 term, the observed ΔH increases rapidly with increasing partial pressure until equilibrium pressure is reached.

An example of partial pressure effects may be described by calculation of the first endothermic kaolinite reaction. Using Arens' specific heat data for kaolinite and calcined kaolinite, 0.201 and 0.428 cal/gm/°C, respectively, and molar heat capacities of 51.1 and 75.0 cal/mole/°C, respectively, the evolution of two moles of water requires another 18.0 cal/mole/°C and the ΔC_p is calculated to be 60 cal/mole/°C. For observed peak shifts at 545 and 620°C (heating rates of 6 and 10°C per min) and a heat of decomposition of 200 cal/gm the preceding equation may be partially evaluated:

$$\log \frac{P_2}{P_1} = -\frac{(200\,\text{cal/gm})(260\,\text{gm/mole})}{2.303(1.987)}\left[\frac{1}{893} - \frac{1}{820}\right] + \frac{60}{2.303(1.987)}\log\frac{893}{820}$$

or

$$\log(P_2/P_1) = 1.1 + 0.5 = 1.6$$

Thus the partial pressure at the higher heating rate is about 40 times higher assuming no permeation.

From the above discussion of reaction models and the reaction heat parameters it should be apparent that in sequential reaction systems a critical interim phase may exist to limit other reactions. For example, the kaolin-metakaolin-mullite sequence has the metakaolin diffusivity as the rate-limiting phase.

By combining DDC measurements and thermal diffusivity measurements it may be possible to predict the limiting heating rate for unidirectional heat flow and reaction. DDC will provide dq/dt as a function of heating rate and reaction rate; from surface temperature and thermal diffusivity the temperature profile, dT/dx, may be computed at any time; after the reaction temperature is reached at any point, the change in local temperature and flux velocity may be determined.

REACTION RATE DETERMINATION

Following the general approach but avoiding the mathematical mistakes of Vold [3] a general expression for heat transfer between a dynamically heated reference and a sample may be derived for DDC measurements. The heat flux *into* respective cells is:

$$\frac{dq_s}{dt} = K_s(T_w - T_s) + \sigma(T_r - T_s) + a_s(T_0 - T_s)$$

$$\frac{dq_r}{dt} = K_r(T_w - T_r) + \sigma(T_s - T_r) + a_r(T_0 - T_r)$$

where s and r refer to sample and reference respectively, K is the heat transfer coefficient between furnace wall and cavity, σ is the heat transfer coefficient between the cavities, and a is the heat transfer coefficient along the thermocouples.

Using the identity

$$\frac{dq}{dt} = \frac{dH}{dt} = \frac{dH}{dT}\frac{dT}{dt} = C\frac{dT}{dt}$$

the heat flux *within* the cells will be

$$\frac{dq_r}{dt} = C_r \frac{dT_r}{dt}$$

and

$$\frac{dq_s}{dt} = C_s \frac{dT_s}{dt} + \Delta H \frac{df}{dt}$$

where f is the reacted fraction at time t. When care is taken to maintain the symmetry of the system

$$K_s = K_r - \delta K$$

$$a_s = a_r - \delta a$$

Equating the internal and external fluxes and solving for

$$\frac{d(T_r - T_s)}{dt} = \frac{dT_r}{dt} - \frac{dT_s}{dt}$$

$$C_s \frac{dT_s}{dt} + \Delta H \frac{df}{dt} - C_r \frac{dT_r}{dt} = K_r(T_w - T_s - T_w + T_r) - \delta K(T_w - T_s)$$

$$+ 2\sigma(T_r - T_s) + a_r(T_0 - T_s - T_0 + T_r)$$

$$- \delta a(T_0 - T_s)$$

$$= K_r(T_r - T_s) + 2\sigma(T_r - T_s) + a_r(T_r - T_s)$$

$$- \delta K(T_w - T_s) - \delta a(T_0 - T_s)$$

and

$$C_s \frac{dT_s}{dt} = (K_r + 2\sigma + a_r)(T_r - T_s) + C_r \frac{dT_r}{dt} - [\delta K(T_w - T_s) - \delta a(T_s - T_0)]$$

$$- \Delta H \frac{df}{dt}$$

$$\frac{dT_s}{dt} = \frac{K_r + 2\sigma + a_r}{C_s}(T_r - T_s) + \frac{C_r}{C_s}\frac{dT_r}{dt} - \frac{[\delta K(T_w - T_s) - \delta a(T_s - T_0)]}{C_s}$$

$$- \frac{\Delta H}{C_s}\frac{df}{dt}$$

$$\frac{dT_r}{dt} - \frac{dT_s}{dt} = -\frac{K_r + 2\sigma + a_r}{C_s}(T_r - T_s) + \left(1 - \frac{C_r}{C_s}\right)\frac{dT_r}{dt} + \frac{1}{C_s}[\delta K(T_w - T_s)$$

$$- \delta a(T_s - T_0)] + \frac{\Delta H}{C_s}\frac{df}{dt}$$

Experimentally dT_r/dt is controlled as a constant, $(T_w - T_s)$ is at least slowly varying, and $(T_s - T_0)$ and $(T_w - T_s)$ appear with δ coefficients so that any change may be neglected.

Introducing more compact nomenclature,

$$y = (T_r - T_s)$$

$$A = \frac{K_r + 2\sigma + a_r}{C_s}$$

$$y_s = \frac{[(C_s - C_r)(dT_r/dt) + \delta K(T_w - T_s) - \delta a(T_s - T_0)]}{(K_r + 2\sigma + a_r)}$$

the integrated form of the general differential at $df/dt = 0$ is

$$y = y_s[1 - e^{-A(t-t_1)}] + y_b e^{-A(t-t_1)}$$

Serving as a boundary condition, $y_b = y$ at $t = t_1$, and y_s is a steady-state value arising from the difference in heat capacities, the heating rate, and the heat transfer coefficients. In other words y_s accounts for the sloping baseline which is characteristic of DDC.

Initially $y_b = 0$ at $t - t_1 = 0$ and $y = 0$; as the system is heated y_s increases. The general differential may be rewritten in the new nomenclature as

$$\frac{\Delta H}{C_s}\frac{df}{dt} = \frac{dy}{dt} + A(y - y_s)$$

Before reaction $df/dt = 0$ and $dy/dt = A(y_s - y)$. Although Vold ([3]) set $dy/dt = 0$ and concluded $y = y_s$, the experimental curve has a positive slope and dy/dt cannot be zero. Since no reaction has occurred up to this point, y must equal zero and $dy/dt = Ay_s$. After reaction df/dt is again zero and $dy/dt = -A(y - y_s)$. Thus $\int_{t_1}^{t_2} A(y - y_s)\,dt$ is the peak area and $dy/dt = (y_2 - y_1)/(t_2 - t_1)$ where y_1 and y_2 are equal to y at t_1 and t_2, respectively.

To determine values of t_1 and t_2, values of $(y - y_s)$ are plotted against time. The points lie on a curve which becomes linear at each time value. The value of A may be obtained from the slopes of the linear portions of curves showing $\ln(y - y_s)$ versus time.

A baseline between t_1 and t_2 is necessary to accurately determine the peak area. Both y_s and A values include the combined heat capacities of the cell and its contents; a weighted baseline must be established. As a first approximation of the relative change in heat capacity, the initial baseline is extended beyond t_2 and its divergence from the past reaction baseline noted. If the lines converge ΔC_s is negligible and the extended line is a suitable base. If the divergence is appreciable the dy/dt expressions for pre- and post-reaction must be equated and A, y, and y_s values used to determine the shape of the base-

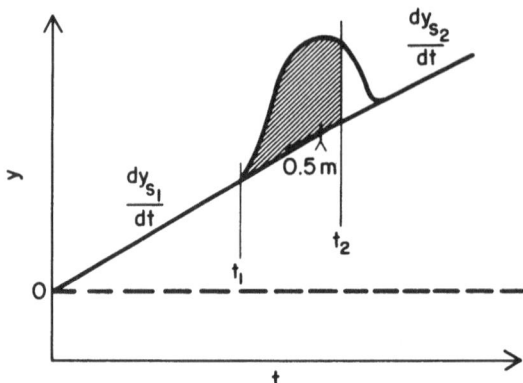

Fig. 4. Schematic of experimental curve.

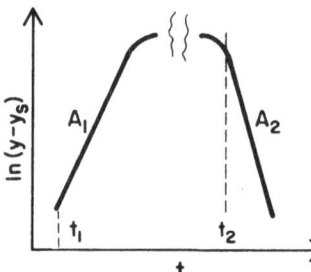

Fig. 5. Analytical plot for the determination of A, t_1, and t_2.

line. Unless a lower entropy reaction-phase or a sequential reaction occurs the post-reaction dy/dt will be of higher magnitude. Consequently the extention of the two baselines will intersect at the point of equal thermal mass of the products and reactants. The thermal mass may be converted into physical proportions by calculations based on the heat capacity of each phase. Also, the time identified by this intersection permits the calculation of df/dt for comparison with the calculated equal thermal mass. The significant analytical curves are shown in Figs. 4 and 5.

Should the post-reaction baseline be nonlinear or too abbreviated for accurate extention, the sample may be cooled and reheated to obtain a full secondary baseline. This experimental approach requires the absence of reversible transitions as well as leaving the system undisturbed between cycles. The combined baseline equation is

$$A_1 y_{s_1} = A_2 y_{s_2} - A_2 y$$

when values between t_1 and t_2 are used to solve for y.

Having determined the ΔH by the above approach the reacted fraction may be calculated for any time and equated to the rate of reaction and rate of heating.

REFERENCES

1. P. L. Arens, *A Study on the Differential Thermal Analysis of Clays and Clay Minerals*, Excelsiors Foto-Offsets, 1951.
2. P. D. Garn, *Thermoanalytical Methods of Investigation*, Academic Press, New York, 1965.
3. M. J. Vold, "Differential Thermal Analysis," in *Anal. Chem.* **21**(6): 683–88 (1949).
4. W. W. Wendlandt, *Thermal Methods of Analysis*, Interscience, New York, 1964.

Chapter 7

Nucleation and Growth Processes in the Dehydration of Salt Hydrate Crystals

W. J. Dunning

Department of Physical Chemistry
University of Bristol
Bristol, England

The changes undergone by the crystals of a salt hydrate during dehydration are discussed. The growth and dispersion of nuclei of the new crystalline phase are examined. Experimental observations of the formation of patches of the dehydrated phase on the surface of dehydrating crystals are reviewed, and the growth and transport of these patches are discussed in terms of nucleation theory.

INTRODUCTION

During the dehydration of a salt hydrate both the structure and the composition of the initial crystal change to produce the new crystalline phase and water vapor. The processes are therefore more complex than the transformation at constant composition of crystals such as tin ([1]), sulfur ([2]), and silver nitrate ([3]).

Let us suppose that the new phase II first appears as a critical nucleus in the interior of the parent phase I. In conformity with the usual theory of homogeneous nucleation ([4]) this nucleus would be considered as being a small particle of phase II which conceptually could be derived from a large crystal of II by subdivision. It would have essentially the same lattice as II though its lattice may be strained or defective.

Since the nucleus is supposed to be in the interior of I, the problem of the relation of its lattice to the parent lattice arises. This relation must be such that the formation of the nucleus is reversible—it may redisperse to reform the original lattice of I; any irreversible changes such as might accompany a cracking away from the parent would need to be excluded in this theory. The nucleus would be visualized as being in topotactic relation to the parent, and any volume change would be accommodated by strain and by defects the formation of which is reversible.

In the theory of the condensation of a supersaturated vapor the growth and dispersion of the embryos are considered, in the simplest case, to take place by the attachment and detachment of single vapor molecules to and from an embryo. The nature of the analogous elementary kinetic processes in the case of crystalline embryos in a crystalline parent phase is vague. There have been suggestions ([5]) that a thin quasigaseous layer exists at the interface into which molecules of the parent phase sublime and from which molecules condense on the surface of the new phase. This model seems an inappropriate one for the transport of salt across the interface, though the objections are not insuperable. We prefer to explore the possibility that the interface is coherent. Movement of the interface in growth or withering takes place then by mechanisms considered in theories of crystal growth ([6]). In the absence of screw dislocations or twin boundaries normal to the interface each lattice plane requires two-dimensional nucleation for its transformation. On the bordering step of a two-dimensional embryo or on steps associated with dislocations there are kinks which are the sites at which the elementary kinetic processes of growth take place. During growth a kink moves discontinuously by jumps along the step and a small region of the parent lattice at the kink changes into a corresponding small region of the new lattice, and these changes are reversible. Such small regions will be referred to as kinetic units; a kinetic unit will be comparable in size to a unit cell.

RATE OF NUCLEATION

With this model in mind we may adapt the Volmer–Stranski–Becker–Döring theory of homogeneous nucleation in a supersaturated vapor to the problem of nucleation of a new crystalline phase II in the interior of an unstable crystalline phase I. Progressive clustering of kinetic units occurs to form embryos:

$$A + A \rightleftharpoons A_2$$
$$A_2 + A \rightleftharpoons A_3$$
$$A_3 + A \rightleftharpoons A_4$$
$$\cdots$$
$$A_{k-1} + A \rightleftharpoons A_k \tag{1}$$
$$\cdots$$
$$A_{m-1} + A \longrightarrow A_m$$

In the unstable phase there are present clusters of all sizes from dimers A_2

up to m-mers. There is a critical size of cluster A_k. the critical nucleus, which is in metastable equilibrium with the unstable phase.

After creating the instability the size distribution of embryos may become steady, and then there will be a steady flow through the critical size. This steady flow is the rate of nucleation J_0 per unit volume per sec,

$$J_0 = n_I g_k \exp(-\Delta G_k/kT) \tag{2}$$

in which n_I is the number of kinetic units per unit volume of phase I (comparable to the number of unit cells per unit volume), g_k is the rate at which kinetic units transform onto the critical nucleus of size k, and ΔG_k is the free energy of formation of a critical nucleus; ΔG_k is equal to $\sum_i \sigma_i O_i/3$, where O_i is the area and σ_i the interfacial tension of the ith interface of the polyhedral nucleus. Assuming that $\sigma_i = \sigma$ for all faces and that $\sum_i O_i = \omega r_k^2$, where r_k is the linear dimension of the nucleus and ω is a shape factor, we obtain

$$\Delta G_k = \frac{4\omega}{3} \frac{\sigma^3 v_{II}^2}{(\Delta\mu)^2} \tag{3}$$

in which $\Delta\mu$ is the thermodynamic driving force and v_{II} the volume of a kinetic unit.

Instead of the thermodynamic driving force $\Delta\mu$ we may define a supersaturation ratio a by

$$a = \exp(\Delta\mu/kT) \tag{4}$$

In the case of a supersaturated vapor $a = p/p_\infty$, where p and p_∞ are the pressures of the supersaturated and saturated vapors. In the case of a supersaturated solution $a = c/c_\infty$, the ratio of the concentrations (or better, activities) of the supersaturated and saturated solutions. In some solid-state transformations the significance of a will be similar—the ratio of the concentrations of a certain species—but in other solid-state reactions it will have a more complex significance.

Using expression (4) the steady-state nucleation rate is given

$$J_0 = n_I g_k \exp\left(-\frac{4\omega}{3} \frac{\sigma^3 v_{II}^2}{k^3 T^3 \log^2 a}\right) \tag{5}$$

When this expression for J_0 is plotted against the supersaturation a it is found that J_0 is extremely small for a wide range of values $a > 1$ and then in the neighborhood of a critical value a_c there is a very sharp increase in J_0 (Fig. 1). This behavior of J_0 on a accounts for the Ostwald metastable limit

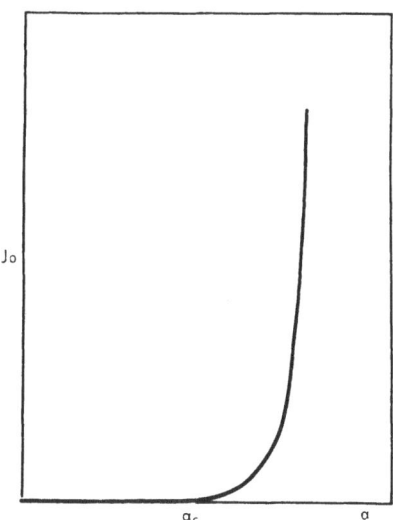

Fig. 1. Steady-state rate of nucleation as a function of the supersaturation ratio α.

in solutions. Naturally the sharpness of the upturn depends on the value of σ, the interfacial tension; if σ is small the upturn is less pronounced.

HETEROGENEOUS NUCLEATION

Nucleation in a crystal may take place at the surface or at dislocations and grain boundaries within the crystal or where these features meet the surface. So long as the embryos and critical nuclei grow three-dimensionally the rate J_0' of such heterogeneous nucleation will be given by an expression of the same form as that for homogeneous nucleation [Eq. (5)]. The shape factor ω will be different for nuclei formed at the surface but essentially unchanged for other cases.

The quantity n_I, the number of kinetic units per unit volume, is to be replaced by n_I' the number of kinetic units per unit area of surface or grain boundary, or per unit length of dislocation. Since $n_I' \ll n_I$, J_0' will be less than the homogeneous nucleation rate J_0 unless the disparity between n_I' and n_I is compensated by other factors. For example, in the case of nuclei formed at the surface the shape factor ω in the exponential may be sufficiently smaller than that for homogeneous nuclei. Also, if the parent phase is unyielding, a change in specific volume will give rise to stresses and strains in homogeneous nuclei, thereby causing an increase in the free energy of formation; such stresses may be absent when the nucleus is formed at the surface or at a dislocation where strains can be relieved by flow more readily. There is another point to be borne in mind; it is possible that the rate of formation of homo-

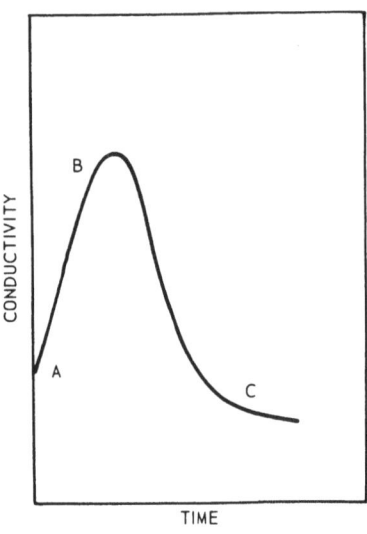

Fig. 2. Electrical conductivity of lithium aluminum hydride as a function of time during thermal decomposition (after Garner).

geneous nuclei is much greater than that of heterogeneous nuclei, but that the rate of growth of the latter to macroscopic size is orders of magnitude higher because of favorable conditions. Heterogeneous nuclei would therefore be more effective in bringing about the overall phase change. Enhanced ease of diffusion in the neighborhood of surfaces, dislocations, and grain boundaries could be such a favorable factor.

BUILDUP OF THE SUPERSATURATION

The creation of a supersaturation takes a finite time. In some cases, for example, the adiabatic expansion of a vapor in a cloud chamber, conditions may be selected so that this time may be neglected. In other cases, for example, the adiabatic expansion of a vapor through a nozzle, the development of the supersaturation down the nozzle must be taken into account ([7]).

As an illustration we may consider the behavior of lithium aluminum hydride, which decomposes above 100°C to give lithium hydride, metallic aluminum, and hydrogen. The conductivity of the $LiAlH_4$ increases initially (Fig. 2, AB) and Garner ([8]) suggests that this increase is probably caused by F′ centers which are produced by loss of hydrogen atoms. Assuming that the change in conductivity reflects the change in supersaturation, Fig. 2 indicates a steady increase AB in supersaturation until a critical supersaturation B, corresponding to the metastable limit, is reached for which the rate of nucleation becomes so large that the supersaturation collapses. Growth of the nuclei reduces the supersaturation to C.

NUCLEATION TIME-LAG

In the steady state under a steady supersaturation clusters or embryos of all sizes will be present. If n_i is the number of embryos of a size corresponding to i kinetic units, a plot of n_i, or more conveniently log n_i, versus i represents the number distribution of size. Figure 3 shows schematically the size distributions for different supersaturations $a = 1, 2, 3$, and 4.

For the saturated system $a = 1$ and the number of embryos decreases continuously with size. For supersaturated systems $a > 1$ there are minima in the distribution curves at $i = k_2, k_3$, and k_4, where k_2, k_3, and k_4 are the numbers of kinetic units in the critical nuclei which are in quasiequilibrium at the supersaturations $a = 2, 3$, and 4. Note that the size k of the critical nucleus decreases regularly from $k_1 = \infty$ for $a = 1$ to k_4 for $a = 4$. In order to represent Szilard's convention of notionally removing nuclei which have grown a little larger than the critical nucleus and returning them as single kinetic units to the system, each of the distribution curves is indicated as a dashed line for $i > k$.

Suppose that starting with a saturated system $a = 1$ a supersaturation is developed. If the rate of development is sufficiently slow the distribution

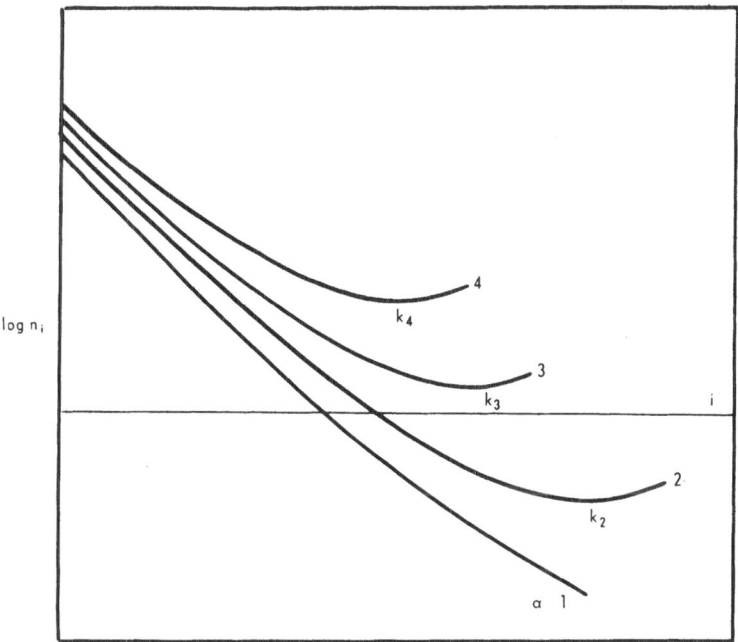

Fig. 3. Representation of the number of embryos (log n_i) as a function of the number distribution of size for supersaturations $a = 1, 2, 3$, and 4.

curves adjust smoothly to curves higher on the diagram, so that the distribution at any time is that which corresponds to the steady state for the supersaturation reigning at that time. The critical supersaturation discussed in the last section is that for which $\log n_{k_c}$ is sufficiently large for $\log J_{0c}$ to be experimentally significant.

Suppose now that the rate of development of supersaturation is very rapid. Start with a saturated system $a = 1$ and suddenly create a supersaturation, say $a = 4$. The distribution of sizes that was characteristic of the saturated state has to alter and readjust itself to a new distribution that is characteristic of the supersaturated state.

This readjustment takes place by the attachment of single kinetic units in accordance with the reaction scheme (1). These net growth processes take time, and the readjustment of the number of critical nuclei of size k_4 takes time. There is therefore a time-lag before a stationary population of critical nuclei is built up. The rate of nucleation $J(t)$ increases during this time-lag until it approaches J_0, the steady-state rate.

A partial differential equation characterizing the nonequilibrium distribution of embryos was derived by Zeldovich [9] and by Frenkel [10] from Becker and Döring's difference equations. After considerable simplifications solutions were obtained by Probstein [11] and by Kantrowitz [12]. The latter obtained

$$J(t) = J_0 e^{-\tau/t} \tag{6}$$

as an approximate solution for small t. In this τ is a time constant given by

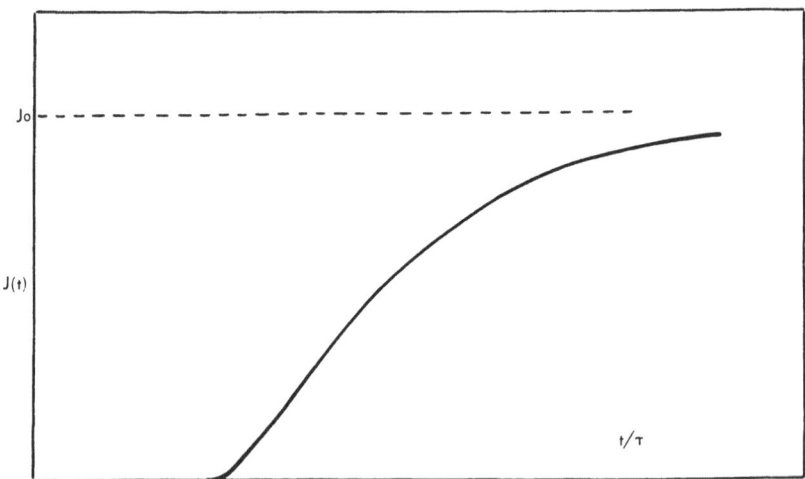

Fig. 4. Rate of nucleation J_t during the incubation period before a stationary population of critical nuclei is achieved as a function of t/τ.

$$\tau = k^2/g_k \qquad (7)$$

where g_k is the rate of growth of the critical nucleus. A plot of $J(t)$ versus t/τ is shown in Fig. 4. For typical cases of vapor condensation the time-lag τ is $\sim 10^{-6}$ sec, but for nucleation in the solid state, where transport processes across the interface are very slow, τ will be much longer.

RATE OF FORMATION OF DEHYDRATION "NUCLEI"

Garner and his collaborators ([13-15]) studied the dehydration of crystals such as $CuSO_4 \cdot 5H_2O$, $NiSO_4 \cdot 7H_2O$ and the alums, e.g., $(NH_4)_2SO_4 \cdot Al_2(SO_4)_3 \cdot 24H_2O$. One of the techniques used was to expose carefully grown crystals to a vacuum and watch the surfaces with a microscope. The new dehydrated phase appeared as patches on the surface which Garner called "nuclei;" we shall refer to them as "patches" and reserve the term "nulei" for critical nuclei.

In the case of $CuSO_4 \cdot 5H_2O$ the patches were star-shaped; on $NiSO_4 \cdot 7H_2O$ they were half-ellipses; on the (001) faces of potassium alum the patches were square; on the (111) faces of this alum they were hexagonal; while on the (111) faces of cesium alum they had trigonal symmetry.

The patches did not appear immediately when the vacuum was applied and Fig. 5 shows how the number of patches on $NiSO_4 \cdot 7H_2O$ increased with time at various temperatures; the curves are parabolic in shape. The number of patches $N(t)$ depends on the time in the following way

$$N(t) = 0 \qquad\qquad t < t_i$$
$$N(t) = c(t - t_i)^2 \qquad t > t_i$$

where t_i is an induction period and c is a constant.

For $CuSO_4 \cdot 5H_2O$ the number of patches showed a different dependence on time (Fig. 6):

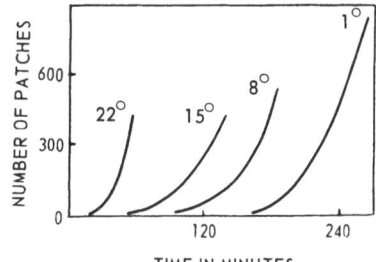

Fig. 5. Increase in number of dehydration patches on $NiSO_4 \cdot 7H_2O$ crystals at various temperatures.

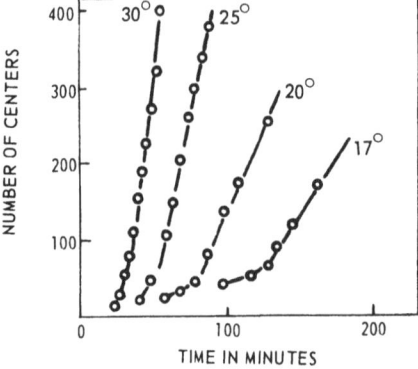

Fig. 6. Increase in number of dehydration patches on $CuSO_4 \cdot 5H_2O$ crystals at various temperatures.

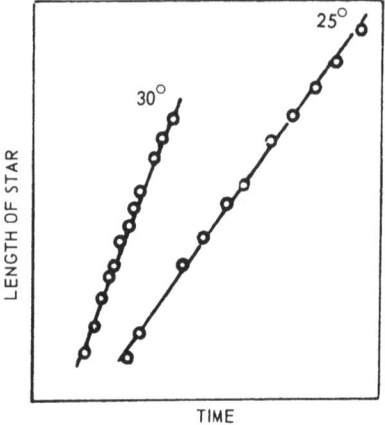

Fig. 7. Rate of growth of dehydration patches on $CuSO_4 \cdot 5H_2O$ as a function of time and temperature.

$$N(t) = 0 \qquad\qquad t < t_i$$
$$N(t) = c(t - t_i) \qquad t > t_i$$

Here a linear dependence on time appears, but note the vestigial feet to the curves. In both cases the induction period t_i decreases as the temperature increases.

Garner also measured the rate of growth of the patches. Figure 7 shows how the length of a star on $CuSO_4 \cdot 5H_2O$ increased with time; the rate of growth is constant with time but varies with the temperature. In Fig. 8a the logarithms of the induction period t_i and of the linear rate of growth are plotted against $1/T$. Both processes have the same activation energy. Again it was found for $NiSO_4 \cdot 7H_2O$ that both processes had the same activation energy (Fig. 8b).

The explanation of the time dependences of the number of patches offered by Garner was the following. He considered that the nucleation of a

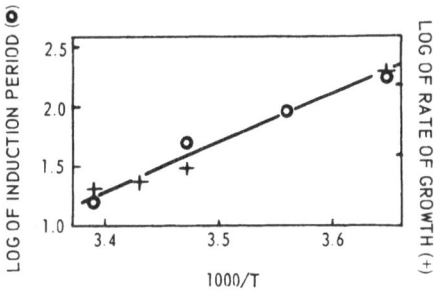

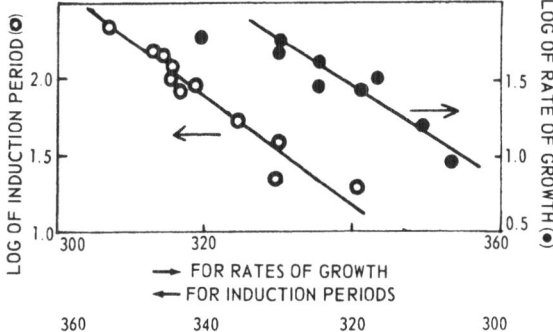

Fig. 8. Corresponding logarithmic plots for the induction period t_i and rate of linear growth as a function of $1/T$ for (top) $CuSO_4 \cdot 5H_2O$ and (bottom) $NiSO_4 \cdot 7H_2O$, illustrating the similarity in activation energy.

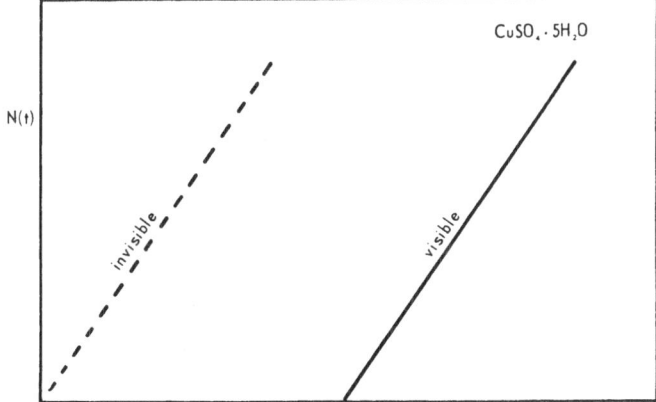

Fig. 9. Empirical lateral shift of the visible rate of growth of nuclei on the assumption that very small nuclei are invisible.

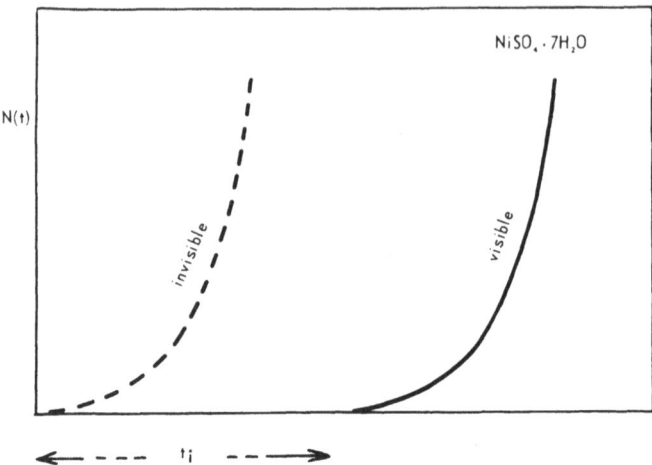

Fig. 10. Curves showing the probability of two independent successive abstractions of water molecules in the formation of hydration patches on $NiSO_4 \cdot 7H_2O$.

patch on the surface of $CuSO_4 \cdot 5H_2O$ depended upon the removal of *one* water molecule from the surface; thus N would increase linearly with time if the removal of one water molecule is the slow rate-determining step. The induction period was accounted for by suggesting that the rate of growth of nuclei to visible size is very slow and the straight line $N(t) = ct$ is shifted laterally, since nuclei are only added to the count when they reach visibility (Fig. 9).

For $NiSO_4 \cdot 7H_2O$ Garner considered that *two* successive abstractions of water molecules independent of each other are necessary to nucleate patches. The probability of two independent abstractions will increase with the square of the time (Fig. 10).

This explanation seems to be *ad hoc*. It is difficult to understand why nucleation depends on one molecule for $CuSO_4 \cdot 5H_2O$ but two molecules for $NiSO_4 \cdot 7H_2O$. Further, to account for the length of the induction period it is necessary to assume that the rate of growth is very much smaller for a patch when it is small compared to the rate of growth when it is of observable size. Yet the activation energies for observable growth and for the induction period are the same.

APPLICATION OF NUCLEATION THEORY TO DEHYDRATION

Instead of relying on *ad hoc* assumptions particular to each case it would be more satisfactory if these observations could be brought into line with and accounted for by nucleation theory. With this in view it is supposed that the exposure of the crystal surface to vacuum evaporates some water molecules, leaving vacant water molecule lattice sites in the hydrated salt. These vacancies diffuse inward from the surface, forming a narrow surface layer in near-equilibrium with the water vapor in the adjacent "vacuum." Nuclei of the new phase II will appear in this vacancy-equilibrated parent phase I, possibly homogeneously and necessarily near the surface, possibly heterogeneously at emergent dislocations. The thermodynamic driving force $\Delta\mu$ for this change is considered in Appendix A.

If it is assumed that the build-up of the supersaturation a is more rapid than the build-up of embryo population and critical nuclei, the number of critical nuclei $N(t)$ at time t is given by

$$N(t) = \int_0^t J(t)\, dt = J_0 \int_0^t e^{-\tau/t}\, dt \qquad (8)$$

This integral is shown in Fig. 11. The lower end of the $N(t)$ curve is very close to parabolic in form and it is suggested that this is the region explored by Garner and Southon [(^{14}), p. 1706] in the case of $NiSO_4 \cdot 7H_2O$. Had they been able to continue the curve it is suggested that the curve would have straightened to a linear portion. It was this linear portion which was found in the case of $CuSO_4 \cdot 5H_2O$. In Fig. 6 the vestiges of the parabolic feet on these curves are discernible.

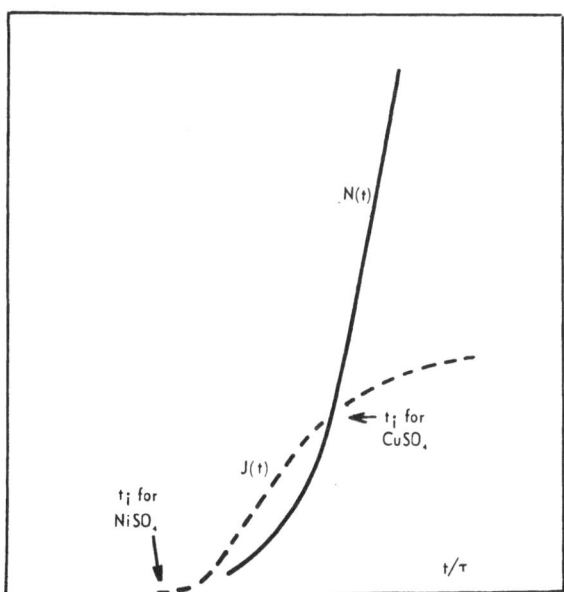

Fig. 11. Number of critical nuclei $N(t)$ as a function of t/τ according to the expression $N(t) = J_0 \int_0^t e^{-\tau/t}\, dt$.

The induction periods t_i observed by Garner *et al.* correspond to points on the curve in the neighborhood of which the course of $J(t)$ changes (Fig. 11). These points occur for certain values of t/τ; for $NiSO_4 \cdot 7H_2O$ $t_i/\tau = 0.2$ or $t_i = 0.2\tau$, and for $CuSO_4 \cdot 5H_2O$ $t_i/\tau = 0.8$ or $t_i = 0.8\tau$. Thus when $\log t_i$ was plotted versus $1/T$ (Fig. 8) it was the same as plotting $\log \tau$ against $1/T$.

Since $\tau = k^2/g_k$ and k is only a slowly varying function of T, the dependence of $-\log \tau$ and therefore $-\log t_i$ is essentially the dependence of $\log g_k$ on $1/T$. A correspondence was found between the activation energies for the induction period and for the linear growth of patches (Fig. 8). This now implies that the activation energy for g_k, the rate of transport of kinetic units across the interface of a nucleus, is closely similar in magnitude to the activation energy for transport across the interface during the growth of a patch. This seems reasonable.

EFFECT OF WATER VAPOR PRESSURE ON THE RATE OF GROWTH OF PATCHES

Garner and his collaborators ([16]) found that the rate of growth of patches on $CuSO_4 \cdot 5H_2O$, ammonium, potassium, and chrome alums was decreased when water vapor was present at a pressure below the equilibrium dissociation

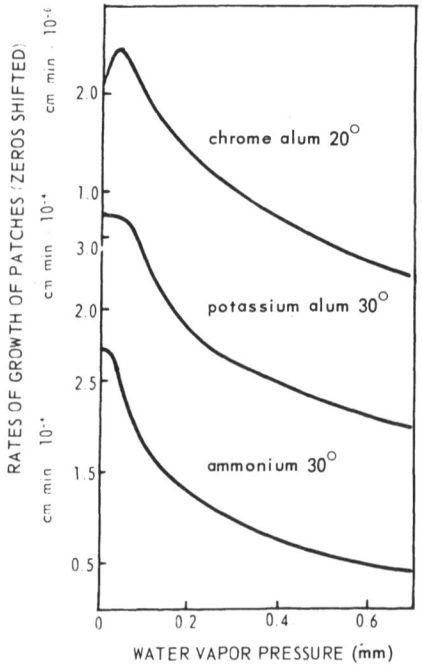

Fig. 12. Rate of growth of dehydration patches on copper sulphate, ammonium, potassium, and chrome alums as a function of water vapor pressure.

pressure (Fig. 12). Garner explained the retarding effect as follows ([16]). Water molecules have to cross the interface between the parent crystal I and the crystals of the patch II. If water molecules are adsorbed at this interface they hinder the passage of the diffusing water molecules out of I; the higher the water vapor pressure, the greater the adsorption and the slower the rates of diffusion and growth.

An alternative and perhaps more natural explanation can be given. It is to be expected that in the absence of other complicating factors the rate of growth should diminish when the thermodynamic driving force is diminished. The driving force is expected to be large when the water vapor pressure is small, i.e., under "vacuum," and the driving force will be small when the water vapor pressure is large, i.e., just less than the equilibrium dissociation pressure of the hydrate.

Let us assume that the growth of a patch proceeds by two-dimensional nucleation of new lattice layers of phase II at the interface between I and II; the rate of growth g of a patch will then be given by

$$g = n_a r \exp(-\Delta G_{k'}/kT) \tag{9}$$

where n_a is the number of kinetic units per unit area of interface, r is the rate of reorganization of kinetic units, and $\Delta G_{k'}$ is the free energy of formation of a two-dimensional nucleus containing k' units. Also

$$\Delta G_{k'} = \pi \rho^2 o_{II}/(-\Delta \mu) \tag{10}$$

in which ρ is the line tension of the periphery of a nucleus, o_{II} is the area of interface occupied by a kinetic unit, and $\Delta \mu$ is the driving force.

In Appendix A the thermodynamic driving force is calculated for an assumed model of the system [Eq. (A16)]. With further assumptions the expression can be simplified to [(A17)]:

$$-\Delta \mu = w_I kT \log\{(K_{II} + p^e)/(K_{II} + p)\}$$

where p^e is the equilibrium dissociation pressure, p is the pressure of water vapor, and K_{II} is the constant for the equilibrium between water vapor, vacant water molecule lattice sites, and occupied water lattice sites in I [see Eq. (A15)]. When K_{II} is smaller than p a further simplification results to give

$$-\Delta \mu = w_I kT \log(p^e/p) \tag{11}$$

Using this expression together with Eqs. (9) and (10) we obtain

$$\log g = \log(n_a r) - \frac{\pi \rho^2 o_{II}}{w_I^2 k^2 T^2 \log(p^e/p)} \tag{12}$$

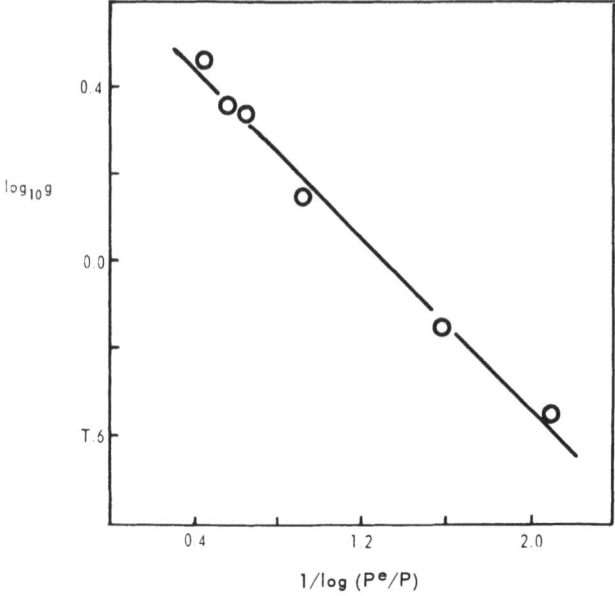

Fig. 13. Influence of partial pressure of water vapor p^e/p on the
rate of growth of dehydration patches.

The value of p^e for ammonium alum has been determined ([16]) and the points
in Fig. 12 for ammonium alum can be replotted as log g versus the reciprocal
of log (p^e/p). Surprisingly a straight line is obtained (Fig. 13), in agreement
with the form of Eq. (12) when r is taken as effectively independent of p.

This agreement may mean that the growth of patches takes place by
Volmer nucleation. Volmer nucleation would only be necessary if the interface
between the patch and the parent crystal were flat and parallel to a low index
plane. Acock et al. [([15]), p. 510] mention that patches on ammonium alum and
other hydrates tend to have crystallographic interfaces. The patches on
chrome alum appear to be hemispherical in shape; two-dimensional nucleation would not be necessary for curved interfaces.

THE ELEMENTARY KINETIC PROCESS

The rate of growth of patches on chrome alum ([16]) goes through a maximum as the pressure of water vapor diminishes (Fig. 12). In the case of ammonium alum ([15]) the rate appears to reach a short horizontal limit at low pressure. These effects may be accounted for by considering the elementary
process in which a kinetic unit of the unstable phase I reorganizes and transforms to a kinetic unit of the stable phase II.

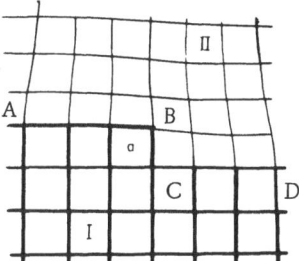

Fig. 14. Section parallel to an interface showing a step site for the elementary process.

In this model process the interface between the two lattices is assumed to be coherent, at least in the neghborbood of that kinetic unit which is transforming, and to be essentially parallel to a low index lattice plane of I. On the interface there are steps (or a step, if it is the periphery of a two-dimensional nucleus) and on the steps there are kinks. In Fig. 14 a section parallel to the interface is shown cutting through a step ABCD with a kink at BC. The two lattices are indicated as being strained in the neighborhood of the kink, the result of a decrease in volume on transformation. Transformation is assumed to occur in the following manner.

In the neighborhood of the interface both phases, although not in equilibrium with each other, are separately in equilibrium with the external water vapor pressure. As mentioned earlier, these vapor-equilibrated interfacial layers may be considered as boundary layers of diffusion fields extending into the interior of phase I. Each layer contains its equilibrium concentrations of vacant water molecule lattice sites and of interstitial sites occupied by water molecules. For simplicity we shall neglect the presence of interstitial water and we shall also assume that in each phase all the water molecule lattice sites are equivalent in energy and that there is no interaction between occupied or vacant sites.

Let the number of water molecule lattice sites in a kinetic unit of I be w_I and in the corresponding kinetic unit of II be w_{II}. There is a certain probability that the kinetic unit a of I adjacent to the kink has a certain number of vacancies q_I. If q_I is equal to or greater than $w_I - w_{II}$ the kinetic unit a of I transforms into a kinetic unit of II with a number q_{II} of vacancies. It is assumed that during this transformation no water molecules are liberated into the vapor phase nor into the surrounding lattices; thus the transformation conserves the number of water molecules in the kinetic unit and involves only the disappearance of vacancies during the rearrangement of the lattice. The number of vacancies disappearing is

$$q_I - q_{II} = w_I - w_{II} \tag{13}$$

Let v_I and v_{II} be the average numbers of vacancies per kinetic unit of

I and II. The probability that the kinetic unit of I contains q_I vacancies is

$$\binom{w_I}{q_I} \left(\frac{v_I}{w_I}\right)^{q_I} \left(1 - \frac{v_I}{w_I}\right)^{w_I - q_I} \tag{14}$$

where the first term is a binomial coefficient. The specific rate at which this kinetic unit transforms in the forward direction (I to II) is $k'_f(q_I)$, where $k'_f(q_I)$ may depend on the number of vacancies in the kinetic unit. In the same way the probability that a kinetic unit of II contains q_{II} vacancies is

$$\binom{w_{II}}{p_{II}} \left(\frac{v_{II}}{w_{II}}\right)^{q_{II}} \left(1 - \frac{v_{II}}{w_{II}}\right)^{w_{II} - q_{II}} \tag{15}$$

and the specific rate at which it transforms backward into I is $k'_b(q_{II})$. The net rate of transformation $r(q_{II})$ of a kinetic unit of I containing q_{II} vacancies in the forward direction is given by

$$r(q_{II}) = k_f \left(\frac{v_I}{w_I}\right)^{q_I} \left(1 - \frac{v_I}{w_I}\right)^{w_I - q_I} - k_b \left(\frac{v_{II}}{w_{II}}\right)^{q_{II}} \left(1 - \frac{v_{II}}{w_{II}}\right)^{w_{II} - q_{II}} \tag{16}$$

where

$$k_f = k'_f \binom{w_I}{q_I} \quad \text{and} \quad k_b = k'_b \binom{w_{II}}{p_{II}}.$$

At equilibrium the water vapor pressure is p^e and $r(q_{II})$ is zero. Hence k_b can be eliminated to give

$$r(q_{II}) = \frac{kp^{w_{II} - q_{II}}}{(p + K_I)^{w_I}} \left[1 - \left(\frac{p + K_I}{p^e + K_I}\right)^{w_I} \left(\frac{p^e + K_{II}}{p + K_{II}}\right)^{w_{II}}\right] \tag{17}$$

where we have put $k = k_f K_I^{w_I - w_{II} - q_{II}}$ and K_I and K_{II} are the constants K_{II} and K_{III} of Appendix A. If excess water is not to be liberated, q_I must be equal to or greater than $w_I - w_{II}$ and this conservation of water in the kinetic unit implies, as mentioned, that

$$q_{II} = w_{II} - w_I + q_I$$

The rates of transformation $r(q_{II})$ for different kinetic units having different values of q_{II} consistent with this condition should be summed to give the total rate of transformation r:

$$r = \sum_{q_{II}} r(q_{II}) = \left[1 - \left(\frac{p + K_I}{p^e + K_I}\right)^{w_I} \left(\frac{p^e + K_{II}}{p + K_{II}}\right)^{w_{II}}\right]$$
$$\times \frac{p^{w_{II}}}{(p + K_I)^{w_I}} \sum_{q_{II}} \frac{k(q_{II})}{p^{q_{II}}} \tag{18}$$

The factor in the brackets is zero when $p = p^e$. Since $w_I > w_{II}$ and K_I is likely to be greater than K_{II}, when $p = 0$ this factor increases monotonically with decreasing pressure to a limiting value

$$1 - \left(\frac{K_I}{p^e + K_I}\right)^{w_I} \left(\frac{p^e + K_{II}}{K_{II}}\right)^{w_{II}} \tag{19}$$

The other factors may increase or decrease with decreasing pressure according to the individual properties of the system considered. A decrease in this factor, combined with an increase in the factor in brackets, could give rise to a maximum in the rate versus pressure curve as found for chrome alum (Fig. 12).

Where the movement of the interface is similar to crystal growth on molecularly rough surfaces or involving screw dislocations the rate of growth of patches will be determined by the rate of transformation r. Where Volmer two-dimensional nucleation on molecularly smooth interfaces is the growth process the rate of transformation r appears as a factor in the rate expression, Eq. (9). It is possible that the short horizontal limit on the rate at low pressures which was found for ammonium alum (Fig. 12) may eventually be accounted for by considering the behavior of r in Eq. (9) at low pressures.

THE ACTIVATION ENERGY FOR THE RATE OF GROWTH OF DEHYDRATION PATCHES

Garner et al. [13,15,16] measured g_0, the rate of growth of patches in a vacuum, as a function of temperature. Expressing this rate in the usual way as $g_0 = B \exp(-E_0/kT)$, they found for ammonium alum that E_0 was (16.5 $\pm$ 0.5) kcal/mole compared with a value of 15.65 for the heat of dissociation. The corresponding figures for potassium alum were 16.6 and 15.7. For copper sulphate pentahydrate E_0 is 18.25 and the heat of dissociation 12.6. In the case of chrome alum $E_0 = 31$ kcal/mole and the heat of dissociation only 10 kcal/mole. Garner considered that the activation energy E_0 in the latter case was "abnormally high." What does the present theoretical treatment predict regarding the relation between E_0 and the heat of dissociation?

If the rate of growth of a patch involves two-dimensional nucleation it will be given by Eq. (9). The activation energy for growth defined as

$$E_g = -R(d \log g_0)/d(T^{-1}) \tag{20}$$

is therefore given by

$$E_g = E_r + E_n \tag{21}$$

where E_r is the activation energy for the reorganization process and E_n the activation energy for the nucleation:

$$E_r = -R(d \log r)/d(T^{-1}) \tag{22}$$

$$E_n = \frac{N_0 p^2 o_{II}}{\log(p^e/p)} \tag{23}$$

where N_0 is Avogadro's number; p has been assumed to be effectively independent of temperature and the thermodynamic driving force (or p^e/p) is assumed to be kept constant. Clearly E_n will be different for different pressure ratios, p^e/p, and, as will be evident from Eq. (27) below, E_r is also dependent on the pressure ratio. There is some evidence from the work of Garner *et al.* that E_g varies with the pressure ratio.

Under vacuum (i.e., the low working pressure p_0 of the experiments mentioned) p^e/p_0 is large and $\Delta G_{k'}$ is small, and it may be assumed in consequence that the reorganization process alone determines the rate under this condition. If this were the case then Garner's activation energy E_0 is equal to E_{r0} obtained from r_0, the reorganization rate at p_0, i.e.,

$$E_0 = E_{r0} = -R(d \log r_0)/d(T^{-1}) \tag{24}$$

In our model of the reorganization process Eq. (18) gives

$$r_0 = \left(\frac{p_0}{K_I}\right)^{w_{II}} \sum \frac{k_f(q_{II})}{(K_I p_0)^{q_{II}}}(1 - L_0) \tag{25}$$

with

$$L_0 = \left(\frac{K_I}{p^e + K_I}\right)^{w_I} \left(\frac{p^e + K_{II}}{K_{II}}\right)^{w_{II}} \tag{26}$$

where it has been assumed that p_0 is negligible compared with K_I. Differentiating we obtain

$$\frac{d \log r_0}{dT} = \frac{d}{dT} \log \sum_{q_{II}} \frac{k_f(q_{II})}{(K_I p_0)^{q_{II}}} - \frac{L_0}{1 - L_0}\left[\frac{w_I p^e}{p^e + K_I}\frac{d \log K_I}{dT}\right.$$
$$\left. - \frac{w_{II} p^e}{p^e + K_{II}}\frac{d \log K_{II}}{dT} - \left(\frac{w_I p^e}{p^e + K_I} - \frac{w_{II} p^e}{p^e + K_{II}}\right)\frac{d \log p^e}{dT}\right] \tag{27}$$

From this we see that Garner's activation energy E_0 is complex. The first term in (27) shows that it bears no simple relationship to the activation energies, $E_f(q_{II}) = kT^2[d \log k_f(q_{II})/dT]$, of the elementary processes of kinetic unit transformation in the forward direction. The second term shows that E_0 bears no simple relationship to the heat of dissociation, $kT^2[d \log p^e/dT]$, but also involves the heats of the vacancy equilibria, $kT^2[d \log K_I/dT]$ and $kT^2[d \log K_{II}/dT]$. It is understandable that the experimental values of

E_0 and the heats of dissociation show different relationships for different reaction systems.

APPENDIX A

Calculation of the Driving Force

Let us consider the change in free energy and hence the driving force when one salt hydrate (phase I) is dehydrated to another hydrate of the same salt (phase II) and water vapor at pressure p. Suppose that in phases I and II there are w_I and w_{II} water molecule lattice sites per unit cell and i_I and i_{II} interstitial sites for water molecules per unit cell. Consider a small homogeneous region of phase I containing n_I lattice cells and assume that its water molecule populations in water molecule lattice sites and in interstitial sites are in equilibrium with the water vapor pressure.

We use the subscripts fl, vl, fi, and vi to designate occupied and vacant lattice and interstitial sites. Thus n_{fl}, etc., and μ_{fl}, etc., are the numbers and chemical potentials of occupied lattice sites. The free energy of n_I unit cells of phase I is

$$n_I G_I = n_I \mu_{AI} + n_{fII} \mu_{fII} + n_{vII} \mu_{vII} + n_{fiI} \mu_{fiI} + n_{viI} \mu_{viI} \tag{A1}$$

where μ_{AI} is the chemical potential of the salt. Using the relations

$$w_I n_I = n_{fII} + n_{vII} \tag{A2}$$

$$i_I n_I = n_{fiI} + n_{viI} \tag{A3}$$

$$\mu_{fII} = \mu_{vII} + \mu_w \tag{A4}$$

$$\mu_{fiI} = \mu_{viI} + \mu_w \tag{A5}$$

where μ_w is the chemical potential of water vapor, we find that the free energy per unit cell of I is

$$G_I = \mu_{AI} + w_I \mu_{fII} + i_I \mu_{viI} + \frac{n_{fiI} - n_{vII}}{n_I} \mu_w \tag{A6}$$

In the same way the free energy per unit cell of II in equlibrium with water vapor is

$$G_{II} = \mu_{AII} + w_{II} \mu_{fIII} + i_{II} \mu_{viII} + \frac{n_{fiII} - n_{vIII}}{n_{II}} \mu_w \tag{A7}$$

When one unit cell of phase I in equilibrium with a water vapor pressure p reacts to form a unit cells of phase II in equilibrium with the same pressure the decrease in free energy per unit cell of I is

$$-\Delta G = G_I - aG_{II} - x\mu_w \qquad (A8)$$

where

$$x = (n_{fII} + n_{fiI} - n_{fIII} - n_{fiII})/n_I \qquad (A9)$$

Inserting (A6), (A7), and (A9) in (A8), using relations (A2)–(A5) and the corresponding relations for phase II together with

$$n_{II} = an_I \qquad (A10)$$

we find

$$-\Delta G = \mu_{AI} + w_I\mu_{vII} + i_I\mu_{viI} - a(\mu_{AII} + w_{II}\mu_{vIII} + i_{II}\mu_{viII}) \qquad (A11)$$

It is convenient to relate the chemical potentials to their values μ^e when the two phases I and II are in equilibrium with each other as well as with the water vapor. For such equilibrium conditions

$$0 = \mu_{AI}^e + w_I\mu_{vII}^e + i_I\mu_{viI}^e - a(\mu_{AII}^e + w_{II}\mu_{vIII}^e + i_{II}\mu_{viII}^e) \qquad (A12)$$

and the driving force per unit cell of I is

$$-\Delta\mu \equiv -\Delta G = \Delta\mu_{AI} + w_I\,\Delta\mu_{vII} + i_I\,\Delta\mu_{viI}$$
$$- a(\Delta\mu_{AII} + w_{II}\,\Delta\mu_{vIII} + i_{II}\,\Delta\mu_{viII}) \qquad (A13)$$

where $\Delta\mu_{AI} = \mu_{AI} - \mu_{AI}^e$, etc.
Since

$$\Delta\mu_{vi} = kT\log(n_{viI}/n_{viI}^e), \text{ etc.} \qquad (A14)$$

and

$$n_{vII}p/n_{fII} = n_{vII}p/(w_In_I - n_{vII})$$
$$= K_{II}, \text{ etc.} \qquad (A15)$$

we have

$$-\Delta\mu = \Delta\mu_{AI} - a\,\Delta\mu_{AII}$$
$$+ kT\log\left(\frac{K_{II}+p^e}{K_{II}+p}\right)^{w_I}\left(\frac{K_{iI}+p^e}{K_{iI}+p}\right)^{i_I}\left(\frac{K_{III}+p}{K_{III}+p^e}\right)^{aw_{II}}\left(\frac{K_{iII}+p}{K_{iII}+p^e}\right)^{ai_{II}} \qquad (A16)$$

Thus a calculation of the thermodynamic driving force requires a knowledge of the values of the equilibrium constants K for equilibrium between vacancies, occupied sites, and the vapor. This information is not available.

Acock et al. ([15]) found that the loss of water from chrome alum when exposed to "hard vacuum" apparently occurs in two stages; the first with a loss of 12 water molecules and the second with a loss of 4. Whether a phase change occurs in each stage is not known; it is possible that the first stage is an equilibration of vacancies. The dehydration of ammonium and potassium alums may be a little more straightforward; the loss of water from these salts increased from 19.8 to 20.6 of their original 24 water molecules as the temperatures was increased from 20 to 40°C.

If it is assumed that this relative constancy with temperature may indicate that $K_{i\text{II}}$ and $K_{i\text{II}}$ are large compared with p^e, the last two factors in the logarithm of (A16) are unity. With the further assumptions that the number of interstitial water molecules in phase I is negligible (i.e., that $K_{i\text{I}}$ is large) and that $\Delta\mu_{A\text{I}}$ and $\Delta\mu_{A\text{II}}$ are negligible, we arrive at

$$-\Delta\mu = w_\text{I} kT \log\left(K_\text{II} + p^e\right)(/k_\text{II} + p) \tag{A17}$$

APPENDIX B

The Heat of Dissociation

Garner measured the dissociation pressure at which water vapor was in equilibrium with both phase I and phase II. When $\log p^e$ for ammonium alum ([15]) was plotted against $1/T$ a good straight line was found. From the slope the heat of dissociation was derived as 15.65 kcal/mole H_2O. What is the significance of this heat?

If we start from Eq. (A12) and adapt it to our simpler model in which only water molecule lattice sites are taken into consideration and interstitial water molecule sites are ignored, we get

$$\mu_{A\text{I}}^e + w_\text{I}\mu_{v\text{II}}^e - a(\mu_{A\text{II}}^e + w_\text{II}\mu_{v\text{III}}^e) = 0 \tag{B1}$$

Using

$$\mu_{v\text{II}}^e = \mu_{v\text{II}}^0 + kT \log(n_{v\text{II}}^e/n_\text{I}) \tag{B2}$$

and a similar expression for $\mu_{v\text{III}}^e$, then differentiating with respect to temperature we obtain

$$w_\text{I}\frac{d\log(n_{v\text{II}}/n_\text{I})}{dT} - aw_\text{II}\frac{d\log(n_{v\text{III}}/n_\text{I})}{dT} = \frac{H_{A\text{I}}^e - aH_{A\text{II}}^e + w_\text{I}H_{v\text{II}}^0 - aw_\text{II}H_{v\text{III}}^0}{kT^2} \tag{B3}$$

where H^0_{vII} and H^0_{vII} are the enthalpies per vacant lattice site in the standard state and H^e_{AI} and H^e_{AII} are the enthalpies of the salt sites per unit all of I and of II. Since

$$n_{vII}/n_I = K_{II}w_I/(p^e + K_{II}) \tag{B4}$$

and

$$n_{vIII}/n_{II} = K_{III}aw_{II}/(p^e + K_{III}) \tag{B5}$$

we find

$$kT^2\left(\frac{w_I p^e}{p^e + K_{II}} - \frac{aw_{II}p^e}{p^e + K_{III}}\right) = kT^2 w_I \frac{d \log K_{II}}{dT} - kT^2 aw_{II}\frac{d \log K_{III}}{dT}$$
$$+ aH^e_{AII} - H^e_{AI} + aw_{II}H^0_{vIII} - w_I H^0_{vII} \tag{B6}$$

and by using (A4),

$\mu_{fII} = \mu_{vII} + \mu_w$ and $\mu_{fIII} = \mu_{vIII} + \mu_w$ and we can show that

$$\frac{d \log p^e}{dT} = \frac{\Delta H}{kT^2} \frac{1}{[w_I p^e/(p^e + K_{II}) - aw_{II}p^e/(p^e + K_{III})]} \tag{B7}$$

Here

$$\Delta H = \{aH^e_{AII} + aw_{II}H^0_{fIII} + (w_I - aw_{II})H^0_w\} - \{H^e_{AI} + H^0_{fII}\} \tag{B8}$$

in which H^0_w is the standard enthalpy per molecule of water vapor and H^0_{fII} and H^0_{fIII} are the standard enthalpies of filled water molecule lattice sites in phases I and II. The term on the right in (B7) can be expressed more simply to give

$$\frac{d \log p^e}{dT} = \frac{\Delta H}{kT^2} \frac{n_I}{n_{fII} - n_{fIII}} \tag{B9}$$

Hence, for this model of the process Garner's heat of dissociation is given by

$$\Delta H \, (n_I/(n_{fII} - n_{fIII}))$$

The linearity of the experimental plot of $\log p^e$ versus $1/T$ may indicate that the concentrations of occupied water molecule lattice sites n_{fII}/n_I and n_{fIII}/n_I in I and II do not vary very much with temperature. In turn this may imply that the concentrations of vacancies in the two phases are small.

REFERENCES

1. W. G. Burgers and L. J. Groen, *Discussions Faraday Soc.* **23**: 183 (1957).
2. C. Briske and N. H. Hartshorne, *Discussions Faraday Soc.* **23**: 196 (1957); N. H. Hartshorne and M. H. Roberts, *J. Chem. Soc.* p. 1097 (1951); N. H. Hartshorne and M. Thackray, *J. Chem. Soc.* p. 212 (1957); R. S. Bradley, N. H. Hartshorne, and M. Thackray, *Nature* **173**: 400 (1954).
3. S. W. Kennedy and P. K. Schultz, *Trans. Faraday Soc.* **59**: 156 (1963).
4. M. Volmer, Kinetik der Phasenbildung, Steinkopf, Dresden and Leipzig, 1939; R. Becker and W. Döring, *Ann. Physik* **24**: 719 (1935).
5. N. H. Hartshorne, *Discussions Faraday Soc.* **5**: 149 (1949); R. S. Bradley, *Proc. Roy. Soc. (London) Ser. A* **205**: 553 (1951).
6. W. K. Burton, N. Cabrera, and F. C. Frank, *Phil. Trans. Roy. Soc.* **A243**: 299 (1951).
7. P. Wegener, *J. Appl. Phys.* **25**: 1485 (1954); P. Wegener and C. Mack, *Advances in Applied Mechanics* (H. L. Dryden, T. L. von Kármán, and G. Kuerti, eds.) Academic Press, New York, 1958, p. 307; K. Oswatitsch, *Z. Angew. Math. Mech.* **22**: 1 (1942); W. J. Dunning, *Discussions Faraday Soc.* **30**: 9 (1960).
8. W. E. Garner, *Chemistry of the Solid State*, Butterworths, London, 1955, p. 236.
9. J. B. Zeldovich, *Acta Physicochim. URSS* **18**: 1 (1943).
10. J. Frenkel, *Kinetic Theory of Liquids*, Oxford University Press, 1946.
11. R. Probstein, *J. Chem. Phys.* **19**: 619 (1951).
12. A. Kantrowitz, *J. Chem. Phys.* **19**: 1097 (1951).
13. N. F. H. Bright and W. E. Garner, *J. Chem. Soc.* p. 1872 (1934).
14. W. E. Garner and W. R. Southon, *J. Chem. Soc.* p. 1705 (1935).
15. G. P. Acock, W. E. Garner, J. Milsted, and H. J. Willavoys, *Proc. Roy. Soc. (London) Ser. A* **159**: 508 (1947).
16. J. A. Cooper and W. E. Garner, *Proc. Roy. Soc. (London) Ser. A* **174**: 487 (1940).

DISCUSSION

P. W. M. Jacobs (University of Western Ontario): It seems that there is a possible test of the theory by measuring t_i as a function of the supersaturation ratio, because τ depends on k, and the shape of the plot of kt against t presumably depends on τ, and therefore on k; k depends on α, the supersaturation ratio. Of course, this is hindsight and I don't suppose it quite explains the problem. Maybe one wants to measure the number of nuclei formed as a function of supersaturation ratio.

Answer: Yes, that would be an interesting investigation. Investigation of the defect structure should also prove fruitful.

Chapter 8

Electronic Mechanisms in Solid-State Photolysis*

P. W. Levy and P. J. Herley[†]

Brookhaven National Laboratory
Upton, New York
and
Explosives Laboratory, Picatinny Arsenal
Dover, N. J.

When some crystalline solids are photolytically decomposed several of the effects which occur indicate that electrons, holes, or excitons are involved in the reaction. In addition the kinetics of many of these reactions appear to be approximately, but not precisely, second order. The nature of these effects suggests that the photolytic reactions are essentially electronic and that they occur at certain crystal sites or defects when these become doubly excited. The results obtained when $NaBrO_3$ is photolytically decomposed, particularly the dependence of the equilibrium rate on light intensity, suggest that the doubly excited sites may also generate additional sites. Furthermore, the data indicates that some decomposition sites may be generated from precursors that are expended during the photolytic reaction. Using these concepts a phenomenological theory has been developed which is consistent with current results. In particular it indicates that the photolytic rate vs. time curves should be resolvable into one constant and two exponential components. It also predicts, in agreement with the data, that the exponential component attributable to the sites generated from precursors should be reduced—even to the point of disappearing—when the material is subjected to ionizing radiation prior to photolysis.

INTRODUCTION

A large number of common crystals can be decomposed by exposing them to intense light, e.g., azides and oxalates ([1]). When this photolytic decomposition process is examined in detail certain general features become obvious. First, in a large fraction of the cases the photolysis-producing light is not absorbed at the crystal surface but in a distribution below the surface. Second, in many instances despite the optical absorption distribution the decomposition process clearly takes place on the surface. Often, while it is apparent that

---

*Research performed primarily at Brookhaven National Laboratory and supported by Picatinny Arsenal and the U. S. Atomic Energy Commission.
†Guest Scientist at Brookhaven National Laboratory, Upton, N. Y.

a large part of the photolysis is occurring directly on the crystal surface, some decomposition ultimately begins to take place in the interior, especially after prolonged exposure to light. Even when decomposition is taking place inside of the crystal it is not uniformly distributed but occurs at well-defined individual locations.

From these observations one may conclude the following: The photolysis process must involve some sort of charge or energy carrier to transport the energy imparted by an individual photon to the point in the crystal where decomposition occurs. The most obvious choices for such carriers are electrons, holes, or excitons. For example, the incident photon may create an electron–hole pair at the point where it is absorbed. While the hole may remain trapped at that point, the (conduction) electron may migrate to an internal or external surface of the crystal. There it may be trapped by a surface state and release enough energy to initiate a decomposition reaction. This electronic carrier point of view is not new. To give just one example it was discussed by Jacobs and Tompkins [2] in 1954. In a sense this paper will describe an extension of this approach to photolytic rate studies.

PHOTON-INITIATED CHARGE AND ENERGY TRANSPORT IN CRYSTALS

When a photon is absorbed by a crystal the absorption process usually involves an interband transition of one or more defect centers. These defect centers may include color centers, trapping centers, impurity ions, etc. When an interband absorption occurs an electron is removed from the valence band and, providing a sufficient amount of energy has been made available by the absorption of the photon, is injected into the conduction band. In the conduction band the electron then migrates until trapped by one of a variety of different kinds of trapping centers. The entity produced by the removal of an electron from the valence band, the well-known (electronic) hole, may remain at the point where it originated or may migrate throughout the valence band until it becomes trapped by a hole trap. Usually the lifetime of the migrating hole is considerably shorter than the lifetime of the migrating electron.

Certain hole traps have an important property. Once having trapped a hole the probability that the center will then trap an electron is relatively large. Consequently the electrons and holes recombine, the charges neutralize one another, and a certain amount of energy is released. Usually this is approximately equal to the energy separating the valence and conduction bands. This important process is termed recombination and the centers at which it occurs are called recombination centers. There are other important absorption processes that may enter into the photolysis process. For example, electrons may be raised from the valence bands to electron traps, or from electron traps to the conduction band. In addition hole centers enter into similar processes.

One additional photon absorption process which may be particularly important in photolytic decomposition is the creation of excitons. An exciton may be thought of as an electron which has been excited sufficiently to produce a bound electron–hole pair or atom. In this case the electron would receive sufficient energy to remove it from the valence band but not enough energy to release it from the Coulomb attraction of the hole. This can be expressed another way; an exciton can be formed when an electron receives sufficient energy to remove it from the valence band but not quite enough energy to elevate it to the conduction band. If this excited electron–hole pair or atom remains fixed at a specific location in the crystal it is usually called a molecular exciton. If, however, the electron–hole pair migrates throughout the crystal it is called a mobile exciton, or simply an exciton. These mobile excitons have the important property that they can transport energy without any charge transport occurring. Once formed the mobile excitons migrate throughout the crystal until they interact with any center or defect which upsets the Coulomb attraction between the two charges. Once this interaction occurs the two charges recombine, or neutralize, with the release of a certain amount of energy. Usually the energy released when an exciton is destroyed is tenths of an electron volt less than the band-gap energy.

Now it is well known that the surfaces of crystals may contain electron traps, hole traps, and/or recombination centers. Clearly the electronic processes occurring at these surface sites or surface states can release sufficient energy to produce reactions at the crystal surface. Some states (^{3}) can be associated with defects or impurities in the crystal lattice or one or more types of atoms chemisorbed on the crystal surface. Other surface states, the Tamm states (^{4}), occur in or on perfect crystals and are a consequence of the quantum mechanical nature of the electronic properties of crystals. Clearly if the surface of a crystal is being eroded by photolytic decomposition there could be ever-present Tamm states on the surface. The more important carriers, surface states, and internal states or traps which are important for photolytic decomposition are summarized in Table I.

It is important to emphasize one other electronic property of crystals which are subject to photolytic decomposition. It has been emphasized that energy release occurs when opposite charges interact with one another. It has also been pointed out that the potential energy associated with an exciton is released when it is sufficiently perturbed by some defect in the crystal or on the crystal surface. This perturbation can be supplied by a trapped electron or hole. Specifically, the diffusing exciton can interact with a trapped electron or hole to release its energy. The probability of such an interaction is propo-

TABLE I

Charge and Energy Carriers in Solids

Carrier	Common symbols	Commonest traps	Carrier reactions	Approximate reaction energy
Electron	$e, e^-, -$	Top of gap	e + trapped h	E_B^*
			e + mobile h	E_B
		Surface state	e + mobile h	$\approx E_B^*/2$
Hole	$h, e^+, +$	Bottom of Gap	h + trapped e	E_B^*
			h + mobile e	E_B
		Surface state	h + mobile e	$\approx E_B^*/2, E_B^*$
Exciton	X, ex., e^0	Trapping of mobile	X + trapped e	$E_B^*, 2E_B^*$
		X at room	X + trapped h	E_B^*
		temperature is	X + perturbation	E_B^*
		unlikely	X + (e, or h in surface state)	$E_B^*, E_B^*/2, 3E_B^*/2$

E_B = band gap = energy separating conduction and valence band. $E_B^* \approx E_B$ less a few tenths eV.

tional to the product of the concentration of diffusing carrier and the concentration of the fixed perturbation. The concentration of electrons, holes, and/or excitons is most likely to be proportional to the intensity of the carrier-producing incident light. Thus the situation we have just described provides a natural explanation for the commonly observed second-order dependence of the photolytic reaction on light intensity. In fact the derivation given below will specifically indicate two relationships. First it will show that light intensity is approximately second-order. Second it will show that deviations from an exact second-order dependence are to be expected and that these deviations may occur in both directions.

THE DECOMPOSITION SITE

In the preceding section the various crystal defects, particularly the recombination centers, that facilitate energy release when charge neutralization occurs have been discussed. When these recombination centers occur either on external or internal crystal surfaces the energy release may facilitate a photolytic reaction. Here the phrase photolytic reaction is used in its most primitive sense. The name decomposition site will be given to any or all traps that facilitate the photolytic reaction. The particular defect, impurity, or surface state acting as a decomposition site need not be identified. However, the electronic processes associated with the decomposition site are quite specific. The first step in the electronic process occurs when the site traps an electron, a hole, or an exciton. In other words the decomposition site is a carrier trap, but it is not necessary to specify exactly what charge it traps. When the decomposition site contains the initially trapped carrier it will be referred to as in the first excited state. The lifetime of this first excited state must be finite and could range from a few nanoseconds to many hours or even days. Having trapped a carrier the decomposition site is capable, as described in detail above, of trapping a carrier whose charge is opposite of the initially trapped charge, or of trapping an exciton and facilitating the release of the exciton potential energy; of course, as described above, the final step occurs when the energy released facilitates a photolytic reaction.

THE GENERATION OF DECOMPOSITION SITES

The energy released in the recombination processes may produce some effects other than photolytic decomposition. In this connection it is useful to very briefly recall some of the effects which occur when alkali halides are subjected to ionizing radiation (^{5}). As is well known these materials become colored when exposed to X rays or gamma rays. Also, some of the alkali halides become colored when exposed to ultraviolet light. This coloring is due to the formation of F-centers and other centers when electrons and holes are trapped on appropriate defects. When certain alkali halides which have been purposely doped with impurities such as calcium are exposed to ionizing radiation the coloring rate is very greatly increased. The room-temperature F-center concentration versus dose curves of calcium-doped alkali halides increase in the low dose region at a much greater rate than similar curves for undoped crystals. The introduction of calcium impurity may be expected to increase the positive ion vacancy concentration in these crystals. However, the F-center is formed when an electron is trapped on a negative ion vacancy. Consequently for this and many other reasons numerous mechanisms have been proposed to account for the generation of negative ion vacancies. Thus it does not seem inappropriate to assume that similar processes may occur in crystals undergoing photolytic decomposition. In particular it seems reasonable to assume that the doubly excited decomposition sites may undergo decomposition reactions under certain circumstances and generate additional decomposition sites under slightly different conditions. To give just a single example of many possibilities consider the following reactions in sodium bromate. Assume that a decomposition site occurs where there is a BrO^- molecule-ion instead of a BrO_3^- ion. When excited the following reaction may occur

$$\text{Excitation} + BrO^- + BrO_3^- \longrightarrow 2(BrO_2^-) \longrightarrow 2Br^- + 2O_2 \qquad (1)$$

In this case the decomposition site is "used up." Alternatively consider

$$2(BrO_2^-) \longrightarrow 2BrO^- + O_2 \qquad (2)$$

In this case both gas evolution occurs and additional sites are generated. Clearly, many different reactions of this type could be proposed. To reiterate, it will be assumed that once a singly excited decomposition site has acquired a second excitation it may enter into a decomposition reaction or it may produce an additional decomposition site without destroying itself.

EXCITED DECOMPOSITION SITES PRODUCED BY EXPENDABLE GENERATORS

The same type of reasoning that led to the assumption that excited decomposition sites may generate additional decomposition sites points to the possibility that excited decomposition sites may be generated from precursors that are used up in the site-creation process. For example, consider a crystal with a substitutional OH impurity. It is conceivable that the OH ion will interact with a BrO_3 ion to form BrO^- by a photolytic process, e.g.,

$$BrO_3^- + OH^- \longrightarrow BrO^- + \tfrac{3}{2}O_2 + H^+ \qquad (3)$$

Most likely this would result from exiton capture or a recombination event. Later we will present data to indicate that a decomposition site-generating process similar to the one we have just discussed appears to be operating in sodium bromate.

Assuming that excited decomposition sites are generated by a process which uses up some sort of precursor it is necessary for the derivation to be given below to specify the rate that precursors are converted to decomposition sites. Lacking specific physical information on such a process one is forced to assume a conversion dependence. The following definitions are used: t is time; S is the concentration of decomposition-site precursors at $t = 0$; $s(t)$ is the concentration of decomposition sites in the singly excited state, s is a function of the time t (s is *not* the concentration of decomposition sites); I is the intensity of photolytic light (the concentration of light-generated carriers is proportional to I); and k_6 is the rate that unconverted precursors are converted to excited decomposition sites per unit time (k_i, $1 \leqslant i \leqslant 5$, defined below).

It is reasonable to assume that the rate at which precursors are converted into excited decomposition sites is proportional to the product of the light intensity and the concentration of unconverted precursors. Thus at time t

the concentration of unconverted precursors is $(S - s)$ and the conversion of precursors into decomposition sites is given by

$$ds/dt = k_6 I(S - s) \qquad (4)$$

The solution of this is

$$s(t) = S[1 - \exp(-k_6 It)] \qquad (5)$$

Then (4) can be written

$$ds/dt = k_6 IS \exp(-k_6 It) \qquad (6)$$

and $k_6 IS \exp(-k_6 It)$, is the rate that excited decomposition sites are generated from precursors.

It must be emphasized that there is no evidence to support the use of this particular conversion dependence. However, it would appear to be a most reasonable one. In addition the discussion in this section has been entirely in terms of the generation of decomposition sites from precursors. The possibility of an expendable precursor-related mechanism for the destruction of decomposition sites cannot be discarded.

In the derivation given below it will be assumed that the excited decomposition sites generated from precursors are the same as those generated during decomposition. Actually this assumption is not necessary. It is conceivable that two or more independent processes are operating. In fact two or more different types of precursors may be involved and each type may or may not be related to the process responsible for the steady-state behavior.

DEPENDENCE OF THE EQUILIBRIUM DECOMPOSITION RATE ON LIGHT INTENSITY

The concepts described above have been used to obtain a theoretical expression for the dependence of the equilibrium photolytic rate on light intensity. This derivation is given in a previous publication[6] but is repeated here inasmuch as it is very short and is an integral part of the extension given below. Originally it was suggested only to explain the dependence of the equilibrium rate on light intensity which was observed when sodium bromate was photolytically decomposed. Three pertinent results emerged from these measurements: (1) The equilibrium photolysis rate was roughly proportional to the square of the light intensity but not precisely so, (2) the deviation from an exact I^2 dependence was markedly increased when the material was exposed to gamma-irradiation prior to photolysis, and (3) the deviation from an exact I^2 dependence was opposite to that found in the majority of similar materials.

The basic assumption that was used is that decomposition occurs, at least part of the time, at decomposition sites when these sites become doubly excited. It is not necessary to specify the excitation process nor is it necessary that the same process produce both excitations. In addition it was assumed that the doubly excited sites may, at least part of the time, generate additional decomposition sites. The following definitions are used: $R_1(t) =$ the decomposition rate $= k_1 Is$, i.e., k_1 is the probability that the photolytic light interacts with existing singly excited sites to produce decomposition; k_2 is the rate of formation of singly excited sites by the photolytic light [$k_2 I$ is the number of singly excited sites formed (per unit time per unit volume) by the photolytic light]; k_3 is the probability that a singly excited site will decay by one or more independent processes that do *not* involve interactions with light-generated carriers, for example, the thermal release of a carrier or, alternatively, recombination with a thermally released charge ($k_3 s$ is the decrease in the singly excited site concentration due to these processes, which will be referred to as spontaneous decay); k_4 is the probability that the existing singly excited sites will produce additional singly excited sites by interaction with the incident light ($k_4 Is$ is the rate of formation of singly excited sites by the interaction of the light and the existing singly excited sites—more than one process may contribute to this mode of formation); k_5 is the probability that existing singly excited sites will be destroyed by interaction with the light flux I, e.g., by electron–hole recombination ($k_5 Is$ is the rate of destruction of singly excited sites by the interaction of the light and the existing singly excited sites).

Using these definitions and with the aid of the following diagram

$$k_2 I \longrightarrow \begin{bmatrix} \text{concentration of} \\ \text{singly excited} \\ \text{sites, } s \end{bmatrix} \begin{array}{l} \longrightarrow k_1 Is \\ \longrightarrow k_3 s \\ \longrightarrow k_5 Is \end{array}$$
$$k_4 Is \longrightarrow$$

one obtains the differential equation for the concentration of single excited sites:

$$ds/dt = -k_1 Is + k_2 I - k_3 s + k_4 Is - k_5 Is$$
$$= k_2 I - [k_3 + (k_1 + k_5 - k_4)I]s \qquad (7)$$

When the decomposition is occuring at a constant rate

$$R(I) = k_1 Is \qquad (8)$$

is constant. This requires that s is constant and that

$$ds/dt = 0 = k_2 I - [k_3 + (k_1 + k_5 - k_4)I]s \tag{9}$$

or

$$s = \frac{k_2 I}{k_3 + (k_1 + k_5 - k_4)I} \tag{10}$$

Combining these equations for the constant rate conditions

$$R(I) = \frac{k_1 k_2 I^2}{k_3 + (k_1 + k_5 - k_4)I} = \frac{(k_1 k_2/k_3)I^2}{1 + [(k_1 + k_5 - k_4)/k_3]I} \tag{11}$$

$$R(I) = \frac{aI^2}{1 \pm \beta I} \approx aI^2(1 \mp \beta I + \cdots) \tag{12}$$

where $a = k_1 k_2/k_3$ and $\beta = (k_1 + k_5 - k_4)/k_3$.

It is unlikely that k_3 would be negative since this would represent spontaneous generation of excited sites. If k_3 is zero, Eq. (11) reduces to one that is linearly dependent on I. Thus k_3 must be greater than zero to obtain an I^2 behavior. Note that a is always positive. However, β may be plus or minus depending on which process is the predominant one; it is unlikely that it will be exactly zero. For β to be less than zero k_4 must be larger than $k_1 + k_5$. Thus a negative sign in the denominator of Eq. (12), i.e., $\beta < 0$, requires that the term representing the generation of additional excited sites be greater than the terms representing the destruction of singly excited sites.

The data obtained for unirradiated $NaBrO_3$ requires that β be slightly negative. With irradiated $NaBrO_3$ the constant β was considerably more negative. Thus this data supports the concept of excited decomposition-site generation.

THE DECOMPOSITION-SITE EQUATION VALID FOR ALL TIME

Using the definitions given above and with the aid of the following diagram:

$$
\begin{array}{c}
k_2 I \longrightarrow \\
k_4 Is \longrightarrow \\
k_6 ISe^{-k_0 It} \longrightarrow
\end{array}
\left[
\begin{array}{c}
\text{concentration of} \\
\text{singly excited} \\
\text{sites, } s
\end{array}
\right]
\begin{array}{c}
\longrightarrow k_1 Is \\
\longrightarrow k_3 s \\
\longrightarrow k_5 Is
\end{array}
$$

It can be seen that the following differential equation describes the concentration of decomposition sites in the singly excited state:

$$ds/dt = -k_1 Is + k_2 I - k_3 s + k_4 Is - k_5 Is + k_6 IS \exp(-k_6 It) \qquad (13)$$

Before considering the solution of this equation it is essential to discuss two mechanisms which may contribute to the excited decomposition site concentration. First there is the possibility that the material contains impurities or defects which facilitate the creation of excited decomposition sites by the direct absorption of incident photons. In a more sophisticated treatment an attempt would have to be made to differentiate between an impurity- or defect-related mechanism and one which depends on the electronic properties of the pure material. However, such a differentiation is not attempted here and all such processes are lumped together in the single rate constant k_2. A second source of excited decomposition sites is the possibility that the crystal may contain some such excited sites prior to photolysis. This possibility is automatically included in the process of solving the differential equation. In order to determine the arbitrary constants that appear in the general solution the obvious boundary condition to specify is the concentration of singly excited sites at $t = 0$. Consequently s_0 is the concentration of singly excited decomposition sites at $t = 0$. Note that if the singly excited sites spontaneously decay, i.e., if k_3 is not zero, it is essential to put s_0 equal to zero. However, k_3 may represent several different decay processes and thus $s_0 \neq 0$ and $k_3 \neq 0$ are not inconsistent.

The solution of Eq. (13) is

$$
\begin{aligned}
s = {} & \frac{SIk_6}{k_3 + (k_1 + k_5 - k_4 - k_6)I} \\
& \times (\exp\{-k_6 It\} - \exp\{-[k_3 + (k_1 + k_5 - k_4)I]t\}) \\
& + \frac{k_2 I}{k_3 + (k_1 + k_5 - k_4)I}(1 - \exp\{-[k_3 + (k_1 + k_5 - k_4)I]t\}) \\
& + s_0 \exp\{-[k_3 + (k_1 + k_5 - k_4)I]t\} \qquad (14)
\end{aligned}
$$

or alternatively

$$
\begin{aligned}
s = {} & \frac{k_2 I}{k_3 + (k_1 + k_5 - k_4)I} \\
& + \left[s_0 - \frac{k_2 I}{k_3 + (k_1 + k_5 - k_4)I} - \frac{k_6 SI}{k_3 + (k_1 + k_5 - k_4 - k_6)I} \right] \\
& \times \exp\{-[k_3 + (k_1 + k_5 - k_4)I]t\} \\
& + \frac{SIk_6}{k_3 + (k_1 + k_5 - k_4 - k_6)I} \exp(-k_6 It) \qquad (15)
\end{aligned}
$$

It is apparent that the concentration of singly excited sites is described by two exponential terms and one constant term. The first term in the latter expression is the one derived above for the special case of equilibrium photolysis. The second and third terms may be ascribed to the transition from the initial state to the equilibrium condition. Using this expression for the excited site concentration the photolytic decomposition rate is given by

$$R_1(t) = k_1 Is \qquad (16)$$

Obviously the photolytic rate curve has the same shape as the excited site curve. It consists of two exponential and one constant components.

SODIUM BROMATE PHOTOLYSIS DATA

The photolytic decomposition of sodium bromate ($NaBrO_3$), both unirradiated and exposed to Co^{60} gamma irradiations prior to photolysis, has been studied. The experimental details are given in ([6]). In these determinations the gas evolution rate is not determined directly. Actually the total gas pressure is determined as a function of time, and to determine the decomposition rate it is necessary to differentiate the pressure vs. time curves. The differentiation is done numerically on a computer. The first step in the differentiation is to make a least square fit to a polynomial (usually second degree) through a specified number of data points. For example, five data points could be used. The second step is to compute the derivative at the center point

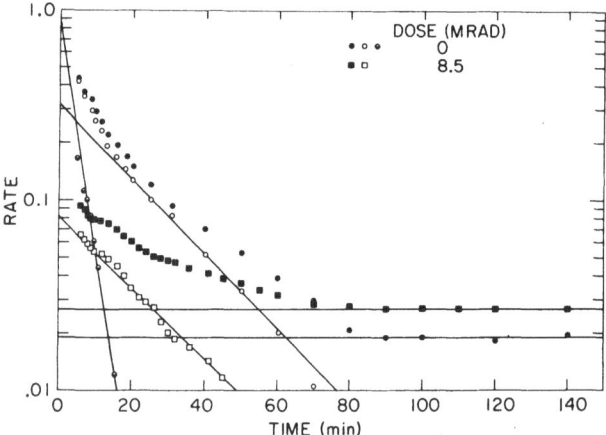

Fig. 1. Rate of gas evolution from powdered $NaBrO_3$ illuminated with 2537-Å light at room temperature. The data from unirradiated material is resolved into one constant and two exponential components. However, the $NaBrO_3$ which was exposed to a 8.5×10^6 rad Co^{60} gamma-irradiation prior to photolysis contains one constant and *only one* exponential component.

in this group of five. The process is then repeated using the next data point and the two points directly below and above the center point. This procedure can be used with any number of points in each cluster and is considered reliable when the derivative curve is found to be independent of the cluster size. Unfortunately the derivative computation is least reliable at the ends of the data curve. For large time there is very little curvature and consequently little or no error is introduced. At the $t = 0$ end of the curve there is relatively little curvature, but the rate is quite high. Consequently at this end, where the data is most interesting, there may be some undetermined but nevertheless small errors in the rate curves.

In addition to the data shown in Fig. 1 curves were obtained from samples subjected to other gamma-ray doses prior to photolysis. All of these curves were analyzed by the methods described above and the pertinent constants are summarized in Table II. The rate of gas evolution, $R_1(t)$, as a function of time t and at a given light intensity I is given by

$$R_1(t) = C_1 + C_2 e^{-a_2 t} + C_3 e^{-a_3 t} \tag{17}$$

where the C_i and a_i are constants. Table II contains values obtained from crystals both unirradiated and irradiated with Co^{60} gamma rays prior to photolysis. The light intensity was constant. The number of data points used to compute $R_1(t)$ from the total pressure versus time curves could be altered without significantly changing the tabulated values. Apparently the procedure used to compute the derivative does not produce exactly equivalent constants when the number of points in each computation cluster is changed. However, the deviations are too small to affect any of the conclusions. The lines which correspond to each of the resolved components are shown in Fig. 2. This figure shows the error associated with the rate determinations. The error in the gamma-ray dose measurements, which is approximately 5%, is not shown. Several conclusions can be obtained from the data. First, the photolytic rate curves can be resolved into one constant and two exponential components as predicted by the theory. Second, Eq. (17) predicts that the exponential con-

TABLE II

Analysis of the NaBrO$_3$ Photolysis Rate Curves

Dose ($\times 10^6$ rad)	C_1	C_2	a_2	C_3	a_3
0	0.021	0.359	0.015	0.95	0.104
5	0.023	0.327	0.019	0.68	0.118
7	0.028	0.204	0.016	0.59	0.136
8.5	0.028	0.081	0.020	—	—
10.0	0.019	0.097	0.021	—	—
20.0	0.025	0.045	0.021	—	—

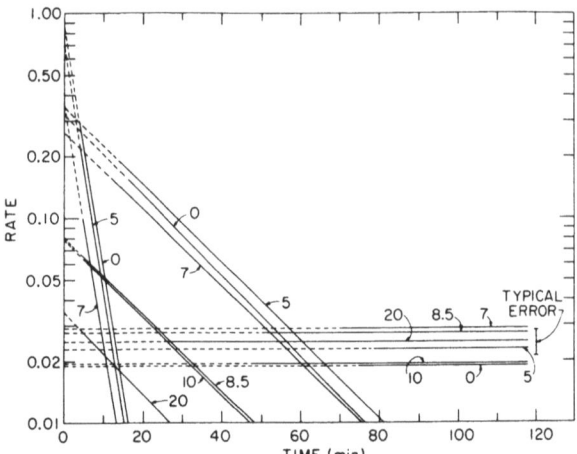

Fig. 2. The exponential and linear components obtained from Fig. 1 and some additional curves. The "numbers" are the gamma-ray dose in (10^6 rad) imparted to the crystal prior to photolysis. In addition to the indicated error in the rate the doses are subject to an error of ± 5 %. For doses greater than 7×10^6 rad the third component is entirely missing.

stants, i.e., the a_1 and a_2 given in Table II, should change if any of the k_i values are modified. As the previous study[6] indicated that k_i did change, the net effect of these changes left the constant term C_1 unchanged. Because of the similarity of the denominator in C_1 to a_2 (in fact $C_1 = k_2 I/a_2$) this exponential should not be modified by radiation if these measurements are to be consistent with the previous work, and this appears to be the case. Third, consider the effects of the radiation on the pre-exponential constants C_2 and C_3. Either of these could be the constant containing S. Assume it is C_3, and unless k_6 is greatly modified by radiation it should remain constant after irradiation for reasons similar to those which explain the constancy of a_2. However, it clearly decreases as the radiation increases. Consequently the possibility that S is decreased by radiation must be included. However, this is in accord with the mechanism postulated for the reaction. It was assumed that the electronic carriers—i.e., electrons, holes, and excitons—generated by the light played an essential role in the photolysis reaction. Since these carriers are generated by gamma-ray irradiation it is not surprising that one observes similar effects from gamma rays and light-generated carriers.

This can be put in another way. It is to be expected that exposure to light will use up the postulated precursors. Since both light-induced and gamma-ray induced reactions involve the same carriers they should produce the same effects. Consequently after exposure to gamma rays crystals might behave like material which had been appreciably photolyzed. The photolytic rate dependence of gamma-ray irradiated material appears to be describable by the Eq. (17) except that t is replaced by $t + t_0$, where t_0 is an "effective time" which reflects the gamma-ray reaction.

Similar arguments can be advanced to explain the behavior of C_2.

REFERENCES

1. D. A. Young, *Decomposition of Solids*, Pergamon, London, 1966, Chapter 4, p. 110.
2. P. W. M. Jacobs and F. C. Tompkins, *Chemistry of the Solid State*, Butterworth, London, 1954, Chapter 3, p. 57.
3. W. Dekeyser, *Reactivity of Solids*, Elsevier, Amsterdam, 1961, p. 376.
4. I. Tamm, Phys. *Z. USSR* **1**: 733 (1932).
5. J. H. Schulman and W. C. Compton, *Color Centers in Solids*, Pergamon, London, 1962.
6. P. J. Herley and P. W. Levy, *J. Chem. Phys.* **46**: 627 (1967).

DISCUSSION

J. B. Wachtman, Jr. (*National Bureau of Standards*): Can you separate the effects of volume and surface processes by studying the effects of specimen thickness on the rate of gas production and has this been attempted?

Answer: In general yes, but we have not yet attempted any such studies on $NaBrO_3$. In considering the possible differences between volume and surface mechanisms the following observation is particularly interesting. If you calculate the distribution of absorbed ultraviolet light in various inorganic solids such as $NaBrO_3$ or KN_3 you will find that the non-negligible absorption may extend as much as a millimeter below the surface. However, microscopic examination of partially reacted crystals indicates that almost all of the decomposition is taking place in rather well-defined locations on the surface, which may be related to the presence of grain boundaries, emerging dislocations, etc. In heavily reacted materials one can detect that a reaction is occurring in the body of the crystal. In this case, too, the reaction seems to occur at rather well-defined internal boundaries.

P. W. M. Jacobs (*University of Western Ontario*): It seems from Fig. 2 that one of the exponential decay processes is removed by gamma-irradiation. Which of the two time-dependent terms in the solution of the differential equation for s corresponds to this process? In other words, has it been possible to identify the process giving rise to the faster decay? Also, is there any hope of obtaining the magnitude of the individual k's?

Answer: Intuitively one would expect that the most rapidly decreasing exponential is the one associated with the "using up" of precursors. Then the remaining exponential would correspond to the approach to equilibrium which would occur after the precursor-related processes are no longer important. This has not been demonstrated in an unequivocal way but all of the available data are in accord with this assignment.

Consider what occurs when the samples are irradiated prior to photolysis. As the dose is increased the rate versus time curves obtained resemble the curves obtained for unirradiated material except that their origins are displaced to a point on the positive time axis. Or, more specifically, if $R_1(t)$ is the curve obtained from unirradiated material the

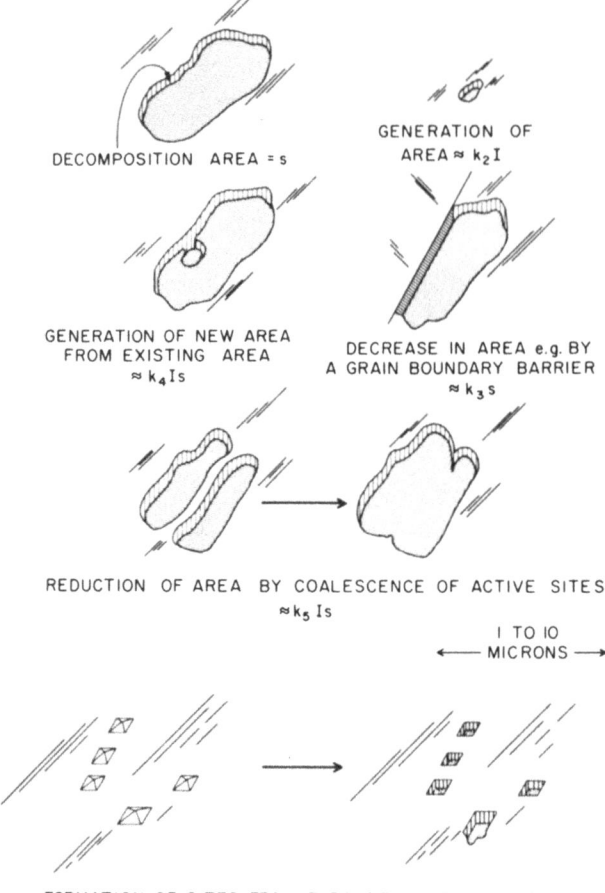

DECOMPOSITION AREA = s

GENERATION OF
AREA ≈ $k_2 I$

GENERATION OF NEW AREA
FROM EXISTING AREA
≈ $k_4 Is$

DECREASE IN AREA e.g. BY
A GRAIN BOUNDARY BARRIER
≈ $k_3 s$

REDUCTION OF AREA BY COALESCENCE OF ACTIVE SITES
≈ $k_5 Is$

I TO 10
◄── MICRONS ──►

FORMATION OF SITES FROM EXPENDED PRECURSORS x $k_6 I$

Fig. 3. The text contains an equation giving the rate of photolytic decomposition vs. time which is based on an "electronic" mechanism. Precisely the same equation may be derived using the assumption that the quantity s (in the text the concentration of singly excited sites) is the total area of the sides of the decomposition sites and the other rate constants refer to the processes indicated here.

curve obtained from irradiated material is given by the same expression with t replaced by $t + t_0$. Furthermore as the dose increases t_0 increases proportionally. Thus the more rapidly changing exponential term is the first to become negligible as the dose increases. This observation would support the assignment of the more rapidly changing exponential to the component associated with precursors.

In regard to the question about the magnitude of the individual k's two comments are required. First, some of the k's are related to measurable quantities. For example, k_6 is the "decay constant" of one or the other of the exponential components; most likely it is the more rapidly changing one. However, some of the k's occur in certain combinations, not as individuals, e.g., $(k_1 + k_5 - k_4)$.

Second, it is necessary to emphasize the phenomenological nature of the theoretical derivation given, and as a consequence of this the phenomenological nature of the various rate constants.

This point can be clearly demonstrated by giving an alternative derivation of the equation in the text based on entirely different physical processes. It is based on the assumption that photolytic decomposition occurs on the sides of the decomposition sites such as those shown by the shaded areas on Fig. 3. Thus s is the total area of the decomposition site, k_1 is the rate that the side progresses during decomposition, k_4 is the rate that new etch pits are created, etc. Thus while it is possible to identify some of the features of the rate vs. time curves with some of the terms of the theoretical expression, and in addition give numerical values to some individual constant or combination of constants, this association must *not* be regarded as a proof that the proposed mechanisms occur as postulated.

Chapter 9

Crystallization and Melting in Glass-Forming Systems

D. R. Uhlmann

Department of Metallurgy and Materials Science
Massachusetts Institute of Technology
Cambridge, Massachusetts

The phenomenology and kinetics of nucleation, crystal growth, and melting in glass-forming systems are discussed. It is shown that studies of transformation kinetics in such systems can help elucidate problems in the field of crystal growth. Consideration is limited to systems which crystallize without change of composition, i.e., pure, single-component materials or congruently melting compunds.

INTRODUCTION

A pure liquid may solidify in either of two ways: it may form a crystalline solid, in which the molecules are regularly arranged on a lattice, or it can form an amorphous solid, called a glass, in which the molecular array is characterized by the absence of long-range order. The former process is termed crystallization from the melt, while the latter is known as glass formation.

At any temperature below the melting point the liquid can transform to a crystalline solid and thereby decrease the free energy of the system. The crystallization process is, however, a discontinuous one and requires the formation of nuclei of the new phase. In this process crystal–liquid interfaces must be created, and their formation represents a barrier to the change of phase. For this reason all liquids can be undercooled, at least to some extent, below their respective equilibrium temperatures without the occurrence of crystallization.

When most liquids are cooled in bulk form the maximum range of undercooling is limited to some small fraction of the melting temperature; upon further cooling nucleation and growth of the crystalline phase is observed. Some liquids, however, can be undercooled indefinitely, assuming reasonable cooling rates, without the occurrence of detectable crystallization. As these liquids are cooled and their viscosity progressively increases a region of temperature will generally be reached where samples may be said to exhibit solid-like behavior and may be termed glasses.

The formation of glasses is widely believed ([1,2]) to be a characteristic of nearly all liquids, providing that they are cooled to a sufficiently low temperature and that crystallization does not intervene. For purposes of the present chapter, however, we shall use the term "glass-forming materials" in a more restrictive but familiar sense, to denote those materials which can form glasses in bulk form when cooled at moderate rates from the liquid state.

In the present chapter we shall be concerned with the kinetics of nucleation and crystal growth from the melt with particular emphasis on glass-forming systems. It will be shown that studies of transformation kinetics in such systems, besides being interesting in their own right, can be important in elucidating some of the central physical problems in the field of crystal growth. In all of our discussion we shall, for simplicity, direct our attention to systems which crystallize without change of composition, i.e., to single-component materials or congruently melting compounds.

NUCLEATION

The theory of nucleation in condensed phases has been well covered in recent reviews ([3-5]) and can be briefly treated here. In this treatment we shall adopt without comment the standard thermodynamic formalism and will not consider questions such as the applicability of bulk thermodynamic quantities for describing very small systems. We shall also neglect any possible modification of the kinetic expressions due to a Lothe–Pound type treatment ([6]) of the translational and rotational free energies of the embryos because any such modification should be quite small for nucleation in condensed phases.

Let us begin with homogeneous nucleation and consider the formation of an embryo of a crystalline phase in a surrounding liquid medium, the system being maintained at constant temperature and external pressure. For simplicity the embryo will be assumed spherical in shape and the crystal–liquid specific surface free energy will be assumed isotropic (independent of orientation).

With these assumptions the free energy of forming the critical nucleus, ΔG^*, may be written ([3-5]):

$$\Delta G^* = \frac{16\pi}{3} \frac{\sigma^3}{(\Delta G_V)^2} \tag{1}$$

and the number of molecules in the critical nucleus:

$$n^* = \left(\frac{32\pi}{3V_m}\right)\left(\frac{\sigma}{\Delta G_V}\right)^3 \tag{2}$$

Here σ is the crystal–liquid specific surface free energy, V_m is the molecular volume, and ΔG_V is the difference between the liquid and crystal phases of the standard Gibbs free energy per unit volume. For small departures from equilibrium this may be expressed as

$$\Delta G_V \approx \Delta H_{fV}(\Delta T/T_E) \tag{3}$$

where ΔH_{fV} is the latent heat of fusion per unit volume, T_E is the equilibrium temperature (melting point), and $\Delta T = T_E - T$ is the undercooling.

For large departures from equilibrium use of this expression for ΔG_V becomes less reliable because account must be taken of the temperature variation of ΔH and ΔS, the differences in enthalpy and entropy between the phases. Using the model suggested by Hoffman [7] to describe these variations ΔG_V may be expressed as:

$$\Delta G_V \approx \Delta H_{fV} \frac{\Delta T}{T_E} \frac{T}{T_E} \tag{4}$$

In estimating the rate of nucleation it is generally assumed that nuclei of critical size form by a series of bimolecular reactions in which a cluster of a given size grows or shrinks by the gain or loss of a single molecule. In the so-called steady-state treatment a steady-state current of molecules through subcritical embryos is assumed; that is, the rate at which embryos containing $n - 1$ molecules change to embryos containing n molecules is assumed equal to the rate at which embryos of n molecules change to embryos of $n + 1$ molecules. With these assumptions, and using the formalism of reaction rate theory [8], Turnbull and Fisher derived an expression for I_V, the rate of homogeneous nucleation per unit volume [9]:

$$I_V = K_V \exp[-(\Delta G^* + \Delta G'_m)/kT] \tag{5}$$

where to a reasonable approximation K_V may be taken as:

$$K_V \approx N_V^0 v_0 \tag{6}$$

Here $\Delta G'_m$ is the activation free energy for motion across the nucleus–matrix interface, v_0 is the number of times per second a molecule attempts this free energy barrier, and N_V^0 is the number of unassociated molecules per unit volume.

For discussing nucleation in glass-forming systems we shall make use of a different expression for the steady-state nucleation rate [10]:

$$I_v \approx \frac{N_v^0 D'}{a_0^2} \exp(-\Delta G^*/kT) \tag{7}$$

Here D' is the kinetic coefficient for molecular transport across the nucleus–matrix interface and a_0 is the molecular diameter. This expression reduces to Eq. (5) when molecular motion at the interface is treated as an activated process:

$$D' = a_0^2 v_0 \exp(-\Delta G_m'/kT) \tag{8}$$

Experimental studies of homogeneous nucleation ([11-14]) have been carried out for a wide variety of materials, including metals, alkali-halides, and organic compounds. The results indicate that the homogeneous nucleation temperature T^*, i.e., the lowest temperature to which liquid droplets can be cooled without crystallizing, for nearly all materials lies between 0.75 T_E and 0.85 T_E. At T^* the free energy of forming the critical nucleus ΔG^* is typically about 50 kT. From such measurements of T^*, values for the crystal–liquid surface free energy have been obtained. When σ is expressed in terms of the molar surface free energy σ_M and compared with the corresponding values for the molar latent heat of fusion ΔH_{fM} it is found that

$$\sigma_M \approx \Delta H_{fM}/2 \tag{9a}$$

for the metals and

$$\sigma_M \approx \Delta H_{fM}/3 \tag{9b}$$

for the semimetals, organic compounds, and alkali halides.

Since independent experimental values of the crystal–liquid surface free energy are generally not available, these correlations based on the magnitude of σ do not represent a check of the theory. The only detailed measurement of nucleation frequency in condensed phases was carried out by Turnbull ([15]) on mercury. This investigator measured the nucleation rate with droplets dispersed in an inert dilatometric medium, and found reasonable agreement (within 7 orders of magnitude†) between theoretical and experimental values of K_V. In the evaluation of K_V from experimental data the surface free energy and the entropy of fusion per unit volume were taken as independent of temperature, and slight variations in these parameters could readily account for the discrepancy in the values of K_V.

In interpreting observations of nucleation phenomena the activation energy for transport across the nucleus–matrix interface $\Delta G_m'$ has generally been equated with that for transport in bulk liquid, ΔG_m. In principle these

---

†The experimental value was 10^{42}, while the theoretical K_V was 10^{35}.

two activation energies can be quite different, but in most instances taking $\Delta G'_m$ as ΔG_m probably represents a rather reasonable assumption. Further, in nearly all cases where the assumption has been applied, transport in the liquid was relatively easy and the kinetic factor, $v_0 \exp(-\Delta G_m/kT)$, was characterized by a value within 2 or 3 orders of magnitude of 10^{12} sec^{-1}.

In contrast, for glass-forming liquids in the high viscosity region, transport in the liquid is difficult and the kinetic factor is quite small. The effect of this lower mobility on the nucleation frequency will be considered later in the present section. It should first be noted, however, that for many glass-forming systems the assumption of a similarity between the transport process at the nucleus–matrix interface and transport in bulk liquid does not sufficiently specify the kinetic factor. The difficulty is related to the lack of reliable measurements of liquid diffusivity and the absence of a satisfactory model for relating this diffusivity to other properties of the materials.

For simple molecular liquids, and even for the liquid metals, the liquid diffusivity D is inversely related to the viscosity η as:

$$D = b/\eta \tag{10a}$$

For these materials the factor b may to a good approximation be taken as the Stokes–Einstein coefficient:

$$b = kT/3\pi a_0 \tag{10b}$$

where a_0 is the molecular or (in the case of metals) the ionic diameter.

In considering more complex systems, such as many of the important glass-formers, an attitude of caution must be adopted. In many cases not only is there uncertainty about the magnitude of b but even the form of relation between D and η is open to question. For purposes of the present discussion, however, we shall assume the relations of Eqs. (10a) and (10b) and merely call attention to the possible error thereby introduced.

Taking then $D' = D = kT/3\Pi a_0 \eta$, let us proceed to estimate the steady-state rate of homogeneous nucleation in glass-forming systems. In this estimate we shall assume the model of Hoffman [Eq. (4)] to evaluate ΔG_V, take σ as independent of temperature and take $\Delta G^*/kT \approx 50$ at $\Delta T/T_E = 0.2$ (consistent with the results on other materials noted above). With these assumptions Eq. (7) becomes:

$$I_V \approx \frac{N_V^0 kT}{3\pi a_0^3 \eta} \exp\left[-\frac{1.024}{T_r^3 \Delta T_r^2}\right] \tag{11}$$

where $T_r = T/T_E$ and $\Delta T_r = \Delta T/T_E$.

Applying this to the nucleation of cristobalite in fused silica as an example we shall use the viscosity data of Fontana and Plummer ([16]), and take

TABLE I

Calculated Nucleation Frequency Versus Temperature for Cristobalite Crystals in SiO_2

T (°C)	ΔT_r	T_r	η (poise)	$\dfrac{N_V^0 kT}{3\pi a_0^3 \eta}$ (cm^{-3} sec^{-1})	$\exp\left(-\dfrac{1.024}{T_r^3\,\Delta T_r^2}\right)$	I_V (cm^{-3} sec^{-1})
1614	0.05	0.95	$3\ \times 10^8$	1.2×10^{23}	10^{-208}	10^{-185}
1514	0.10	0.90	$2\ \times 10^9$	1.7×10^{22}	10^{-61}	10^{-39}
1415	0.15	0.85	1.5×10^{10}	2.1×10^{21}	$2\ \times 10^{-32}$	10^{-11}
1316	0.20	0.80	1.6×10^{11}	1.9×10^{20}	5.6×10^{-21}	1
1217	0.25	0.75	2.2×10^{12}	1.3×10^{19}	7.9×10^{-16}	10^4
1117	0.30	0.70	4.5×10^{13}	5.8×10^{17}	3.2×10^{-14}	10^4

$T_E = 1713°C$, $a_0 = 2.5$ Å, and $N_V^0 = 2 \times 10^{22}$ moles/cm³. With these values the expected nucleation frequencies may readily be calculated, with the results shown in Table I.

It is apparent from this table that the steady-state rate of homogeneous nucleation will be relatively small for a material, such as SiO_2, having a viscosity which is high at the melting point and increases strongly with falling temperature below the melting point. The largest values tabulated in Table I are, in fact, not much above the lower limit of observability (about 1 cm^{-3}sec^{-1}). For comparison the nucleation frequency observed for a molten metal at a relative undercooling (ΔT_r) of 0.2 is about 10^{13} cm^{-3}sec^{-1}, some 13 orders of magnitude larger than the expected value for SiO_2 at the same relative undercooling.

In taking σ as independent of temperature we have neglected a consideration which could materially reduce the expected nucleation frequency. In particular, the crystal–glass surface energy may be comparable to a grain boundary energy ([10]), while a crystal–liquid surface energy is usually only about half the grain boundary energy. Such an increase in σ as the material becomes more rigid could serve to make the nucleation frequency unobservably low at any temperature.

At temperatures below the glass transition temperature T_g the difference in Gibbs free energy between crystal and glass can well be approximated by the difference in enthalpy. Then if the specific heats of crystal and glass are similar—as is often the case—the motivating free energy for nucleation will not change appreciably with temperature. Hence if the crystal–glass surface energy is also independent of temperature, $\Delta G^*/kT$ will not change appreciably over a range of temperature below T_g. Further, since the time scale for molecular rearrangement is large below T_g, the viscosity of a typical (unstabilized) glass sample is not expected to change appreciably in reasonable times as it is cooled below T_g. Consequently the steady-state rate of homogeneous nucleation might be expected to show little variation over a range of temperature below T_g.

It should be noted that in using Eqs. (5) or (7) to describe the nucleation frequency we have assumed a steady-state concentration of subcritical embryos and have ignored the effect of transients during which such concentrations are established. Such neglect is generally justified whenever the time required to establish the steady-state nucleation rate is small relative to the total transformation time and to the time scale of the experiment—conditions which may not be fulfilled for some transformations in condensed phases.

The transient nucleation behavior of a condensed system is usually approximated by an expression of the form ([3,4]):

$$I_t = I_V \exp(-\tau/t) \tag{12}$$

where I_t is the nucleation frequency at time t and the transient time τ is given to perhaps order-of-magnitude accuracy† as:

$$\tau \approx (n^*)^2/N_s v' \tag{13}$$

Here N_s is the number of molecules in the surface of the critical nucleus, $v' = D'/a_0^2$ is the frequency of molecular transport at the nucleus–matrix interface and n^* is the number of molecules in the critical nucleus given by Eq. (2).

Applying this to SiO_2 as an example, using Eqs. (9b) and (4) to estimate the surface energy and motivating potential, again taking $D' = kT/3\Pi a_0 \eta$ and utilizing the same values for the various parameters as before, we obtain the results shown in Table II.

It is apparent from Table II that the transient time can be quite large for

---

†A more satisfactory treatment of the transient problem has recently been carried out by Professor K. C. Russell of M.I.T., who found for the case considered (nucleation without change of composition) a transient time within about an order of magnitude of that given by Eq. (13).

TABLE II

Calculated Transient Time Versus Temperature for Cristobalite in SiO_2

T (°C)	$(n^*)^2/N_s$	τ (sec)
1415	1040	1×10^4
1316	420	4.5×10^4
1217	225	3.5×10^5
1117	140	1.5×10^7

materials like SiO_2. This period, during which the steady-state concentrations of embryos are established, increases as the viscosity increases and becomes longer than usual experimental times as the glass transition is approached. For comparison the transient time expected for a molten metal at its homogeneous nucleation temperature ($\Delta T_r \approx 0.2$) is less than a microsecond.

It should be emphasized again that the magnitude of the transient period must be viewed in the perspective of the total transformation time and the time scale of the experiment. Transient effects are most likely to be significant when the barrier to nucleation is large, when the kinetic factor for transport across a macroscopic crystal–liquid interface (in crystal growth) is materially larger than that for transport across the nucleus–liquid interface, or when D' itself is very small (smaller than 10^{-15} cm²/sec, for example).

In all the above discussion we have considered only homogeneous nucleation. It is recognized, of course, that most nucleation takes place heterogeneously at external surfaces, foreign particles, and imperfections. It is also recognized that little can be said *a priori* about the number or potency of nucleating heterogeneities associated with a given sample of a given material.

Experience with glass-forming materials indicates that crystal nucleation almost invariably takes place at external surfaces, and sometimes ([17]) but not always ([18]) at interior bubble surfaces. Nucleation at external surfaces of SiO_2 has been associated ([19]) with superficial condensed phase impurities, and the nucleation sometimes observed at interior bubble surfaces may well have a similar origin (see the section below on melting). While internal nucleation is seldom observed in glasses, it has occasionally been noticed on long runs at relatively low temperatures. The resulting crystallization has been observed in the form of rosettes ([19]) and elongated striae ([20]), and may be associated with nonuniform concentrations of impurities in the melt.

When many glass-forming materials are cooled into the glassy state and then reheated to a temperature T between T_g and T_E copious crystallization is frequently observed. In contrast, when a sample of the same material is cooled directly to T from a temperature above the melting point the sample may remain free of visible crystallization for an extended period. The origin of this difference in behavior is presumably associated with the process of nucleation, but such observations cannot of themselves be taken as evidence for homogeneous nucleation. Indeed the homogeneous nucleation of crystals in glass-forming liquids may never have been observed and at least does not seem to have been established. As we have noted, internal crystallization is itself rarely observed in such systems and in all or nearly all cases it seems to be associated with heterogeneities in the melt.

This is not to say that homogeneous nucleation could not be observed in a glass-forming material. There is, of course, a spectrum in the glass-forming capabilities of various materials, from those which will not form glasses even when cooled rapidly in droplet form to those which will form glasses even when cooled slowly in bulk. Within the latter category, which we have designated glass-forming materials, homogeneous nucleation can be observed only if crystallization on heterogeneities can be avoided or its manifestations identified and conceptually eliminated.

The latter alternative is made possible by the fact that these materials are generally characterized by a relatively high viscosity at a given relative undercooling and a single nucleation event need not result in rapid growth and crystallization of the entire sample. With this approach a sample of large dimensions should be used so that nucleation can (hopefully) be observed in its interior before the crystallization front, advancing from the external surfaces, reaches the central region. To the present writer's knowledge, however, experiments of this type have yet to yield fruitful results. The observed internal crystallization seldom if ever occurs in the numbers or distribution expected for a homogeneous nucleation process.

In all of the above discussion we have been concerned with systems which crystallize without change of composition, i.e., pure materials or congruently melting compounds. In the more general case, with composition as an additional variable, the techniques of controlled nucleation and crystal growth have been fruitfully employed in recent years. Applied in various contexts they form the basis of several important commercial processes ([21,22]).

In many of these applications the phenomenon of liquid–liquid immiscibility occurs as a precursor to the desired crystallization process. Consider as an example the crystallization of a LiO_2–Al_2O_3–SiO_2 glass containing about 5 wt.% TiO_2 (Corning Code 9608). It has recently been reported ([23]) that production of this material in the form of a fine-grained glass-ceramic body proceeds in the following sequence: on cooling the glass liquid–liquid immiscibility takes place, resulting in a phase-separated structure with a characteristic dimension of about 50 a.u. On reheating a Ti–Al crystalline phase (perhaps $Al_2Ti_2O_7$) forms, also on a scale of about 50 a.u. These crystals apparently serve as nucleating agents for β-eucryptite ($Li_2O \cdot Al_2O_3 \cdot 2SiO_2$) crystals, whose growth results in the formation of a highly crystalline material. At sufficiently high temperatures the β-eucryptite transforms to β-spodumene ($Li_2O \cdot Al_2O_3 \cdot 4SiO_2$), and other minor crystalline phases appear. The final product—a uniform, fine-grained, largely crystalline material—could not be obtained without heterogeneous catalysis, here provided by the Ti–Al crystalline phase (for reasons which should be apparent from the previous discussion). In turn, the formation of this phase in a finely-dispersed form seems to depend upon the phenomenon of liquid–liquid immiscibility.

The process of phase separation and its effects on the properties of glasses are the subjects of intense investigation at the present time. The separation phenomenon is expected to affect crystallization behavior through its effects on the motivating potential for crystallization, on the surface energy barrier to nucleation, and on the material transport process. Although these questions are of considerable interest they lie outside the purview of the present chapter and will not be considered further here. The interested reader is referred to Hillig [24] and Cahn [25].

CRYSTAL GROWTH

The current status of crystal growth from the melt has been summarized in a recent paper [26], which may be used as ancillary material for the present discussion. Below we shall be specifically concerned with the phenomenon of crystal growth in glass-forming systems and will indicate how kinetic studies in such systems can provide important information for elucidating some central problems in the field of crystal growth. As before we shall direct out attention to single-component materials and congruently melting compounds.

Let us begin with the standard expression relating the rate of advance of a crystal–liquid interface to the motivating potential for crystallization. For unit cross section of the interface this is [27]:

$$u = \frac{fD''}{a_0}[1 - \exp(-V_m \Delta G_V / RT)] \tag{14}$$

Here u is the growth rate in cm/sec, f is the fraction of sites at the interface where atoms can preferentially be added or removed, D'' is the kinetic coefficient for transport across the crystal–liquid interface (having dimensions cm^2/sec), a_0 is the molecular diameter, V_m is the molar volume, and ΔG_V is the free energy change per unit volume accompanying crystallization (the motivating potential).

If D'' is taken as inversely proportional to the melt viscosity ($D'' = b/\eta$) and Eq. (3) is used to express the motivating potential in terms of the undercooling Eq. (14) becomes:

$$u = \frac{bf}{a_0\eta}[1 - \exp(-\Delta H_{fM} \Delta T / RTT_E)] \tag{15}$$

For discussing crystallization kinetics it is useful [28] to rewrite this expression in terms of a reduced growth rate u_R which is proportional to f:

$$U_R = \frac{u\eta}{[1 - \exp(-\Delta H_{fM} \Delta T / RTT_E)]} = \frac{b}{a_0}f \tag{16}$$

Provided only that the transport process at the interface varies with temperature in the same way as the viscosity[†] the u_R versus ΔT relation should provide a useful representation of the temperature dependence of f.

Most of the kinetic data in the literature are discussed in terms of three models:

1. Growth at all sites on the interface [29,30]: Here $f = 1$, and a plot of u_R versus ΔT should be a horizontal line. For small departures from equilibrium the growth rate is proportional to the undercooling:

$$u \approx \frac{D'' \, \Delta H_{fM}}{a_0 R T T_E} \Delta T \tag{17}$$

This is known as Wilson–Frenkel or linear kinetics.

2. Growth at steps provided by screw dislocations [31]: Here f should be proportional to ΔT, as:[‡]

$$f = \frac{a_0 \, \Delta H_{fM}}{4\pi\sigma T_E V_M} \Delta T \tag{18}$$

and a plot of u_R versus ΔT should be a straight line of positive slope. For small departures from equilibrium the growth rate is proportional to $(\Delta T)^2$.

3. Growth at steps provided by two-dimensional nuclei on the interface [32]: Here an evaluation of the number of growth sites depends critically on an evaluation of cluster distributions on the interface. With the standard assumption of a Boltzmann distribution the growth rate should vary exponentially with undercooling:[§]

$$u = A D'' \exp[-B/(T \, \Delta T)] \tag{19}$$

Again taking $D'' = b/\eta$ a plot of $\ln(u\eta)$ versus $1/(T \, \Delta T)$ should be a straight line of negative slope. For a nucleus in the form of circular pillbox of height a_0 using an equilibrium treatment to estimate the rate of formation of such nuclei

$$B = \pi a_0 V_m T_E \sigma_E^2 / k \, \Delta H_{fM} \tag{20}$$

where σ_E is the specific edge surface free energy of the nucleus.

[†]The plausibility of this assumption may be enhanced by noting that the temperature dependences of both shear and volume viscosity have been found similar for the glass-forming materials investigated to date.

[‡]At large departures from equilibrium Eq. (18) might better be replaced by $f = (a_0 \, \Delta H_{fM} T_r / 4\pi\sigma T_E V_M) \Delta T$. For purposes of the present discussion, however, this modification will be neglected.

[§]Several different models have been proposed to describe growth by a surface nucleation mechanism. As discussed in [28,33,34] these models differ somewhat in the form of A and B.

It has been emphasized ([26]) that the validity of these kinetic models as descriptions of crystal growth has yet to be provided. The advance of a crystal–liquid interface should depend critically on the detailed structure (on a molecular scale) of the interface. In considering crystal growth it is in general important to know the number and distribution of solid-like atoms on an interface plane as well as the population and form of molecular groups in the liquid. Unfortunately such structural details are very difficult to calculate on a statistical basis because the relevant statistics relate to small cluster size and limited numbers of configurations.

The importance of a bulk thermodynamic property, the entropy of fusion, in the crystallization process has been emphasized by Jackson ([35,36]). According to his criterion for metals, SiO_2, and other materials characterized by low entropies of fusion ($\Delta S_{fM} < 2R$), even the most closely-packed interface planes should be rough and the growth rate anisotropy should be small. In contrast, for most organic compounds, B_2O_3, and other materials characterized by large entropies of fusion ($\Delta S_{fM} > 4R$) the most closely packed faces should be smooth, while the less closely packed faces should be rough and the growth rate anisotropy should be large.

The predictions based on this model have been well confirmed by experimental observations. Materials characterized by large entropies of fusion freeze with a faceted interface morphology, while materials with low entropies of fusion exhibit under similar conditions the nonfaceted interface morphologies characteristic of nearly isotropic growth. Examples of such morphological differences are shown by Jackson in Chapter 12 of this volume.

Extrapolating from this model, one might expect that for materials with rough interfaces the fraction of growth sites on the interface should be of the order unity, and while it will in general depend on orientation, it should not vary strongly with undercooling. For smooth-interface materials, on the other hand, growth is expected to be affected significantly by defects. For sufficiently perfect crystals nucleation barriers to the formation of new layers may well be noted. The magnitude of these barriers may, however, be poorly represented by the relation of Eq. (20).†

In discussing theories of crystal growth we have been concerned with the relation between the growth rate and the undercooling at the interface. Unfortunately, this relation is often difficult to obtain because of the difficulty in determining the interface temperature. In many experimental situations the growth rate is limited not by interface kinetics but by the rate at which the latent heat of fusion, generated in the freezing process, can be

---

†Growth rates proportional to $\exp[-a/(T\,\Delta T)]$ have been observed in salol for undercooling between 1.6 and 2.5°C, in gallium for undercoolings between 0.5 and 0.76°C, and in the growth of ice normal to the basal plane for undercoolings between 0.03 and 0.07°C. See([26]).

removed from the freezing front, or even by the rate of motion of a furnace with respect to the sample.

For materials of high fluidity a number of techniques have been employed to circumvent this problem ([26]). In no case, however, has it been possible to obtain data over a wide range of undercooling because the fluidity of the melt results in large freezing rates. In contrast, the large viscosities of glass-forming materials result in growth rates (and hence rates of latent heat evolution) which can be small even at large undercoolings. The interface temperature can then be well taken as the bath or furnace temperature and the growth rate can be measured over a wide range of undercooling.

Because of the rapid increase in viscosity which accompanies decreasing temperature in glass-forming materials the "quenched-in" interface morphology provides a valid representation of the morphology at the temperatures of crystallization. For materials with glass transitions above room temperature the interface morphology can conveniently be studied by microscopic observations at ambient temperature. Examples of morphologies observed in this way are shown in Figs. $1a$ for $Na_2O \cdot 2SiO_2$ and $1b$ for SiO_2

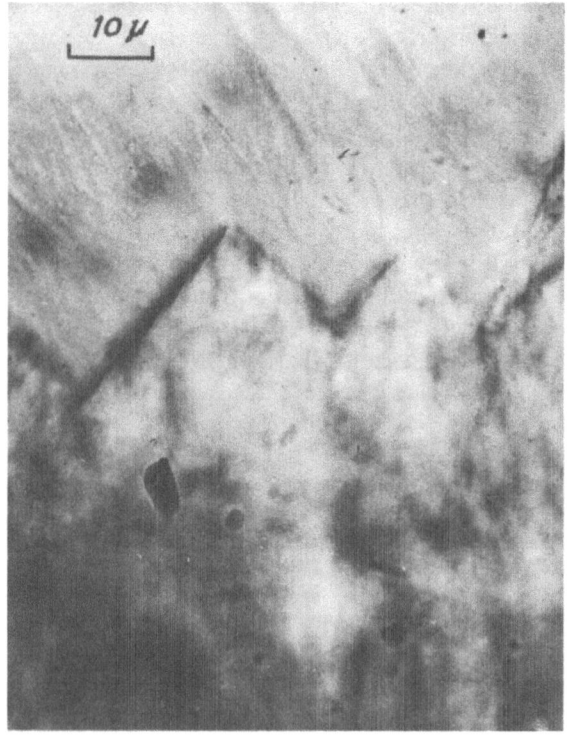

Fig. $1a$. Interface morphology of $Na_2O \cdot 2SiO_2$ crystal growing at $\Delta T = 244°C$.

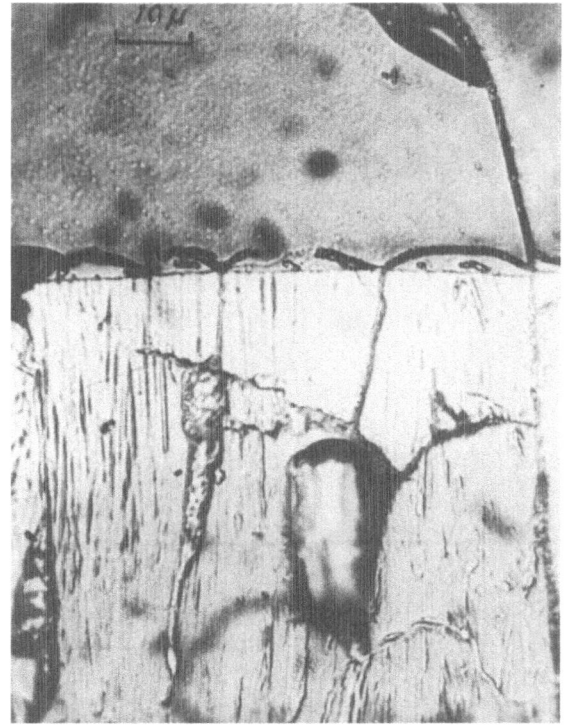

Fig. 1*b*. Interface morphology of SiO_2 crystal growing at $\Delta T = 250°C$.

crystallized at similarly large undercoolings. Taking the formula unit as the appropriate unit of composition for evaluating the entropy of fusion, these morphological observations are in accord with the predictions of Jackson.†

To illustrate the expected form of the growth-rate–temperature relation for a glass-forming system Fig. 2 shows the expected variation for SiO_2. This was calculated using Eqs. (15) and (10*b*) with the viscosity data of [16] and taking $f = 1$, $a_0 = 2.5$ Å and $\Delta H_{fM} = 1835$ cal/mol. As seen in the figure the expected growth rate is low at small undercoolings, rises to a maximum, and then decreases as the fluidity decreases with further increases in undercooling. In the region of the melting point the growth rate is said to be undercooling-limited, while at large undercoolings it is said to be viscosity-limited.

Curves of this form have been observed for a number of materials, e.g.,

---

†The entropy of fusion of SiO_2 is about 0.5 R[37], while that of $Na_2O \cdot 2SiO_2$ is about $4R$ [38].

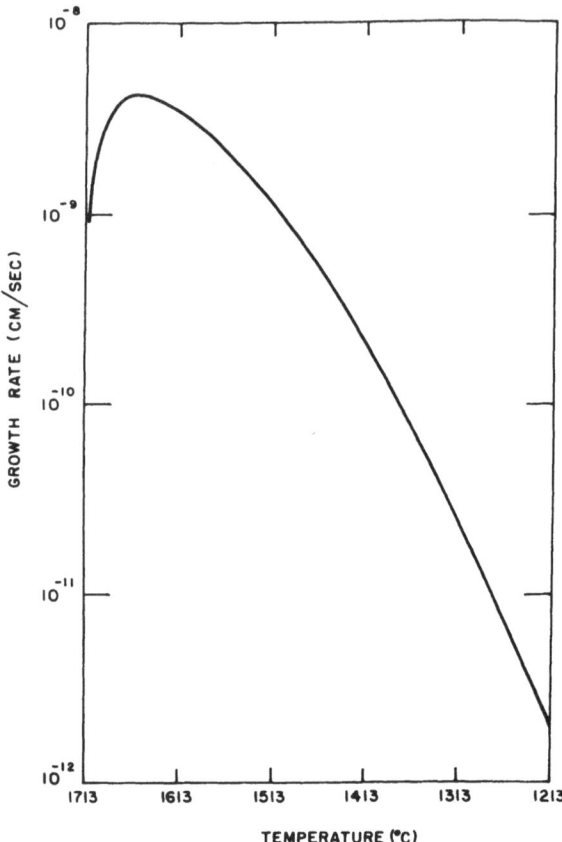

Fig. 2. Logarithm of growth rate versus temperature for
the crystallization of cristobalite from fused silica, calculated from Eq. (15) with assumptions outlined in text.

glycerine ([39]), salol ([40]), orthoterphenyl ([41]), $K_2O \cdot 2SiO_2$ ([42]), $Na_2O \cdot 2SiO_2$
([38,43,44]), and tri-α-naphthyl benzene ([45]). Only in the last two cases, however,
have observations of growth rate, viscosity, and interface morphology been
carried out by the same investigators on the same material. In the other cases,
where concurrent growth rate and viscosity measurements were not carried
out, questions regarding the similarity of material studied and uniformity
of environmental conditions cannot be eliminated with any confidence.
Consequently we shall consider in detail only the studies of $Na_2O \cdot 2SiO_2$
and tri-α-naphthyl benzene.

.   In the former study ([38,44]) the growth rates were measured by heat treating samples for different periods of time at a given temperature. The crystallization was observed to proceed uniformly in from the free surfaces and the

growth rates were given by the slopes of the linear extent of crystallization versus time plots. By carrying out such heat treatments at a variety of temperatures the growth rate–temperature relation shown in Fig. 3 was obtained. The interface morphology was observed to be faceted over the entire range of undercooling investigated ($\Delta T = 6$–$298°C$). A typical example of this morphology was shown in Fig. 1a. Combining the growth rate and viscosity data, the reduced growth rate versus undercooling relation shown in Fig. 4 was obtained. The form of the relation suggests that the fraction of sites on the interface where atoms can preferentially be added increases with increasing undercooling. The linear relation for undercoolings greater than about 60°C may be suggestive of growth by a screw dislocation mechanism, but the experimental slope is larger than that calculated from Eqs. (16), (18), and (10b) by a factor of about 150. It has been pointed out ([28]), however, that D'' may exceed the diffusion coefficient for transport in bulk liquid—which Eq. (10b) was intended to describe—by a factor of 10 or 100. In any case there is no *a priori* justification for assuming the particular relation of Eq. (10b).

The growth rate of tri-a-naphthyl benzene ([45]) was measured by directly observing interface motion using a microscope. Consistent with the predictions of Jackson, the interface showed marked faceting under all conditions where the morphology could be observed in detail ($\Delta S_{f_M} \approx 10.7R$). With

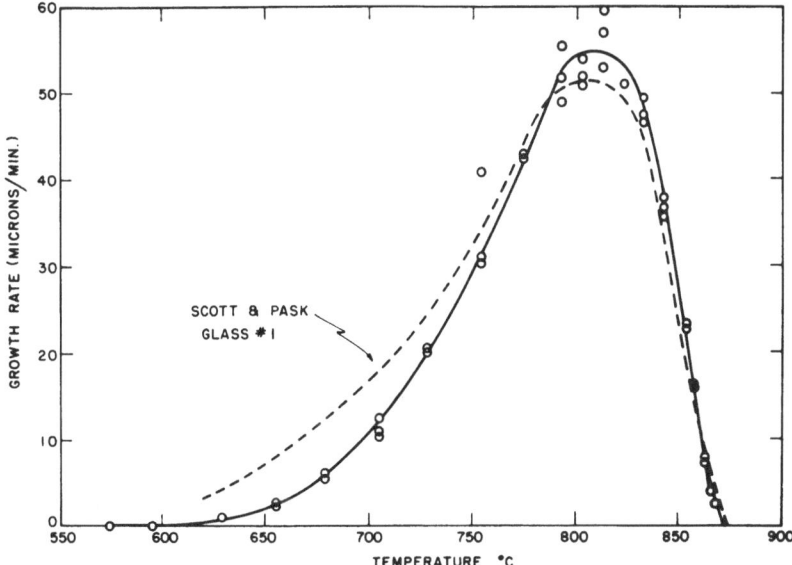

Fig. 3. Observed growth rate versus temperature for the crystallization of $Na_2O \cdot 2SiO_2$ [after Meiling and Uhlmann ([44])].

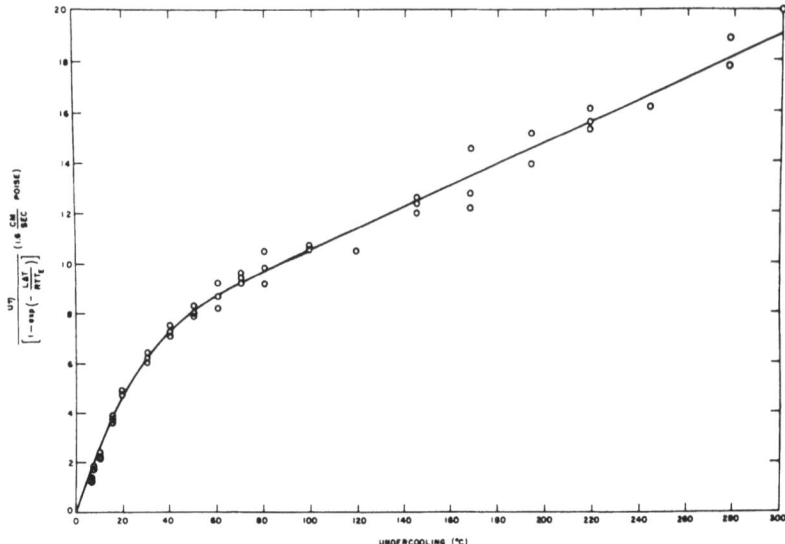

Fig. 4. Reduced growth rate versus undercooling relation for $Na_2O \cdot 2SiO_2$ [after Meiling and Uhlmann ([44])].

increasing undercooling the morphology changed from a single-crystal habit (at $\Delta T < 40°C$) to a radiating array of crystals (at $\Delta T > 60°C$). Combining growth rate with viscosity ([46]) data, the log ($u\eta$) versus $1/T\,\Delta T$ relation shown in Fig. 5 was obtained. As shown in this figure the data may be described by two straight lines of different slope which intersect at about $\Delta T = 50°C$ (the region of the change in interface morphology). The form of this relation and the absence of observable growth for $\Delta T < 2°C$ are suggestive of growth by a surface nucleation mechanism. The slope at large undercoolings is, however, smaller by about a factor of 2 than that predicted by Eq. (20) with Eq. (9b).

In both these materials, then, the interface morphology seems indicative of anisotropic growth, and the fraction of preferred growth sites seems to increase with increasing undercooling. In both cases care was taken to obtain pure starting materials, but in neither case could purity levels greater than 99.9% be claimed. While no evidence for impurity effects was indicated, neither could the absence of such effects be demonstrated. More generally, this question of purity is one which often arises in studies of glass-forming materials because they generally are not obtained in ultrapure condition and they are often sensitive to dissolved and atmospheric impurities. In some cases the problem could conveniently be investigated by comparing results for starting materials of different purity such as those prepared by normal procedures, by zone refining, and by doping with selected impurities. Such investigations have not yet been carried out.

In discussing the fraction of preferred growth sites it might be noted that for no material in any range of undercooling has this fraction been found independent of undercooling. From the suggestions outlined above such an independence seems most likely to be demonstrated by materials characterized by small entropies of fusion. Within the class of glass-forming materials SiO_2 and GeO_2 seem well suited in this regard. Their entropies of fusion are low and their crystal–liquid interfaces have already been shown to be non-faceted (see Fig. 1b, for example). Both materials are presently under investigation in this laboratory. In both cases the crystallization kinetics are quite sensitive to dissolved impurities, atmospheric impurities (viz., water and oxygen), and to the stoichiometry of the melt ([19,47-52]).

MELTING

From considerations similar to those outlined above in the section on nucleation it is expected that considerable superheating above the melting point would be required for the homogeneous nucleation of melting in a crystal. Such nucleation has, however, not yet been observed. When a crystal is heated above its melting point liquid is invariably observed to form at the free surfaces at negligible departures from equilibrium. The melting process

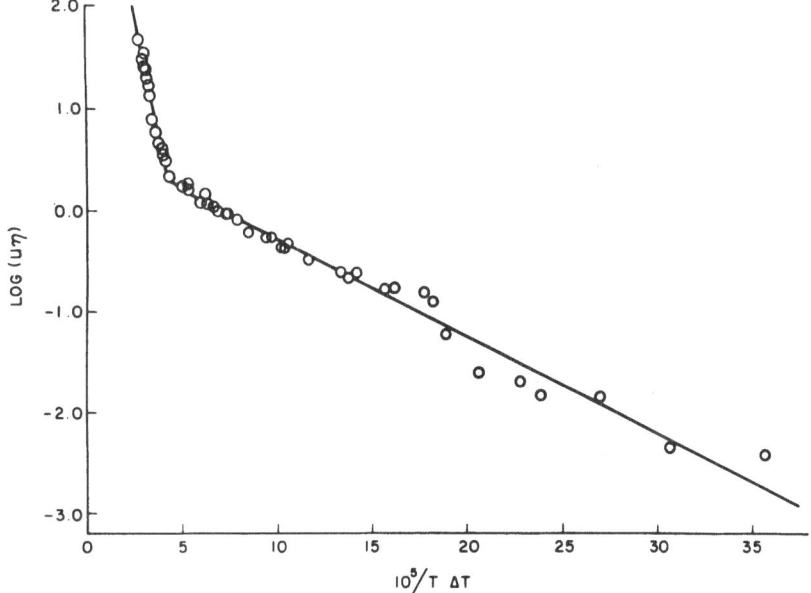

Fig. 5. Logarithm of growth rate times viscosity versus $1/(T \Delta T)$ for tri-a-naphthyl benzene [after Magill and Plazek ([45])].

then takes place by the inward propagation of the liquid–crystal interface. In most materials the rate of this interface advance is limited by the rate of supply of heat, and significant superheating of the interior cannot be obssrved.

For many glass-forming materials, on the other hand, the viscosity of the liquid at the melting point is quite high (perhaps 10^5 poise or more) and the rate of interface advance is limited by interface kinetics rather than by the flow of heat. Under these circumstances it should be possible to superheat the crystals by substantial amounts. Such superheating has in fact been observed. As early as 1905 the superheating of albite ($NaAlSi_3O_8$) crystals by more than 100°C was reported [53]. More recently crystals of quartz were superheated by as much as 350°C [54].

In neither of these studies, nor in any other investigation where superheating has been described, has there been evidence for homogeneous nucleation of melting. As indicated above melting invariably initiates on the free surfaces and proceeds inwards; it is observed in the interior of crystals only along cracks or similar imperfections.†

From the fact that melting occurs heterogeneously at free surfaces at negligible departures from equilibrium we may infer a relation between the various surface energies:

$$\sigma_{cv} \geqslant \sigma_{cl} + \sigma_{lv} \qquad (21)$$

where σ_{cv}, σ_{cl}, and σ_{lv} are the crystal–vapor, crystal–liquid, and liquid–vapor specific surface free energies, respectively.

If we assume that this relation applies as well at temperatures below the melting point, we may infer that free surfaces of liquids should not themselves act as sites for the heterogeneous nucleation of crystals. This may readily be seen from Fig. 6, where it may be noted that such nucleation would imply

$$\sigma_{cv} \leqslant \sigma_{cl} + \sigma_{lv} \qquad (22)$$

From this result we may infer that the preferential nucleation of crystals at external surfaces of glasses like SiO_2 is very likely associated with second-phase particles on the surfaces. Further, the preferential nucleation sometimes

---

†We are neglecting the case where radiation may be focused at a point in the interior of a crystal, resulting in local melting.

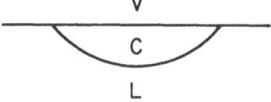

Fig. 6. Heterogeneous nucleation of crystals at liquid–vapor interface (schematic).

seen at interior bubble surfaces of these glasses seems likely to have a similar origin (assuming that the relation of Eq. (21) remains valid with the vapor of the bubbles).†

The kinetics of melting have been reported for two materials, P_2O_5 [55] and $Na_2O \cdot 2SiO_2$ [44]. In both cases the melting always occurred at the free surfaces, and the propagation of the liquid–crystal interfaces was limited by interface kinetics rather than by heat flow. In both cases the extent of melting at a given temperature was found to be a linear function of time until just prior to the disappearance of the crystals, when the rate of melting was observed to increase.

While some scatter is seen in the data on P_2O_5, the curves of growth rate and melting rate versus motivating potential seem to be continuous with the same slope through the melting point (Fig. 7a). With $Na_2O \cdot 2SiO_2$, on the other hand, a pronounced asymmetry is observed in the curves of growth rate and melting rate versus motivating potential in the vicinity of the melting point. As shown in Fig. 7b at a given departure from equilibrium, melting is observed to take place at a faster rate than crystallization.

By the principle of microscopic reversibility if molecules are added to and taken from similar sites (in freezing and melting respectively) the curves

---

†The plausibility of this assumption is enhanced by observations that the nucleation of melting requires negligible superheating in any atmosphere studied.

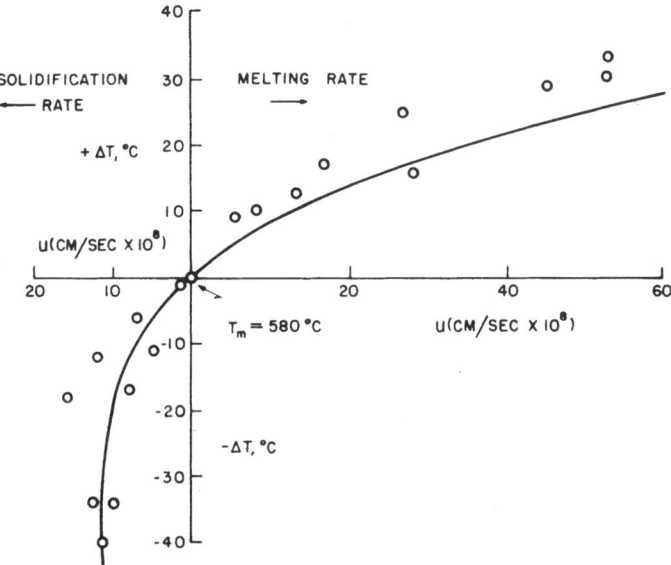

Fig. 7a. Melting and crystallization rates versus motivating potential for P_2O_5 [after Cormia et al. [55]].

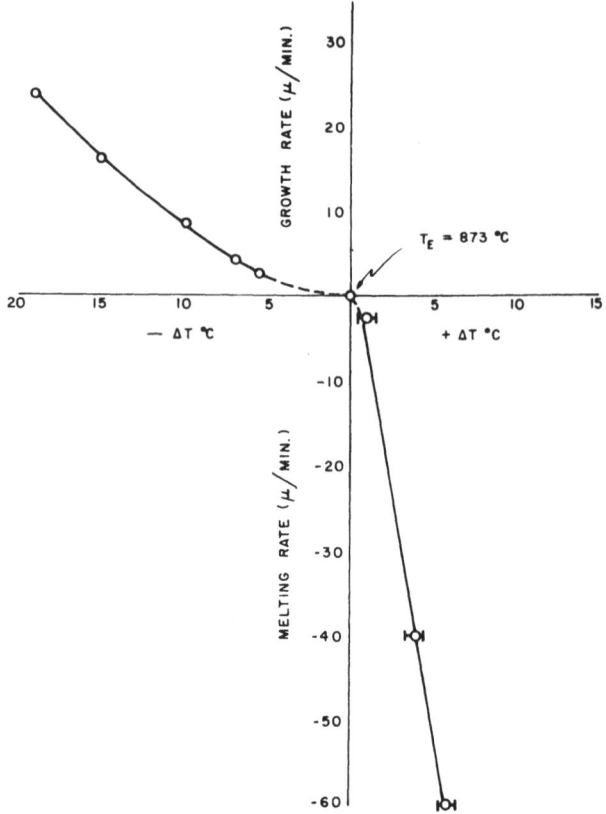

Fig. 7*b*. Melting and crystallization rates versus motivating potential for $Na_2O \cdot 2SiO_2$ [after Meiling and Uhlmann ([44])].

of growth rate and melting rate versus motivating potential should be continuous with the same slope through the equilibrium temperature. However, the extent of applicablility of the principle is expected to be strongly dependent upon the fraction of preferred sites for growth or melting.

For materials with rough solid–liquid interfaces, which exhibit the features of nearly isotropic growth, where the interface morphologies are similar in growth and melting and where the fraction of growth or melting sites is expected to be of the order unity, the rates of crystallization and melting may well be equal out to appreciable departures from equilibrium.† In contrast, for materials with smooth solid–liquid interfaces, which exhibit

—————————

†The domain over which an equality of the rates might be expected will also depend upon the variation with temperature of the mobility. This factor may be included by considering the product of the rate times the viscosity, $u\eta$, rather than the rate u.

faceted interface morphologies in growth but rounded morphologies in melting and where growth is expected to take place at steps provided by screw dislocations or nuclei formed on the interface, but where melting may take place by the removal of atoms from high-energy corner sites on the atom layers, marked asymmetry could be observed in the rates of freezing and melting at equal small motivating potentials.

The results obtained for $Na_2O \cdot 2SiO_2$ (Fig. 7b) have been interpreted in these terms ([44]). It seems somewhat premature, however, to attempt a more detailed discussion until other studies ([47,52,56]) are completed.

CONCLUSIONS

In the present chapter we have considered the phenomenology and kinetics of nucleation, crystal growth, and melting in glass-forming systems. We have discussed some of the advantages and problems of working with these systems, and have indicated some of the areas where studies of glass-forming materials may contribute materially to our understanding of the phenomena.

In closing this discussion we should emphasize again the striking paucity of reliable kinetic data on the crystal–liquid transition and express confidence that investigations of crystal growth and melting phenomena will provide stimulating activity for many years to come.

ACKNOWLEDGMENTS

The author is happy to acknowledge stimulating discussions of these topics with Professor David Turnbull of Harvard University and Dr. K. A. Jackson of the Bell Telephone Laboratories. Financial support for the work was provided by the Atomic Energy Commission.

REFERENCES

1. M. H. Cohen and D. Turnbull, *J. Chem. Phys.* **31**: 1164 (1959).
2. T. P. Seward, D. R. Uhlmann, and D. Turnbull, *Vapor Deposition of Homogeneous Glasses*, in press.
3. D. Turnbull, in: *Solid State Physics*, Vol. 3, Academic Press, New York, 1956.
4. J. W. Christian, *The Theory of Transformations in Metals and Alloys*, Pergamon Press, New York, 1965.
5. D. R. Uhlmann and B. Chalmers, in: *Nucleation Phenomena*, American Chemical Society, Washington, 1966.
6. J. Lothe and G. M. Pound, *J. Chem. Phys.* **36**: 2080 (1962).
7. J. D. Hoffman, *J. Chem. Phys.* **29**: 1192 (1958).
8. S. Glasstone, K. J. Laidler, and H. Eyring, *The Theory of Rate Processes*, McGraw-Hill, New York, 1941.

9. D. Turnbull and J. C. Fisher, *J. Chem. Phys.* **17**: 71 (1949).
10. D. R. Uhlmann, J. F. Hays, and D. Turnbull, *Phys. Chem. Glasses* **7**: 159 (1966).
11. D. Turnbull and R. E. Cech, *J. Appl. Phys.* **21**: 804 (1950).
12. D. G. Thomas and L. A. K. Staveley, *J. Chem. Soc.* p. 4569 (1952).
13. J. H. DeNordwall and L. A. K. Staveley, *J. Chem. Soc.* p. 224 (1954).
14. E. R. Buckle and A. R. Ubbelohde, *Proc. Roy. Soc. London Ser. A* **259**: 325 (1961).
15. D. Turnbull, *J. Chem. Phys.* **20**: 411 (1952).
16. E. H. Fontana and W. A. Plummer, *Phys. Chem. Glass* **7**: 139 (1966).
17. C. G. Bergeron and J. P. DeLuca, *J. Am. Ceram. Soc.* **50**: 116 (1967).
18. J. E. Neely and F. H. Ernsberger, *J. Am. Ceram. Soc.* **49**: 396 (1966).
19. N. G. Ainslie, C. R. Morelock, and D. Turnbull, in: *Symposium on Nucleation and Crystallization in Glasses and Melts*, American Ceramic Society, Columbus, 1962.
20. G. S. Meiling and D. R. Uhlmann, *Low Temperature Crystallization of SiO₂*; in press.
21. S. D. Stookey and R. D. Maurer, in: *Progress in Ceramic Science*, Vol. 2, Pergamon Press, New York, 1962.
22. P. W. McMillan, *Glass-Ceramics*, Academic Press, New York, 1964.
23. P. E. Doherty, D. W. Lee, and R. S. Davis, *J. Am. Ceram. Soc.* **50**: 77 (1967).
24. W. B. Hillig, in: *Symposium on Nucleation and Crystallization in Glasses and Melts*, American Ceramic Society, Columbus, 1962.
25. J. W. Cahn, *Thermodynamic Relations between Phase Separation and Crystallization*, in press.
26. K. A. Jackson, D. R. Uhlmann, and J. D. Hunt, *J. Cryst. Growth* **1**: 1 (1967).
27. D. Turnbull, *J. Phys. Chem.* **64**: 609 (1962).
28. J. W. Cahn, W. B. Hillig, and G. W. Sears, *Acta Met.* **12**: 1421 (1964).
29. H. A. Wilson, *Phil. Mag.* **50**: 238 (1900).
30. J. Frenkel, *Z. Sovjetunion* **1**: 498 (1932).
31. W. B. Hillig and D. Turnbull, *J. Chem. Phys.* **24**: 914 (1956).
32. M. Volmer and M. Marder, *Z. Physik. Chem.* **154**: 97 (1931).
33. W. B. Hillig, in: *Kinetics of High Temperature Processes*, Wiley, New York, 1959.
34. W. B. Hillig, in: *Growth and Perfection of Crystals*, Wiley, New York, 1958.
35. K. A. Jackson, in: *Liquid Metals and Solidification*, American Society for Metals, Cleveland, 1958.
36. K. A. Jackson, in: *Growth and Perfection of Crystals*, Wiley, New York, 1958.
37. F. C. Kracek, *J. Am. Chem. Soc.* **52**: 1436 (1930).
38. G. S. Meiling and D. R. Uhlmann, in: *Crystal Growth*, Pergamon Press, New York, 1967.
39. M. Volmer and M. Marder, *Z. Physik. Chem.* **154**: 97 (1931).
40. K. Neumann and G. Micus, *Z. Physik. Chem.* **2**: 25 (1954).
41. R. J. Greet, *J. Cryst. Growth* **1**: 195 (1967).
42. A. Leontijewa: *Acta Physicochim. URSS* *16*: 97 (1942).
43. W. D. Scott and J. A. Pask, *J. Am. Ceram. Soc.* **44**: 181 (1961).
44. G. S. Meiling and D. R. Uhlmann, *Phys. Chem. Glasses* **8**: 62 (1967).
45. J. H. Magill and D. J. Plazek, *J. Chem. Phys.* **46**: 3757 (1967).
46. D. J. Plazek and J. H. Magill, *J. Chem. Phys.* **45**: 3038 (1966).
47. P. J. Vergano and D. R. Uhlmann, in: *Reactivity of Solids, VI*, Wiley, New York, 1969.
48. S. D. Brown and S. S. Kistler, *J. Am. Ceram. Soc.* **42**: 263 (1959).
49. F. E. Wagstaff, S. D. Brown, and I. B. Cutler, *Phys. Chem. Glasses* **5**: 76 (1964).
50. F. E. Wagstaff and K. J. Richards, *J. Am. Ceram. Soc.* **48**: 382 (1965).
51. F. E. Wagstaff and K. J. Richards, *J. Am. Ceram. Soc.* **49**: 118 (1966).

52. D. R. Uhlmann and D. Turnbull, *Crystallization and Melting Kinetics of SiO₂*, to be published.
53. A. L. Day and E. T. Allen, *Carnegie Inst. Wash. Publ.* p. **31** (1905).
54. N. G. Ainslie, J. D. Mackenzie, and D. Turnbull, *J. Phys. Chem.* **65**: 1718 (1961).
55. R. L. Cormia, J. D. Mackenzie, and D. Turnbull, *J. App. Phys.* **34**: 2239 (1963).
56. K. A. Jackson, *Crystal Growth Kinetics*, in press.

DISCUSSION

E. F. Riebling (Corning Glass Works): From our work with GeO_2 we know that trace impurities play a tremendous role in changing, for instance, the viscosity. It changes drastically with just a fraction of a per cent of impurity. Would you expect a correspondingly large effect on the crystallization behavior?

Answer: GeO_2 is an interesting case, where impurities, including departures from stoichiometry, can strongly affect both the viscosity and the crystallization kinetics. Recent work by Vergano in our laboratory has provided a striking example of impurity effects on the crystallization behavior. He prepared one glass by melting in vacuo for 4 hr at about 1200°C and another by melting at 1300°C for 1 hr in vacuo and 1 hr in a nitrogen atmosphere. He then crystallized (devitrified) samples of the two glasses under identical conditions at 795°C and observed that the overall rate of crystallization of the glass melted at 1200°C exceeded by some two orders of magnitude that of the glass melted at 1300°C. These observations, which are presently being extended by Vergano, are very likely associated with departures of the melt from stoichiometry. A marked effect of melting conditions on subsequent crystallization behavior has also been observed in the case of SiO_2. In comparing these systems it might be noted that much more pronounced departures from stoichiometry can be observed with GeO_2 than with SiO_2. The solubility of Ge in GeO_2 seems appreciably greater than that of Si in SiO_2.

More generally impurities of various types can strongly affect the observed crystallization behavior, and care should be taken to isolate and eliminate impurity effects in all studies of crystallization kinetics whether in glass-forming or nonglass-forming systems. As was noted, little systematic attention has been directed to these effects in the past, and their study might well be a fruitful endeavor in its own right.

Riebling: There are a number of molten salt hydrate systems which form glasses. Angell and others have been working on these materials. They might be more conveniently studied, the temperatures being perhaps a little bit lower than your silicates.

Answer: That is an interesting suggestion. For our present purposes, however, such systems present a number of possible drawbacks. First, the compositional regions of bulk glass formation in many of these systems do not include the compositions corresponding to congruently melting compounds. In such cases, where the crystallization process takes place with a change of composition, a substantial part of the driving force may be used in driving the transport process. Indeed, the crystallization may well be controlled by this transport rather than by interface kinetics.

Second, relatively few salt hydrate systems which melt congruently are characterized by small entropies of fusion ($\Delta S_f < 2R$). Those which do have small ΔS_f are generally characterized by low viscosities at the melting point. As indicated, the most promising area of investigation at the present time seems to be crystallization kinetics in glass-forming systems with small entropies of fusion. Two such materials, SiO_2 and GeO_2, have already been shown to be characterized by nonfaceted interface morphologies and for the reasons outlined above seem likely to exhibit linear crystallization kinetics. Glass-forming fused salt systems, with congruently melting compounds characterized by small extropies of fusion which do not suffer from hydration effects, would likewise be interesting to investigate.

W. D. Scott (University of Washington): From your work so far do you think that the decreasing growth rate with increasing undercooling which you mentioned as being viscosity-controlled is in fact viscosity-controlled? If so, then it seems that the site factor f would not be very important.

Answer: When we speak of the growth rate as being undercooling-limited or viscosity-limited we refer to the dominant factor contributing to the variation of growth rate with temperature. In the vicinity of the melting point the variation of viscosity with temperature is small in comparison with the variation of the driving force, while at large departures from equilibrium the change of driving force with temperature is small relative to the change in viscosity.

It should be noted that even at large undercoolings, where the temperature dependence of the site factor is not so important as that of the viscosity in determining the overall variation of growth rate with temperature, the site factor versus undercooling relation is still important in elucidating the nature of the growth process. And this relation can be evaluated from concurrent determinations of growth rate and viscosity on the same material under similar conditions.

There is, of course, a possible problem in evaluating the site factor at a given temperature. If one says, for the sake of discussion, that the measurements of viscosity could be in error by a factor of 2 and that the growth rate measurements might be in error by a similar factor, then one could be significantly in error in evaluating the site factor. Further, there is an uncertainty of this magnitude or more in assuming a particular proportionality coefficient relating D'' to η. The advantages of considering the site factor versus undercooling relation, rather than merely the magnitude of the site factor at a given temperature, should therefore be apparent.

In our study of sodium disilicate uncertainties from the first two sources were, to the best of our knowledge, insignificant. Three sets of samples were run at each temperatures and, except in the vicinity of the maximum in the growth rate curve, there was essentially no scatter in the data. The viscosity data, obtained by two techniques on the same material used in the crystallization study, fitted well with the data of other investigators in different temperature ranges.

Scott: I was thinking that it would be possible to have a large surplus of growth sites and the kinetics controlled strictly by the sites.

Answer: The great utility of the reduced growth rate versus undercooling relation is the information which it provides about the fraction of preferred growth sites on the interface. If there were a suplus of growth sites, or in general if the site fraction were independent of temperature, then the u_R versus ΔT relation would be a horizontal line. Let me emphasize again the assumption made in using this relation, viz., that the kinetic coefficient for transport at the interface varies with temperature in the same way as the viscosity. With this assumption one can evaluate the temperature dependence of the site factor; and if there were a surplus of sites, one wouldn't expect to see the observed temperature dependence of this factor.

W. J. Dunning (University of Bristol): In studying the effect of temperature on these plots did you use one crystal for all temperatures or different crystals at different temperatures?

Answer: In our study of sodium disilicate we prepared a large glass sample, cut many plates from it for the growth rate measurements, and used the remainder for chemical analysis, viscosity measurements, and the like. The growth rate data were obtained by running different sample plates at different temperatures, and in fact, different samples at different times at a given temperature. At each temperature, as mentioned previously, three different sets of samples were run, and essentially no scatter was seen in the results.

Dunning: There is one test of whether it is dislocation growth or nucleation growth or proof of what the process might be. If there is a wide scatter with respect to the dislocations, so that each individual crystal has its individual characteristics, one might expect growth rates which vary from one crystal to another and hence significant scatter in the results.

Answer: You are quite right. Perfect and imperfect crystals show markedly different crystallization kinetics at least at small undercoolings for smooth interface growth. In this range of small undercoolings the crystallization behavior would similarly show a sensitivity to damage of the crystals. Such a sensitivity would not be expected for rough interface materials. Finally, as noted above, the growth rate may also be strongly sensitive to small amounts of impurities, and these may likewise introduce scatter into the results.

Incidentally, some rather interesting observations on the variability in growth rate for a smooth-interface material (salol) were recently reported ([26]). In that case significant scatter was found in the growth rate observed at a given undercooling for different crystals, different facets of a given crystal, and a given crystal at different stages of growth.

Chapter 10

Role of Impurities in Precipitation of Potassium from Supersaturated KCl:K Systems

D. G. Muth and G. C. Kuczynski

Physics Department and Radiation Laboratory*
University of Notre Dame
Notre Dame, Indiana

Experiments are described which add evidence to the hypothesis that impurities introduced from the ambient atmosphere are necessary for the nucleation of the precipitation of the excess metal in superstaturated solutions of alkali halides. Potassium chloride single crystals were used, and the effects of impurities on X- and F-center formation are examined.

INTRODUCTION

An alkali halide crystal can be saturated by alkali metal by heating it in the alkali metal vapor. The crystals thus treated are colored due to the electrons trapped in the halogen ion vacancies and exhibit the characteristic F-center absorption band. When these crystals are cooled to the temperature at which the solution is supersaturated the excess alkali metal atoms separate out and the bands characteristic of a new phase appear. In potassium chloride crystals at relatively low temperatures ($T < 350°C$) the first absorption band which appears has a peak at 735 mμ. This is the so-called X band. After prolonged heating the peak of the X band decreases in intensity, and a new band peaking around 760 mμ appears. During the entire process the intensity of the F band decreases. These two bands are characteristic of the products of precipitation of the excess metal from the supersaturated solution. These results were interpreted by Scott and co-workers [1,2] as a heterogeneous equilibrium between small colloidal metal particles very much like the equilibrium between solid and vapor phases. Shatalov [3,4] was first to point out that this interpretation may be incorrect for several reasons. In the first place the X band does not appear in the scattering spectrum, indicating that if it is due to colloidal particles they would have to have diameters smaller than 10^{-7} cm. More impor-

---

*The Radiation Laboratory is operated by the University of Notre Dame under contract with the Atomic Energy Commission. This is AEC Document number COO–38–543.

tant, however, is the fact that broadening of the X band has a dependance on temperature very similar to that of the bands which are known to be due to the atomically or molecularly dispersed centers such as the F and M centers, although the value of 10^{13} sec^{-1} measured for the vibrational frequency of X centers in both NaCl and KCl is much too high. These observations led Shatalov to assign the F_2 structure to the X center. This assignment is obviously wrong because the indentity of the F_2 with the M center is well established. Recent investigations (^{5}) seem to indicate that the X center is due to the K_2 molecules precipitated out in the crystal cavity.

Compton (^{6}) was first to discover that H_2O or OH$^-$ impurity is required to form the X band, and this work was followed by Etzel (^{7}). These studies lead them to believe that oxygen may help nucleation of the centers responsible for X band absorption.

This report describes a series of experiments which add new evidence to the hypothesis that impurities introduced from the ambient atmosphere are necessary for nucleation of the precipitation of the excess metal in supersaturated solution of alkali halides.

EXPERIMENTAL PROCEDURE

Potassium chloride single crystals used in these experiments were obtained from the Harshaw Chemical Company. Single-crystal disks approximately 2.6 mm thick and 13 mm in diameter were additively colored by the van Doorn technique (^{8}). A stainless steel tube was used to contain the potassium vapor and the KCl crystals. The pressure of the potassium vapor surrounding the crystals was controlled by first introducing nitrogen gas into the system at a desired pressure prior to coloration. Crystals were colored at temperatures ranging from 650°C to 710°C and at times from 10 min to 90 min. At the end of coloration the tube was quenched into a water bath. After coloration the crystals contained approximately 3.2×10^{17} F centers per cm^3. These crystals were annealed at 225°C for 24 hr or longer in a vacuum of 5×10^{-2} mmHg. The absorption spectra of the crystals after annealing were taken on a Cary Model 14 R spectrophotometer. The optical densities were measured as a function of the thickness of the crystals. The crystals were thinned by polishing with a water–methanol mixture (equal parts) on a polishing cloth, rinsed in anhydrous ether, and air dried.

EXPERIMENTAL RESULTS AND DISCUSSION

Under the proper condition of coloration it was possible to obtain a crystal in which the interior was immune to X-center formation during thermal annealing or optical bleaching of the F centers. This is shown in Fig. 1, in which the X-center absorption coefficient and the F-center concentration are plotted as functions of the distance from the surface of the crystal. The X centers formed only near the surface of the crystal, and the F center concentration in this region was close to the solubility concentration observed by Scott and Smith ([1]). Near the center of the crystal no X bands could be observed, and the F center concentration equaled the original concentration before annealing. The distribution of the X centers and F centers did not change after 48 or 72 hr of annealing. After the crystal was thinned down to 0.7 mm this central portion was reannealed at 300°C for 2 hr with no change in the F-center concentration.

The "front" of the region in which the X centers formed, denoted by X_1 in Fig. 1, varied with the time and temperature of coloration. The results of coloration for various times and temperatures are shown in Fig. 2. The

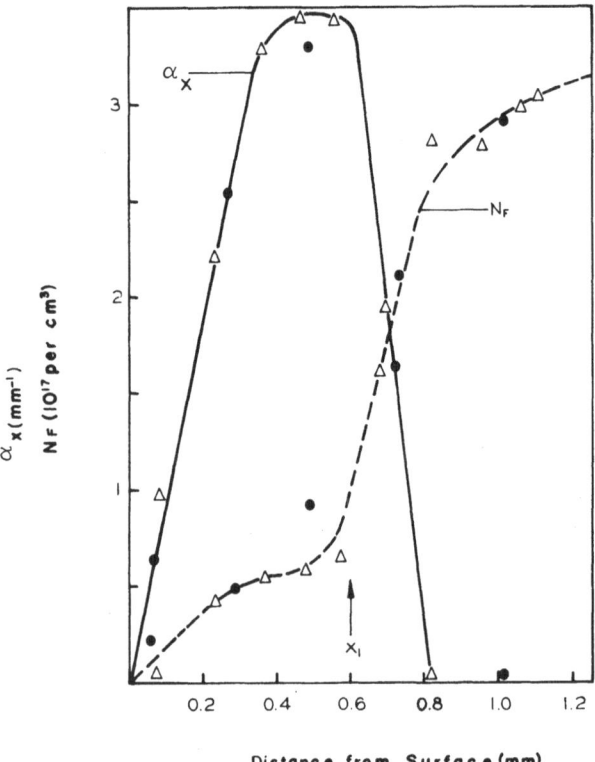

Fig. 1. Distribution of X centers (α_X) and F centers (N_F) from the surface of a colored KCl crystal after annealing. Original thickness was 2.44 mm.

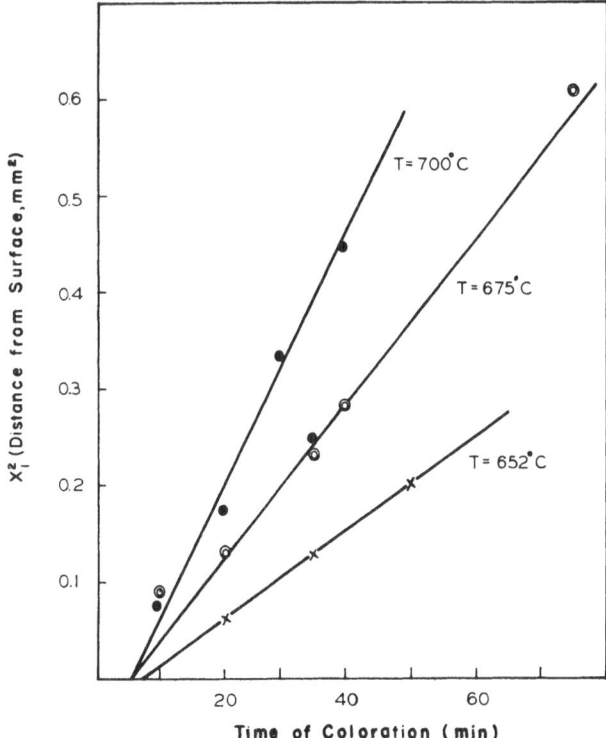

Fig. 2. Dependance of the distance of the X-center front (X_1)
on the time of coloration for three coloration temperatures.

fact that X_i^2 is proportional to the time of coloration indicates that diffusion
of some impurity into the crystal during coloration is responsible for X-center
formation. From the slopes of the curves in Fig. 2 a diffusion coefficient of
this impurity could be estimated,

$$X_i^2 = 4Dt$$

where t is the time of coloration, X_1 is the distance from the surface of the
crystal to the "front", and D is the diffusion coefficient. It has been found
that

$$D = 5 \times 10^2 \exp(-1.75 \, eV/kT) \quad \text{cm}^2/\text{sec} \tag{1}$$

Within 0.1 mm of the surface the intensity of the X bands was reduced
due to a low concentration of F centers existing in this region prior to annealing. In this same region the absorption at 204 mμ, due to the hydroxyl ion
(OH⁻), was greatly enhanced. Apparently water vapor in the atmosphere

surrounding the crystal during coloration bleached the F centers and formed
OH⁻ centers near the surface. For distances greater than 0.2 mm from the
surface the OH⁻ concentration was the same as that in an uncolored crystal
and did not change during annealing. Hence the hydroxyl ions were not
responsible for nucleation.

Control of the impurity concentration in the coloration atmosphere
could not be achieved. In fact the results described above could only be
obtained in crystals colored in a stainless steel tube the interior of which had
been previously exposed to air at 700°C. Normally precipitation occurred
throughout the entire crystal. However, even in the latter case the distribution of X centers and F centers was not always uniform. Three different dis-

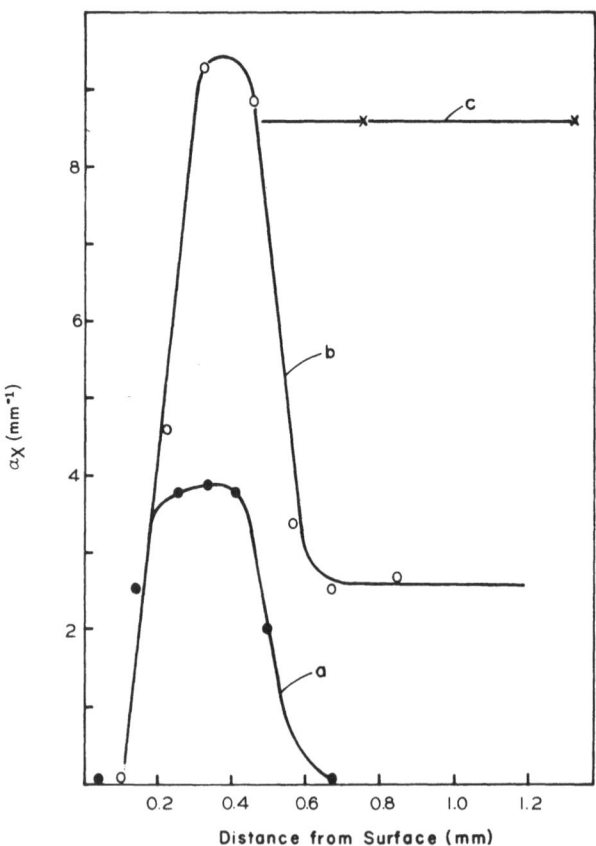

Fig. 3. Distribution of X centers from the surface of annealed
crystals. Time and temperature of coloration and annealing are
identical for each curve. Original thicknesses were approximately
2.6 mm.

tributions were observed in crystals which had been colored at 675°C for 35 min at a pressure of 10 mm of potassium, but apparently with differing impurity concentrations in the ambient atmosphere. These distributions are shown in Fig. 3. The absorption coefficient of the X centers, which is proportional to their concentration, is plotted against the distance from the surface of the crystal. The crystals were colored separately at 675°C for 35 min and originally contained 3.2×10^{17} F centers per cm³. Each crystal was then annealed for 24 hr at 225°C in vacuum.

Curve (a) of Fig. 3 represents the results discussed previously, i.e., the central portion of these crystals were immune to X-center formation. In specimens represented by curve (b) of Fig. 3, which was the case most frequently observed, precipitation occurred throughout the crystal. However, as indicated by the slopes of the curve the concentration of X centers was lower in the interior of the crystal in comparison with that near the surface. The F-center concentrations followed this change in X-center concentration. Thus even in this case the crystal showed two differing regions of precipitation. For distances less than approximately 0.7 mm from the surface the X-band absorption coefficient was higher, its peak wavelength was 735 mμ, and the F-center concentration was 0.8×10^{17} per cm³. For distances greater than 0.8 mm from the surface the absorption coefficient of the X band was reduced, the peak wavelength had shifted to 760 mμ, and the F-center concentration approximately 2.0×10^{17} per cm³. Only in specimens characterized by curve (c) of Fig. 3 does the X-center concentration appear to be uniform throughout the crystal. The peak position of the X band was at 735 mμ and the F-center concentration was 0.8×10^{17} per cm³. This case could not be reproduced in a consistent manner, but nonetheless it occurred with some regularity. Apparently only in this last case was the impurity concentration uniform throughout the crystal after coloration.

These experimental results indicate that some impurity is necessary to nucleate an X center. The source of this impurity or impurities is the ambient atmosphere of the coloration. The impurities enter at the gas–crystal interface and diffuse inside along with F centers, but at slower rate. Indeed, the best measured value for the diffusion coefficient of F centers in KCl recently obtained by Byun ([9]) is

$$D_F = 10^2 \exp(-1.3 eV/kT) \tag{2}$$

much larger than that of the impurity as given by Eq. (1). Potassium oxide is an unavoidable impurity in the technique used for coloration. This, together with the Etzel's observation that heating of the colored crystals in hydrogen atmosphere prior to annealing, prevents X-center formation, strongly indicates that oxygen is the necessary agent for the precipitation of excess metal.

Whatever is the model of precipitation, a necessary step toward formation of a colloidal metallic particle (the ultimate product of decomposition of a supersaturated solution of F centers in alkali halide crystals) is the formation of an aggregate of diatomic potassium molecules. This is difficult to visualize in a perfect lattice or even in a dislocation, as such an aggregate would resemble an M center rather than a K_2 molecule. The problem is similar to that of nucleation of the latent image in the photographic process, which is ascribed to the formation Ag_2S molecules at the surface of AgBr crystallites. In KCl crystals a K_2O molecule may form and thus provide a sufficiently asymmetrical site in the lattice to act as a center of formation of other K_2 molecules.

REFERENCES

1. A. B. Scott and W. A. Smith, *Phys. Rev.* **83**: 982 (1951).
2. A. B. Scott, W. A. Smith, and M. A. Thompson, *J. Phys. Chem.* **57**: 757 (1953).
3. A. A. Shatalov, *Soviet Phys—JETP* **2**: 725 (1956).
4. A. A. Shatalov, *Bull. Acad. Sci. USSR Phys. Ser.* **22**: 1315 (1958).
5. D. Muth, Ph. D. thesis, University of Notre Dame (1967).
6. W. D. Compton, *Phys. Rev.* **107**: 1271 (1957).
7. H. W. Etzel, *Phys. Rev.* **118**: 1150 (1960).
8. C. Z. van Doorn, *Rev. Sci. Inst.* **32**: 755 (1961).
9. J. Byun, Ph. D. thesis, University of Notre Dame (1967).

DISCUSSION

P. W. Levy (Brookhaven National Laboratory): Would you tell me how you established the Fermi level in the KCl crystals? Secondly, could you tell me how you picture this K_2 to be located in the crystal?

Answer: Using simple theory of semiconductors and knowing the F-center concentration one can estimate the position of the Fermi level with respect to the conduction band. I do not determine the Fermi level experimentally, but by calculation from known F-center concentration. I believe that the K_2 molecules are contained in small cavities in the crystal. The volume of a cavity containing one K_2 molecule should essentially be equal to the combined volumes of two K^+ and two Cl^- ions, because K_2 is formed from two cations and two F centers. Some K_2 molecules exist by themselves and others condense into larger cavities as in the formation of a droplet of liquid in a supersaturated vapor.

Levy: If you have K_2 in the crystal in a cavity, which essentially permits it to exist as a gas molecule, then you ought to be able to identify this by the infrared absorption of such molecules.

Answer: Yes, but the molecules in their cavities are highly compressed; hence we should expect that their spectrum would be modified.

Levy: Are you saying that the X-absorption band then is essentially that of the K_2 molecule modified by the presence of the lattice?

Answer: Yes, the absorption band of K_2 in the visible part of spectrum is shifted to shorter wavelengths by the effect of compression. Drickamer, in a series of beautiful experiments, has demonstrated such a shift of F band caused by high pressure. Unfortunately, his experiments on X bands are not complete; therefore we do not have a good experimental evidence to refer to in this case.

Chapter 11

Nucleation and Kinetics of Ferroelectric Domains

H. F. Kay

University of Bristol
H. H. Wills Physics Laboratory
Bristol, England

The relation between the reversal of ferroelectric domains and the change of phase by nucleation and growth of other systems is pointed out. The suggestion is made that the relative simplicity of the ferroelectric system together with the large number of observational techniques available make it suitable for testing nucleation theories. An outline is given of the current methods of observation and understanding of the nucleation and kinetics of domains in single crystals, with particular reference to the ionic materials; a discussion is then given of the relation of this to the technologically important ceramic materials, about which much less is understood.

INTRODUCTION

The similarity of the processes of ferroelectric domain reversal to many solid–solid phase change phenomena is considerable and yet not generally recognized. Moreover, the ferroelectric analog is relatively simple and as such appeals as a potentially suitable system in which to test general theories of nucleation and growth.

Figure 3 reveals several stages in the reversal of the polar direction P_0 of a ferroelectric crystal by the application of a reversing field E. This consists of the formation of a dynamic or static precursor, its growth in size, and its eventual extension to include the whole crystal. If now we consider the original direction of polarity as constituting one phase and the new induced direction of polarity as constituting the other phase, the similarity to general nucleation growth processes becomes quite obvious. The special simplifications of this system over most others are noteworthy:

1. The initial and final phases are chemically and structurally identical. This simplification means that no atoms change their relative position during the process, i.e., no bonds are broken. In addition this means that the surface energy contribution to the process are singular and symmetric to the two phases.

2. The method of transport from one phase to another is clearly defined as a simple atomic shift of distance much smaller than any cell parameter and usually no more than a few 0.1 Å. The resistance to this shift is reasonably open to calculation and observation. Thus there exist no problems of volume, surface, or grain boundary diffusion transport mechanisms nor of heat diffusion, problems which beset most systems studied.

3. The driving force of the reaction is clearly defined as the electric field, analogous to supersaturation in other systems. This field can be easily applied to any degree at any rate of change; also, it can be oscillated at will about the equilibrium condition.

4. The kinetics of the phase change can be monitored by a greater number of techniques than is available to most other systems. The individual phases can usually be seen by optical microscope, at least in single crystals. The amount of second phase can be measured instantaneously as the total charge that has passed, while the reaction rate is measured by the instantaneous current. Observations on ferroelectric specimens of incremental permittivity, initial permittivity, and hysteresis constants such as remanence, coercivity, and squareness are all useful measures of the reaction kinetics.

5. The problem of the surface shape of each phase, while still present to some extent, especially in considering the precursors of size below instrumental resolving power, is much reduced in the ferroelectric case because phase (domain) boundaries can be observed and the shape obtained quite directly. Furthermore, different ferroelectrics provide essentially different nucleus shapes which provide further tests of the theories.

There are some features which are of similar complexity in all systems, although sometimes the geometrical simplicity of the ferroelectric system is more amenable to calculation. For example:

6. The ingression of expanding nuclei and consequent exhaustion of material remaining for conversion to the second phase is a common problem. The technique of Avrami and others are used to calculate this aspect.

7. The volume change occuring on phase change is included due to the existence of piezostriction which occurs in all ferroelectrics. However, in the case of 180° domain reversal this contribution is at least minimized and fairly easily calculable.

A few features of ferroelectrics are unique to this system, but are, in principle, soluble by classical electromagnetic theory, although in practice this sometimes proves difficult:

8. There is no analog in other systems except ferromagnetics for the long-range ordering, disordering, or depolarizing effect by which a portion

of phase two can react on both itself and on another, separate, portion of phase two by the action of its own electric field.

In the comparison of ferroelectric or ferromagnetic systems as suitable for testing nucleation theory it is clear that the latter have several disadvantages. For example, in the case of ferromagnetics domains are only visible at the specimen surface. If they are transparent, then they are difficult oxides to produce. Furthermore, their low magnetostriction relative to ferroelectrics makes their boundaries less visible and their orientation freedom greater. Each different orientation is analogous to a different phase and the inherent simplicity of the system is thereby reduced.

It would appear then that the reversal of 180° domains in a ferroelectric might be a simple system to use as a test of solid–solid nucleation theories without so many of the complexities of nucleus shape, transport mechanism, surface energy, and volume change found within most other systems. It is therefore intended to outline the current state of knowledge of ferroelectric domains and their nucleation and kinetics, with particular reference to ionic materials.

FERROELECTRIC DOMAINS

The intense development of studies of ferroelectrics over the last 20 years has made it necessary to subdivide the subject. This is done most naturally either by substance (ferroelectric types) or by property (usually electrical). The subject of ferroelectric domains, their nucleation and movement under external forces, is a division of the subject well suited to this text; the background material concerning general crystallography, the origin and theories of ferroelectricity, and general electrical, mechanical, and thermal behavior can be found in reviews by Megaw ([1]), Känzig ([2]), Jona and Shirane ([3]), Merz and co-workers ([4,5]), and Martin ([6]). Most of the work to be discussed here relates to the "hard" ferroelectrics such as $BaTiO_3$ and similar oxides, and it is just these materials which relate most closely to the subject of ceramics. By reasons of the mechanical strength, essential cyrstallographic simplicity, and greatest application of the $BaTiO_3$ type materials they have been studied more extensively than the others. Nevertheless, reference to other materials will be made in passing where there is some important point of relevance.

As in the majority of solid state studies most information is obtainable from experiments with single crystals. The results are not always easily related to the observed properties of the polycrystalline agglomerate so often of such immediate practical importance. Even so, without detailed single-crystal knowledge it is even more difficult to explain the agglomerate, and much of our discussion will relate to single crystals, essential for understanding of the basic phenomena and later for interpreting the properties of ceramics of the oxide type.

It is not intended to present a detailed bibliography but only to enable the reader to enter areas of the subject of relevance here. Precedence will therefore often pass unacknowledged.

Ferroelectric domains are those macroscopic (greater than 20 Å) regions in a solid having natural crystal polarity which is constant, but whose direction of polarity can be altered irreversibly by a practically imposable electric field. Implied in this definition is the existence within the material of a spontaneous electrical dipole moment hence the existence of piezoelectricity, and the reversing effects of applied fields hence the existence of hysteresis, and of the probability that at some temperature before complete crystal disruption the spontaneous polarization will be destroyed or directionally randomized on an atomic scale. At this temperature the dipolar instability between the decision of order or disorder will usually produce a high peak value of incremental permittivity.

The analogy with ferromagnetic domains is obvious and leads to the name ferroelectrics, even though very few contain atoms usually associated with major magnetic characteristics. However, it is the differences between them that need most constant reminder. Ferroelectrics are different in that they may contain free electric charges which can compensate the surface charges of a domain; the electrical polarity exists in almost any chosen atomic region of a crystal, so that dipole location becomes less specific; the long-range coupling of Lorenz contrasts with the short-range magnetic coupling; the larger coupling forces of dipole to crystal lattice (crystal anistropy forces) ensure greater electric (piezo) strain compared with magnetic strain; ferroelectric domains are effectively voltage operated rather than current operated as in the magnetic case.

Reversal of polarity of a crystal occurs by the nucleation and growth in volume of domains having an orientation and shape within the crystal which is of lower energy than its neighboring domains, which similarly contract in volume. The low-energy domain is separated from its neighboring domains by a thin boundary or wall in which the polarity changes direction over a distance of a few unit cells. This nucleation and growth, hence boundary growth or movement, involves no local chemical change involving the breaking and reforming of bonds associated with the usual kinetic mechanisms of diffusion, dislocation movements, or separate phase nucleation and growth. However, there are analogies with other crystal growth processes. The driving force of domain rotation is that of applied electric field analogous to supersaturation; growth shapes are controlled by a domain wall energy analogous to surface energy of separate phase growth; the speed of growth is controlled by depolarizing effects and charge-migration time constants analogous to diffusion in phase growth. In both cases there are the mechanical restrictions of adjacent materials and the nucleating or interaction effects of dislocations and chemical defects.

Domains are defined usually by the direction of the polar axis in relation to the principal axes of the unit cell or the macrocrystal. The possible domain orientations are governed by the crystal symmetry and the possibility of changing the orientation of this symmetry system by small-scale movement of the atoms. Further limitations are imposed on the whole domain assembly by the necessity of avoiding high concentrations of charge, e.g., "head-to-tail" arrangements, and high mechanical hindrance with neighbors having different orientations. This means that there are severe limitations to the so-called "easy directions" of polarization, and most frequently the angle between adjacent polar directions is 180° or 90°, giving rise to separating domain walls designated by the angle involved.

The very existence of domains and their behavior under applied stress has considerable bearing on all those properties of ferroelectrics which have currently been found useful. These include optical properties important for the application to modulators and for the basic study of the ferroelectric mechanism on an atomic scale; electrical properties important in applications as dielectric, resistance, parametric, or memory elements; and mechanical properties in relation to piezoelectric transducers and as potential materials having special high-temperature strength characteristics. In all these applications the special limitations of high applied stress fields are important especially in relation to the properties of coercivity, loss, saturation, and aging, all of which are intimately related to domain nucleation and kinetics.

METHODS OF OBSERVATION

The properties affected by the existence of domains are naturally those utilized to study them.

Optical Method. This is the most frequently used technique and utilizes either the strain imposed on the crystal by the domain boundaries or the differences of birefringence imposed by different orientations of the polar axis. The sign of polarity is observed by the application of small perturbing electric fields, which cause wall motion or enhanced wall contrast by small angular rotation of the polar direction away from an "easy" crystallographic direction. Most difficulties are usually experienced in the observation of 180° domains, for in this case the mechanical distortion of boundary cells is smallest and birefringence differences are zero because the angle between polar axes is an exact multiple of $\pi/2$. However, careful phase-contrast microscopy will render such 180° boundaries visible if viewed along the polar axis also parallel to a small applied field. Visibility results from surface strains due to the different piezostriction occurring on either side of the boundary.

Etching Method. This provides a surface picture of emergent domains. Weak acids have been used for hard ferroelectrics such as oxides ([7]) and even water for soft water-soluble materials such as triglycine sulfate (TGS). The method differentiates the polarity by virtue of differences in solution rates but is slowly destructive. Although repeated etching will give indications of domain progress the patterns soon become confused, but the initial resolution is high, especially for 180° domains.

Bitter Pattern Method. This is the exact analog of the method found so successful for ferromagnetics in which sprinkled powders or colloidal suspensions of high susceptibility material are placed on the surface to decorate the domain boundaries, or in which somehow charged particles preferentially decorate one sign of surface emergent domains ([8]). Suspensions of Pb_3O_4, TiO_2, or S in hexane, the deposition of crystals of $PbCl_2$ from saturated solutions, and the preferential deposition of evaporated Cd have all been used with success. The exact mechanism is not established in detail in all cases but is probably associated in the one case with the attraction of charged particles to the emergent surface charge of domains, possibly encouraged by minute temperature changes providing a pyroelectric charge, or in the other case by the well-known attraction of high susceptibility particles to regions of field gradient occurring at emergent domain boundaries.

Electron Microscope Method. Specimens thin enough to allow the transmission of the electron beam will have their domains revealed either by the breakdown of Friedel's law at 180° domain walls ([9]) or by the surface charges resulting from beam heating. Contrast has also been observed due to the small angular differences occurring in some cases between adjacent domains, and surface replicas of fractured oxide ceramics also show domain patterns due to the influence of the slight disorientations of adjacent domains on the fracturing process.

X-Ray Method. For small crystals totally irradiated in the beam the existence and relative proportion of domains can be assessed depending on sensitivity of the instrumental resolution and the angular differences between adjacent domains. This varies from a few minutes of arc by normal single-crystal techniques ([10]) to a few seconds by improved versions of the Berg-Barrett technique [([11]), also standard texts on X-ray diffraction techniques]. By suitable choice of X-ray wavelength it is also possible to distinguish between 180° domains by utilizing the breakdown of Friedel's law ([12]).

Pyroelectric Method. Since domains are strongly pyroelectric it is possible to measure the surface electric charge induced by local heating. Pulsed infrared radiation ([13]) and scanning electron beams ([14]) analogous to the microbeam probe analyzer method have been used.

Hysteresis Method. In this method details of the individual domains are lost, and only average polarization, hence domain orientation effects, are observable at each point in relation to an alternating applied field. Relatively simple apparatus ([15]) can be used, but the method is more useful if coupled with simultaneous measurements of the piezoelectric strain ([16]), which then gives incremental piezoelectric as well as dielectric constants providing a useful byproduct to assess the domain kinetics.

Switching Method. Sharp step functions or slow-rising sawtooth electrical pulses are applied to a ferroelectric specimen and the total switching current is measured as a function of time. For step-function inputs the resultant current pulse is a measure of the nucleation and growth of reversal domains, while for slow sawtooth inputs the current is usually a series of small pulses analogous to Barkhausen pulses in ferromagnetics. The results provide information regarding domain sizes and incremental volumes swept up by the jerky movement of domain walls.

DOMAIN ENERGY

As in ferromagnetic materials the domain array adopted by a ferroelectric crystal is not unique for any one crystal under given environmental conditions, but is such as to reduce the free energy components to a minimum:

1. Interaction energy is associated with the work required to disrupt the natural tendency of the atomic dipoles to align parallel, a situation induced by a local field enhanced by little more than the long-range Lorenz factor $4\pi P_0/3$.

2. Crystal anisotropy energy is associated with the work required to move the polar direction away from an easy crystal direction; this is really the variation of the interaction energy with orientation relative to the crystal lattice.

3. Electrostrictive energy is associated with change of interaction energy with mechanical strain, which may in turn be related to elastic constants.

4. Electrostatic self energy is associated with the work in producing an external electric field due to dipole surface charges.

5. Domain wall energy is associated with the energy of formation of the dividing domain walls; a surface tension, comprised of components 1–4 giving a resultant energy per unit area of wall.

CALCULATION OF DOMAIN WALL ENERGY

Calculations of this energy are still limited somewhat by lack of clarity of atomic detail in the domain wall. Energy calculations for $BaTiO_3$ 180° walls, including only the simplest approximations for interaction and electrostrictive terms, gives the thickness in unit cells of $N = 1$ and the wall energy $\sigma = 10$ ergs/cm². More extensive works give $N = 0$–4 and $\sigma = (1.5$–$10)$ ergs/cm², with somewhat higher values of $N = 3$–10^3 and $\sigma = (4$–$65)$ ergs/cm² for 90° walls in the same material ([17-20]), although electron micrographs ([9]) of the latter seems to provide an upper limit of $N = 50$. Direct observations of σ are difficult to make. Carefully annealed crystals under insulating conditions show a tendency for an equilibrium domain size for a definite crystal thickness which value gives a measure of σ ([21]). Measurement of coercivity of crystals will also give an estimate of σ if sufficiently correct assumptions can be made regarding the domain shape ([22]), but observations to date are rare. It can safely be said that wall energy values, either from calculation or measurement, are still very approximate, no doubt in part due to the variation in σ with crystal direction and the various other factors controlling domain shapes.

DOMAIN WALL INTERACTION AND MOVEMENT

The 90° domains and walls usually nucleate at one surface as small wedges whose constituent walls grow roughly normally to their surface so that the wedge extends into the crystal (Fig. 1). The coercive field is high (2.0 kV/cm in $BaTiO_3$) due to the larger strain (1 %) on a 90° direction change of P_0. The maximum velocities involved are therefore near the velocity of sound.

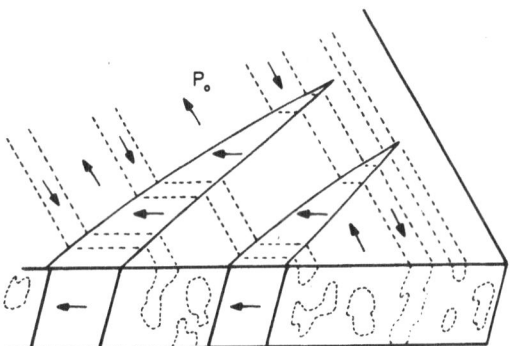

Fig. 1. Diagram of 90° domain wedges in a portion of a $BaTiO_3$ crystal bounded by $\langle 100 \rangle$ axes. The 180° domains are also shown in dotted outline.

Alternating low-frequency electric fields applied along $\langle 100 \rangle$ causes domain walls to oscillate back and forth at the same frequency, and in the case of 90° walls this movement has superimposed an asymmetric drift whose magnitude and direction is adequately explained in terms of the concentration of 180° domain walls also present in the system ([23]) as depicted by the dotted lines in Fig. 1. Incremental hysteresis loops between applied field and 90° wall displacements are reasonably explained in terms of an average separation of lattice imperfections which provide a series of energy barriers ([24]), while the observed reduction of the boundary displacement with increase of frequency is accounted for by including a dissipative term dependent on wall velocity, although the exact origin of this term is not yet clearly understood.

The 180° domains and walls also nucleate usually near a crystal surface (especially the cathode) and extend as spikes along the P_0 direction as in Fig. 2. They usually have approximately cylindrical or square symmetry for those materials of high crystal symmetry perpendicular to P_0 such as $BaTiO_3$, lenticular symmetry about P_0 for moderate crystal anisotropy perpendicular to P_0 such as triglycine sulfate (TGS), and plane or bladelike symmetry for high crystal anisotropy perpendicular to P_0 as in Rochelle Salt (RS). The acicular nature of 180° domains is attributable to strong dipole coupling along the polar direction, while their lower coercivity (zero to a few volts/cm) and greater maximum wall velocities (10^7 cm/sec or more) are due to the fact that 180° domains cause little mechanical disturbance, particularly if the crystal symmetry in the plane normal to P_0 is high as in $BaTiO_3$. This makes 180° domain reversal the most important mechanism controlling the elec-

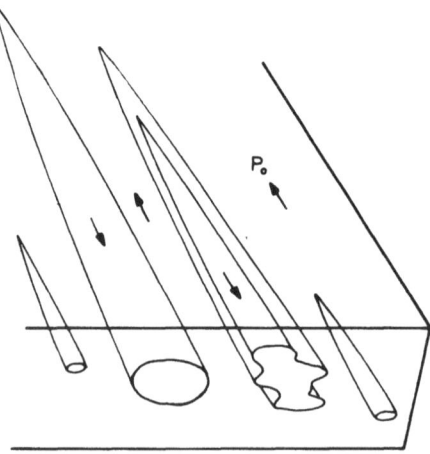

Fig. 2. Diagram of 180° domain spikes in a portion of a $BaTiO_3$ crystal bounded by $\langle 100 \rangle$ axes. The cone angle is much exaggerated.

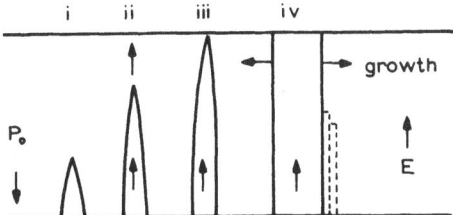

Fig. 3. Diagram of the four phases of 180° nucleation and growth in BaTiO$_3$. (i) Primary nucleation of critical size. (ii) Longitudinal growth through the crystal. (iii) Contact of domain with far electrode. (iv) Lateral growth of cylinder.

trical properties of oxide ceramics, but 90° domain mechanisms the most important contribution to the electromechanical and aging properties.

DOMAIN NUCLEATION AND GROWTH

The recognizable stages of 180° reversal for discussion consists of (1) primary nucleation and achievement of a critical size, (2) longitudinal growth through the crystal, (3) contact of the domain with the distant electrode, and (4) lateral growth of the domain boundary, as depicted in Fig. 3.

Clearly these stages are not completely independent, as some sideways growth will also occur during longitudinal growth. Also, there must be a distinction between freely nucleating new domains and the nucleation of secondary layers (dotted lines in Fig. 3) on an already existing wall, which together with their rapid transit through the crystal thickness constitutes the most probable mechanism of lateral wall growth.

Primary Nucleation. The nucleation rate for new domains is found ([25]) to be constant, with R produced per unit area of crystal and per unit time. This rate is dependent on the applied field such that

$$R \propto E^{3/2} \tag{1}$$

reaching a rate as high as 10^{13} sites/cm^2-sec at 450-kV/cm applied field. Explanation of this field dependence is currently not agreed upon. The usual starting point for nucleation is that the change in energy U forming the new reversed nucleus should be a minimum. For 180° domains this statement takes the form:

$$U = 2P_0 EV + \sum_1 \sigma_i A_i + U_d \tag{2}$$

where the first term represents the reduction in energy constituting the driving force for reversal due to applied field E acting on the reversed volume V, the second term represents a resistance to growth produced by the domain wall energy σ summed over all the particular areas A_i and orientations involved, and U_d represents a resistance to growth due to the self-depolarizing field produced by surface charges on the new domain.

This type of equation predicts a critical energy barrier, representing a nucleus of critical dimensions, which has to be overcome by thermal fluctuations before the domain can grow. This critical value is strongly dependent on nucleus shape and it is impossible to determine this from the equation, so that starting assumptions must be made based on observations. Assuming the nucleus is a half prolate spheroid of base radius r at the crystal surface and length l, then the critical values r^*, l^*, and U^* can be calculated [26] for the critical nucleus, and using for $BaTiO_3$ the values of P_0, σ, and ϵ measured from separate experiments one finds that at room temperature with an applied field of 500 V/cm, which easily causes reversal,

$$r^* \propto 1/E \approx 7 \times 10^{-6} \text{ cm}, \qquad l^* \propto 1/E^{3/2} \approx 7 \times 10^{-4} \text{ cm},$$
$$U^* \propto 1/E^{5/2} \approx 10^6 kT \tag{3}$$

Thus the critical nucleus dimensions are spatially practicable in experimental crystals ($\sim 10^{-2}$ cm) but the probability of activation is infinitesimally small. There are several escapes from this dilemma but the one actually operative is not yet firmly established.

1. Dislocations terminating at the surface and of various lengths l_D parallel to the polar direction provide, by their elastic distortion, cores of reversed or unstable material such that spheroidal nuclei of length l_D sited on these dislocations need negligible energy to be produced. Then for any applied field E the critical length from Eq. (3) is l^*, and dislocations of greater length produce reversed domains, those shorter dislocations require energy

$$U^* \propto (l^* - l_D)/E$$

and for one cell parameter difference this becomes $\sim 12\,kT$ at the field considered, giving $R \propto \exp(-\alpha/E)$. If higher fields are used and multiple parameter differences are included, then an $R \propto E^n$ with $n > 1$ can be expected. This suggestion has some further merit in that a dependence of nucleation rate on crystal thickness is involved owing to the fact that dislocations cannot be of greater length than this. Observations indicate that in some $BaTiO_3$ crystals dislocations mostly terminate within 6×10^{-3} cm from the surface [27].

2. Thin cylindrical reversed regions present at all stages, possibly in dislocations extending right through the crystal, are too small to be seen but expand by lateral growth of the cylinder. As it stands this idea fails, since the usual crystal values for BaTiO$_3$ at applied field of 500 V/cm again give

$$r^* \propto 1/E \approx 5 \times 10^{-8} \text{ cm}, \qquad l^* = d \approx 10^{-2} \text{ cm},$$

$$U^* \propto d/E \approx 4 \times 10^4 kT \tag{4}$$

This again is too large for activation and requires that a large background energy be available in the cylindrical core. Also, the model does not readily fit the observed travel of reverse domain spikes through the crystal, and the predicted thickness dependence operates in a direction opposite to that observed in practice, i.e., greater thickness causes smaller activation as judged by greater nucleation rate.

Longitudinal Growth. Once the critical size of the reversed spike has been surpassed it continues its growth through the crystal until it reaches the other surface; it then reduces its energy by achieving a cylindrical shape having its base radius, thereby reducing wall energy and making its electrostatic energy more negative. The depolarizing energy is completely absorbed by the electrode.

Transit velocities of 5×10^6 cm/sec have been observed at high fields of 450 kV/cm, which is 10 times the velocity of longitudinal sound waves c_0, but this only requires a lateral growth of the walls of a 2° apex cone of 10^5 cm/sec or $0.2c_0$, so that acoustic impedance is probably still not operative.

If the voltage is removed before the switching is complete, i.e., intermittent voltage pulsing, then some domains will be caught in transit. When the domain is fairly near the other surface observations have revealed a slight speedup owing to image force attraction of the far electrode, but if the domain tip is well short of the far electrode then the domain will tend to retreat to its point of origin. This occurs because on reduction of the field the domain energy U can become strongly positive owing to surface energy σA, depolarizing energy U_d, and the backfield of similarly oriented adjacent domains. The result is a certain amount of backswitching, making intermittent pulses less effective than continuous pulses of the same total duration for switching a crystal ([28,29]).

The high mobility of domains in transit also accounts for the large increase of incremental permittivity ϵ_i observed in ferroelectrics ([30,31]) and a large increase in second harmonic generation ([32]). The increase of ϵ_i depends on the frequency and varies with $1/\nu$ or $1/\nu^2$ for low and high frequencies, respectively. This behavior can be reasonably explained ([33]) using normal theories of damped forced oscillation of a domain wall of an effective mass per unit area having a restoring force due to its elastic tension.

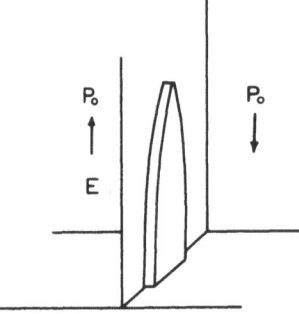

Fig. 4. Diagram of growth of laminar sheet of reverse 180° material on an existing 180° wall.

Lateral Growth. Once the small reversed domain is cylindrical it then grows in extent laterally. Observations show that up to about 2 kV/cm the wall velocity $v \propto \exp(-\alpha/E)$ and above this field value it tends toward $v \propto E^{1.5}$, the same field dependence as for the nucleation rate of new domains [Eq. (1)]. This similarity of field dependence of both R and v strongly suggests similar mechanisms, but the subject has not been developed sufficiently to show this in detail. There are strong theoretical reasons ([18]) to suggest that a bodily sideways increment of the entire cylindrical wall does not occur as a whole, based on the large energy density of the wall (1 erg/cm² in one lattice thickness compared with similar wall energy in 100 lattice thicknesses in ferromagnetic ion). It now seems certain that the wall moves laterally by the addition of thin bladelike layer nuclei reaching from one or both crystal surfaces and lying flat on the domain wall already present (Fig. 4). This is rather like the growth of new layers on a crystal growing from solution, but whose origin is the local field and piezoelectric stress concentrations at the domain wall crystal–surface intersection rather than growth steps forced by emergent dislocations. Again one assumes the lowest energy form would be either a flat wedge or ellipse of base width w of one cell thickness to ensure minimum addition to the wall energy. In this case there is little positive contribution to U resulting from the depolarizing effect and calculations involving several similar shapes give for the same conditions of field as before, the critical values

$$w^* \propto 1/E \approx 5 \times 10^{-6} \text{ cm}, \qquad l^* \propto 1/E \approx 7 \times 10^{-6} \text{ cm},$$
$$U^* \propto 1/E \approx 9kT \tag{5}$$

which is reasonable for thermal activation and requires very little more energy than that calculated to advance the wedge leading edge by one unit cell once the critical nucleation has occurred:

$$w^* \propto 1/E \approx 10^{-5} \text{ cm}, \qquad l^* = a \approx 2 \times 10^{-8} \text{ cm}, \qquad U^* \approx 9kT \tag{6}$$

The probability of wall nucleation and its velocity of advance is then proportional to $\exp(-\alpha/E)$, as observed, and at higher fields above about 2 kV/cm, when new blades two and three lattice constants thick begin to contribute to the velocity, it can be shown that this approaches a velocity proportional to $E^{1.5}$, also as observed.

Eventually the domains lose their cylindrical shape and take up that shape governed by the variation of σ with lattice direction (i.e., become squared in $BaTiO_3$, lenticular in TGS, etc.). As they grow in size they eventually touch and coalesce; the rapid filling up of the reentrant regions near points of contact have been definitely associated with the larger Barkhausen pulses ([34,35]).

SWITCHING CURVES

The relative proportion of nucleation and domain wall lateral growth is expected to be different for different fields and different materials, but this proportion is extremely difficult to observe directly and most of such detail has been deduced from the switching curve. This is the variation of switching current i_s with time when a step field is applied and fairly typically is as depicted in Fig. 5(a). The current observed is the added effect of all the nucleation and wall growth processes which occur between the electrodes. Nuclea-

a)

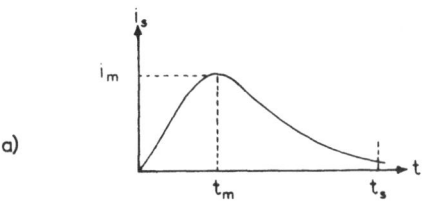

b)

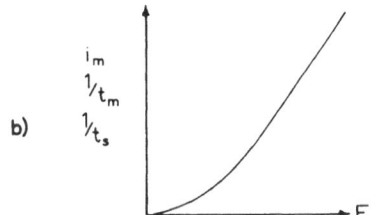

Fig. 5. Typical results of switching curves (a) showing output pulse characteristics usually measured and (b) showing typical plots of characteristics against applied field E.

tion occurs at different places in the crystal at different times depending on field, and the general geometry of domains will be different for different materials. Hence it is necessary to assume a nucleation model utilizing all known features of domain geometry, etc., and then to develop the expected i_s/t curve and see if the assumed model is indicated.

Figure 5(a) shows the measurements characterizing the switching pulse. Included are the current i_m and time t_m of the maximum current, the time t_s at which a definitely large proportion of the crystal has switched, and also the variation of these characteristics with applied field, as in Fig. 5(b). Another characteristic which is sometimes measured is the curve symmetry defined in several different ways—as an example, by the ratio of crystal polarization switched at i_m to the total switched polarization $(P_m/2P_0)$, which can vary from zero to 0.5 in theory. The total area under the curve is, of course, $2P_0V$, where V is the crystal volume between the electrodes. Several theories have been developed based on different nucleation models, and even the simple ones are frequently very effective in explaining observations:

1. It is assumed there are a fixed number of nucleation sites independent of whether the contribution is due to primary initiation or to wall growth, each site developing through the same growth path. This approach has some physical significance in that the direct observations on $BaTiO_3$ indicate some similarities of characteristics, e.g., voltage dependence, for both mechanisms. This type of theory is based on the simplest nucleation equation:

$$dn/dt = \sigma(n_0 - n) \tag{7}$$

where the rate of switching of sites dn/dt is proportional to a probability or rate factor σ and the unreversed sites $(n_0 - n)$, thus accounting directly for exhaustion of available sites. As it stands this provides a switching curve of exponential decay having a maximum very close to the origin of t, and relates only to those particular cases where initial nucleation contributes far more to the reversal process than growth of the nuclei by lateral motion, e.g., guanidinium aluminum sulfate hexahydrate (GASH).

2. The same conditions as (1) are assumed but with a rising time dependence of σ. This allows for an increased nucleation rate due to already formed nuclei providing their own preferential sites at the walls. Depending on the domain geometry and the law accepted or observed for the growth velocity of domain walls, this can give $\sigma \propto t^n$ with n varying between about 0.5 and 2.0. For example, a constant wall velocity as observed for $BaTiO_3$ ([35,36]) gives $\sigma \propto t$, providing the finite size of initial nuclei is neglected. Such an approach has been found very successful. Solution of Eq. (7) now gives

$$i_s = 2P_0(t/t_m^2) \exp[-\tfrac{1}{2}(t/t_m)^2] \tag{8}$$

and also constancy of the product $i_m t_m$ independent of field [see Fig. 5(b)], as has been observed for many substances such as $BaTiO_3$ [37] and TGS and its isomorphs [38]. This theory is particularly useful in these materials, where lateral growth is the major factor in reversal.

3. Separate initial and wall nucleation rates are assumed. This naturally produces descriptive equations of somewhat greater complexity. It is assumed that the initiation of primary nuclei is controlled by a constant rate factor k and that n_0 sites are capable of nucleating before exhaustion. This gives, by integrating Eq. (7), the number of primary domains n which have nucleated:

$$n = n_0(1 - e^{-kt}) \tag{9}$$

Adding the fact that new wall sites are available depending on the velocity of lateral growth v gives an equation of the type:

$$i_s = (Avn_0/k)(2P_0 - P)(kt - 1 + e^{kt}) \tag{10}$$

a result derived in several different ways by various authors [39,40] and checked experimentally for colemanite, for example, by Wieder [43]. The effect of field intensity on v and k can be inserted if such information is available, or to obtain the best fit between theoretical and experimental switching curves. Field dependence of nucleation on powers of E up to $E^{3/2}$ can be understood as mentioned earlier, but the occurrence of higher powers such as $E^{5/2}$ for trihydrogen selenites or $E^{7/2}$ for tetramethylammonium trichloromercurate (TTM) are still unexplained.

4. The two time constant theory assumes switching is controlled by two time constants. These may be loosely associated first with the transit time τ_d of growth of nuclei across the crystal and contributing a current pulse $i_0 \exp(-t/\tau_d)$ and second with the exhaustion decay of these sites τ_n. This theory seems most applicable when little wall growth occurs (small domain nucleus interaction) or at high fields, where this approximation is perhaps more acceptable. The equation describing such a process is:

$$i_s = i_0 N(e^{-t/\tau d} - e^{-t/\tau n})/1/\tau_n - 1/\tau_d) \tag{11}$$

giving a risetime dependent on the smaller time constant (more rapid process) irrespective of whether it corresponds to nucleation or transit times. Several interesting experimental correlations with this type of theory have been found, in TGS, for example [41].

COERCIVITY

The coercive field E_c of many ferroelectrics including $BaTiO_3$ tested with sinusoidal AC shows a linear relationship $1/E_c \propto -\log w$, the applied angular frequency over several orders of frequency([37,42,43]). This relationship can be derived fairly simply from Eq. (8) and can be considered to represent satisfactory agreement between theory and experiment. However, there is also observed([44]) a strong dependence on crystal thickness, which for BaTiO$_3$ taking the form

$$E_c = E_\infty + A/d \tag{12}$$

where $A \approx 2.2$ V and E_∞ is the limiting value of coercive field at infinite thickness of crystal. Unfortunately, in the case of BaTiO$_3$ there is strong evidence in many investigations ([17,45-47]) of the existence of a surface electrical double layer ~ 1000 Å thick with about 1 V across it. This might be expected to give thickness effects depending on the significance of this layer relative to the bulk crystal. However, other crystals not so subject to this particular fault e.g., TGS, still show a strong thickness dependence of coercivity such that

$$E_c \propto 1/d^{2/3} \tag{13}$$

where here the coercivity is defined as the starting field for the beginning of rapid reversal ([22,48]). The explanation of this lies in the details of the variation of critical size and shape of nuclei as the applied field increases. For convenience, consider these nuclei to be half prolate spheroids, although rather more probable from the activation point of view are flattened versions of this shape growing on existing walls. At low fields the length l^* of the critical nucleus is greater than the crystal thickness and therefore the domain cannot form because the required activation is too great or no imperfections exist to provide a core of this length. As the applied field increases the critical length l^* becomes equal to the crystal thickness and nuclei can form stretching to the far electrode and then grow by first becoming cylindrical and finally expanding laterally. Equating l^* to d in Eq. (3) and arguing that this situation corresponds to the coercive field gives immediately the required dependence of E_c on $1/d^{2/3}$.

EFFECT OF DOMAINS IN PIEZOELECTRIC STRAIN

The most useful observations are those in which piezoelectric strain x and displacement D are measured simultaneously, as depicted in Fig. 6, which shows typical observation made for a ceramic specimen. The results for single crystals are more squared off due to the smaller range of field over which the majority of switching occurs rather suddenly. Single-crystal results also show hysteresis or crossover effects in the D/x characteristic due to the sudden change in sign of the piezostriction upon sudden change in the sign

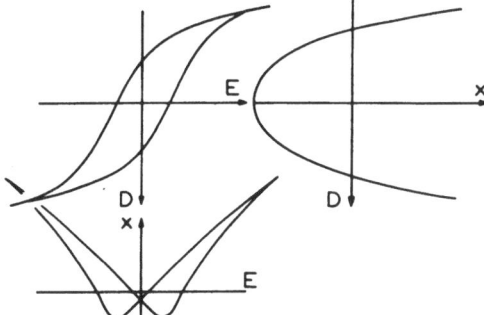

Fig. 6. Typical results of piezostrain measurements x with simultaneous values of displacement D and applied field E.

of D on 180° rotation. This is reduced in ceramics due to clamping and the direction averaging, both of which reduce the piezostriction and also make the 180° domain reversals occur over a greater range of field. In the case of BaTiO$_3$ crystals, with fields applied along $\langle 100 \rangle$, domain reorientation can occur by two mechanisms, 180° reversal and 90° reorientations.

If the relative proportions of each possible reversal mechanism occurring at each particular coercive field were known, then the expected D/E, D/x, and x/E characteristics could be roughly predicted. However, it is not so easy to ascertain these proportions and coercive fields from the observations of the electrical characteristics themselves. Nevertheless, many features of behavior of crystals or ceramics can be obtained from an analysis of the various possibilities that can occur for any particular materials and comparing these possibilities with experimental results. Three basic mechanical strain processes can then occur for BaTiO$_3$ and similar materials:

1. Rotation through 180° by growth of 180° domains and associated change of piezostrain.

2. Rotation through 90° by growth of 90° domains and associated change of lattice constant and piezostrain.

3. Changes of piezoelectric strains without rotation.

Table I shows details of the relative magnitudes of mechanical strain change Δx and electrical strain change ΔD and the ratios of these variables insofar as they can be reasonably assessed for the $\langle 100 \rangle$ axes of a perfect BaTiO$_3$ crystal if each of these three processes were to occur alone. Where necessary, for purposes of comparison a field of 1.5 kV/cm and a piezoelectric constant d_{33} of 4×10^{-6} esu are used to calculate relative magnitudes. The shear effect d_{51} is ignored as being clamped. The major strain mechanism is obtainable from the measured characteristics from this table, especially if

TABLE I

Mechanism	Δx	ΔD	$\Delta x/\Delta D$	$\Delta D/\Delta E$	$\Delta x/\Delta E$
180° Rotation	$2d_{33}E$	$8\pi P_0$	—	$8\pi P_0/\Delta E$	$2d_{33}$
	4×10^{-5}	9.4×10^5	4.2×10^{-11}	$\sim 3.1 \times 10^5$	8×10^{-6}
	Small	Large	Small	Large	Small
90° Rotation	$c - a$	$4\pi P_0$	—	$4\pi P_0/\Delta E$	$(c - a)/\Delta E$
	10^{-2}	4.7×10^5	2.1×10^{-8}	$\sim 1.5 \times 10^5$	$\sim 2 \times 10^{-3}$
	Large	Large	Large	Large	Large
Piezostriction	$d_{33}E$	$\epsilon_3 E$	—	ϵ_3	d_{33}
along c axis	2×10^{-5}	10^3	3.3×10^{-8}	200	4×10^{-6}
	Small	Small	Large	Small	Small

one mechanism takes precedence in any particular range of applied field. However, in $BaTiO_3$-type materials the complication arises that two quadra-phase 90° rotations of equal volume occurring at the same time cannot be distinguished from one 180° rotation, so that only the uncompensated 90° rotations will contribute to the strain. Measurements on thin crystals having P_0 perpendicular to the plate and with the field applied parallel to P_0 show that 180° rotation is the major process, but as little as 0.5 % of crystal volume passing through the 90° position is detectable ([49]).

An important feature of the strain characteristics of $BaTiO_3$ is that as the temperature is raised above its Curie temperature the piezoelectric constants disappear because the symmetry becomes cubic. The D/x curve thus becomes truly parabolic ($\propto P^2$), while at lower temperatures the strain is related to even powers of the polarization by a relation of the type

$$x = Q_1 P^2 + Q_2 P^4 \tag{14}$$

This is clearly the type of relation to use with the ceramic, which is rendered isotropic to some extent by general crystallite orientation and clamping effects, both of which reduce the relative effectiveness of Q_2. The values of $Q_1 = 7.7 \times 10^{-13}$ esu and $Q_2 = 8.4 \times 10^{-23}$ have been measured ([50]) and found to be roughly independent of temperature.

DOMAINS IN CERAMICS

The progress made to date in the application of domain information from oxide single crystals to the behavior in the ceramic is severely limited. The differences are essentially connected with the spread of spontaneous polarization and cell anisotropy due to clamping, impurity segregation at crystallite boundaries, and the effects of limited crystallite size. In the ideal circumstance of no crystallite clamping the observable saturation polarization in $BaTiO_3$ ceramic would be only 0.89 P_0 if both 90° and 180° rotations could freely occur, or 0.5 P_0 if only 180° rotation could occur. Values greater than this are rarely found in ceramics. The latter is more likely because of clamping of adjacent crystallites, and X-ray observation ([51]) has revealed no evidence of 90° rotations. Polished and etched ceramic surfaces ([52]) have shown a majority of 180° domain walls, but they are lamella-shaped instead of the cylindrical type found in single crystals, no doubt owing to the anisotropy of local stresses in the ceramic. The stability of domains in ceramics is shown by the lack of disturbance of the pattern during polishing, a feature certainly not observed in single crystals. Of importance is the observation of the existence of quite numerous 90° domains, which are expected to oscillate slightly with applied AC field but not to move radically because of the steric hindrance involved.

Attempts have been made to interpret the dielectric loss of D/E loops of ceramics to various domain movements and to the inhibiting effect of small crystallite size on the number of domain walls, but much more work is required to elucidate these processes in detail.

· Ceramics also demonstrate considerable aging effects, i.e., change of properties with time during the application of a field or after the field has been relaxed. These effects have been attributed to low-energy arrangements of domains forming closure groups which are formed essentially by the clamping effects of neighbors when the material is cooled below the Curie temperature. These can be altered by application of a bias voltage in addition to the measuring AC field, or by large AC measuring fields themselves. Once altered by essentially a 90° rotation of some of the unfavorably oriented components of the closure group they relax, at least partly, to the original format. The existence of regions of domain closure can be used to explain the variations of ultrasonic attenuation ([55]), the aging under bias ([56]), the slow changes in the D/E hysteresis loops on suddenly increasing the applied field ([54,57]), and the reduction of loss and permittivity on storage ([58]). It is established that these aging effects can be reduced by any mechanism which inhibits the elastic relaxation of 90° boundaries. This includes impurities, e.g., $MgSnO_3$, and exposure to intense γ or X rays.

No direct observations of domain movements in ceramics have been made, and it remains to be seen whether some of the speculation concerning loss and aging effects in ceramics can be confirmed more directly in the future.

REFERENCES

1. H. D. Megaw, *Ferroelectricity in Crystals*, Methuen, London, 1957.
2. W. Känzig, "Ferroelectrics and Antiferroelectrics," in: *Solid State Physics* Vol. **4**, Academic Press, New York, 1957.

3. F. Jona and G. Shirane, *Ferroelectric Crystals* Pergamon, New York and London, 1962.
4. W. J. Merz, *Progress in Dielectrics* Vol. 4, Heywood and Co., London, 1962.
5. E. Fatuzzo and W. J. Merz, *Ferroelectricity*, North Holland Pub. Co., Amsterdam, 1967.
6. J. J. Martin, *Die Ferroelectricker*, Akademische Verlags-gesellschaft, Leipzig, 1964.
7. J. A. Hooton and W. J. Merz, *Phys. Rev.* **78**: 409 (1955).
8. G. L. Pearson and W. L. Feldman, *J. Phys. Chem. Solids* **9**: 28 (1959).
9. M. Tanaka and G. Hanjo, *J. Phys. Soc. Japan* **19**: 954 (1964).
10. H. F. Kay, *Acta. Cryst.* **1**: 229 (1948).
11. C. B. Busquet, M. Lambert, A. M. Quittet, and A. Guinier, *Acta. Cryst.* **16**: 989 (1963).
12. N. Niizeki and M. Hasegawa, *J. Phys. Soc. Japan* **19**: 550 (1964).
13. A. G. Chynoweth, *J. Appl. Phys.* **27**: 78 (1956).
14. J. C. Burfoot and R. V. Latham, *Brit. J. Appl. Phys.* **14**: 933 (1963).
15. J. K. Sinha, *J. Sci. Inst.* **42**: 696 (1965).
16. G. Schmidt, *Z. Phys.* **145**: 534 (1956).
17. W. J. Merz, *Phys. Rev.* **95**: 690 (1954).
18. W. Kinase and H. Takahasi, *J. Phys. Soc. Japan* **12**: 464 (1957).
19. V. A. Zhirnov, *Soviet Phys.—JETP* **8**: 822 (1959).
20. L. N. Bulaevski, *Soviet Phys.—Solid State* **5**: 2329 (1964).
21. C. Kittel, *Rev. Mod. Phys.* **21**: 541 (1949).
22. H. F. Kay and J. W. Dunn, *Phil. Mag.* **7**: 2027 (1962).
23. J. Fousek and B. Brèzina, *Czech. J. Phys.* **B117**: 344 (1961).
24. J. Fousek and B. Brèzina, *J. Phys. Soc. Japan* **19**: 830 (1964).
25. H. L. Stadler and P. J. Zachmanidis, *J. Appl. Phys.* **34**: 3255 (1963).
26. R. Landauer, *J. Appl. Phys.* **28**: 227 (1957).
27. S. Waku, *J. Phys. Soc. Japan* **17**: 1068 (1962).
28. K. Husimi and K. Kataoka, *J. Appl. Phys.* **29**: 1247 (1958).
29. K. Husimi and K. Kataoka, *J. Appl. Phys.* **30**: 323 (1959).
30. M. E. Drougard, H. L. Funk, and D. R. Young, *J. Appl. Phys.* **25**: 1166 (1954).
31. M. Prutton, *Proc. Phys. Soc.* **70**: 702 (1957).
32. K. Husimi, *J. Appl. Phys.* **30**: 978 (1959).
33. E. Fatuzzo, *J. Appl. Phys.* **32**: 8 (1961).
34. R. C. Miller and A. Savage, *Phys. Rev.* **112**: 755 (1958).
35. R. C. Miller and A. Savage, *Phys. Rev.* **115**: 1176 (1959).
36. R. C. Miller and A. Savage, *J. Appl. Phys.* **31**: 662 (1960).
37. C. F. Pulvari and W. Kuebler, *J. Appl. Phys.* **29**: 1315 (1958).
38. C. F. Pulvari and W. Kuebler, *J. Appl. Phys.* **29**: 1742 (1958).
39. H. H. Wieder, *J. App. Phys.* **31**: 180 (1959).
40. E. Fatuzzo, *Phys. Rev.* **127**: 1999 (1962).
41. E. Fatuzzo, *Helv. Phys. Act.* **33**: 21 (1960).
42. D. S. Campbell, *J. Electronics and Control* **3**: 330 (1957).
43. H. H. Wieder, *J. Appl. Phys.* **28**: 367 (1957).
44. W. J. Merz, *J. Appl. Phys.* **27**: 938 (1956).
45. W. Känzig, *Phys. Rev.* **98**: 549 (1955).
46. A. G. Chynoweth, *Phys. Rev.* **102**: 705 (1956).
47. M. E. Drougard and R. Landauer, *J. Appl. Phys.* **30**: 1663 (1959).
48. V. Janovec, *Czech. J. Phys.* **8**: 3 (1958).
49. H. L. Allsop and D. F. Gibbs, *Phil. Mag.* **4**: 359 (1959).
50. M. Madden, Doctorate Thesis, University of Bristol (1959).
51. J. W. Dunn, "Domains in BaTiO₃" Reports I, II. Contract Mos/7/GEN/1408/PR3 Min. of Aviation, 1958.
52. V. J. Tennery, and F. R. Anderson, *J. Appl. Phys.* **29**: 755 (1958).

53. B. Lewis, *Proc. Phys. Soc.* **73**: 17 (1959).
54. W. Heywang and R. Schöfer, *Z. Angew. Phys.* **8**: 209 (1956).
55. T. F. Heuter and D. P. Neuhaus, *J. Acoust. Soc. Am.* **27**: 292 (1955).
56. W. P. Mason, *Acustica* **4**: 200 (1954).
57. H. L. Allsop, *Phil. Mag.* **2**: 1100 (1958).
58. Z. Pajak and J. Stankowski, *Proc. Phys. Soc.* **71**: 1144 (1958).

DISCUSSION

D. F. Riebling (Sullivan Research Center, Corning Glass Works): You state "ceramics also demonstrate considerable aging effects," and then "these have been attributed to low-energy arrangements of domains forming closure groups which are formed essentially by the clamping effects of neighbors when the material is cooled below the Curie temperature." Why is it that ceramics (I know this is a broad definition) would be more susceptible to this clamping and aging than would other materials?

Answer: By ceramic material I really mean polycrystalline material. On cooling such material through the Curie temperature polarization has to occur with its associated 1 % extension of the unit cells along any one of the cubic directions. For a crystallite to do this as a whole along one particular direction would create excessive mechanical strain energy because of the spatial restriction imposed by differently oriented adjacent crystallites. However, the crystallite can polarize along three orthogonal directions in both a positive and negative sense, and therefore it can absorb the relatively large 1 % extension isotropically and maintain its original shape and very nearly its original volume by the formation of domains equally distributed in all allowable directions. To reduce the free field energy these will take up a head-to-tail arrangement as nearly as possible. One arrives naturally at the conclusion of electrically "closed" domain arrays of the picture-frame variety, although there is little direct evidence of the existence of such domains in the individual crystallites.

Single crystals, on the other hand, unless very large or perhaps strained by impurity gradient arising during growth, usually transform into simpler domain arrays because the possible polarization directions are parallel throughout the whole crystal and also because one external dimension is often sufficiently small to ensure considerably reduced resistance to polarization along that direction. In such a case there can be arrangements composed of only 180° domains, which will accommodate the 1 % strain simply by extending to the crystal surface. Nevertheless, in some large single crystals, domain arrays have been observed which could well be described as closure domains, but they are relatively rare compared with the large volume of domain material which does not provide complete electrical "closure." It is therefore reasonable to expect that such domains could also exist in crystallites.

Since the breakup of these closure domains involves encouraging the 1 % strain which they are designed to overcome, one can say with reasonable certainty that they are electrically more stable, i.e., are less easily aligned by an electric field. They will align only if the strain can be other wise relieved, i.e., by dislocation movement and its associated long relaxation times, which are associated with one aspect of the aging of electrical properties.

D. R. Uhlmann (Massachusetts Institute of Technology): What is not clear is the effect of ceramic modifiers, improvers, and the like on the domain and hence electrical behavior. Surely this must seriously affect any kinetic process, especially a situation where compensation of electrical charge is important?

Answer: You are absolutely right. However, if the crystals are clear enough for optical study, their electrical resistance is usually so high that only small variations of domain behavior are to be expected due to difference in charge compensation ability, dislocation of impurity mobility, etc. If the crystals are very unsuitable for optical study, as so many practical materials are, then these impurities will have a more profound effect. Much is still to be learned about these electrical and chemical defect conditions.

D. F. Riebling: Would you say if anything were to be gained by producing these materials in the glassy phase and if you could predict the behavior of such material?

Answer: It is possible to extrapolate part of the way toward an understanding of what might be expected from a glassy phase by consideration of the behavior of very fine polycrystalline materials. The properties tend toward a limit as the particle size reduces. For example, the electromechanical properties tend toward a truly parabolic stress/strain relation and hysteresis due to domain wall movement decreases, as the latter cannot be maintained in ultrafine particles. However, such materials are still polycrystalline in that the majority of atoms have ordered neighbors, and reasonable diffraction patterns are obtainable therefrom.

In the glassy state significant fractions of the atoms have neighbors of incoherent placing. This usually necessitates the addition of special modifiers to encourage the glassy phase, and so the electrical effectiveness is reduced even assuming optimum density, which must be less than the crystalline equivalent. This also reduces the effectiveness of the cooperate electrical coupling acting from one polar grouping of atoms to another.

Thus the electrical properties such as dielectric and piezoelectric constants will be reduced, but so also will the temperature variation of these properties and the dielectric losses. The optimum balance will depend on the application intended and will lie between the known extremes of a large-grain polycrystalline material and an isotropic dielectric material of value around 150.

P. W. M. Jacobs (University of Western Ontario): It would seem that the current depicted in Fig. 5(a) is the analog of the chemical reaction rate. Thus one might expect the function $i(t)$ to obey the differential form of the Avrami equation. Is this so, and what is the value of n?

Answer: It is true that in Fig. 5(a) the value of the current at any time $i(t)$ is analogous to the chemical reaction rate. Therefore the integral under this curve represents the proportion of material that has changed its phase at any time and is therefore the same as the sigmoid curves described in your presentation (Chapter 3) of the various nucleation theories. The flattening off in the sigmoid curve is due to exhaustion of material either due to the formation of sufficient nuclei of finite volume or to the growth of these nuclei, or to both. Simple mathematical values can be given to both these processes except toward the final stages, when one nucleus grows into another, especially when the latter process is predominant. In the case of random nucleation this feature may be corrected for by a fairly simple theory due to Avrami. Such overlap correction has been applied with success to obtain better correlation between observed and theoretical integrated sigmoid curves for ferroelectrics using the value of n expected for the observed two-dimensional growth and overlapping of cylinders of new phase which penetrate the full thickness of the crystal.

Chapter 12

Crystal Growth Kinetics and Morphology

K. A. Jackson

Bell Telephone Laboratories, Inc.
Murray Hill, New Jersey

The growth kinetics and morphology of crystals are related to the change in entropy during a phase transformation. It is shown that for small entropy changes the crystal growth kinetics are relatively rapid and isotropic. In these circumstances diffusional processes dominate growth rate. Conversely when the entropy change is large growth kinetics are slow and very anisotropic. Simulation of crystal growth processes is exemplified using low-melting, transparent organic compounds, adjusting composition and conditions of crystal growth in an appropriate manner. Technique is discussed and typical systems illustrated.

Recent work ([1]) has shown that there are transparent organic compounds which solidify in the same way that metals do. These organic compounds have been used as analogs to study the solidification behavior of metals ([2-4]). The ability to observe directly the crystal growth processes in these transparent materials, together with their relatively low melting points, have permitted the elucidation of some of the phenomena which occur in metals and alloys.

In fact there is a very large number of organic compounds, so that a model for most crystal growth processes found in other materials can also be found among them. The organic compounds can be used to study a variety of phase transformation processes, with the advantages of transparency and low transformation temperature. These permit the direct observation of the phenomena with high-powered optical equipment.

In this chapter I will review briefly and illustrate some of the work that has been done to date with transparent organic materials, but first I will indicate the theoretical basis for the similarities and differences in the growth of various materials.

INTERFACE STRUCTURE AND CRYSTAL GROWTH
KINETICS

The structure of the interface on an atomic scale determines the kinetics of the crystal growth processes, and indeed of phase transformations in general. The details of the atomic motions which constitute the phase change are dependent on the environment of the atoms involved. In first-order phase transformations these atomic processes take place at an interface separating the two phases. Motion of the interface propagates the phase transformation.

Let us consider an initially flat, atomically smooth interface separating two phases, and ask the question: What is the lowest free energy configuration of such an interface? The simplest model for the structure of the interface is to consider extra atoms added to a plane interface. Using Bragg–Williams statistics the change in total free energy of the interface ΔF_S is given by [5,6]:

$$\Delta F_S/NkT_E = \alpha x(1 - x) + x \ln x + (1 - x) \ln(1 - x) \qquad (1)$$

where N is the total number of surface sites, k is Boltzmann's constant, T_E is the equilibrium temperature between the two phases, x is the fraction of the N surface sites which are occupied, and α is given by

$$\alpha = (L/kT_E)\xi \qquad (2)$$

where L is the heat of transformation and ξ will be discussed below. Equation 1 is shown in Fig. 1 for various values of α. For α less than 2 the surface free energy has a minimum for $x = \frac{1}{2}$, that is, with half the surface sites filled. For α greater than 2 the surface has a minimum free energy with a small fraction of the surface sites filled and a small number of holes. For large α the fraction of extra atoms is $e^{-\alpha}$ and the fraction of holes is $e^{-\alpha}$.

The term α is made up of two factors [Eq. (2)]: ξ, which depends on the structure of the solid and on the orientation of the interface, and L/kT_E, which is the entropy change associated with the transformation. The term ξ is always less than unity, and is closest to unity for the most closely packed planes of the structure. It decreases progressively for less closely packed faces of the crystal.

The entropy change, L/kT_E, is about unity for metal crystals growing from their melts, for some ionic materials growing from their melts, and for a few organic compounds growing from the melt. It is somewhat higher, 2 or 3, for a few near-metals and semiconductors, and for some ionic and organic materials growing from the melt. It is still larger, 4 to 8, for most molecular materials and some ionic materials growing from the melt. The term L/kT_E is about 10 for many materials (including metals) growing from their vapors.

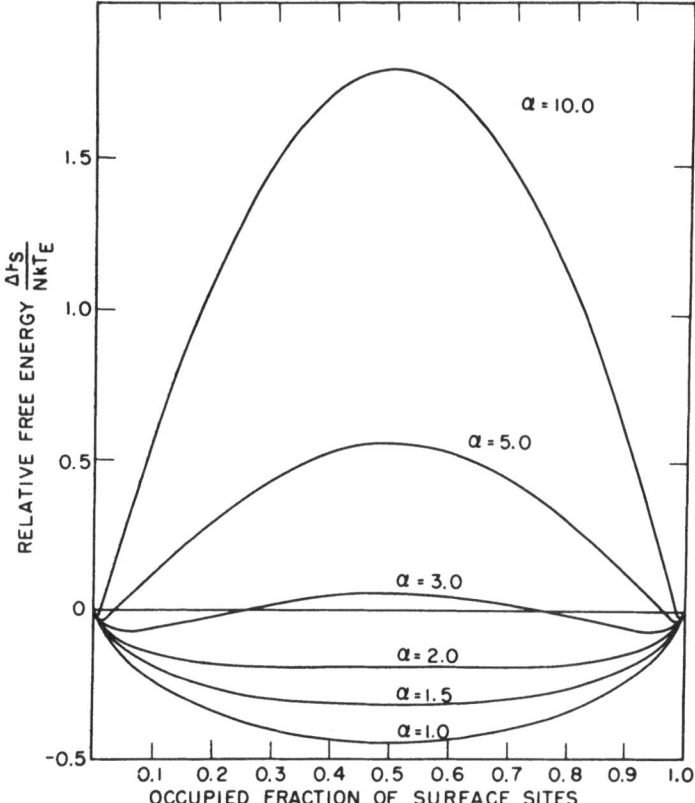

Fig. 1. Free energy of an interface versus occupied fraction of surface sites.

For growth from solution the entropy change varies over a wide range, being similar to that for melt growth if a small amount of solvent is present, and larger for crystals growing from dilute solutions ([7]). For materials with high molecular weights, such as very large organic molecules, including the polymers, the entropy of fusion can be very large.

Crystal faces with low α factors (any face of a low entropy of fusion material such as the metals or the high index faces of materials with a higher entropy of fusion) are rough on an atomic scale. These interfaces will be rougher than indicated by the single-layer model used above [Eq. (1)], and there may be a few or even several atom planes parallel to the interface in which some of the atoms are solid atoms and some are liquid atoms. The growth of the crystal on such faces should be relatively rapid.

For high-α faces, such as the close packed faces of materials having a large entropy change, there may be difficulty in initiating new layers. Surface nucleation or growth on dislocations may be operative. For these faces the growth rate should be much slower than on rough faces.

The growth rate for low-entropy-change growth should be relatively isotropic, since any bounding face of the crystal will be rough. For a given atomic mobility the growth rate should be relatively rapid at small undercoolings.

Materials which have high entropy changes during growth will form a solid bounded by the slowest growing, highest α-factor faces. The growth rate of these materials will be markedly anisotropic, and for a given atomic mobility will be relatively much slower than for a low entropy change. The higher the entropy change the more difficult it will be to form new layers. The anisotropy of growth will increase and the growth rate will decrease for higher entropy change on crystallization.

Some crystals have very anisotropic structures. These materials may be bounded by some high- and some low-α ($\alpha < 2$) faces. In these crystals the growth rate will be very anisotropic.

The growth characteristics of a material should depend on the entropy change, the crystallographic orientation of the interface, and the atomic mobility, but not on whether the material is metallic or organic, molecular or ionic. In the next section photographs of materials in several of these categories are presented.

CRYSTAL GROWTH MORPHOLOGIES

The morphology of a growing crystal depends in part on the kinetics of the growth process, which were discussed in the preceding section. The morphology also depends on transport processes, such as diffusion and convective mixing of matter and heat. Convective mixing usually occurs on a fairly large scale, and for our present purposes we will consider only diffusion as a mixing process.

The photographs presented here were taken on a temperature gradient microscope stage ([8]) shown schematically in Fig. 2. The sample contained in a thin glass cell can be moved at a controlled rate across the gap between a hot and a cold plate. The temperatures of the two plates can be adjusted to give the desired temperature gradient, with the transformation occurring between the two temperatures.

The growth morphology for several organic materials having various entropies of fusion are shown in Figs. 3–7. The effects of diffusion on each are shown by the addition of various amounts of soluble impurities. These growth forms are typical for the various values of entropy change.

The lowest entropy of fusion range is typified by CBr_4 ($L/kT_E = 0.8$) as shown in Figs. 3a–3c. The growth is isotropic, or almost so (Fig. 3a).

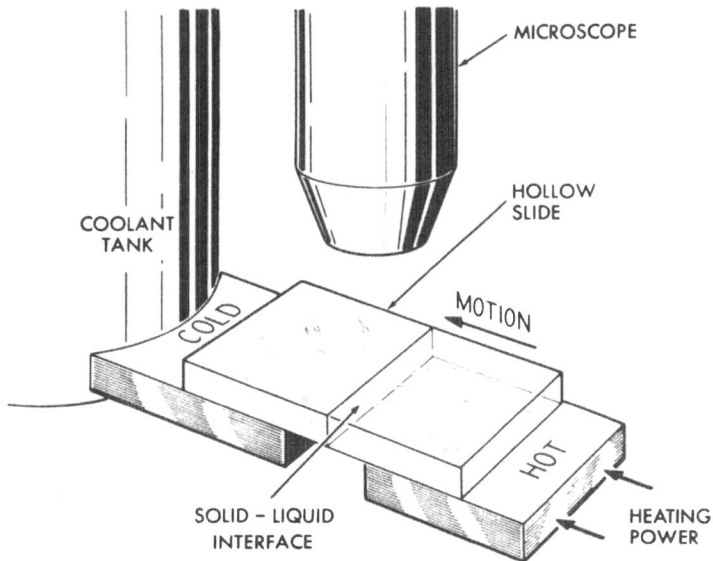

Fig. 2. Schematic drawing of temperature gradient microscope stage.

The interface is parallel to an isotherm. The addition of a small amount of impurity causes instability of the planar isothermal interface (Fig. 3b). The impurity collects in the cell boundaries. This form of growth has been studied extensively in metals ([9-11]). Several percent impurity results in dendritic

Fig. 3a. Planar interface growth in carbon tetrabromide (L/kT_E = 0.8) growing from the melt between glass cover slide approximately 25 μ apart. Photograph taken using a temperature gradient microscope stage (150 $\times$).

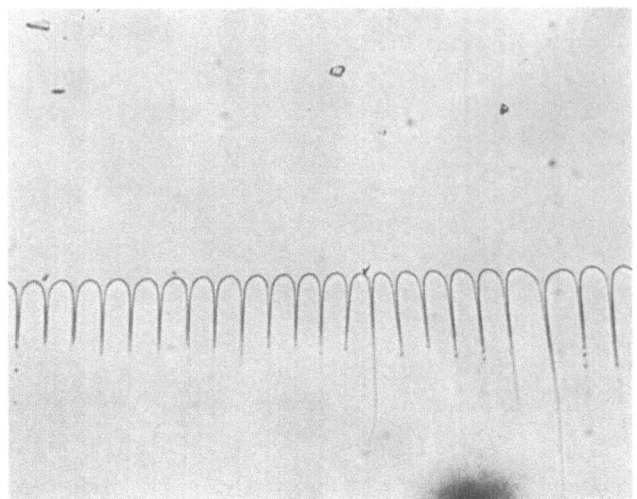

Fig. 3*b*. Same conditions as Fig. 3*a*. Cellular growth with a small amount of impurity (150 ×).

growth (Fig. 3*c*). Dendritic growth can also occur into a thermally super-cooled melt. In each of these photographs the shape of the interface is determined almost entirely by diffusion. In Fig. 3*a* heat flow dominates. In Fig. 3*b* solute diffusion is important on a microscopic scale. In Fig. 3*c* the dendritic growth is a result of solute diffusion. There is no evidence of low index faces forming the interface in any of these photographs.

Figures 4*a* and 4*b* show the growth of *t*-butyl alcohol ($L/kT_E = 2.6$). This material grows with a faceted interface when it is "pure" (Fig. 4*a*). The addition of several percent impurity results in a pseudodendritic growth form (Fig. 4*b*), where the crystal, in spite of having a somewhat dendritic shape, is nevertheless bounded by segments of low index faces. The diffusion processes are important here, but the anisotropic growth rate is evident.

Figures 5*a* and 5*b* show the growth of benzil ($L/kT_E = 6$). Figure 5*a* shows "pure" benzil and Fig. 5*b* shows the benzil–azobenzene eutectic. Both phases are present in Fig. 5*b* and both are bounded by low index faces, despite the diffusion processes involved. The growth here is quite anisotropic.

For higher entropy of fusion materials the growth becomes progressively more "spikey." These materials also require high undercooling for growth. Quite imperfect crystals form at these large undercoolings. The crystals change orientation during growth, and even renucleate on themselves ([12]), as shown in salol (Fig. 6), which is growing at a large undercooling ($\sim 25°$). Very high entropy-change growth is often spherulitic, as shown in Fig. 7, a photograph of tristearin ($L/kT_E = 62$). This form of growth is common among the polymers.

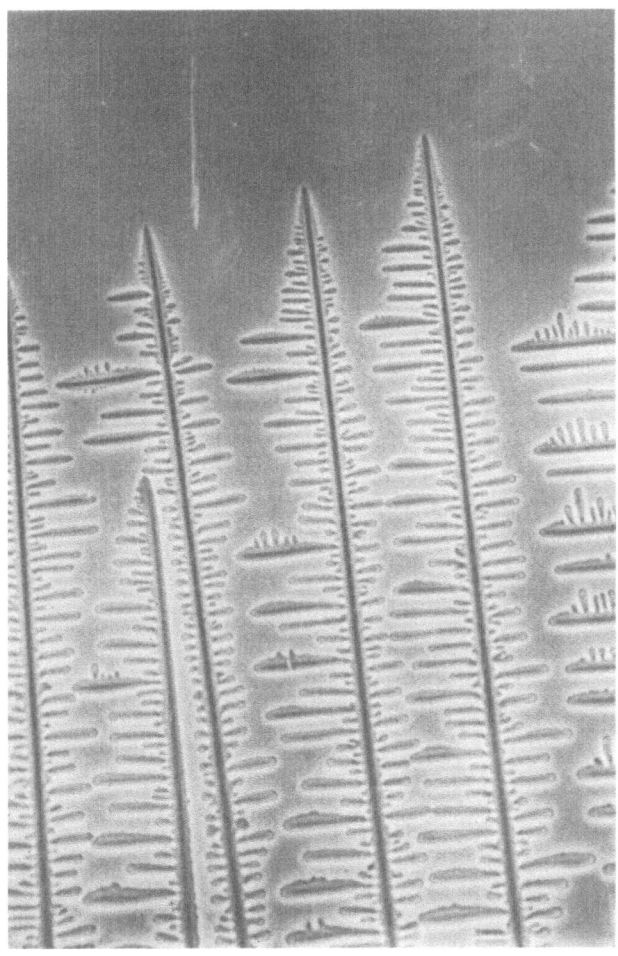

Fig. 3c. Same conditions as Fig. 3a. Dendritic growth with several percent impurity (150 ×).

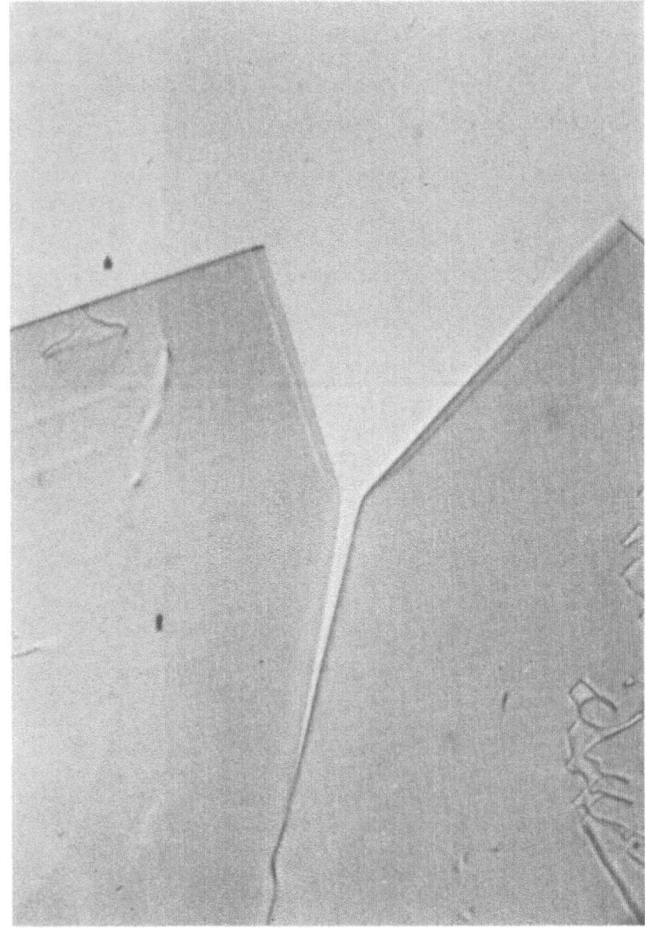

Fig. 4a. Faceted growth in t-butyl alcohol ($L/kT_E = 2.6$) growing from the melt as in Fig. 3a ($150 \times$).

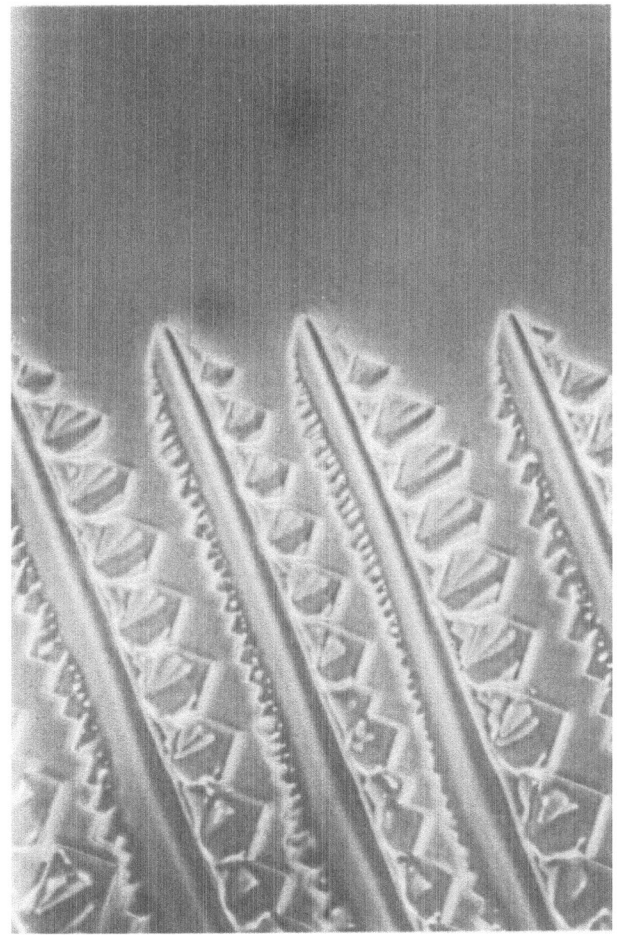

Fig. 4b. Same conditions as Fig. 4a. Pseudo-dendritic growth (150×).

Fig. 5a. Benzil ($L/kT_E = 6$) growing from the melt as in Fig. 3a
($150\times$).

Fig. 5b. Benzil–azobenzene eutectic growing from the melt as in Fig. 3a (150×).

Fig. 6. Salol ($L/kT_E = 7$) growing from the melt as in Fig. 3a. Notice the "autonucleation" of crystals of a different orientation which occurs only at large undercoolings in salol ($150 \times$).

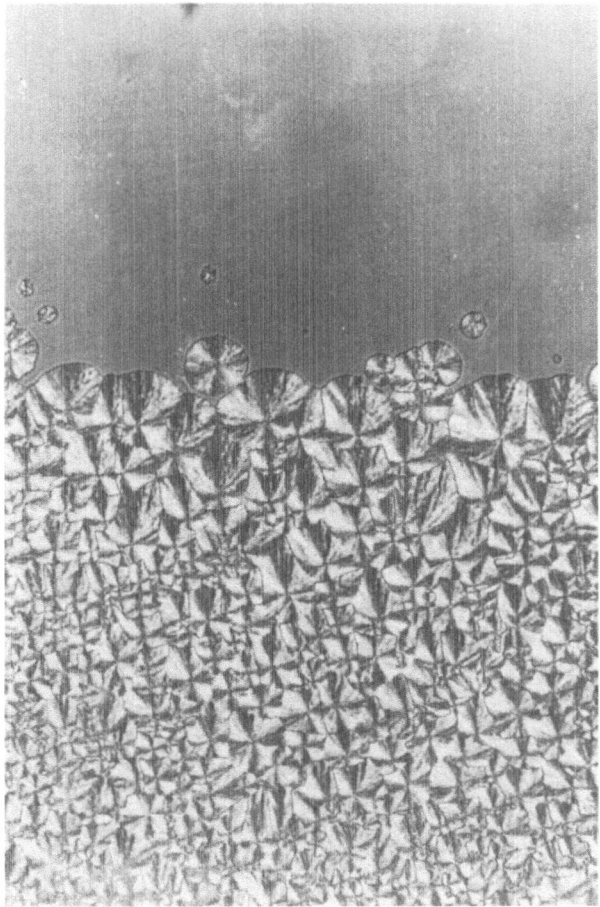

Fig. 7. Tristearin ($L/kT_E = 63$) growing from the melt as in Fig. 3a (150×).

The observations reported here are quite general, and similar observations have been made on a wide variety of materials. In general the expectations outlined in the preceding section are verified. Isotropic growth occurs for low entropy-change materials, and the anisotropy of growth increases for higher entropy change. Diffusion can dominate the morphology for the low entropy-change cases, but has little effect on morphology for high entropy change. Measurements of growth rate also follow the expected pattern ([13]): for a given mobility the growth rate on the closest packed faces decreases for high entropy-change transformations.

EUTECTIC GROWTH

The above classification scheme can also be applied to two-phase growth ([3]). When both phases are low entropy-change the growth kinetics are unimportant and diffusion dominates. This leads to rod or lamellar eutectic growth forms, as shown in Fig. 8. Figure 9 shows the calculated shape of such an interface compared to the observed shape for various growth conditions ([4]). These growth forms are common in metallic and some ionic

Fig. 8. The interface on a growing carbon tetrabromide–hexachloroethane eutectic (1000×). (Phase contrast.)

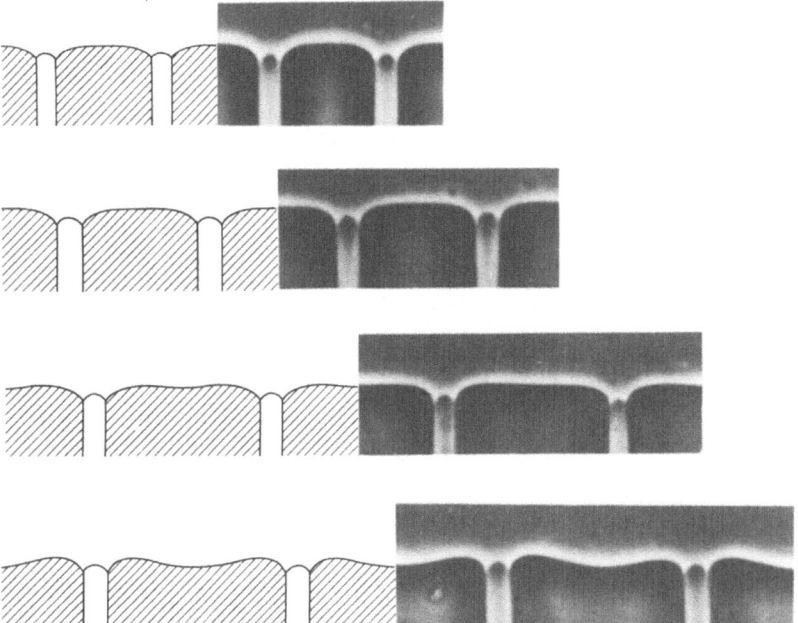

Fig. 9. Comparison of calculated and observed eutectic interface shapes. Carbon tetrabromide–hexachloroethane eutectic growing as in Fig. 3a (1000×). (Phase contrast.)

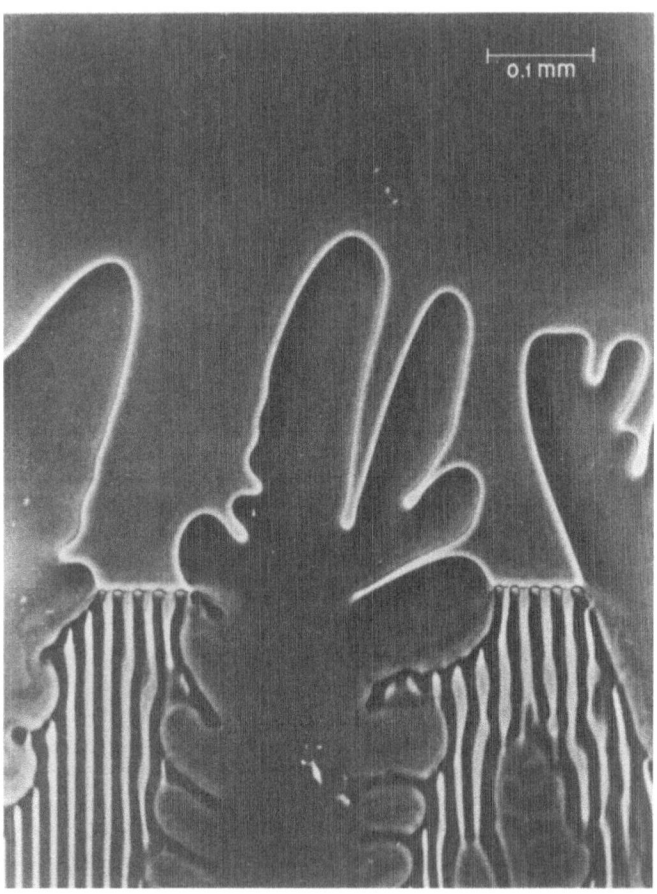

Fig. 10. Dendrites growing ahead of an eutectic interface in a carbon tetrabromide-rich alloy of carbon tetrabromide–hexachloroethane. (Phase contrast, 150×.)

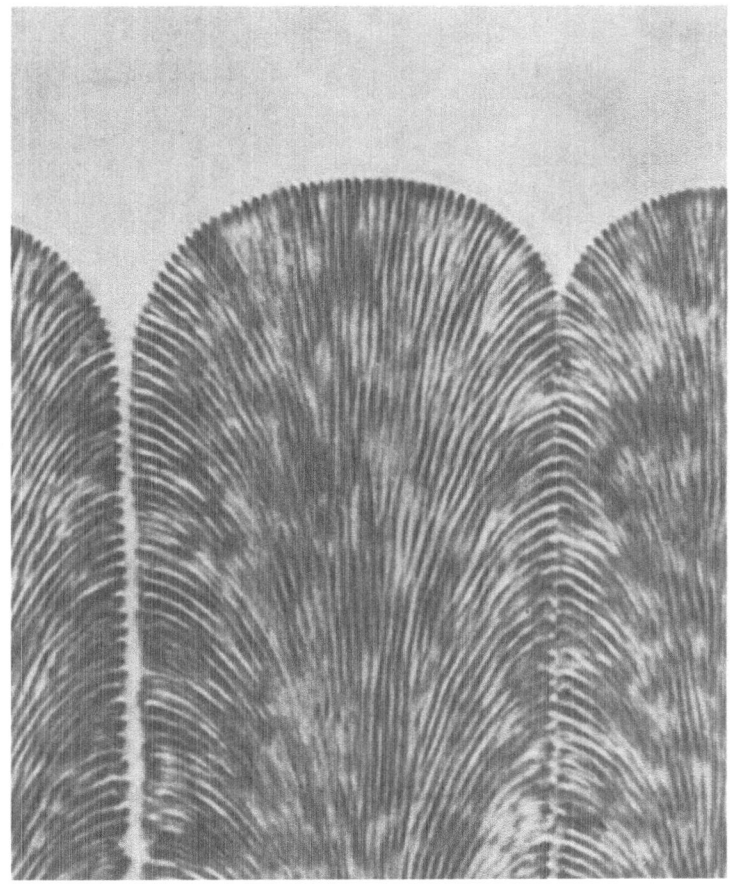

Fig. 11. All structures produced in camphor–succinonitrile eutectic containing a third component impurity. (Phase contrast, 750×.)

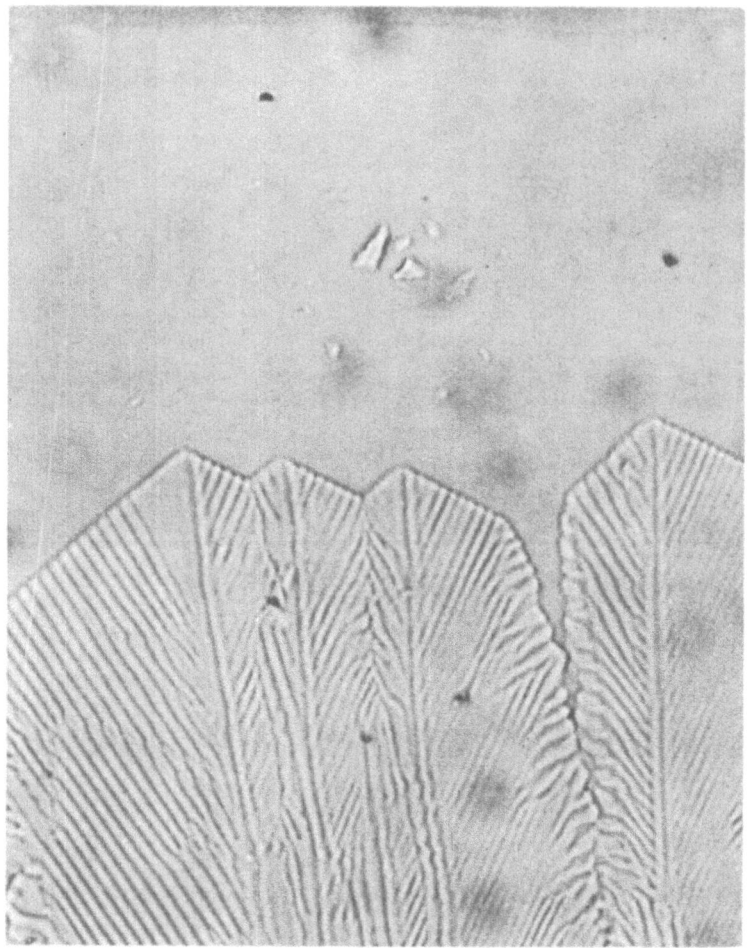

Fig. 12. Shape of the solid–liquid interface during the growth of cyclohexane–camphor eutectic. (Phase contrast, 1000 ×.)

eutectic system (³). A mixture growing off-eutectic composition grows with dendrites protruding into the liquid ahead of the eutectic interface (Fig. 10). A third component impurity added to a eutectic causes cellular growth (Fig. 11) similar to that shown in Fig. 3b.

Figure 12 shows a eutectic in which one phase has a high entropy change and the other a low one. The interface shape is determined by the high entropy-change phase, and the low entropy-change phase forms in regular alternate layers because of the diffusion requirements. This mode of growth has been observed in eutectics between metals and semiconductors. Other growth forms are also observed in eutectic systems with one high and one low entropy-change phase ([3]). Figure 5b shows a eutectic between two high entropy-change materials. The growth is uncoupled: the two phases grow independently. This type of growth is characteristic of organic eutectics.

SUMMARY

The growth kinetics and morphology of crystals depends on the change in entropy that occurs with the transformation. For small entropy change the growth kinetics are relatively rapid and isotropic, so that growth is likely to be dominated by diffusion. For large entropy change the growth kinetics are relatively slow and very anisotropic. Diffusion may be important but the crystals will be bounded by slow growing faces.

Almost any crystal growth process can be simulated by an appropriate low-melting transparent organic material, with the composition and growth conditions adjusted appropriately.

REFERENCES

1. K. A. Jackson and J. D. Hunt, *Acta. Met.* **13**: 1212 (1965).
2. K. A. Jackson, J. D. Hunt, D. R. Uhlmann, and T. P. Seward, III, *Trans. Met. Soc. AIME* **236**: 149 (1966).
3. J. D. Hunt and K. A. Jackson, *Trans Met. Soc. AIME* **236**: 843 (1966).
4. K. A. Jackson and J. D. Hunt, *Trans. Met. Soc. AIME* **236**: 1129 (1966).
5. K. A. Jackson, in: *Liquid Metals and Solidification*, American Society for Metals, Metals Park, Ohio, 1958.
6. K. A. Jackson, in: *Growth and Perfection of Crystals* (R. H. Doremus *et al.*, eds.), John Wiley, New York, 1958.
7. V. V. Voronkov and A. A. Chernov, in: Crystal Growth, Supplement to *J. Phys. Chem. Solids* (H. S. Peiser, ed.) p. 593 (1967).
8. J. D. Hunt, K. A. Jackson, and H. Brown, *Rev. Sci. Inst.* **37**: 805 (1966).
9. J. W. Rutter and B. Chalmers, *Can. J. Phys.* **31**: 15 (1953).
10. D. Walton, W. A. Tiller, J. W. Rutter, and W. C. Winegard, *Trans. AIME* **203**: 1023 (1955).
11. E. L. Holmes, J. W. Rutter, and W. C. Winegard, *Can J. Phys.* **35**: 1223 (1957).
12. K. A. Jackson, in: Crystal Growth, Supplement to *J. Phys. Chem. Solids* (H. S. Peiser ed.), p. 17 (1967).
13. K. A. Jackson, D. R. Uhlmann, and J. D. Hunt, *J. Cryst. Growth* **1**: 1 (1967).

DISCUSSION

Burmann (Westinghouse): The problem we have had among the observations we have made with nuclear ceramic fuels during radiation is the development toward the center of a cylindrical rod of columnar grains which we currently believe to be due to the migration of pores up a thermal gradient. Material vaporizes from the hot side of the pore and deposits on the cold side and behind the pore as it moves, and a columnar grain grows. It is important for us to understand this process and study it. Attempts to study it by simulating the conditions have been very disappointing. We have run tungsten rods through the center and tried to pour very large currents through them. We cannot maintain the situation for long enough times. We have been seriously considering the possibility of simulating it using more amenable organic materials that can be investigated at lower temperature and perhaps observing it by means of photography. Now it becomes obvious that it is not enough just to simulate the UO_2 by selecting an amenable organic that has a reasonable vapor pressure at a reasonable temperature. It will also be necessary to simulate such things as the change in entropy on crystallization.

Answer: I agree that the choice of a suitable organic for your modeling purposes should take into account the entropy change on crystallization, as well as other properties, such as the vapor pressure, that are important in the process you want to study. We have often observed the migration of liquid inclusions, which contain soluble impurities, through a crystal in the direction of increasing temperature. This process has been studied for brine inclusions in ice by J. D. Harrison [14].

ADDITIONAL REFERENCES

14. J. D. Harrison *J. Appl. Phys.* **36**: 326 (1965).

Chapter 13

Crystal Growth in Ceramic Powders

H. J. Oel

Max Planck Institut für Silikatforschung
Würzburg, Germany

The importance of crystal size as opposed to particle size in powders and in polycrystalline solids is discussed. Crystal growth may be followed throughout a complete ceramic fabrication process. Optical methods for determining crystal size distributions are reviewed and an X-ray method is described in detail. The importance of determining distributions instead of average values for studying the process of crystal growth is emphasized. Experimental results for crystal growth in powders (tempering, calcination) and in powder compacts and polycrystalline bodies (sintering) are compared with the rate law as derived from thermodynamic and kinetic considerations. In all cases which were investigated agreement was good. It may be concluded that crystal growth in crystalline systems is governed by the theoretical rate law [Eq. (24)] with the kinetic factor u_k depending on the material, the area of contact, and on temperature.

INTRODUCTION

Particle size and crystal size are two important parameters in characterizing a powder. Both quantities seem to be easily definable at first sight. In the case of fine powder particles, however, this is not the case. When a dry powder is mixed with water the number of individually identifiable particles may alter considerably because water is an effective dispersing medium for electrically charged particles. In air the moisture content of a powder leads to agglomeration.

Crystal size is usually less sensitive and is independent of environment. Heat treatment does, of course, cause an increase in crystal size, and crystal growth may be followed all the way through a complete ceramic fabrication process. Thus while a particular oxide or metal powder produced by a chemical process may have a crystallite size of 10–100 Å, subsequent heat treatment (calcination) by which the powder is prepared for slip-casting, dry pressing, etc., may increase the crystal size to about 0.01–0.1 μ, and from the final sintering process a crystal size of about 100 μ may result.

Products for which crystal size is important include catalysts, pigments, and inorganic fillers for rubber and plastic materials. Crystals may also grow from a glass matrix. Well known examples are the undesired and often disastrous devitrification on the one hand and the lucrative fabrication process of glass-ceramics on the other hand.

EXPERIMENTAL

Methods for Crystal Size Determination

Particle sizes above 1 μ may be determined by sedimentation and optical microscopy and smaller particles by centrifugation and electron microscopy.

Crystal sizes in polycrystalline materials and glass may also be measured under the microscope since there the crystals are distinguishable. For powders, however, the method of direct observation is limited to monocrystalline and a few other exceptional coarse powders since in most cases, even given the broad imagination of an experienced electron microscopist, one is incapable of identifying the crystals. This gap is to some extent filled by an X-ray line-broadening method.

All methods employed should obviously aim at determining size distribution functions because these hold much more information than simple averages.

Microscopy

Etched polished sections (Figs. 1–4) are directly observed under the optical microscope while replicas of such sections or of fractured surfaces (Figs. 5 and 6) serve for the electron microscope.

If the crystals have no systematic elongation or flattening and no preferential orientation they may be considered to be spheres and their cross sections in a polished section to be circles. The radii of these are measured with a Zeiss Particle-Size Analyzer. This is a semiautomatic device which allows for classification of 48 different radii. The sizes thus obtained are not the real crystallite size, since a crystallite (sphere) with radius r in the polished section may be cut so as to represent a circle with a smaller radius ρ. For obvious reasons this problem is referred to in the German literature as the "Tomatensalat Problem" (Tomato Salad Problem). A simple mathematical solution which is especially adapted for a digital computer was developed in a recent publication ([1]). Figure 7 shows the evaluation of the true crystallite-size distribution $P(r)$ from the measured distribution $q(\rho)$.

It should be noted that such a transformation $q(\rho) \rightarrow P(r)$ is only sensible when a sufficiently large number of crystals have been measured (about 5000 per sample in the present work). For details see ([1]).

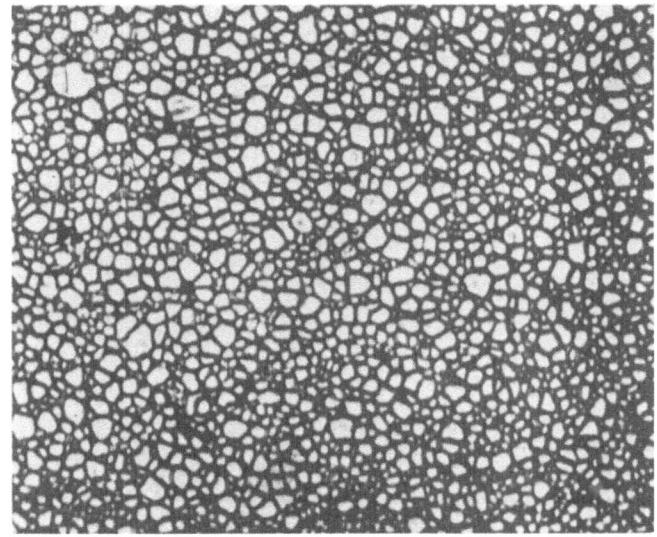

Fig. 1. ThO$_2$ compacts sintered at 1900°C for 20 min. 600×.

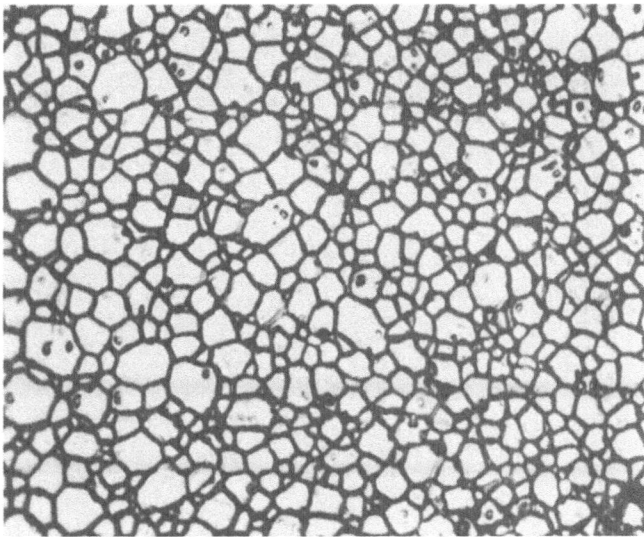

Fig. 2. ThO$_2$ compacts sintered at 1900°C for 1 hr. 600×.

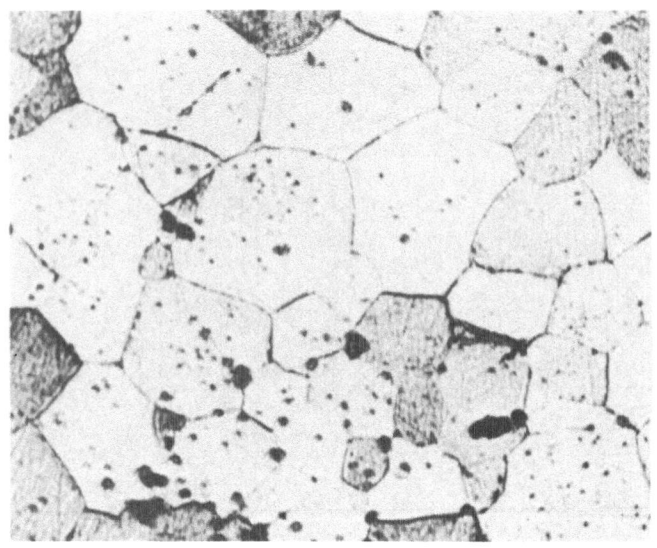

Fig. 3. UO$_2$ compacts sintered at 1600°C for 7 hr. 600×.

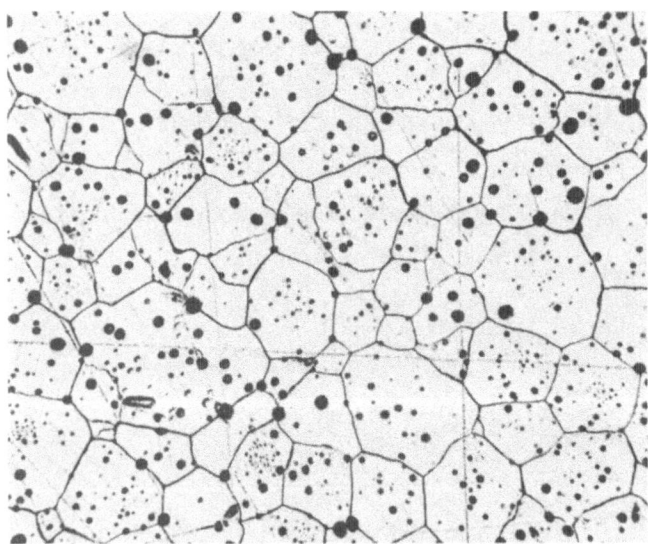

Fig. 4. CeO$_2$ compacts sintered at 1800°C for 1 hr. 200×.

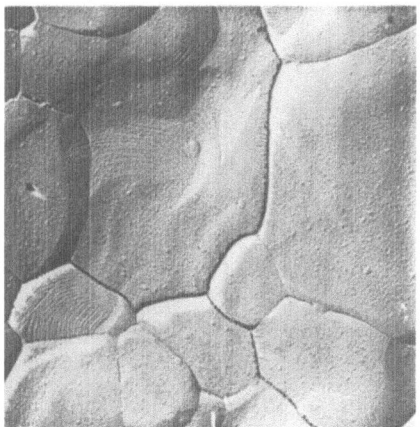

Fig. 5. UO$_2$ compacts sintered at 1550°C for
5 hr. 12,000×.

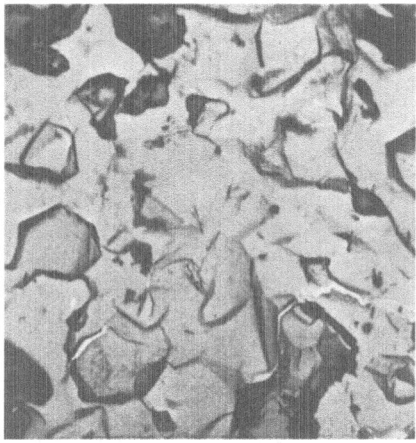

Fig. 6. UO$_2$ compacts sintered at 1650°C for
10 min. 10,000×.

Crystal size determinations from fractured surfaces are problematic since it is not clear how the fracture propagates with regard to the grains or crystals. Sometimes evaluation is impossible (Fig. 6).

Photomicrographs of monocrystalline powders taken with the electron microscope are directly evaluated with the Zeiss Analyzer (Figs. 8, 9). On ordinary polycrystalline powders crystal size measurements are sometimes difficult (Fig. 10) and mostly impossible (Figs. 11 and 12).

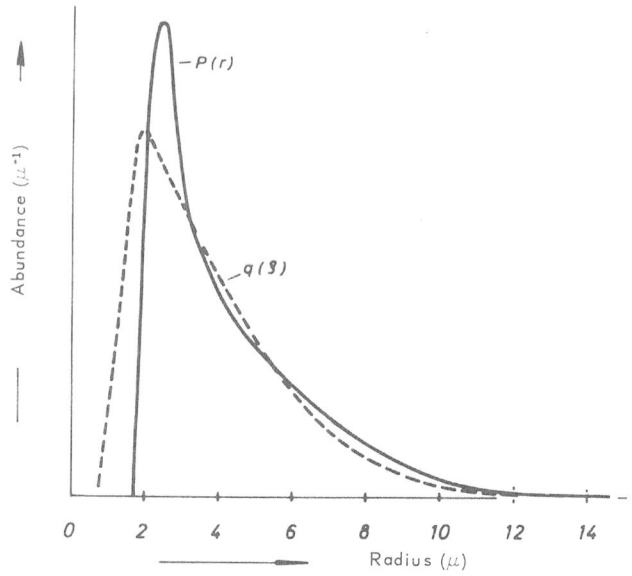

Fig. 7. True crystal size distribution $P(r)$ and measured distribution $q(\rho)$.

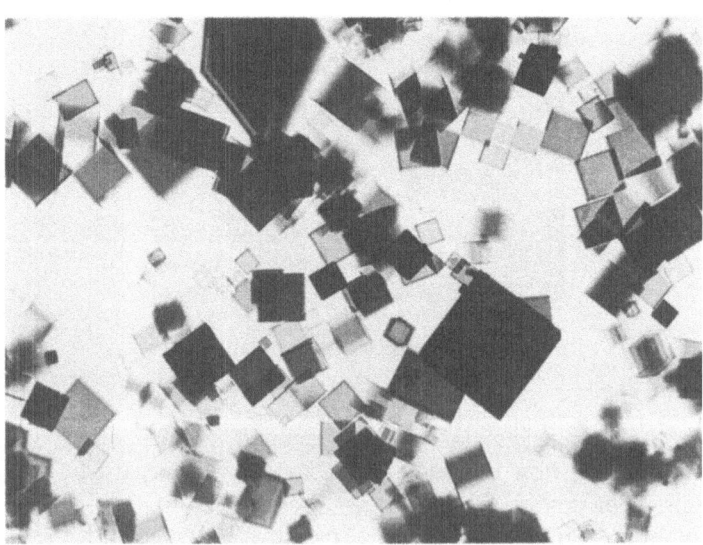

Fig. 8. MgO powder (smoke). $73,600 \times$, reduced 35 percent for reproduction.

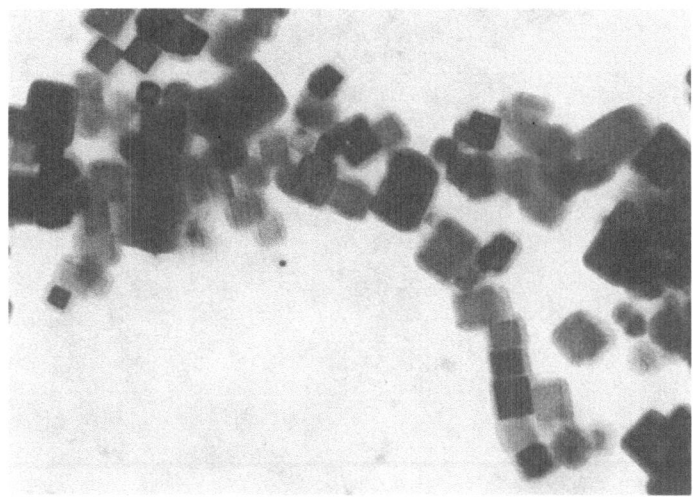

Fig. 9. ZrN powder (Zr + NH₃) 73,600 × , reduced 35 percent for reproduction.

Fig. 10. MgO powder 30,000 × , reduced 35 percent for reproduction.

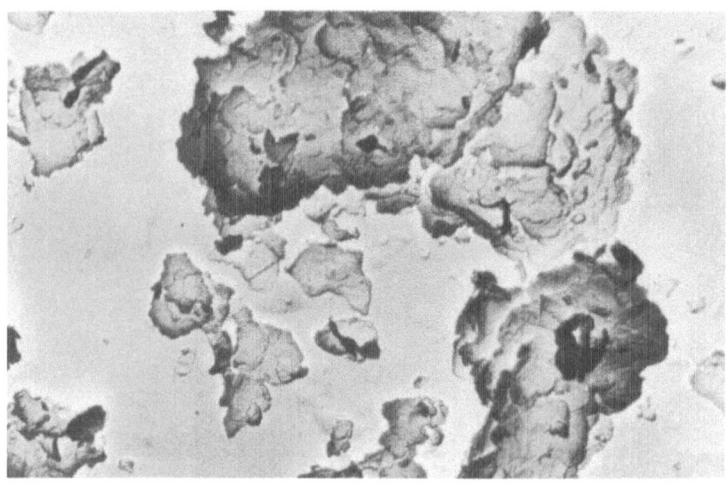

Fig. 11. MgO powder 80,000 ×, reduced 35 percent for reproduction.

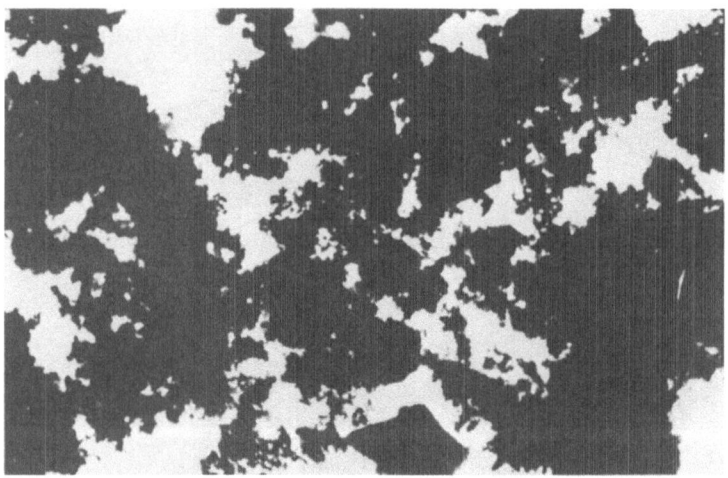

Fig. 12. UO₂ powder 15,000 ×, reduced 35 percent for reproduction.

X-Ray Line-Broadening

Evaluation of the Pure Broadening Due to Crystal Smallness. The profile of X-ray diffraction lines for crystalline powders or polycrystalline solids is given by

$$h(x) = \int_{-\infty}^{+\infty} f(u)g(x - u) \, du$$

where $h(x)$ is the measured profile, $x \equiv 2(\theta - \theta_0)$, θ is the incidence angle, and θ_0 the Bragg angle of the particular line observed. The function $g(x)$ is called instrumental broadening and depends on the experimental setup. It is measured as the line profile of the same substance which has been heat-treated so as to grow crystals large enough ($> 1 \, \mu$) not to show broadening due to crystal smallness. The term $f(x)$ is then the pure broadening due to the finite crystal size, i.e., their smallness. By the general method of Stokes the desired function $f(x)$ is obtained from the measured functions $h(x)$ and $g(x)$ by Fourier transform. The method is described in detail in textbooks on X-ray diffraction ([2]).

Diffraction by Small Crystals. As Fig. 13 shows, the amplitude of a diffracted ray will be maximum when the path difference Δl between rays reflected from succession atomic lattice planes with interplanar distance d is equal to a multiple of the wavelength λ:

$$\Delta l = 2d \sin \theta_0 = n\lambda \tag{1}$$

When the glancing angle differs from θ_0 by an amount $x/2$ the path difference

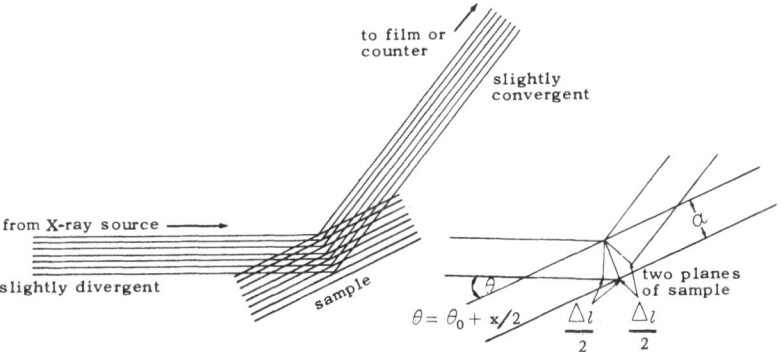

Fig. 13. Schematic illustration of amplitude of a diffracted ray as a function of path difference.

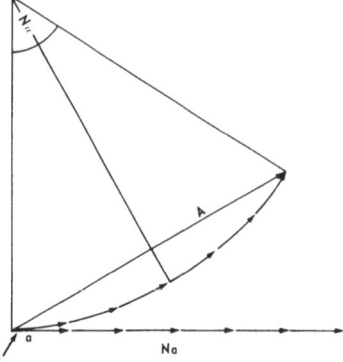

Fig. 14. Schematic illustration of amplitude summation.

becomes

$$\Delta l = 2d \sin (\theta_0 + \tfrac{1}{2}x)$$
$$= 2d(\sin \theta_0 \cos \tfrac{1}{2}x + \cos \theta_0 \sin \tfrac{1}{2}x). \qquad (2)$$

With Eq. (1) and x small

$$\Delta l = n\lambda + xd \cos \theta_0 \qquad (3)$$

This corresponds to a phase difference for the reflected waves of

$$\alpha = (2\pi/\lambda)\Delta l = 2\pi xd \cos \theta_0/\lambda \qquad (4)$$

from adjacent planes.

If a is the amplitude reflected from one plane (see Fig. 14) the total amplitude A reflected from N planes is the sum of N equal vectors a which differ successively in phase by α.* From geometrical considerations which are easily verified (Fig. 14) we have $A = 2p \sin (N\alpha/2)$, and since $pN\alpha = Na$,

$$A = Na\frac{\sin(N\alpha/2)}{N\alpha/2} \qquad (5)$$

---

*This simplified derivation is based on Bragg's reflection analogy. Its main insufficiencies are: (1) While the number of planes N corresponds to the dimensions of the crystal perpendicular to these planes, no account is taken of the lateral extent of the planes, that is, the dimensions of the crystal parallel to the planes. (2) Absolute values for the reflected intensity can not be calculated (a is unknown). For the present calculation this is, however, of minor importance since a cancels out. The same final formula could be derived by a more exact but less convenient derivation ([3]).

Introducing the value for α from Eq. (4) one obtains

$$A = aN\frac{\sin[(\pi/\lambda)Ndx \cos \theta_0]}{(\pi/\lambda)Ndx \cos \theta_0} \qquad (6)$$

and for the intensity

$$i_N(x) = a^2N^2\frac{\sin^2[(\pi/\lambda)Ndx \cos \theta_0]}{[(\pi/\lambda)Ndx \cos \theta_0]^2} \qquad (7)$$

For the maximum intensity

$$i_N(0) = a^2N^2 \qquad (8)$$

since $\lim \sin^2 \epsilon/\epsilon^2 = 1$ for $\epsilon \rightarrow 0$. At half maximum intensity one can then write

$$\frac{i_N(x_{1/2})}{i_N(0)} = \frac{\sin^2[(\pi/\lambda)Ndx_{1/2} \cos \theta_0]}{[(\pi/\lambda)Ndx_{1/2} \cos \theta_0]^2} = \frac{1}{2} \qquad (9)$$

Since from $\sin^2 \delta/\delta^2 = \frac{1}{2}$ follows $\delta = \sqrt{2} \sin \delta$ and from that $\delta = 1.4$

$$(\pi/\lambda)Ndx_{1,2} \cos \theta_0 = 1.4 \qquad (10)$$

Introducing the half maximum breadth of the reflex $\beta_{1/2} = 2x_{1/2}$ and the crystal diameter $D = Nd$ one obtains

$$N_{1/2} = \frac{0.89\lambda}{d\beta_{1/2} \cos \theta_0} \quad \text{or} \quad D = \frac{0.89\lambda}{\beta_{1/2} \cos \theta_0} \qquad (11)$$

the well-known Scherrer equation ([2]).

Size Distributions. If $P(N_i)$ is the total area of platelets consisting of N_i planes and the material investigated contains crystal sizes with $N_1, N_2, \ldots,$ N_m planes, the total intensity will be the sum of all intensities i_N derived according to Eq. (7):

$$J(x) = \sum_{i=1}^{m} P(N_i)a^2N_i^2\frac{\sin^2[(\pi/\lambda)N_idx \cos \theta_0]}{[(\pi/\lambda)N_idx \cos \theta_0]^2} \qquad (12)$$

and the maximum intensity for $x = 0$

$$J(0) = \sum_{i=1}^{m} P(N_i)a^2N_i^2$$

With

$$\bar{N}^2 \equiv \sum_{i=1}^{m} P(N_i)N_i^2$$

one obtains

$$f(x) = \frac{J(x)}{J(0)} = \sum_{i=1}^{m} \frac{P(N_i)}{\bar{N}^2} \frac{\sin^2[(\pi/\lambda)N_i dx \cos \theta_0]}{[\pi/\lambda)dx \cos \theta_0]^2} \tag{13}$$

Taking values of $f(x)$ for discrete values of x and writing C for $(\pi/\lambda)d \cos \theta_0$ the following system of equations is obtained beginning with $x_0 = 0$ and proceeding toward larger values of x (Note that $f(0) = 1$):

$$f(0) = P(N_1) \quad \frac{N_1^2}{\bar{N}^2} \quad + P(N_2) \quad \frac{N_2^2}{\bar{N}^2} \quad + \cdots + P(N_m) \quad \frac{N_m^2}{\bar{N}^2}$$

$$f(x_1) = P(N_1)\frac{\sin^2(N_1 Cx_1)}{\bar{N}^2 C^2 x_1^2} + P(N_2)\frac{\sin^2(N_2 Cx_1)}{\bar{N}^2 C^2 x_1^2} + \cdots + P(N_m)\frac{\sin^2(N_m Cx_1)}{\bar{N}^2 C^2 x_1^2}$$

$$f(x_2) = P(N_1)\frac{\sin^2(N_1 Cx_2)}{\bar{N}^2 C^2 x_2^2} + P(N_2)\frac{\sin^2(N_2 Cx_2)}{\bar{N}^2 C^2 x_2^2} + \cdots + P(N_m)\frac{\sin^2(N_m Cx_2)}{\bar{N}^2 C^2 x_2^2}$$

$$\cdot \quad = \quad \cdots \quad + \quad \cdots \quad + \quad \cdots$$

$$f(x_n) = P(N_1)\frac{\sin^2(N_1 Cx_n)}{\bar{N}^2 C^2 x_n^2} + P(N_2)\frac{\sin^2(N_2 Cx_n)}{\bar{N}^2 C^2 x_n^2} + \cdots + P(N_m)\frac{\sin^2(N_m Cx_n)}{\bar{N}^2 C^2 x_n^2}$$

$$\tag{14}$$

The N_i in the above system are arbitrary at first, but only for an appropriate choice of N_i does a solution exist. A good choice is easily obtained by taking $N_{1/2}$ from Eq. (11) as a medium value and then selecting about 10 equally spaced classes with N_m twice $N_{1/2}$ to start with. Then the $P(N_i)$ are evaluated. They show whether the choice of the N was good or not. If the classes near N_m are empty (e.g., $P(N_m)$, $P(N_{m-1})$, $\ldots$, $= 0$) N_m was too large. If they are too full, N_m was too small. If necessary the calculation must then be repeated with a new set of N_i. Obviously there must not be more N_i than x_i, otherwise the $P(N_i)$ cannot be determined unambiguously. It should be realized that the above system can never be quite exact since the values $f(x)$ are obtained from measurements which are subject to error. In solving the system, therefore, this has to be considered ([3]). Once the $P(N_i)$ have been obtained these may be used to calculate crystallite size distributions. Crystals without preferential directions may be treated as cubes. Their diameter being $2r = Nd$, the size distribution function $W(r_i)$ is given by

$$W(r_i) = P(N_i)/r_i^2 \tag{15}$$

To comprehend the validity of this formula one has only to remember that $P(N_i)$ may be visualized as the total area covered by platelets with the thickness $2r_i$.

Though the whole method is quite straightforward, for practical application the numerical calculations involved are rather tedious. It was therefore programmed for the computer Zuse Z 22 ([3]).

Limits of the X-Ray Method. The method just described is limited with regard to larger crystals because for them the broadening effect is too small to be clearly separated from instrumental broadening. For too small crystals the method is also limited because the broadening caused by them disappears in the background. These and other limitations are discussed elsewhere ([3,4]).

For monocrystalline powders (Figs. 8 and 9) the X-ray method may be checked against direct measurement taken from photomicrographs obtained with the electron microscope. With standard equipment, that is, a Geiger counter goniometer as commercially available, the agreement is good for distributions with maxima between 20 and 200 Å ([4]) (Fig. 15).* Between 200 and 500 Å the X-ray method does not give the complete distribution function but can still be applied (Figs. 16 and 17).* With special equipment such as a monochromator and a high-resolution camera these limits may be extended somewhat. For larger crystals no comparison is possible, since monocrystalline powders of sizes above 500 Å are not available. It should be borne in mind that an agreement is not to be expected. With the electron microscope it is the size of crystals which is measured. From X-ray line-broadening, however, one obtains the size of regions that cause coherent X-ray diffraction, i.e., dislocations, impurities, and other defects also contribute to line broadening. For crystal sizes below 200 Å the size effect is predominant. Above 200 Å this is no longer so. (The limit of 200 Å given here will of course vary from

---

*Note that the class sizes and the scale are different for the two distribution functions in one figure. Therefore they seem different at first sight although they are very similar.

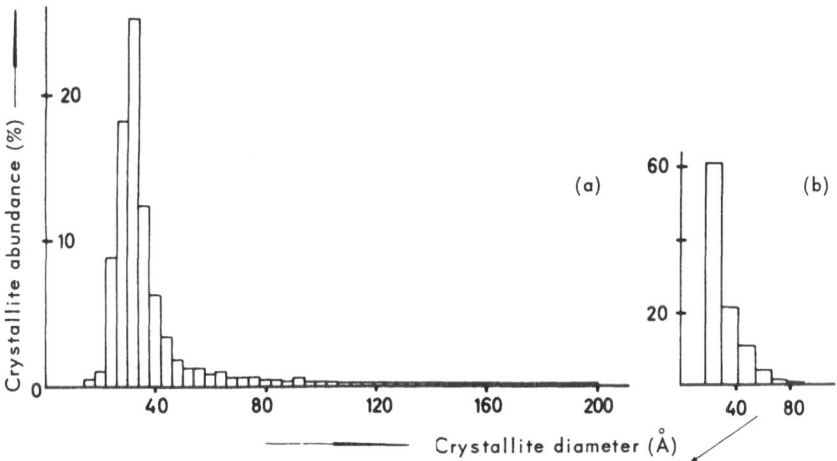

Fig. 15. Crystal size distribution in colloidal silver (*a*) from electron micrograph; class breadths, 4 Å; (*b*) from X-ray line-broadening of the (111) reflection; class breadths, 12 Å.

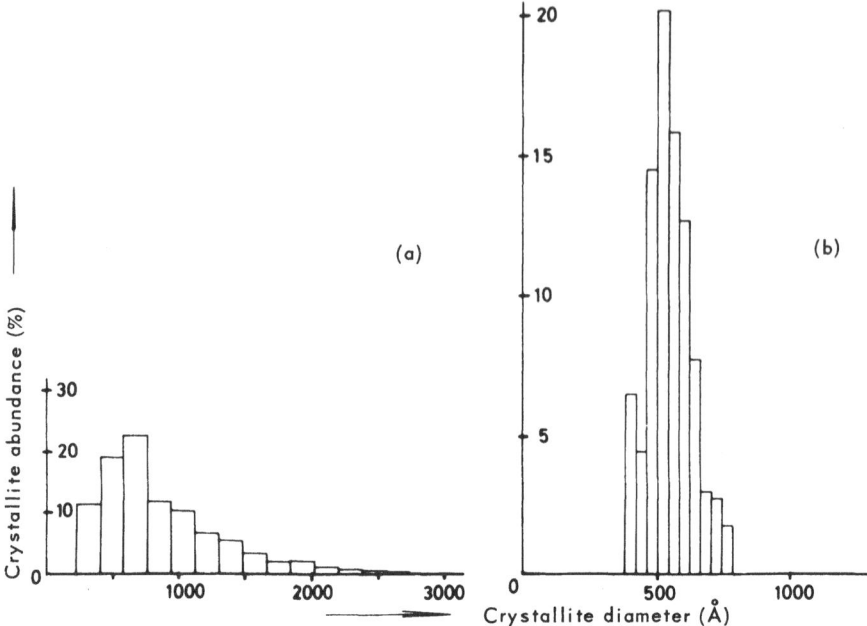

Fig. 16. Crystal size distribution in ZrN prepared from Zr + NH₃, 1 hr, at 1200°C (*a*) from electron micrograph (Fig. 9), class breadths, 180 Å; (*b*) from X-ray line-broadening of the (200) reflection, class breadths, 40 Å.

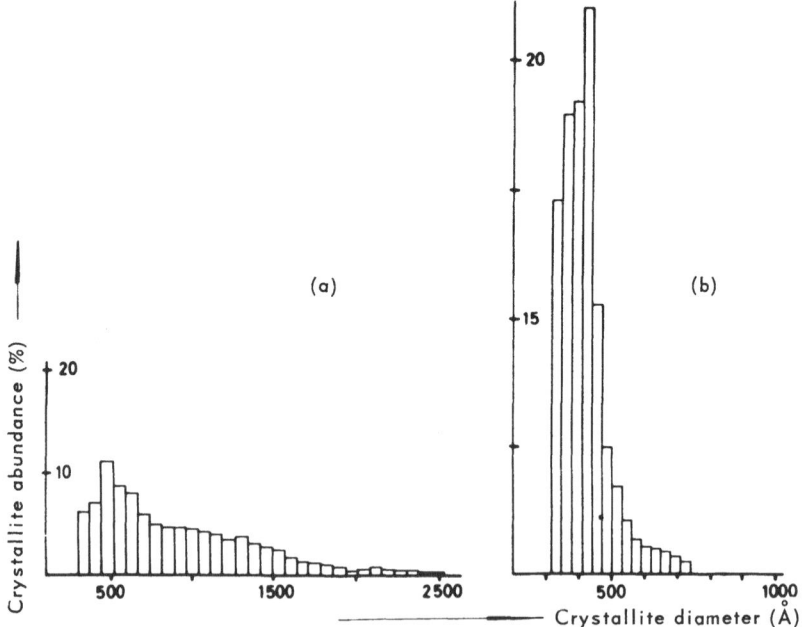

Fig. 17. Crystal size distribution in MgO powder (*a*) from electron micrograph (Fig. 8), class breadth, 75 Å; (*b*) from X-ray line-broadening of the (200) reflection; class breadths 30 Å.

powder to powder.) In applying X-ray line-broadening to larger crystals one therefore measures the faultiness of the material in general, i.e., the deviation from an ideal crystal.

Rate of Crystal Growth As Determined Experimentally

When crystals grow in glass one may assume that after having reached a sufficient size they all grow at the same rate, and this can then be determined by the optical microscope. The classical work on devitrification proceeds this way.

Crystal growth in polycrystalline materials, however, is characterized by the fact that one group (the larger crystals) grows while another group (the smaller) decreases in size and finally disappears, the total amount of crystalline material remaining constant. (In glass devitrification, however, the amount of crystalline material increases.) A thorough study will therefore consider this difference and determine growth rate as a function of crystal size. The impossibility of following the life cycle of individual crystals necessitates reliance on a statistical approach.

Let the size distribution function $W(r_i)$ be known as a function of time $W(r_i, t)$. Since for $r > 0$ no crystal appears or disappears suddenly, $W \, dr/dt$ (which is the number of crystals moving from some value r_i to larger values r_{i+n}) must be equal to $(d/dt) \int_{r_i}^{\infty} W \, dr$, which is the overall change in number for this part of the distribution function. Therefore

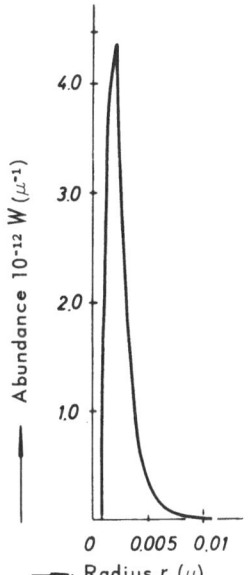

Fig. 18. Crystal size of ThO_2 original powder (from X-ray line-broadening).

$$\frac{dr}{dt} = \frac{1}{W}\frac{d}{dt}\int_{r_i}^{\infty} W\, dr \tag{16}$$

holds and plots of W versus r (Figs. 18–27) may be evaluated graphically to obtain values for dr/dt, the rate of growth.

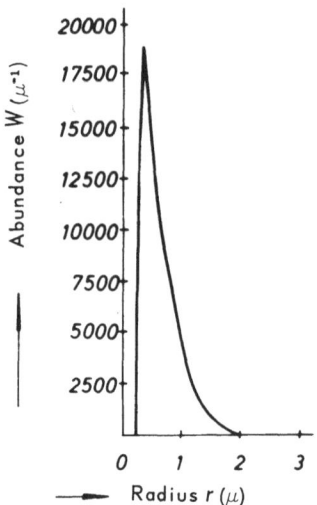

Fig. 19. Crystal size of ThO_2 compacts sintered at 1800°C for $\frac{1}{3}$ hr (from polished sections).

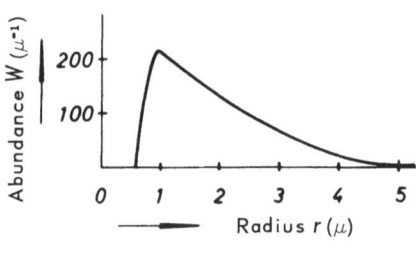

Fig. 20. Crystal size of ThO_2 compacts sintered at 1800°C for 1 hr (from polished sections).

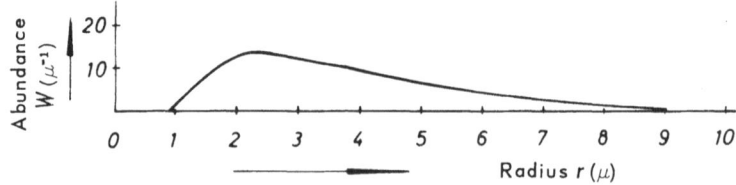

Fig. 21. Crystal size of ThO_2 compacts sintered at 1800°C for 3 hr (from polished sections).

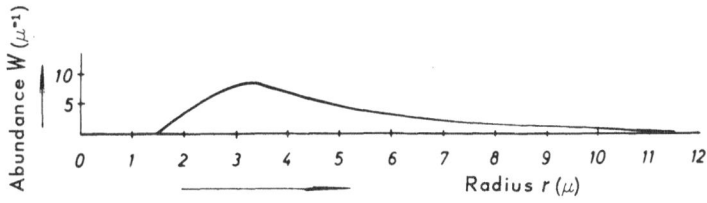

Fig. 22. Crystal size of ThO_2 compacts sintered at 1800°C for 9 hr (from polished sections).

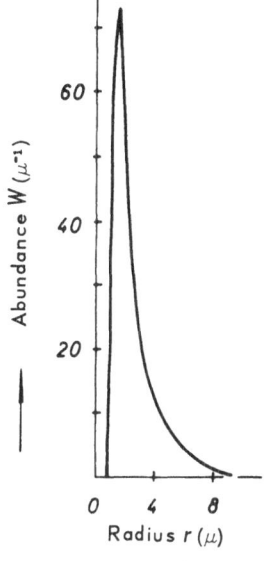

Fig. 23. Crystal size of MgO compacts sintered at 1800°C for $\frac{1}{3}$ hr (from polished sections).

Fig. 24. Crystal size of MgO compacts sintered at 1800°C for 1 hr (from polished sections).

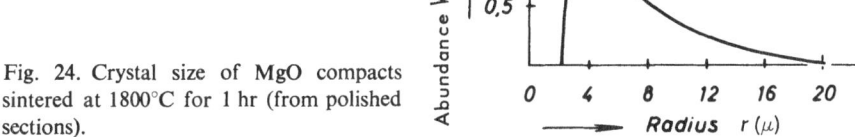

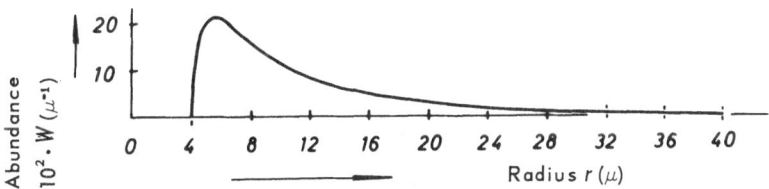

Fig. 25. Crystal size of MgO compacts sintered at 1800°C for 3 hr (from polished sections).

Abundance $10^3 \cdot W \; (\mu^{-1})$

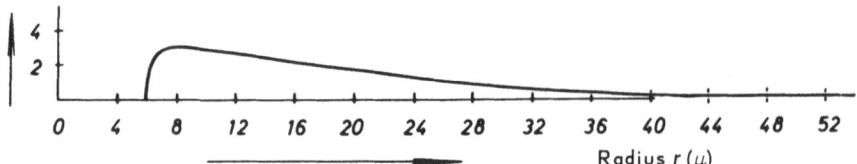

Fig. 26. Crystal size of MgO compacts sintered at 1800°C for 8 hr (from polished sections).

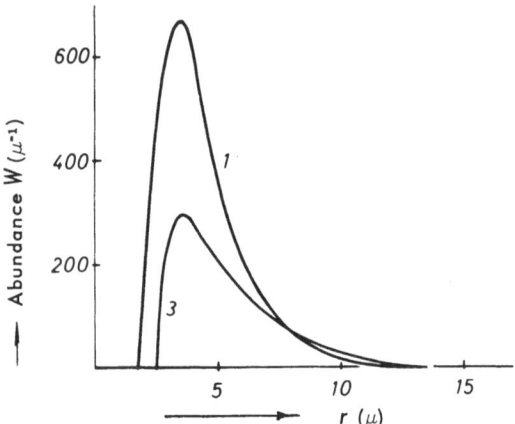

Fig. 27. Crystal size of VC compacts sintered at 2200°C for 1 and 3 hr (from polished sections).

THEORETICAL CONSIDERATIONS AND COMPARISON WITH EXPERIMENTAL VALUES

Consider two crystals with radii r_1 and r_2 in contact. Material of volume dV is now transferred from one to the other so that the radii will change.

$$dV = 4\pi r_1^2 \, dr_1 = -4\pi r_2^2 \, dr_2 \qquad (17)$$

The surface energy will change by

$$dg = \gamma 8\pi (r_1 \, dr_1 + r_2 \, dr_2) \qquad (18)$$

where γ is surface energy per cm², and together with Eq. (17)

$$dg = 2\gamma \left(\frac{dV}{r_1} - \frac{dV}{r_2} \right) \qquad (19)$$

and per mole, with V_m the molar volume,

$$\Delta g = 2_\gamma V_m\left(\frac{1}{r_1} - \frac{1}{r_2}\right) \tag{20}$$

In a polycrystalline material the crystal under consideration is in contact with many crystals of different radii. Therefore one derives the following equation for the crystal under observation:

$$\Delta g = \sum_i 2\gamma_i V_m\left(\frac{1}{r_i} - \frac{1}{r}\right)\frac{f_i}{\sum_i f_i} \tag{21}$$

where f_i is the contact area between the observed crystals and crystals with radius r_i.

Replacing the sums in Eq. (21) by integrals one obtains, assuming a constant value for γ_i,

$$\Delta g = 2\gamma V_m\left(\frac{1}{r_n} - \frac{1}{r}\right) \tag{22}$$

with

$$f_i/\sum_i f_i = Wr^2\,dr\bigg/\int_0^\infty Wr^2\,dr \quad\text{and}\quad r_n \equiv \int_0^\infty Wr^2\,dr\bigg/\int_0^\infty Wr\,dr \tag{23}$$

It now seems reasonable to write

$$dr/dt = u_k 2\gamma V_m\left(\frac{1}{r_n} - \frac{1}{r}\right) \tag{24}$$

for crystal growth, meaning that it will be proportional to the change of free energy involved. Here u_k may be considered as a transport coefficient for the materials transfer between two adjacent crystals.

Figures 28–33 show a comparison between the values for dr/dt obtained from measurements by application of Eq. (1) (points indicated as ××× or ooo) and from the theoretical Eq. (24) (solid lines). Since independent values for u_k are not available this was adapted to a best fit. The agreement between theory and experiment with regard to the dependence of dr/dt on r is as good as can be expected considering the many simplifications involved in the derivation of Eq. (24).

It is realized experimentally that crystal size distributions are conveniently expressed as logarithmic Gauss functions ([5]):

$$W(r, t) = C(t)\exp\left\{-a^2\left[\ln\frac{r}{r_0(t)}\right]^2\right\} \tag{25}$$

where r_0 is the value of the maximum and changes with time. The factor $C(t)$

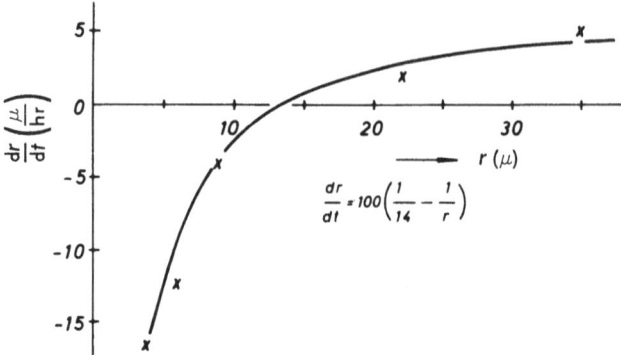

Fig. 28. Rate of growth for MgO compacts sintered at 1800°C for 2 hr.

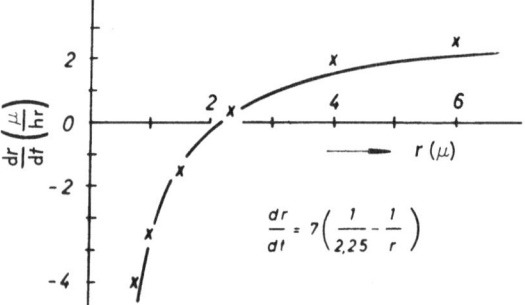

Fig. 29. Rate of growth for ThO$_2$ compacts sintered at 1900°C for 1 hr.

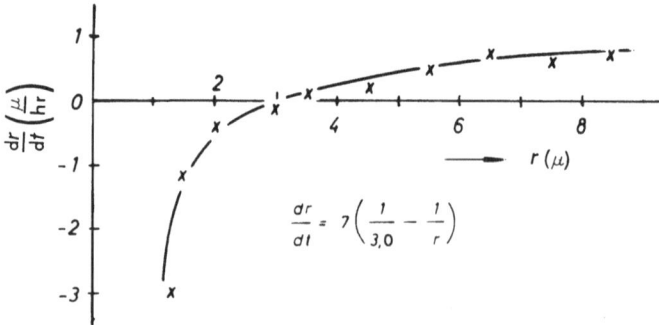

Fig. 30. Rate of growth for ThO$_2$ compacts sintered at 1900°C for 3 hr.

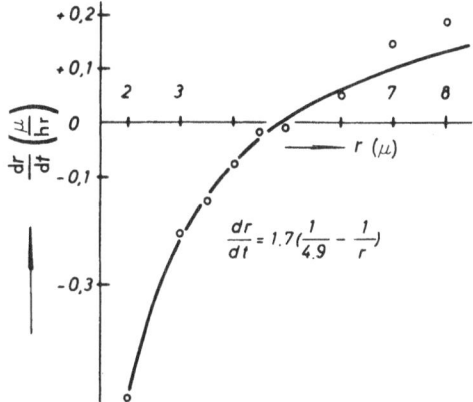

Fig. 31. Rate of growth for VC compacts sintered at 2000°C for 5.5 hr. [Data taken from Torkar et al. (6)].

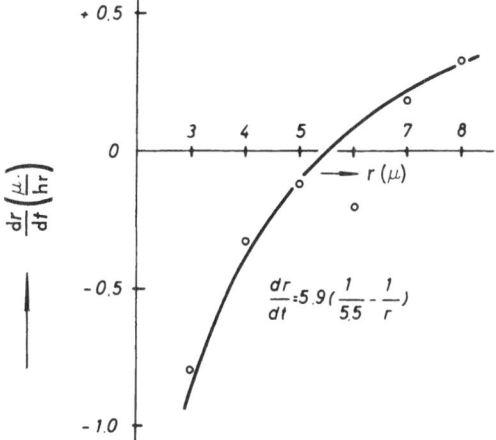

Fig. 32. Rate of growth for VC compacts sintered at 2200°C for 2 hr. [Data taken from Torkar et al. (6)].

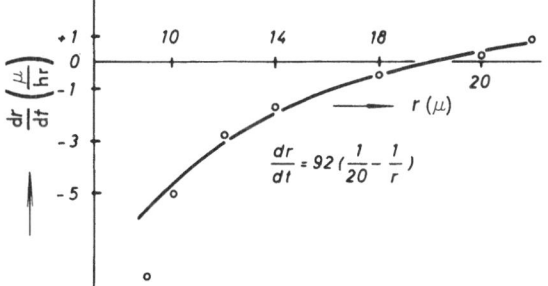

Fig. 33. Rate of growth for VC compacts sintered at 2400°C for 2 hr. [Data taken from Torkar et al. (6)].

is omitted in literature and in fact usually does not need consideration. When one wants to calculate rate of growth, however, it is essential. For crystal growth in polycrystalline material the condition

$$\int_0^\infty W(r, t)r^3 \, dr = \text{const} \tag{26}$$

means that the total amount of crystalline material remains the same, i.e., it holds. Putting the constant equal to unity it follows

$$C^{-1} = \int_0^\infty \exp\{-a^2[\ln(r/r_0)]^2\} \, r^3 \, dr \tag{27}$$

Thus the distribution function correctly normalized for crystal growth in crystalline material reads

$$W = \frac{\exp\{-a^2[\ln(r/r_0)^2]\}}{\int_0^\infty \exp\{-a^2[\ln(r/r_0)]^2\} \, r^3 \, dr} \tag{28}$$

This value for W may now be put into Eq. (16) and one obtains

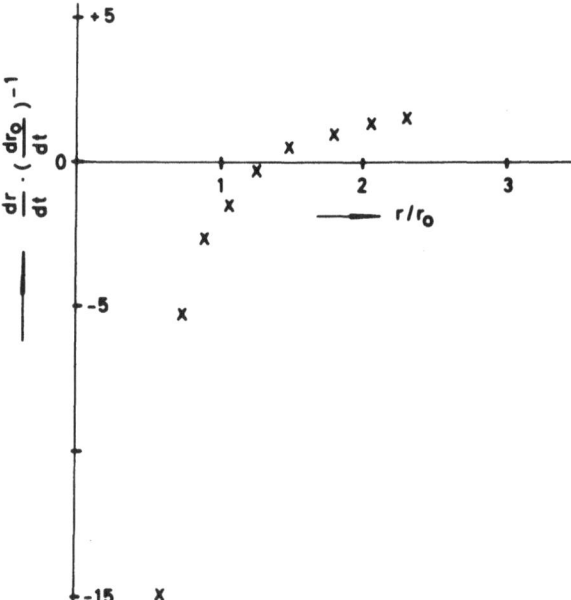

Fig. 34. Rate of growth evaluated from a series of Gauss distribution functions [Eq. (29)].

$$\frac{dr}{dt} = \left(1 - \frac{3\sqrt{\pi}}{2a}e^{-y^2}[1 - \phi(y)]\right)\frac{r}{r_0}\frac{dr_0}{dt} \tag{29}$$

with $y \equiv a\ln(r/r_0) - (1/2a)$ and $\phi(y) \equiv$ error curve. A plot of Eq. (29) (Fig. 34) shows that dr/dt thus obtained is in qualitative agreement with Eq. (24). A series of logarithmic Gaussian functions as obtained during tempering of powders [5] therefore corresponds to the rate law of growth as expressed in Eq. (24).

ACKNOWLEDGMENTS

Credit is due to A. Krauth for supplying data on powders and to K. Doerbecker for checking the mathematics.

REFERENCES

1. H. J. Oel, "Bestimmung der Grössenverteilung von Teilchen, Kristalliten und Poren," *Ber. Deut. Keram. Ges.* **43**(10): 624–641 (1966).
2. H. P. Klug and L. E. Alexander, *X-ray diffraction procedures or polycrystalline and amorphous materials* John Wiley, New York, 1954, p. 495.
3. F. Hossfeld and H. J. Oel, "Die Bestimmung der Kristallitgrössenverteilung aus der Röntgenlinienverbreiterung." *Z. Angew. Phys.* **20**(6): 493–498 (1966).
4. A. Krauth, F. Hossfeld, and H. J. Oel, *Röntgenographische und elektronenmikroskopische Bestimmung von Kristallitgrössenverteilung*, in press.
5. A. Krauth and H. J. Oel, "Kennzeichnung der Eigenschaften von Oxidpulvern," *Ber. Deut. Keram. Ges.* **43**(4): 264–270 (1966).
6. K. Torkar, H. J. Oel, and A. Illigen, "Selbstdiffusion und Sinterkinetik an Vanadiumkarbid," *Ber. Deut. Keram. Ges.* **43**(2): 162–172 (1966).

DISCUSSION

Alan D. Franklin (ARPA): X-ray line-broadening arises from several sources, strain and stacking faults in particular. Have you taken these into account in your treatment? Detailed analyses have been worked out in the past by Bertaut, by Warren and his students, and recently by Bienenstock.

Answer: We are of the opinion that the form of our analysis constitutes an improvement over these earlier works, especially in that ours is more readily adaptable to computer solution. The details of this comparison have been discussed in an earlier work [3].

To guard against error in the theoretical and experimental treatment of the X-ray line-broadening method we also examined our powder with the electron microscope. The close agreement between the particle-size distribution data from the two methods in the applicable particle-size range is reassuring.

R. Coble (MIT): It would appear that integration of your general growth equation would give $r^2 \sim t$. The Greenwood, Wagner, and Lifshitz–Slyozow models would give $r^3 \sim t$ owing to introduction of an additional r term to convert the free energy difference to a gradient. Have you measured the time dependence for increase in $\bar{r}_0$?

Answer: The models which you cite apply to the case of the growth of particles in a continuum. This is not the case with which we are concerned here. In order to adapt the continuum model to the sintering case it is necessary to carry out an averaging process [Eqs. (21–23)]. In this way the r^3 dependence of the continuum model degenerates to the r^2 dependence.

Experimental data for the time dependence of $\bar{r}_0$ appears in Fig. 34. More detailed information may be found in an earlier paper ([5]).

J. E. Burke (General Electric): Do you feel that experimental results support the conclusion that the shape of the size distribution curve becomes invariant with time as the crystal grows larger?

Answer: We have no data regarding particle-size distributions other than logarithmic Gaussian distributions. These are the ones normally encountered, however. With them the tendency to remain logarithmic Gaussian in character is well supported by experiment. Figure 34 is representative of a number of distribution functions from an earlier paper ([5]).

Chapter 14

Impurity Effects in Surface Diffusion on Aluminum Oxide

W. M. Robertson and F. E. Ekstrom

Science Center, North American Rockwell Corporation
Thousand Oaks, California

Surface diffusion on aluminum oxide was studied by measuring the growth rate of grain boundary grooves at high temperatures. Three materials were studied, high-purity aluminum oxide crystals containing low-angle subgrain boundaries and two grades of polycrystalline sintered alumina having different impurity contents. The rate of grooving on the higher-purity surfaces was $\frac{1}{3}$ to $\frac{1}{2}$ that on the lower-purity surfaces. The activation energy for surface diffusion on the high-purity crystals and on one of the sintered materials is about 130 kcal/mole, compared to 75 kcal/mole for the less-pure sintered material. Intentional impurity additions to the surfaces indicate that small amounts of impurities can appreciably change the rate of groove growth.

INTRODUCTION

Surface diffusion on oxide materials can contribute to processes such as sintering, condensation, and phase changes. The amount of information available about surface diffusion on oxides is rather sparse. Robertson and Chang [1] reported a study of surface diffusion on aluminum oxide and Robertson [2] has studied surface diffusion on magnesium oxide. The present study is an extension of this previous work. Surface diffusion was studied by measuring the rate of growth of grain boundary grooves forming on polished surfaces of polycrystalline aluminum oxide samples. Three different sources of materials have been used, each having somewhat different kinds and amounts of impurities. These different materials do not behave the same, and the differences in behavior seem to be related to the different impurity contents.

The object of this investigation was to examine the effect of impurity level on surface diffusion in aluminum oxide and, hopefully, to better understand the mechanism by which surface diffusion occurs. It will become evident that our understanding is by no means complete.

MATERIALS AND EXPERIMENTAL PROCEDURE

Grain boundary groove formation was studied on three different aluminum oxide materials. The first, from the Linde Division of Union Carbide Corporation, was a sapphire boule which contained a number of lineage boundaries. X-ray analysis indicated that the c axis was about 55° from the boule axis. The misorientations across the lineage boundaries were a maximum of about 2° of tilt between grains. Small samples were cut from the boule and surfaces perpendicular to the boule axis were polished for thermal grooving.

The second material used was Lucalox alumina from General Electric Corporation. The material was purchased as disks $\frac{1}{16}$ in. thick by $\frac{1}{2}$ in. diameter. One face was polished for thermal etching.

The third material used was Triangle RR alumina from Morganite, Inc. This is the material used in the previously reported work ([1]). It was purchased as disks $\frac{1}{8}$ in. thick by about 1 in. diameter. Pieces $\frac{1}{8}$ in. $\times$ $\frac{1}{4}$ in. $\times$ $\frac{1}{4}$ in. were cut from the disks for use as samples. The Lucalox and the Morganite aluminas were sintered materials with moderately fine grain size. The Linde and Lucalox were polished to smooth, pit-free surfaces using a diamond abrasive. Because of considerable porosity, the best polished surfaces of Morganite still contained numerous pits and holes.

The compositions of the three materials as determined by spectrographic analysis are given in Table I. The spectrographic analyses were performed on powders obtained by grinding portions of each of the materials in a steel mortar and then removing iron particles with a magnet. This procedure may have introduced some iron into the powders, so that the reported iron contents may be greater than the amount originally present in the samples.

TABLE I
Composition (in wt. %) of the Aluminas Used

	Linde alumina	Lucalox alumina	Morganite alumina
Iron	0.028	0.009	0.024
Silicon	0.0075	0.016	0.019
Manganese	0.0027	<0.002	<0.002
Magnesium	0.00043	0.034	0.0012
Copper	0.000084	0.0020	0.000072
Calcium	0.00080	0.0022	0.013
Gallium	<0.001	<0.001	0.0062
Nickel	<0.0005	<0.0005	0.0012
Titanium	<0.001	0.0014	<0.001
Alumina	remainder	remainder	remainder
	No other elements detected		

The only departure from the experimental procedure previously described was that the heating furnace temperature was controlled by a temperature controller and saturable core reactor to within $\pm2°C$ over long periods of time. The measurements of groove widths were made on interferographs of the thermally etched surfaces.

RESULTS

The data consisted of values of grain boundary groove widths at several times at each of several temperatures. Mullins' theory of grooving by surface diffusion ([3]) predicts that the groove width w will grow with time t according to the relation

$$w = 4.6(Bt)^{1/4} \tag{1}$$

with

$$B = D_s \gamma \Omega^2 n / kT \tag{2}$$

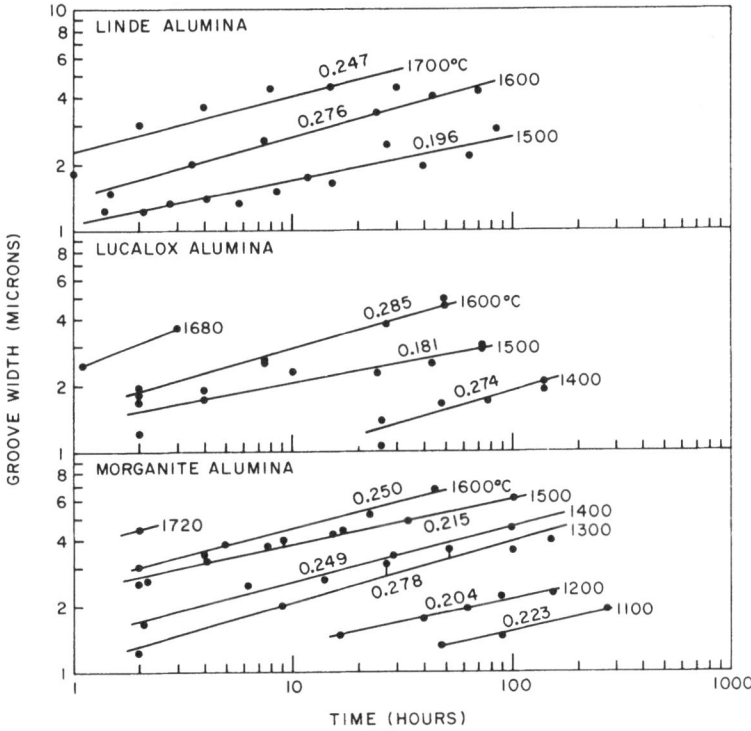

Fig. 1. Groove widths as a function of time for the three different aluminas used.

where D_s is the surface diffusion coefficient, γ the surface free energy, Ω the molecular volume, n the surface concentration of the diffusing species (assumed to be $\Omega^{-2/3}$), k Boltzmann's constant, and T the absolute temperature. A plot of log w versus log t yields a straight line of slope $\frac{1}{4}$ and the intercept yields B. Another possible mechanism of grooving is volume diffusion ([4]), which gives $\frac{1}{3}$ for the exponent of time. Since in every case in the present experiments the exponent of the time was close to $\frac{1}{4}$, surface diffusion was taken to be the grooving mechanism.

Figure 1 gives plots of log w versus log t for the three materials. The lines were fitted to the experimental points by the method of least squares. Each point is the average of the groove widths measured on at least four different interferographs for each annealing time. The attached numbers give the slope of the line through the points. It is apparent that the slopes are close to $\frac{1}{4}$, in accordance with the surface diffusion mechanism.

It can be seen that the groove widths at a given temperature are narrower for the Linde and Lucalox materials than for the Morganite alumina. The data for the three materials at 1600°C are compared in Fig. 2. At lower temperatures the difference becomes greater.

The groove width data were used to calculate surface diffusion coefficients from Eqs. (1) and (2), using the following constants: $\gamma = 905$ ergs/cm^2, $\Omega = 2.11 \times 10^{-23}$ cm^3/atom, $n = 1.31 \times 10^{15}$ atoms/cm^2, and $k = 1.38 \times 10^{-16}$ ergs/°K. The calculated diffusion coefficients are shown in Fig. 3. The ratio of two to three in the groove widths among the materials causes the diffusion coefficients to differ by one to two orders of magnitude.

The grooves were so small at 1400°C on the Linde material that it was not possible to obtain accurate width measurements. The calculated D_s

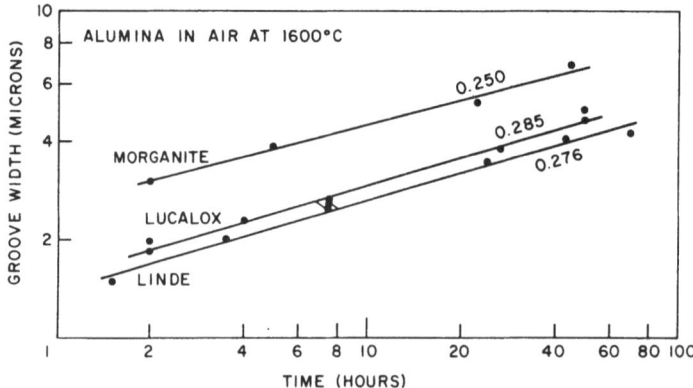

Fig. 2. Comparison of groove growth on the three materials at 1600°C.

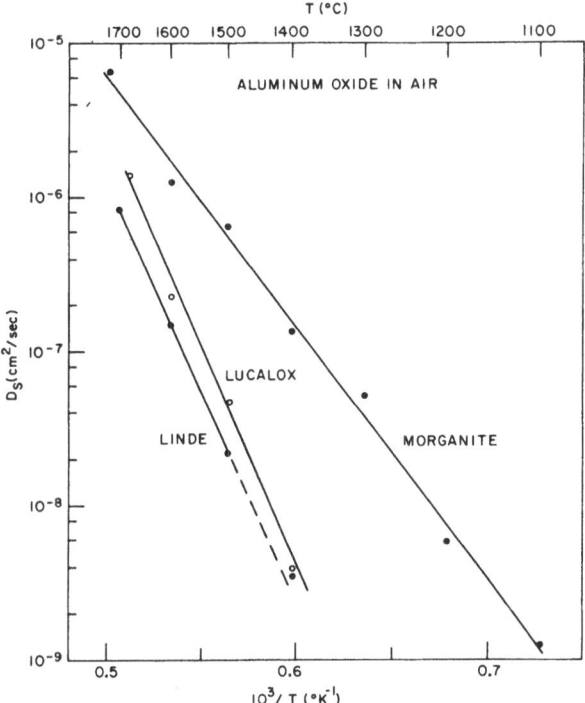

Fig. 3. Surface diffusion coefficients for aluminum oxide in air.

for this material at 1400°C is the maximum value estimated from the data. The data given for the Morganite alumina are those given previously ([1]) plus a few additional points. However, the values of D_s were recalculated and a small numerical error of the previous work corrected, so these values are larger by about 10%. The activation energy is unchanged.

Lines were fitted to the points by the method of least squares according to the relation

$$D_s = D_0 \exp(-Q_s/RT)$$

For the Linde alumina the point at 1400°C was not included in the above line, though it is evident that it falls close to the calculated lines. The values of D_0 and Q_s obtained from the least squares fits are given in Table II. The Morganite alumina had an activation energy for surface diffusion of 75 kcal/mole, while that for the Lucalox and Linde aluminas is 50–60 kcal/mole larger.

In an attempt to determine if the above difference is due to impurities

TABLE II

Surface Diffusion on Alumina

$$D_s = D_0 \exp(-Q_s/RT)$$

Material	D_0 (cm²/sec)	Q_s (kcal/mole)
Linde	1×10^8	128
Lucalox	8×10^8	133
Morganite	9×10^2	75

various oxides were added to the Linde samples as powders sprinkled on the polished surfaces. The oxides added were SiO_2, Na_2O, 70% SiO_2–30% Na_2O, and 60% SiO_2–40% CaO. Grooving was carried out at 1500°C. In every case the powder formed small particles on the surfaces, which made the samples appear much rougher than before. The SiO_2 and Na_2O additions had little effect on the rate of grooving of the alumina. The SiO_2–CaO addition gave somewhat variable results. The mixture of these two oxides gives a liquid at 1500°C. In places where the grain boundary had been completely covered by the liquid at temperature quite wide grooves formed. In some areas rather narrow grooves formed, little if any wider than in the absence of the impurity. In other areas the grooves were of intermediate width. Figure 4 shows examples of the grooving occurring with the SiO_2–CaO additions. In some cases after considerable heating times the liquid spread along the grain boundaries at the surface. The results of these experiments on the addition of impurities are somewhat ambiguous. Where liquid formed the grooving was rapid. But in the absence of liquid the impurities did not always greatly

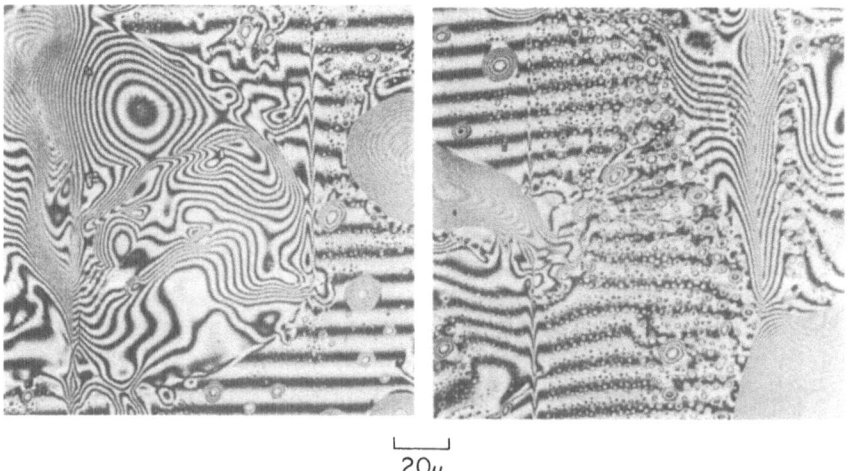

L_____J
20μ

Fig. 4. Grain boundary grooving on Linde alumina at 1500°C with the addition of 60% SiO_2–40% CaO to the surface.

accelerate grooving. The experimental results on the effects of impurities on grooving are not very clear-cut, so it is difficult to ascribe the observed differences in grooving behavior entirely to differences in impurity levels. As will be noted in the discussion, the differences may be due to the particular impurities present.

In the previous work on Morganite alumina ([1]) there seemed to be little variation of the grooving rate with surface orientation. The same general conclusion also holds in the present work, though the grooving rate did appear to be somewhat different from the average in a limited number of cases. An appreciable number of grooves on the Lucalox alumina were asymmetric, with a well-developed hump on one side of the groove and very little hump on the other. A possible cause of this is that the surface diffusion coefficients on the two grains may be different; however, it could also be because the boundary was not normal to the surface, causing boundary movement to occur. In addition in a few cases a groove was considerably smaller at some particular grain boundary than at most of the other boundaries. In some of these cases this was obviously because the boundary had moved, so that the groove had not grown as long as the others; but in some cases it appeared that the boundary had not moved. These observations indicate that there may be some orientations on which surface diffusion is appreciably less than on most orientations.

DISCUSSION

The most notable effect that this work demonstrates is that there is a marked difference in the thermal grooving behavior of different aluminas under similar experimental conditions. Hence we must look for differences in the materials to be able to understand the differences in behavior. The composition of the materials is the obvious place to look for differences, and here we immediately run into problems of interpretation. The total impurity content for the two sintered materials are about equal and about twice that of the Linde alumina. Hence the total impurity content does not rationalize the fact that the Linde and Lucalox behave similarly while the Morganite grooves more rapidly. We must then look for differences in the particular impurities present in order to rationalize the observed behavior.

Carniglia ([5]), in a review of the mechanical properties of refractory oxides, has emphasized that small amounts of impurities can have pronounced effects on the material behavior when they segregate to grain boundaries and surfaces. The present experiments are likely to be affected by precisely this type of segregation. The impurity levels in all three of the materials are such that we cannot be assured that we are measuring the intrinsic behavior of alumina on any of them. However, it is possible that we may be approaching intrinsic behavior in the Linde alumina, which has a purity of about 99.99 %

if we discount the iron which may have been introduced during the grinding process prior to analysis.

The Lucalox alumina has magnesium oxide as its principal impurity. Jorgenson ([6]) has demonstrated that magnesium oxide reduces the sintering rate of alumina in the early stages of sintering, and he interpreted this as being due to a decrease in the volume diffusion coefficient. This suggests that a similar decrease due to magnesium oxide might also be observed in surface diffusion experiments. Thus the Lucalox alumina, despite its higher impurity content, would have a low effective surface diffusion coefficient. The Morganite alumina has as principal impurities silica and calcium oxide. These two oxides tend to form a low-melting liquid when mixed together. It is suggested that in this material the SiO_2 and CaO segregate to the surfaces, forming a region that has a relatively high atomic mobility, though not necessarily forming a liquid film. This high-mobility region would give a relatively large effective surface diffusion coefficient and a rapid rate of thermal grooving. The experiments on the intentional additions of SiO_2–CaO mixtures indicated that in some cases the rate of grooving was appreciably accelerated.

The activation energy for surface diffusion on the Linde and Lucalox aluminas is about 130 kcal/mole, compared to about 75 kcal/mole for the Morganite alumina. The accompanying values of D_0 are about 10^8 cm²/sec and 10^3 cm²/sec, respectively. These values suggest that the differences are due to different mechanisms of surface diffusion rather than just to differences in the energy barrier for diffusion owing to changed impurity or defect levels. This value of 130 kcal/mole is similar to the activation energy for volume self-diffusion of aluminum ions in alumina, 114 kcal/mole ([7]), and approaches that for oxygen ion diffusion in alumina, 152 kcal/mole ([8]). However, if one uses the measured aluminum or oxygen ion diffusion coefficients to calculate the rate of grooving, the calculated rate is smaller than the measured rate by at least an order of magnitude. This reinforces the conclusion reached earlier, based on the time exponent of groove growth, that surface diffusion is the grooving mechanism.

In a previous discussion of the mechanism of surface diffusion ([2]) it was suggested that the activated species in the diffusion process was a partially evaporated oxygen ion with one or two associated metal ions. This model predicted that the activation energy for surface diffusion should be about two-thirds of the heat of sublimation of the material; taking 120 kcal/mole as the heat of sublimation of alumina, this model predicted 80 kcal/mole for the activation energy for surface diffusion, in reasonable agreement with the measured result on Morganite alumina. The present results, as measured on Linde alumina, suggest that the activation energy for surface diffusion on pure alumina is near 130 kcal/mole rather than 75 kcal/mole. Thus the above model does not give good agreement unless the heat of sublimation is of the order of 190 kcal/mole, which seems too high to be reasonable. We are left in the situation that we do not have a good model to explain the mechanism by which surface diffusion occurs on alumina. We could consider the energetics of the formation of several different diffusing species containing aluminum and oxygen ions, but the present uncertainty about the properties of suboxides of aluminum appears to make this line of speculation not worthwhile.

CONCLUSIONS

1. Thermal grooving on alumina occurs by a mechanism of surface diffusion.

2. In reasonably high-purity materials the activation energy for surface diffusion is about 130 kcal/mole and the preexponential factor is about 10^8 cm^2/sec. Some impurity additions can appreciably raise the apparent surface diffusion coefficient, with a decrease in both the activation energy and the preexponential factor for surface diffusion.

3. Some impurity additions can appreciably raise the apparent surface diffusion coefficient, with a decrease in both the activation energy and the preexponential factor for surface diffusion.

ACKNOWLEDGMENTS

We are indebted to the Linde Division of Union Carbide Corporation for the gift of an alumina boule containing numerous subgrain boundaries. The X-ray determinations of orientation were performed by D. Swarthout of this laboratory. Numerous discussions with Dr. H. Wiedersich of this laboratory are gratefully acknowledged.

REFERENCES

1. W. M. Robertson and R. Chang, in: *Materials Science Research*, Vol. 3 (W. W. Kriegel and H. Palmour, III, eds.), Plenum Press, New York, 1966, pp. 49–60.
2. W. M. Robertson, in: *Sintering and Related Phenomena*, (G. C. Kuczynski, N. A. Hooton, and C. F. Gibbon, eds.), Gordon and Breach, New York, 1967, pp. 215–32.
3. W. W. Mullins, *J. Appl. Phys.* **28**: 333–39 (1957).
4. W. W. Mullins, *Trans. Met. Soc. AIME* **218**: 354–61 (1960).
5. S. C. Carniglia, in: *Materials Science Research*, Vol. 3 (W. W. Kriegel and H. Palmour, III, eds.), Plenum Press, New York, 1966, pp. 425–71.
6. R. J. Jorgenson, *J. Am. Ceram. Soc.* **48**: 207–210 (1965).
7. A. E. Paladino and W. D. Kingery, *J. Chem Phys.* **37**: 957–62 (1962).
8. Y. Oishi and W. D. Kingery, *J. Chem. Phys.* **33**: 480–86 (1960).

DISCUSSION

J. E. Burke (General Electric): In the sapphire boules I would anticipate that only low-angle grain boundaries, or highly symmetrical grain boundaries, could exist, because thermal stresses will lead to fracture along high-angle boundaries. In the polycrystalline specimens higher-angle boundaries are encountered. This difference might account for the observed differences between the Linde specimens and the Lucalox and Morganite aluminas. This effect could show up as a variation in groove width from one grain boundary to another in the polycrystalline specimens, with certain cusp orientations or low-angle boundaries showing narrower grooves. Was much variation in groove width observed?

Answer: On the Morganite polycrystalline material a small fraction of the grooves were narrower than the average. Generally though the grooving was quite uniform over the sample surface. The same was true of the Lucalox material—the grooving was uniform over the surface. We concluded that for the range of orientations exposed in the polycrystalline materials grooving was independent of orientation.

The boundaries in the Linde material were quite low-angle boundaries. The largest grain rotations were about 2°, as measured by X-ray methods. All of the surfaces observed on the Linde material were about the same orientation, with the normal to the surface about 55° from the c axis of the crystal.

R. Coble (M.I.T.): Have you looked at the groove angles as a function of temperature?

Answer: No.

Coble: Could some temperature dependence arise from changes in the surface energy due to segregation or desegregation of impurities?

Answer: Yes. The experimental measurement yields the product $(D_s \gamma \Omega^2 n)$. It was assumed that γ, Ω, and n are independent of temperature, so all the temperature dependence is thrown into D_s. If γ varies with temperature due to segregation, this would have to be measured by a separate experiment and the temperature-dependent γ used in the calculation of D_s.

P. E. Morgan (Franklin Institute): It has been reported that sulfur (in the form of sulfate, presumably) is a major impurity in Linde A powder from which the Linde boules and the Lucalox material are derived. Small amounts of covalently bonded material at surfaces and interfaces appear from many sources to greatly reduce diffusion values. The fact that your Linde and Lucalox values approach the behavior of the Morganite material at higher temperatures suggests that the sulfate is being progressively driven off and that the Morganite material is in fact showing more nearly the intrinsic behavior with the more nearly correct activation energy and D_0 value.

Answer: We have not looked for sulfur in our samples. This is something that needs to be done.

P.W.M. Jacobs (University of Western Ontario): Dr. Morgan's proposal that the Morganite alumina data gives the intrinsic surface diffusion coefficient is supported by the pre-exponential (D_0) factors in Table II. A quick calculation shows that the entropy of activation is about 1×10^{-3} eV/deg, which, though high, is not impossibly so. The value for Lucalox alumina would be twice as large, corresponding to an unbelievably large vibrational frequency.

Answer: This calculation from D_0 assumes a jump distance of one interatomic distance. In trying to rationalize observed large D_0 values in surface diffusion on metals ([9]) it has been suggested that rather than having such large entropies and vibrational frequencies the jump distance of the diffusing species is several tens or even hundreds of interatomic distances. The same argument could perhaps be applied to the present case.

Hayne Palmour, III (North Carolina State University): What are the angular misorientations in the polycrystalline Lucalox and Morganite aluminas?

Answer: We didn't measure the grain orientations of the polycrystalline materials. It would be expected that the materials would contain a large range of grain misorientation and a large range of orientations of exposed surfaces. For a given polycrystalline material there did not appear to be much variation in grooving behavior across the surface of the sample.

Palmour: Were second phases visible in the boundaries of the Lucalox and Morganite? We have been studying pure and doped alumina bicrystals with 10° twist misorientation about [11$\bar{2}$3]. Boundaries doped with SiO_2 (<1000 Å thick) develop thickened boundaries containing second phases presumed to be mullite and siliceous glass. Those doped with MgO did not show boundary phases, and some evidence indicates that magnesia diffuses into the adjacent alumina crystals. Microhardness studies on polished surfaces normal to the bicrystal boundary plane show decreasing hardness beginning about 50 μ from the boundary. In pure and SiO_2-doped bicrystals no other variation in the normal hardness of the crystal is observed, but in the MgO-doped specimens the microhardness increases above the normal average in regions adjacent to the boundary and then drops sharply at the boundary proper. It appears that the magnesia could be altering the behavior in the region of the boundary.

Answer: We did not see anything that looked like a second phase in any of the boundaries. However, in some studies of the Morganite material at about 1800°C the groove profile showed a hump, as if something was being extruded out of the boundary. This was not very reproducible. This may indicate that the material in the boundary is softer than the rest of the crystals. But this softer material did not appear to be a second phase.

C. L. Morgan (Oak Ridge National Laboratory): Did you note any difference from one impurity ion to another in the effect on surface diffusion? Did it give any clue as to how the effect actually worked?

Answer: We did attempt to calculate the net charge at the surface, assuming all the impurities are actually at the surface. The results were inconclusive. The Linde and Lucalox materials, which had about the same grooving behavior, appeared to have opposite signs of the charge at their surfaces. The Morganite material, which grooved quite a bit differently from the Linde, had about the same net charge as the Linde. We didn't seem to be getting anywhere, so we didn't follow it further.

G. Kuczynski (Notre Dame): Does the pore structure within the sample affect your surface diffusion measurement? This is the most obvious difference between the three materials.

Answer: There is a difference in the internal pore structure of the materials. The effect of the pores would show up as deeper, wider grooves at the surface just above the pore. The surface diffusion measurement assumes all of the material moves along the surface to form the groove. A subsurface pore on the grain boundary could act as a sink for atoms, and the groove could enlarge by material diffusing down the boundary to the pore, shrinking the pore. Only pores within a distance of the surface of about the groove width would be likely to be effective sinks for atoms. This would cause the grooving to be irregular over the sample surface. Effects of this sort have been observed on metals [10]. We did not see any indications of this effect, so we believe it is not important in our experiments.

ADDITIONAL REFERENCES

9. J. Y. Choi and P. G. Shewmon, *Trans. Met. Soc. AIME* **230**: 123–132 (1964).

10. Y. Y. Geguzin and L. N. Paritskaya, *Phys. Metals and Metallog.* **13**(4): 102–108 (1962).

Chapter 15

Diffusion Path Networks in Oxides*

R. H. Condit

Lawrence Radiation Laboratory, University of California
Livermore, California

An understanding of diffusion mechanisms in oxides, particularly cation diffusion, must take account of the various possible paths which an atom may follow in going from one normally occupied site to another equivalent site. Such paths may be branched and will likely include stops at metastable points along the way. This chapter describes methods whereby path networks within oxides can be characterized, and discusses diffusion in the structures NaCl, ZnS, FeS, and BeO.

INTRODUCTION

A common introductory description of the diffusion process treats it in terms of an atom leaving a normal site in a crystal and entering a neighboring site by a direct jump between the two. A large number of such random jumps results in a migration over distances which can be described in terms of a diffusion coefficient and a duration of diffusion. The coefficient D is frequently given in the form

$$D = \tfrac{1}{6}\nu\lambda^2 \tag{1}$$

where ν is the jump frequency and λ is the jump distance.

Detailed consideration will reveal that with the exception of perhaps the simplest crystal structures the "jump" must involve a sequence of momentary arrests at intermediate stops, metastable configurations of the system, with probable changes in direction of movement before completion of the jump. These stops may be interstitial positions for the moving atom or other configurations of the system in which one or more atoms are on sites which are different from their normal ones. In addition to a path being zigzag it may branch before completing the connection between normal sites. Also, more than one type of path may be available for diffusion, and the origin

---

* Work performed under the auspices of the U.S. Atomic Energy Commission.

and terminus of a given path may not always be immediately neighboring sites.

If diffusion in ceramic systems is to be understood in terms of the individual atomic migrations, it is necessary that the jump possibilities be first enumerated and their relationships be articulated. It will be the purpose of this chapter to introduce methods for analyzing these problems. It will lead us to the point where we can express a diffusion coefficient in terms of the component intermediate jump frequencies between each of the possible intermediate stops. The application of absolute rate theory to the study of these individual frequencies will not be taken up. However, techniques for deriving these frequencies from measured migration rates will be suggested.

The method of approach will be akin to that used in the analysis of electrical networks where current to and from nodes and along individual circuit branches can be calculated. In fact, a principal purpose of this chapter will be to introduce the concepts of electrical network circuit analysis to the study of diffusion mechanisms. The physical picture to which this discussion will relate throughout the chapter is that for self-diffusion in a crystallographically perfect, homogeneous material. Thus, no attention will be given to grain boundary or surface diffusion.

DERIVATION OF THE DIFFUSION EQUATION

We will first consider the motion of a single atom between a series of points, although the mathematical formalism can be applied equally well to the more physically realistic, formal analysis which treats the system as a whole. Such a statistical mechanical model recognizes that there must be displacements of a number of atoms if one atom is to transfer between normal sites. However, for the moment this sophistication would not serve a useful purpose.

It will be convenient to derive an expression for the diffusion coefficient by the comparison of four basic equations. The first, Fick's equation, defines how the coefficient is measured as a function of the flux J of a selected species of atoms and the gradient in their concentration C along a direction paralleling the x axis,

$$J = -D \, dC/dx \qquad (2)$$

There are several restrictions on the application of this equation which are usually introduced in self-diffusion studies; the diffusion coefficient is supposed to be independent of concentration, concentration gradient, and position in the material. In addition, it will be assumed below that: (1) the gradient in concentration is small, so that the fractional difference in concentration

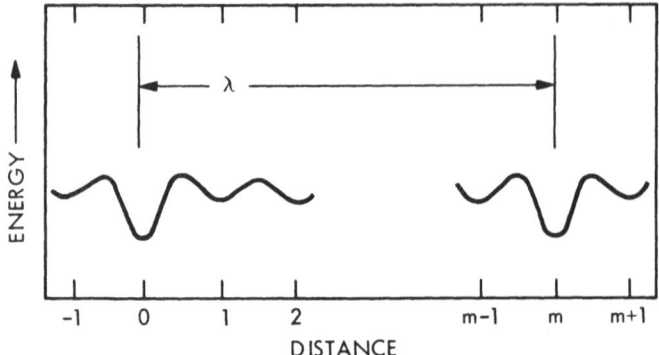

Fig. 1. Energy diagram for a succession of sites.

between neighboring planes of atoms is small; (2) the gradient does not change appreciably over the mean distance between the origin and terminus of possible paths between normal sites; and (3) the gradient does not change during the time an atom is in transit.

The macroscopic description of the concentration can be related to the microscopic description in a second equation. We will picture the atoms in a lattice as lying in planes normal to the diffusion direction x, with the distance between successive planes being λ. The difference in concentration between planes will be $\lambda(dC/dx)$. Let us designate successive planes of normal sites as $0, m, 2m, \ldots$, as has been done in Fig. 1. For the moment consider that normal sites are connected by simple chains of interstitial sites which are designated $1, 2, \ldots m - 1, m + 1, m + 2, \ldots$. Let there be N normal sites per unit area on a plane. If we define the fraction of the kth sites occupied by atoms as F_k, the concentration of atoms at the 0th and mth planes may be accounted as

$$C_0 = \frac{N}{\lambda} \sum_{k=0}^{m-1} F_k \quad \text{and} \quad C_m = \frac{N}{\lambda} \sum_{k=m}^{2m-1} F_k \tag{3}$$

It will be convenient to group together the fractions of interstitial site occupancies,

$$r = \frac{1}{F_0} \sum_{k=1}^{m-1} F_k \tag{4}$$

where $F_0(1 + r)$ is the Maxwell–Boltzmann partition function, $1/(1 + r)$ is the fraction of the total number of atoms on normal sites, and $r/(1 + r)$ is the fraction on interstitial sites. The second relation then becomes

$$\frac{dC}{dx} = \frac{C_m - C_0}{\lambda} = \frac{N}{\lambda^2}(F_m - F_0)(1 + r) \tag{5}$$

The third basic equation is an expression for the flux in terms of the frequencies of jumps. Let the frequency with which an atom on the kth site leaves it in the forward and backward directions be $\phi_{k,k+1}$ and $\phi_{k,k-1}$, respectively. The total departure rate from the kth site will then be $F_k(\phi_{k,k+1} + \phi_{k,k-1})$, and a similar expression will apply for the $(k+1)$th site. The net flow between these two sites along the path joining them will be

$$j = F_k\phi_{k,k+1} - F_{k+1}\phi_{k+1,k} \tag{6}$$

The macroscopic flux will be the sum of these microscopic fluxes. In the example being used at the moment, Fig. 1, the number of paths is equal to the number of normal sites per unit area of a plane, and

$$J = Nj \tag{7}$$

The fourth and final equation will give us the concentrations of atoms on each of the types of sites in a lattice. Under the steady-state conditions which we postulate the number of atoms arriving at a particular type of site will be equal to the number departing,

$$F_k(\phi_{k,k+1} + \phi_{k,k-1}) = F_{k-1}\phi_{k-1,k} + F_{k+1}\phi_{k+1,k} \tag{8}$$

By solving the series of simultaneous equations which result from writing down this equation for each of the sites $k = 0, 1, 2, \ldots, m-1, m, \ldots, 2m$ it is possible to find expressions for F_k for each of the interstitial sites in terms of F_0, F_m, F_{2m}, and the values of the frequencies ϕ. Combination of the four equations, (2), (5), (7), and (8), will then give the diffusion coefficient in terms of the frequencies ϕ and the interplanar spacing λ.

The use of this technique can be conveniently illustrated for the case where there are two metastable sites between each normal site, i.e., values of m will be $0, 3, 6, \ldots$ (see Fig. 2). Equation (5) will become

$$dC/dx = (N/\lambda^2)(F_3 - F_0)(1 + r) \tag{9}$$

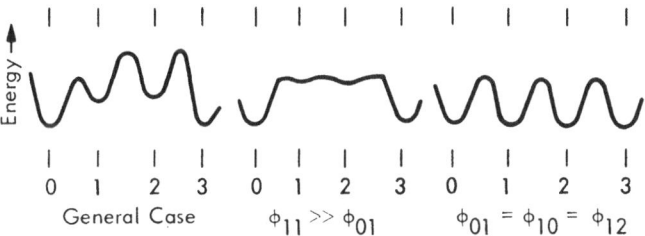

Fig. 2. Energy diagram for a network with three types of sites in a chain.

and Eq. (8) will read

$$F_1(\phi_{10} + \phi_{12}) = F_0\phi_{01} + F_2\phi_{21} \tag{10a}$$

$$F_2(\phi_{21} + \phi_{23}) = F_1\phi_{12} + F_3\phi_{32} \tag{10b}$$

For simplicity assume $N = 1$, and then

$$J = F_0\phi_{01} - F_1\phi_{10} \tag{11}$$

A value for F_1 derived from (10) can be substituted in (11), and use may be made of the relationship,

$$\phi_{01}\phi_{12}\phi_{23} = \phi_{32}\phi_{21}\phi_{10} \tag{12}$$

a result of the fact that sites 0 and 3 are equivalent, and therefore that a transit in one direction will occur with a probability which is equal to that for the reverse direction. This is an expression of the principle of microscopic reversability. It will then be found that

$$J = (F_0 - F_3)\Phi \tag{13}$$

where

$$\Phi = \frac{\phi_{01}\phi_{12}\phi_{23}}{\phi_{12}\phi_{23} + \phi_{10}\phi_{23} + \phi_{21}\phi_{10}} \tag{14}$$

and this will be spoken of as the composite jump frequency. Comparing Eqs. (2), (9), and (13) gives what is actually a general equation,

$$D = \lambda^2\Phi/(1 + r) \tag{15}$$

In practice Φ and r may not be separable unless r is large enough to be revealed by structure studies. The term here, $\Phi/(1 + r)$, is equivalent to the v in Eq. (1).

Discussion of this example will serve to bring out some important consequences of this detailed examination of the diffusion process and may help to make its physical meaning more apparent. To recapitulate, Eq. (15) applies to a case depicted in Fig. 2, where an atom in going from one normally occupied site to a neighbor and must pass through two interstitial positions. If the stopping points 1 and 2 were very shallow traps or actually unstable positions, one could have the case where ϕ_{01} and ϕ_{32} were equal and very much smaller than any of the other jump frequencies. For the moment take the other frequencies, ϕ_{12}, ϕ_{21}, ϕ_{23}, and ϕ_{10}, as equal, and it will follow that

TABLE I

| Number of jumps | Probability of the atom being in the given site | | | |
	Site 0*	Site 1†	Site 2†	Site 3*
0	1	—	—	—
1	0	1	0	0
2	$\frac{1}{2}$	0	$\frac{1}{2}$	0
3	$\frac{1}{2}$	$\frac{1}{4}$	0	$\frac{1}{4}$
4	$\frac{1}{2} + \frac{1}{8}$	0	$\frac{1}{8}$	$\frac{1}{4}$
5	$\frac{1}{2} + \frac{1}{8}$	$\frac{1}{16}$	0	$\frac{1}{4} + \frac{1}{16}$
...	...	...	...	...
∞	$\frac{2}{3}$	0	0	$\frac{1}{3}$

*Normal.
†Interstitial.

$$r = 0 \tag{16}$$

and

$$\Phi = \tfrac{1}{3}\phi_{01} \tag{17}$$

The coefficient $\frac{1}{3}$ is a consequence of the fact that the atom stops at site 1 and from this can continue on to 2 or return to 0, and when at 2 it may complete its transit to the next normal site by continuing to 3, or may return to 2. We can tabulate the probability of position as a function of the number of jumps taken (Table I) remembering that after the first jump the sites 0 and 3 are to be regarded as permanent traps. All this says is that the designation of sites 1 and 2 as stopping points for the atom implies that the atom comes to local equilibrium and then has no memory about the direction from which it arrived. Nothing has been said about the actual energy of the atom in the sites. If it were to be carrying appreciable energy as it crossed the energy saddle point into the site so that it tended to continue its direction of motion, the formalism which is being taken in this chapter would require modification.

In another case described by Eq. (14) the sites 0, 1, 2, and 3 might all be energetically equivalent, and the individual jump frequencies would be equal. As before,

$$\Phi = \tfrac{1}{3}\phi_{01} \tag{18}$$

but

$$r = 2 \tag{19}$$

and

$$D = \tfrac{1}{9}\lambda^2 \phi_{01} \tag{20}$$

This would be equivalent to saying that the basic jump distance was $\lambda/3$, which would give the same result for the diffusion coefficient,

$$D = (\lambda/3)^2 \phi_{01} \tag{21}$$

A matter of nomenclature should be brought out at this point. The discussion has turned on the value of the 0, 1 jump, whereas the frequency with which an atom on the 0-site leaves is really the sum of 0, 1 and 0, -1 jumps. In a typical symmetrical case these might be described as jumps from a normal to an interstitial site. Then the frequency of jump from the normal will be

$$\phi_{ni} = \phi_{01} + \phi_{0-1} \tag{22}$$

and in the case of negligible interstitial population described by Eqs.(16) and (17) we would have

$$\Phi = \tfrac{1}{6}\phi_{ni} \tag{23}$$

An interesting feature of the composite jump frequency, Eq. (14), is its possible nonsimple dependence on temperature. Individual jump frequencies may be expected depend upon temperature in the usual way,

$$\phi = \phi_0 e^{-\Delta Q/RT} \tag{24}$$

but since the denominator is the sum of such terms, the composite frequency

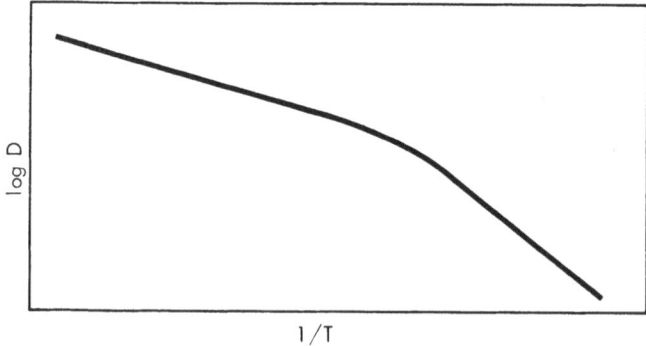

Fig. 3. Possible dependence of the diffusion coefficient on temperature.

may not necessarily have this simple, conventional dependence on temperature. Thus it might happen that the enthalpy for the jump 1, 2 might be lower than for 2, 3, while the entropy of activiation for the jump 1, 2 might be higher than for 2, 3. This could have the consequence that at low temperatures the rate-limiting step for the overall transit would be the 2, 3 jump while at high temperatures it might be the 1, 2 jump. Then the graph of the log D versus $1/T$ would have a rollover from a region of low activation energy at high temperatures to a region of high activation energy at low temperatures (Fig. 3).

DIFFUSION IN SOME OXIDE STRUCTURES

We will now consider diffusion within the interstice network of the two close-packed structures, face-centered cubic and hexagonal close-packed. The networks are shown in Figs. 4 and 5.

In the cubic system we will consider the two structures, NaCl and zinc blende. These have the common feature that the anions are in close packing. The cations may be regarded as occupying the interstices between the usually

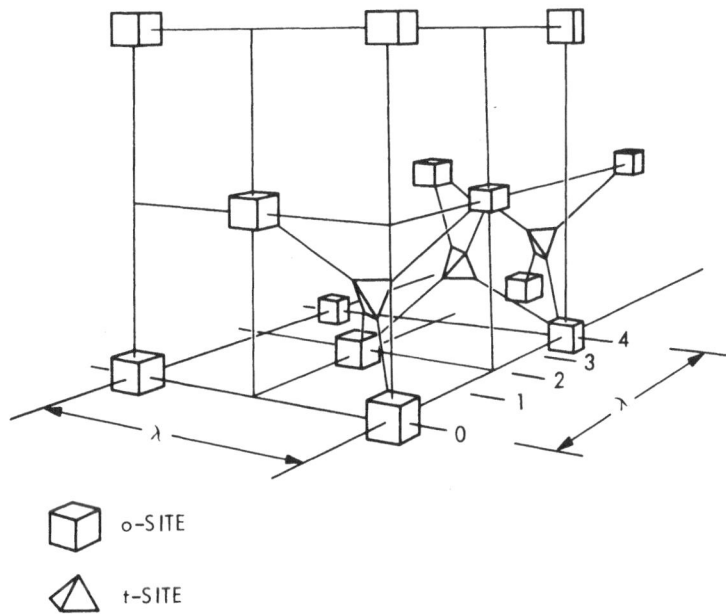

o-SITE

t-SITE

Fig. 4. The interstice network within the f.c.c. structure. Only the tetrahedral interstices within the lower, right-hand quarter of the unit cell are shown. The octahedral interstices on the front and bottom are shown. In the zinc blende structure every other tetrahedral interstice is normally occupied.

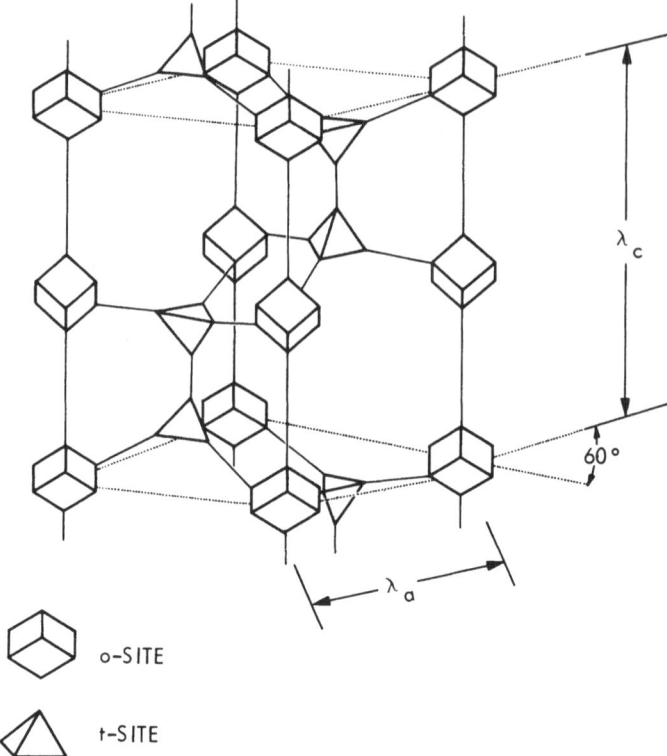

o-SITE

t-SITE

Fig. 5. The interstice network within the h.c.p. structure. A unit
cell is shown containing a two *o* sites and four *t* sites. In the
würtzite structure the upper one of the pair of tetrahedral interstices is normally occupied.

much larger anions. In these structures there are four anions per unit cell,
four interstices wherein a cation would have octahedral coordination with the
six anions about it, and eight tetrahedral interstices wherein a cation would
find itself sorrounded by four anions. In the NaCl structure the cation is
normally within the octahedral site, hereafter to be called the *o* site, while in
the zinc blende structure the cation normally resides within the tetrahedral
sites, *t* sites, although only half of these sites are actually filled.

It is reasonable to hypothesize that a magnesium ion moving through
the MgO (NaCl-type) lattice, to take an example, would not make direct
jumps between normal cation sites. For the cation to do so it would have to
pass through the point where two oxygens are almost in direct contact.
Instead, the magnesium would more likely try to go around the oxygen ions.
In leaving its *o* site the magnesium would probably take a path through an
orifice formed by three of the oxygen ions touching on its *o* site. This jump

would bring it into a site where it was in tetrahedral coordination with the anions. It could not continue in the direction of motion which brought it into this t site because it would then be heading directly toward the center of the anion on the far side. It would then be expected to leave by another jump through an orifice like that through which it entered, and any of these would lead to another o site like that from which it originally departed. In any actual process the completion of a transit from one normal site to another will be contingent upon that second one being vacant if the diffusion is by a vacancy mechanism, or if it is by an interstitial or interstitialcy mechanism the presence of such point defects will be necessary for migration. For the present the migration mechanism will be ignored and attention will only be concentrated on the properties of the networks themselves. The processes in some actual crystals will be considered later.

DIFFUSION IN THE FACE-CENTERED CUBIC STRUCTURE

We will now consider atom migration between interstitial sites in the face-centered cubic lattice. Each o site has eight paths leading away from it toward a t site and each t site has four paths leading to an o site. Therefore, of the possible o–t jumps from an o site only $\frac{1}{8}$ will be along a given path, and of the possible t–o jumps from a t site only $\frac{1}{4}$ will be along a given path from that site. Inspection of the interstice network, Fig. 4, will reveal that there are eight paths reaching away to the right from the cube face of a unit cell. The flux will then be

$$J = (1/\lambda^2)8j \tag{25}$$

where j is the flux along a single path channel and λ is the cell dimension. Taking the o sites on the surface to be plane number 0 and the adjoining t-sites on plane 1, with the next plane of o sites being plane 2, one can designate the individual channel flux as

$$j = F_{o_0}(\phi_{ot}/8) - F_{t_1}(\phi_{to}/4) \tag{26}$$

where F_{o_0} and F_{t_1} are the fractions of atom occupancy on the o_0 site and t_1 site, respectively, and ϕ_{ot} and ϕ_{to} are the frequencies of jump per atom from the two types of sites.

The value of F_{t_1} can be determined by assuming that it is regulated by the condition of dynamic equilibrium,

$$F_{t_1}\phi_{to} = 2F_{o_0}(\phi_{ot}/8) + 2F_{o_2}(\phi_{ot}/8) \tag{27}$$

Substitution of this and (26) into (25) gives

$$J = (1/2\lambda^2)\phi_{ot}(F_{oo} - F_{o2}) \tag{28}$$

The concentration gradient will be

$$\frac{dC}{dx} = \frac{C_2 - C_0}{\lambda/2} = \frac{4}{\lambda^4}[(2F_{o2} + 4F_{t3}) - (2F_{oo} + 4F_{t1})] \tag{29}$$

By using Eq. (27) this can be reduced to

$$\frac{dC}{dx} = \frac{8}{\lambda^4}(F_{oo} - F_{o2})\left(1 + \frac{\phi_{ot}}{\phi_{to}}\right) \tag{30}$$

and by substituting this and (28) into Fick's Eq. (2) it follows that

$$D = \frac{\lambda^2}{16}\frac{\phi_{ot}\phi_{to}}{\phi_{ot} + \phi_{to}} \tag{31}$$

Under the normal condition, $\phi_{to} \gg \phi_{ot}$, this becomes

$$D = (\lambda^2/16)\phi_{ot} \tag{32}$$

This value for the diffusion coefficient would apply to the case of interstitial diffusion in a metal having its atoms in cubic close-packing. It should also apply to the case of cation diffusion in the NaCl structure if the diffusion is by a vacancy mechanism. However, this calculated cation diffusion coefficient would have to be reduced by the correlation factor. It would probably not be applicable to interstitialcy diffusion, because the sequence of energy barriers to be surmounted for diffusion by that mechanism would probably differ from those proposed here.

Diffusion in the zinc blende structure can be treated in the same way as for the NaCl structure, but a distinction must now be made between the two tetrahedral interstices, since half are normally occupied while the other half of the t sites and all of the o sites are normally unoccupied. The arrangement of the normally occupied sites, hereafter called n sites, is such that they are in tetrahedral arrangement about the o sites and are also in tetrahedral arrangement about the anions. This may also be regarded as a structure related to the diamond when the bonding between cations and anions is more covalent than ionic.

The flux equation analagous to (25) must now take the form

$$J = (1/\lambda^2)(4j_{ot} + 4j_{on}) \tag{33}$$

It should be noted too that from an o site there are now four paths leading to t sites and four leading to n sites, rather than eight o–t paths as in the

NaCl case. With recognition of these differences it will be found that

$$D = \frac{\lambda^2}{16} \frac{\phi_{no}\phi_{to}(\phi_{on} + \phi_{ot})}{\phi_{no}\phi_{to} + \phi_{no}\phi_{ot} + \phi_{to}\phi_{on}} \tag{34}$$

It can be assumed in those compounds where the intrinsic disorder is not large that $\phi_{no} \ll \phi_{on}$, and the diffusion coefficient then simplifies to the form

$$D = \frac{\lambda^2}{16}\phi_{no}\left(1 + \frac{\phi_{ot}}{\phi_{on}}\right) \tag{35}$$

Thus, if after jumping from a normal site into an octahedral site, the atom generally makes its next jump into a tetrahedral site, $\phi_{ot} \gg \phi_{on}$, the migration will be an interstitial diffusion mechanism, and the atom may travel quite a distance before jumping into another normal site again. On the other hand if $\phi_{ot} \ll \phi_{on}$, it will be found that the rate-limiting step is simply the jump from the normal to the o site, and

$$D = (\lambda^2/16)\phi_{no} \tag{36}$$

Equations (35) and (36) should be expected to apply to those zinc blende structures which are primarily ionic and in which there is not a directional bond between atoms. The diffusion mechanisms in the latter case would probably have to be described by a different network, since the metastable configurations would probably differ from those envisioned above.

DIFFUSION IN THE HEXAGONAL CLOSE-PACKED STRUCTURE

We now turn to the network of interstices in the hexagonal close-packed structure. It will be found (Fig. 5) that each o site is connected to six t sites and two other o sites, while each t site is connected to three o sites and to another t site. In a unit cell whose base area is $(\sqrt{3}/2)\lambda_a^2$ the flux in the axial direction can be seen to be

$$J_c = (2/\sqrt{3}\,\lambda_a^2)(j_{oo} + j_{tt}) \tag{37}$$

The concentrations on the t_0 and t_1 sites are given by the equations

$$F_{t_0}(\phi_{to} + \phi_{tt}) = 3F_{oo}(\phi_{ot}/6) + F_{t_1}\phi_{tt} \tag{38}$$

and

$$F_{t_1}(\phi_{to} + \phi_{tt}) = 3F_{o_1}(\phi_{ot}/6) + F_{t_0}\phi_{tt} \tag{39}$$

These equations and the expression for the concetration gradient,

$$\frac{dC}{dx} = \frac{8}{3\lambda_a^2\lambda_c^2}[(F_{o_1} - F_{oo}) + 2(F_{t_1} - F_{to})] \qquad (40)$$

will lead to a value for the diffusion coefficient,

$$D_c = \frac{\lambda_c^2}{4} \frac{1}{1 + (\phi_{ot}/\phi_{to})}\left[\frac{1}{2}\phi_{oo} + \frac{1}{2}\phi_{ot}\phi_{tt}/(\phi_{to} + 2\phi_{tt})\right] \qquad (41)$$

The diffusion coefficient in the basal plane direction can be calculated by analagous methods, with the result

$$D_a = \frac{\lambda_a^2}{6} \frac{\phi_{ot}\phi_{to}}{\phi_{ot} + \phi_{to}} \qquad (42)$$

These values for D_a and D_c in the hexagonal close-packed network should be completely applicable to diffusion of interstitial atoms within a metal. Within oxides, however, their use will be limited in the above form because correlation and near-neighbor effects may be important. These will be mentioned below in the discussion of FeS, which has the NiAs structure, the hexagonal equivalent of the NaCl structure and the one to which (41) and (42) apply.

The hexagonal equivalent of the zinc blende structure will now be discussed, the ZnO wurtzite structure. The distinction between t sites which are not normally occupied and tetrahedral sites which are occupied, n sites, will be made as before. It turns out that the derivation of the diffusion coefficient for the axial direction is no more difficult than in any of the cases mentioned so far, and it need merely be stated as

$$D_c = \frac{\lambda_c^2}{4} \frac{1}{1 + (\phi_{no}/\phi_{on}) + (\phi_{nt}/\phi_{tn})}\left[\frac{\phi_{oo}\phi_{no}}{2\phi_{on}} + \frac{\phi_{no}\phi_{ot}\phi_{tn}}{\phi_{nt}\phi_{ot} + \phi_{tn}\phi_{on} + \phi_{to}\phi_{on}}\right] \qquad (43)$$

For the calculation of the coefficient for diffusion in the basal plane direction it will be convenient to refer to Fig. 6, where the separation of successive planes is $\lambda_a/2$, the width of a unit cell face for present purposes will be taken as $\sqrt{3}\,\lambda_a$, and the height of the cell is λ_c. The flux will then be

$$J_a = (2/\sqrt{3}\,\lambda_a\lambda_c)(j_{no} + j_{on} + j_{to} + j_{ot}) \qquad (44)$$

where the values for the flow along each channel will be

$$j_{no} = \tfrac{1}{3}(F_{no}\phi_{no} - F_{o_1}\phi_{on}) \qquad (45a)$$

$$j_{on} = \tfrac{1}{3}(F_{oo}\phi_{on} - F_{n_1}\phi_{no}) \qquad (45b)$$

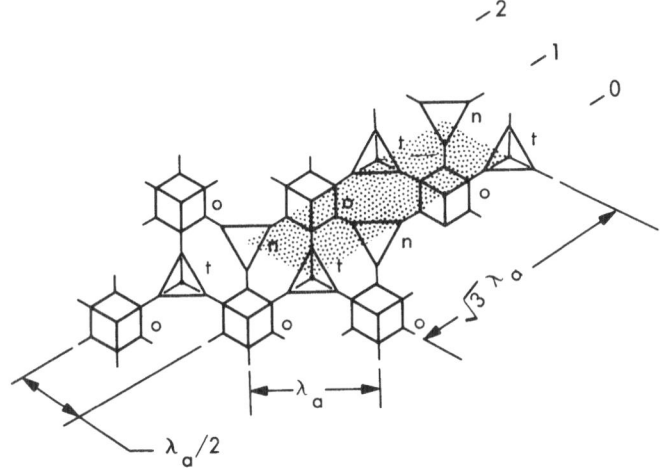

Fig. 6. A view of interstitial sites in the h.c.p. structure, top view of the basal plane. The tetrahedral interstices have been separately identified as normally unoccupied t sites and normally occupied n sites, as they are in the wurtzite lattice.

$$j_{to} = \tfrac{1}{3}(F_{to}\phi_{to} - F_{o_1}\phi_{ot}) \tag{45c}$$

$$j_{ot} = \tfrac{1}{3}(F_{o_0}\phi_{ot} - F_{t_1}\phi_{to}) \tag{45d}$$

To solve these equations we will need values for the fractions F_{o_0}, F_{o_1}, F_{t_0}, and F_{t_1}. To find these it is necessary to solve four simultaneous linear equations which can be conveniently presented in matrix form:

Coefficients:

F_{o_0}	F_{o_1}	F_{t_0}	F_{t_1}		
$(\phi_{ot} + \phi_{on})$	0	$-\phi_{to}$	0	$= \phi_{no}F_{n_0}$	(46a)
0	$(\phi_{ot} + \phi_{on})$	0	$-\phi_{to}$	$= \phi_{no}F_{n_1}$	(46b)
$-\phi_{ot}$	0	$(\phi_{to} + \phi_{tn})$	0	$= \phi_{nt}F_{n_0}$	(46c)
0	$-\phi_{ot}$	0	$(\phi_{to} + \phi_{tn})$	$= \phi_{nt}F_{n_1}$	(46d)

It turns out in this case that Eqs. (46a) and (46c) can be paired off and solved together independently of the remaining pair, and this simplifies the arithmetic. It is also convenient to use the relationship,

$$\phi_{to}\phi_{on}\phi_{nt} = \phi_{tn}\phi_{no}\phi_{ot} \tag{47}$$

which is an expression of the fact that there is no net circularity to the migration of atoms within a structure, i.e., a jump from an n site to a t site to an o site and finally a return to the n site is as probable as a jump from an n site to an o site to a t and to the n site again. It will be found that

$$j_{no} = j_{on} = \tfrac{1}{3}(F_{n_0} - F_{n_1})\phi_{no} \tag{48a}$$

$$j_{to} = j_{ot} = \tfrac{1}{3}(F_{n_0} - F_{n_1})(\phi_{nt}\phi_{to}/\phi_{tn}) \tag{48b}$$

The concentration gradient will then be

$$\left(\frac{dC}{dx}\right)_a = \frac{8}{\sqrt{3}\,\lambda_a^3\lambda_c}(F_{n_1} - F_{n_0})\left(1 + \frac{\phi_{no}}{\phi_{on}} + \frac{\phi_{nt}}{\phi_{tn}}\right) \tag{49}$$

Combining (2), (44), (48), and (49) gives the diffusion coefficient

$$D_a = \frac{\lambda_a^2}{6} \frac{\phi_{no} + (\phi_{to}\phi_{nt}/\phi_{tn})}{1 + (\phi_{no}/\phi_{on}) + (\phi_{nt}/\phi_{tn})} \tag{50}$$

As has been pointed out before this equation and Eq. (43) for D_c in würzite have been derived for diffusion of an isolated atom in a network. The atom has been pictured as free to move from normal to metastable to other normal sites without interference from atoms on other neighboring sites. In some cases this picture may be a satisfactory description of the actual process within a lattice, but in many instances it will be necessary to modify the equations to provide a better description of next-nearest neighbor' interactions.

DIFFUSION IN REAL OXIDES

Two examples of diffusion will be selected to illustrate how real materials may be more complicated in their behavior than has been pictured in the discussion to this point: diffusion of iron in ferrous sulfide and diffusion of beryllium in beryllium oxide. The first will serve to bring out the possible dominance of the correlation factor in a diffusion process, and the second will draw attention to the probable interaction between a migrating atom and the neighboring filled n sites.

In ferrous sulfide, which has the nickel arsenide structure, the sulfide ions are in hexagonal close-packing and the iron ions occupy the octahedral interstices. Thus the equations for diffusion within the h.c.p. structure, (41) and (42), might be expected to apply. It has been found in ferrous sulfide ([1]) that the ratio of diffusion coefficients in the two directions is

$$D_c/D_a = 1.7 \tag{51}$$

a relationship which is independent of temperature and also of deviation from stoichoimetry in the iron-deficient structure, FeS. This suggests that the rate-limiting step for diffusion in the two directions is the same, and that it must therefore be the o–t jump, with the o–o jump being of minor importance.

It is possible, however, that the o–o jump is the most frequent one in the FeS structure. That is, a vacancy on an o-site could most likely be filled by the jump of an iron ion on a neighboring o site making a direct jump of the o–o type. The vacancy could be highly mobile in this way, but its random migration would not lead to a net movement of the iron ions. The number of vacancies passing a particular site along one of these axial channels of the crystal in one direction would be equal to the number passing in the reverse direction, and there would be no net migration of iron ions. In other words the correlation factor would be zero. For an interchange between iron ions to take place it would be necessary for one of them to jump aside, an o–t jump, to allow the other to pass. Thus the rate-limiting step for diffusion in the two directions might be the same even though the value of ϕ_{oo} was very much larger than ϕ_{ot}.

In beryllium oxide, which has the wurtzite structure, the diffusion of beryllium has been studied [2]. Here again it appears that the ratio of diffusion,

$$D_c/D_a = 1.3 \tag{52}$$

is independent of temperature and concentration of impurities which govern the vacancy concentration. The rate-limiting step which is common to diffusion in the two directions, Eqs. (43) and (50), is the n–o jump. Close inspection of the structure reveals, however, that if there is a vacancy in an n site and it is about to be filled by a jump from an adjoining o site, the beryllium ion in the o-site is rather close to the two other n sites which would likely be filled. The electrostatic repulsion between this o-site beryllium and the two nearby n-site ions might make the entrance into the o site a highly unfavored step in the first place. This difficulty could be avoided if one n-site beryllium were to shift momentarily to its adjoining t site. The sequence of atomic configurations might be of the sort tabulated below, with reference to Fig. 7:

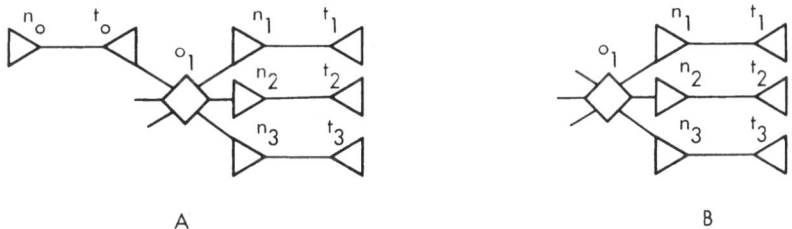

A   B

Fig. 7. A two-dimensional representation of the interstice network in the wurtzite structure.

For diffusion in the c *direction* [Fig. 7(a)],

Step 1.	Be$_1$ ion in n_0	Be$_2$ ion in n_1	vacancy in n_2
Step 2.	t_0	n_1	n_0, n_2
Step 3.	t_0	t_1	n_0, n_1, n_2
Step 4.	o_1	t_1	n_0, n_1, n_2
Step 5.	n_2	t_1	n_0, n_1
Step 6.	n_2	n_1	n_0

For diffusion in the a *direction* [Fig. 7(b)]

Step 1.	Be$_1$ ion in n_1	Be$_2$ ion in n_2	vacancy in n_3
Step 2.	n_1	t_2	n_2, n_3
Step 3.	o_1	t_2	n_1, n_2, n_3
Step 4.	n_3	t_2	n_1, n_2
Step 5.	n_3	n_2	n_1

It can be seen that there is a step in these sequences which is common to migration in both directions. In the c direction there is a step 1, 2—an n–t jump, which occurs again in step 2, 3. For diffusion in the a direction step 1, 2 also involves an n–t jump. Thus it might be that the diffusion in the two directions is limited by this n–t jump rather than by an n–o jump.

This sequence of configurations, which involves two ions at a time rather than just the one which we have been discussing so far, is susceptible to the same network analysis as already outlined. The points in this new network would now represent more complicated configurations than just single atom displacements, but the sequence of jumps between points and the branching between possible connecting paths could be analyzed in basically the same way.

DISCUSSION AND CONCLUSION

The purpose of this chapter has been to draw attention to the possibility of using a network analysis technique for treating diffusion in oxides. Such analyses should be able to articulate possible relationships between individual jumps within a structure and measured values of diffusion coefficients. The determination of the values of individual jump frequencies may then be approached in a number of different ways, and the following have some prospect of being useful:

1. Calculations of the energy of ions at various metastable points within a structure can be carried out using crystal field theory and by the same general routines as are used in calculations of the electrostatic energies of ionic solids.

2. In anisotropic crystals the value of the diffusion coefficient in several directions can be measured, and the ratio of the values as a function of temperature can be determined.

3. Possible deviations from a simple Arrhenius-type dependence of $\log D$ on $1/T$ may be revealing.

4. Studies of the effect of variations in stoichiometry and doping on diffusion rates should be of particular value.

5. Determination of the correlation factor by comparisons between ionic conductivity and self-diffusion measurements or by measurements of the mass-dependent isotope fractionation in tracer diffusion should provide clues about the mechanisms.

6. Studies of dielectric and stress relaxation rates have helped in the identification of atomic jump mechanisms. This has proven useful in the analysis of carbon interstitial diffusion in iron, for example [3], where the jump frequencies have been identified.

In summary the mathematical approach introduced here has not attempted to treat the theory of jump frequencies. Of course, the study of diffusion in ceramics will only begin reach a good level of sophistication when these can be calculated and found to be in agreement with experiment. This chapter has attempted, however, to point a way toward the construction of a bridge which can lead from diffusion measurements to the determination of individual frequencies so that the relationships between theory and experiment can then be assessed.

ACKNOWLEDGMENTS

This work was begun while the author was a candidate for the Ph.D. degree at Princeton University, and Prof. C. E. Birchenall made many helpful suggestions at that time.

REFERENCES

1. R. H. Condit and C. E. Birchenall, "Self-Diffusion of Iron and Sulfur in Ferrous Sulfide," Air Force Office of Scientific Research Technical Note 60–245, Contract No. AF 49(638)–533, Princeton University. Also University Microfilms, Ann Arbor, Michigan, L. C. Card No. Mic. 60–4974.
2. R. H. Condit and Y. Hashimoto, "Self-Diffusion of Beryllium in Polycrystalline Beryllium Oxide," *J. Am. Ceram. Soc.* **50**: 425–432 (1967).
3. R. H. Condit and D. N. Beshers, "Interstitial Diffusion in the B.C.C. Lattice," *Metallurg. Trans. AIME* **239**: 680–683 (1967).

DISCUSSION

R. J. Bratton (Westinghouse Research Labs.): The models assume that the anions are essentially stationary. However, at elevated temperature, where both anions and cations are mobile, does the movement of the anions influence the cation diffusion paths and mobilities?

Answer: The intrinsic concentration of point defects which participate in migration is small for both the cations and anions. It seems that the random probability of finding a cation–anion defect pair is very small. Binding of such a pair should be negligible at normal diffusion temperatures, and I do not see any reason to expect that such a pair would be mobile. Therefore the contribution of such interaction to cation diffusion might be expected to be small. I am assuming that the cations are small and can move between the large anions without displacing them greatly. A vacancy on the cation site is likely to be well shielded from neighboring cation sites by the large, somewhat polarized anions. Anion migration, on the other hand, might be a different situation. The removal of an oxide ion from MgO would leave six magnesium ions facing one another, and it may be that the energy of this system will be much less if there is a cation vacancy associated with the anion vacancy. The defect pair might be the mobile unit in anion diffusion. If this is in fact the case it might be that in oxides the anion diffusion rates are less sensitive to doping than are cation diffusion rates, because while doping might increase one defect concentration while suppressing the other, their product would not be greatly changed. This would be the concentration term for the defect pair.

A. Franklin (ARPA): The important advance here seems to be the focusing of attention on the details of the path taken by an ion moving from one "normal" (crystallographically equivalent) site to the next. Diffusion treatments normally use an effective jump frequency, which is perfectly proper. To relate this to jump frequencies calculated by lattice dynamic methods the details of the path must be considered. You have shown that where intermediate metastable sites exist a reflection coefficient must be included and can be calculated on the basis of detailed models. This assumes some intermediate sites are metastable. For instance, in the NaCl structure the interstitial site normally has a zero Madelung potential and will present an effective negative Madelung potential to a diffusing cation. Electrostatic effects may combine with repulsive terms to make this site metastable, making your treatment important for this case.

C. H. Greene (State University of New York, Ceramic College): Does not the actual diffusion mechanism involve some cooperative phenomena? Before a cation can jump from one normal site to a second normal site the cation occupying this second site must be removed in some way. It is likely that the rate at which cations jump out of such destination sites is increased by the presence of jumping cations at intermediate non-normal tetrahedral or octahedral sites.

Answer: The major part of the mathematical treatment in this chapter has dealt with the case where a single atom is regarded as moving through a network of interstitial positions, or voidal positions, as L. Azaroff proposed to call them a few years ago (4). It has been proposed that one can ask the question, "If there is a vacancy on a cation site, how will that vacancy be filled?," and then answer this question by using the network analysis directly. Your question deals with a slightly different mechanism, essentially an interstitialcy mechanism where an atom on an interstitial site drives an atom from a normal site into another interstitial site while it replaces it. Certainly, as you say, the probability of displacement of the second ion from the destination site must be enhanced by the presence of the first ion in its neighborhood. The exact sequence of metastable configurations might be fairly complicated, however, and to map out the paths whereby these configurations may go from one to another will probably be quite complicated, and the solution of the flow problem between points in this network will probably require the services of a computer.

Paul Levy (*Brookhaven Data Labs.*): I wonder if it is realistic to consider this multiple lattice site configuration which Dr. Franklin described in terms of the fact that the rate-controlling step is from one of the circle points over the large cell points.

A. Franklin (*ARPA*): If you look at reaction theories, the jump rate is some function including geometrical factors and an exponential term which contains the free energy of motion over kT and a transition probability as a sort of a reflection coefficient. Dr. Condit is inferring that if you do not have a perfectly rounded potential function, if for some reason there is a reflection possibility at the top of the barrier (and Dr. Condit took this into account in his treatment), then you must modify the jump frequency by an additional term. That is exactly what Dr. Condit has introduced by having the atom pause and decide that it can go either way, thereby introducing a reflection coefficient. The activation energy will be controlled by the highest activation energy in the passage, but the reflection coefficient can be introduced separately.

Answer: I have not tried to take into account the possibility that the ion may remember to some extent the direction from which it arrived into a site. The notion of a reflection coefficient referred to by Dr. Franklin is an interesting one, and I will want to look into it further.

R. Coble (*M.I.T.*): I might comment further about other qualitative or semiquantitative considerations of the importance of the occupancy of the non-normally occupied sites during diffusion of cations in close-packed oxygen matrices. In MgO there are no normally empty octahedral interstices, hence Schottky disorder is required. The observed activation energies in this case are less than those observed for cations in M_2O_3 compounds, in which non-normally occupied sites ($\frac{1}{3}$ of the octahedral holes) might provide paths for cation diffusion with no Schottky disorder. However, because the observed energies are higher (and the diffusivities lower), it is tentatively concluded that Schottky disorder is also required for cation diffusion in M_2O_3 compounds. There is still the question of whether intrinsic diffusivity has been observed in any of these material.

Answer: I think that we might agree that the presently available experimental data in MgO and such M_2O_3 compounds as Al_2O_3 cannot help us out very much at this time, because intrinsic diffusion coefficients have probably not yet been measured. The network calculation for the diffusion coefficient for the corundum lattice turns out to be quite involved, and I have not yet been able to solve the set of simultaneous equations which comes forth when you account for all of the possible points in the network for this structure.

ADDITIONAL REFERENCES

4. L. Azaroff, *J. Appl. Phys.* **32**: 1658–65 (1961).

Chapter 16

Grain Boundary Grooving at a Solid-Solid Interface

A. H. Feingold*, J. M. Blakely, and Che-Yu Li

Department of Materials Science and Engineering
Cornell University, Ithaca, New York

Grain boundary grooving at an internal solid–solid interface has been studied in the Ni–Al$_2$O$_3$ system. Data obtained from the kinetics of growth of Al$_2$O$_3$ grain boundary grooves indicate that at high temperatures the process is controlled by diffusion of aluminum in the nickel phase. The Ni–Al$_2$O$_3$ interfacial tension has been calculated from the kinetics constant using measured solubilities and published values of the diffusivities and it agrees well with the value obtained from equilibrium shape measurements. The relevance of these results to considerations of the thermal stability of dispersed systems are discussed.

INTRODUCTION

In a system consisting of a dispersed phase in a solid matrix the stability of the system with respect to both particle shape and size distribution will be determined by the magnitudes of interfacial tensions, solubilities, and volume or interfacial diffusivities. A number of analyses have been made of the process of precipitate aging in solids ([1-4]). These are primarily concerned with (1) the problem of the compatibility of the various atomic fluxes and the constraints on the system, and (2) the statistical problem associated with describing the material flux among particles for some assumed initial size and spatial distribution. Since this distribution is rather difficult either to control or measure in practice, the validity of a particular analysis is somewhat difficult to test in any rigorous way. For this reason it is desirable to study the transport processes occurring at solid–solid interfaces in situations for which a rigorous solution can be found for interface shape development with time. The measurement of such shape changes can then be used to derive values of the effective diffusivities and interfacial tensions which are also appropriate to more complex geometrical situations.

It is a common feature of diffusional mass transport phenomena involving shape changes that when local equilibrium exists at the interface the rate

---

*Present address engineering materials laboratory, engineering research division, E. I. Dupont de Nemours and Co. (Inc.), Wilmington, Delaware.

is determined by the product of an interfacial tension, a concentration, and a diffusion coefficient ([5]). In a one-component metal system, for example, the diffusion coefficient is usually taken as the tracer diffusivity corrected for correlation effects. For a multicomponent system the effective diffusivity appropriate to the transport process is a combination of the diffusivities of the separate components and their solubilities ([6,7]). The experimental work to be discussed in this chapter concerns the study of the development of grain boundary grooves at the Ni–Al_2O_3 interface.

For the one-component case Mullins ([8]) has shown that the distance W between the maxima of the groove profile may be expressed in terms of the isothermal annealing time t as

$$W = (K't)^{1/3} \tag{1}$$

where

$$K' = 125(nD_V V^2 \gamma_S/kT) \tag{2}$$

with n the number of atoms per unit volume, D_V the mass transport volume diffusivity, V the atomic volume, and γ_S the surface tension. For the case of groove development at a Ni–Al_2O_3 interface, where volume diffusion in the metal is the controlling process, expressions similar to Eqs. (1) and (2) apply provided that the appropriate solid–solid interfacial tension is used and D_V is replaced by an effective diffusivity ([6,7]):

$$K = 125(nD)_{eff} V_{mol}^2 \gamma_i/kT \tag{3}$$

where γ_i is the specific interfacial free energy, V_{mol} is the volume of an Al_2O_3 molecule in Al_2O_3, and the term representing the effective diffusivity is given by

$$\frac{(nD)_{eff}}{kT} = \frac{n_{Ni}B_{Ni}(nB)^{mol}V_{Ni}^2}{n_{Ni}B_{Ni}V_{Ni}^2 + (nB)^{mol}V_{mol}^{*2}} \tag{4}$$

in which

$$(nB)^{mol} = \frac{B_{Al}B_0 n_{Al} n_0}{\beta^2 B_{Al} n_{Al} + \alpha^2 B_0 n_0} \tag{5}$$

with n_{Ni}, n_{Al}, and n_0 the numbers of atoms per unit volume of nickel, aluminum, and oxygen, respectively, in the nickel phase[†]; V_{Ni} is the atomic

---

†Equations (3)–(5) correspond to the result of the discussion of Eqs. (21)–(33) in ([7]). Equations (25a) and (26) of ([7]) represent an arbitrary separation of the product of concentrations and mobilities which appear in the flux equations. This allows a convenient definition of $B_{A_aB_b}$ and $D_{A_aB_b}$ with the correct dimensions.

We have chosen to define $(nB)^{mol}$ to represent the term multiplying $\nabla \mu_{A_aB_b}$ in Eq. (25) of ([7]), emphasizing that it always appears as a product containing the dimensions of concentration and mobility and that only the flux of the molecules is well defined. The $(nD)_{eff}$ is defined similarly for the case where both volume and concentration constraints exist.

volume of nickel and V_{mol}^* the difference in volume of Al_2O_3 in the oxide and in the metal; B_{Ni}, B_{Al}, and B_0 are the mobilities of nickel, aluminum and oxygen in the nickel lattice. For Al_2O_3 we have $\alpha = 2$ and $\beta = 3$.

In the present work the measured values of the kinetics coefficient K are compared with the theoretical predictions from Eqs. (3), (4), and (5). The solid–solid interfacial tension can then be calculated from the kinetics constant using measured values for the solubilities and published data for the diffusivities. Another method will also be described for the determination of the solid–solid interfacial tension from the equilibrium shape established at the solid–solid interface. The values obtained from these two methods will be compared.

EXPERIMENTAL PROCEDURE

The system $Ni–Al_2O_3$ was chosen for study primarily because it is of interest as a practical composite system ([9]), but also because single-crystal aluminum oxide, sapphire, is transparent in the visible region. The latter feature allows direct observation of internal interface morphologies without any disturbance to the system.

The arrangement for sample preparation is shown schematically in Fig. 1. High-purity (99.999%) nickel from the United Mineral and Chemical Company was used. Single crystals of Al_2O_3 in the form of disks 0.5 in. in diameter and 0.020 in. thick were obtained from the Linde Division of Union Carbide Corporation. These were oriented with the c axis at 60° to the surface

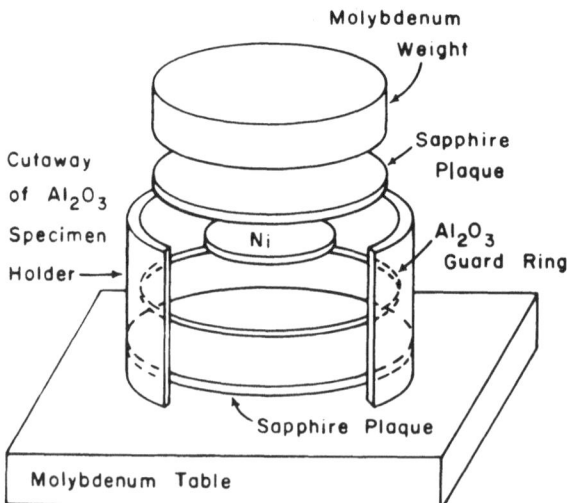

Fig. 1. Schematic of sapphire–Ni sandwich fabrication.

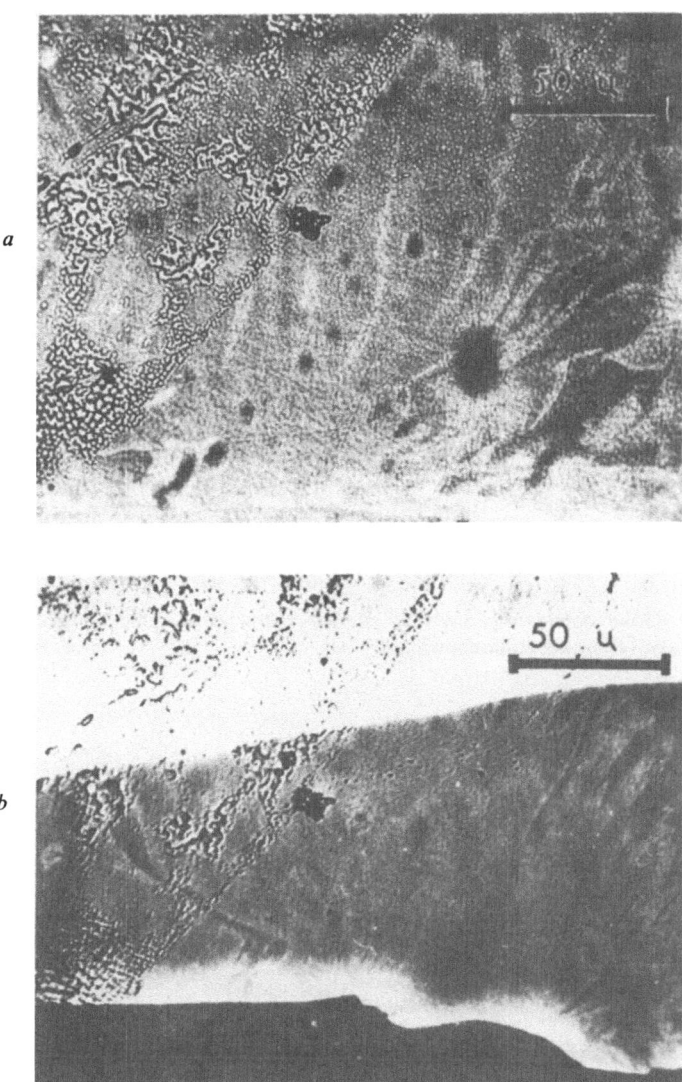

Fig. 2. Delineation of low-angle boundaries by polarization. Sapphire plaque as seen with (*a*) ordinary transmitted light; (*b*) polarized transmitted light.

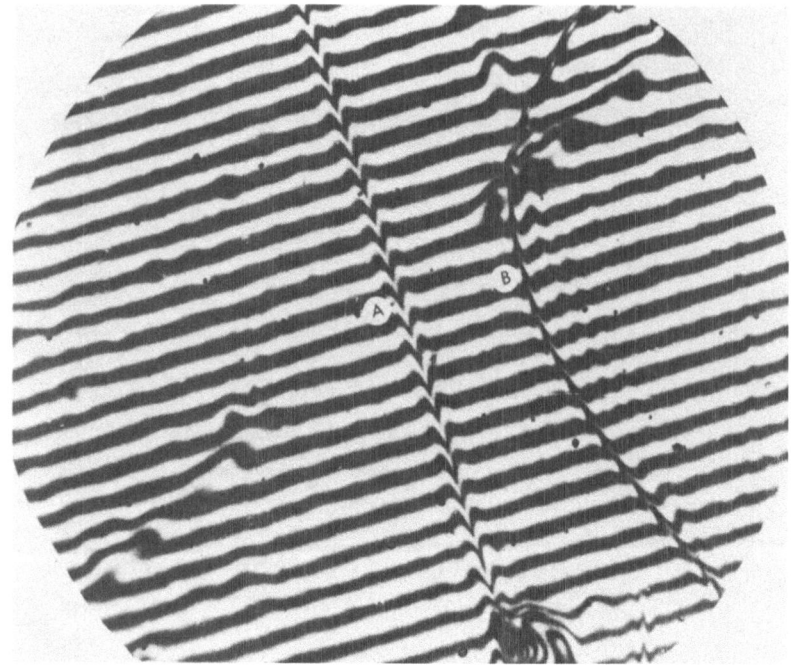

Fig. 3. Grain boundary grooves at internal interface: (A) aluminum oxide grain boundary, (B) nickel grain boundary.

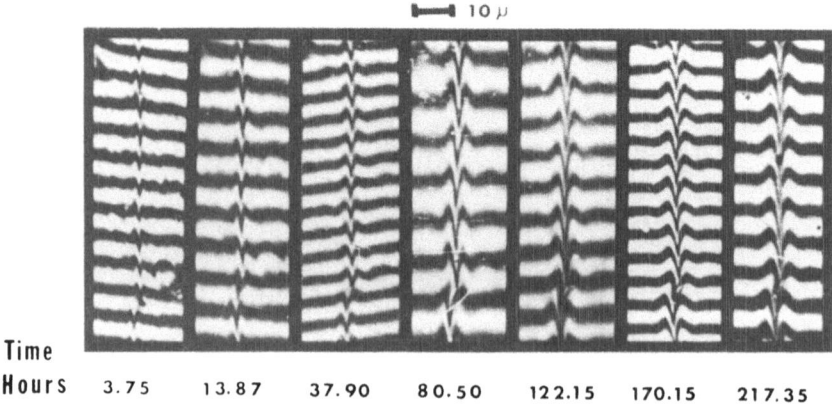

Fig. 4. Grain boundary growth shown by interferograms of groove annealed at 1400°C for increasing times.

normal. Chemically polished nickel disks (0.25-in. diameter × 0.030 in.) were sandwiched between sapphire crystals and melted under a helium atmosphere in an inductively heated molybdenum tube. A molybdenum weight served to compress the sandwich while the nickel was molten. This ensured a large area of solid–solid interface for observation.

The theory outlined in the introduction applies to the development of grooves at the interface due to grain boundaries either in the nickel or in the Al_2O_3. After fabrication both types were visible, but considerable migration of the nickel boundaries occurred during subsequent annealing. Kinetic studies were thus confined to grooves formed at Al_2O_3 boundaries. These boundaries were of small angle misorientation ($1°$–$5°$) and were identified before sample preparation by observation between crossed polarizers (Fig. 2). The samples were then annealed in an N.R.C. tantalum resistance furnace under a dynamic vacuum of less than 10^{-5} mm Hg for various annealing times at temperatures of 1350, 1375, 1400, and 1425°C. After each annealing period the interface was examined and photographed with a Zeiss interference microscope. Figure 3 is an interferogram of an interface and shows profiles of both aluminum oxide (A) and nickel (B) grain boundary grooves. The continuity of the interference fringes across both boundaries illustrates the intimate contact at the metal–ceramic interface. Figure 4 shows a typical series of interferograms taken after successive anneals at 1400°C.

The interfacial tension γ_i was determined from measurements of the

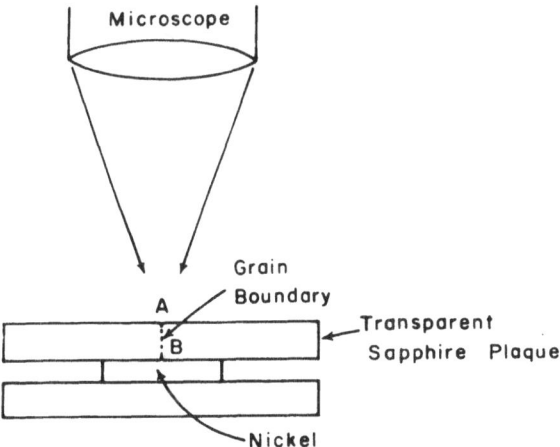

Fig. 5. Schematic of groove observation technique with Zeiss interference microscope. Focusing at A yields the external dihedral angle, which depends on the ratio of surface to grain boundary tension. Focusing at B yields the internal dihedral angle, which depends on the ratio of interfacial to grain boundary tension.

equilibrium dihedral angles of the grooves formed at the intersections of a particular Al_2O_3 grain boundary with the metal–ceramic interface and with external Al_2O_3 surface (Fig. 5). A combination of these two measurements leads to a value of the ratio of interfacial tension γ_i to Al_2O_3 surface tension γ_S. The latter quantity has been measured previously by Kingery[10] at 1850°C in a helium atmosphere, so that the interfacial tension may be derived. In the present experiment the grooves on the Al_2O_3 surface were formed initially at 1500°C in helium during sample preparation. No systematic changes in the equilibrium dihedral angles at the surface were observed after subsequent vacuum anneals.

RESULTS AND DISCUSSION

Data on the grain boundary groove angles and the resulting values of the ratio γ_i/γ_S are summarized in Table I. The true dihedral angles (2θ) of the grooves are deduced from the angles (2α) measured directly from the interferograms using the expression

$$\tan \theta = (n2L/\lambda M) \tan \alpha \qquad (6)$$

where L is the fringe spacing, λ the wavelength of the light used (Th, 0.5 μ), and M the linear magnification. For grooves at the external Al_2O_3 surface n the refractive index is that of air and is taken as unity. For the internal grooves n is the refractive index of sapphire, 1.77. Due to the small dimensions of the grooves the ratio of γ_i and γ_S is accurate only to about $\pm 20\%$. The

TABLE I
Solid–Solid Interfacial Tensions

| Temperature (°C) | Dihedral Angle | | γ_i/γ_S | γ_i * |
	Al_2O_3–nickel interface (deg)	Al_2O_3 surface (deg)		
1350	87.4	81.9	3.17	2870
1375	87.3	83.7	2.39	2160
1375	87.2	83.7	2.29	2070
1400	86.2	80.0	2.59	2340
1400	86.5	79.5	2.94	2660
1400	86.7	80.6	2.80	2530
1400	85.6	79.0	2.50	2260
1425	87.3	80.9	3.31	2990

*Assuming $\gamma_S = 905$ ergs/cm². The γ_i average ≈ 2500 ergs/cm²

data are thus not sufficiently accurate to detect any significant trend with temperature. Using the values from Table I and Kingery's value of 905 ergs/cm^2 for γ_S we obtain an average value for the interfacial tension of 2500 ergs/cm^2. We have adopted this value at all temperatures for the purpose of comparing the measured and predicted growth rates. It may be noted that the sum of the surface tension of nickel ([11]) and of Al$_2$O$_3$, $\sim$2750 ergs/cm^2, is not very much larger than our measured interfacial tension. However, the value reported for the solid Al$_2$O$_3$–liquid Ni specific interfacial free energy is 1750 ergs/cm^2 ([12]). Considering the fact that the solid–solid interfacial tension should contain a contribution from strain energy, our value of 2500 ergs/cm^2 is reasonable.

The measurements on the time dependence of the internal grain boundary groove widths at 1400°C is shown in Fig. 6. Due to the method of fabricating the samples the width of each groove is not zero at the start of the annealing treatment but is equal to W_0. In this case Eq. (1) may be generalized to

$$W^n - W_0^n = Kt \tag{7}$$

where $n = 3$ for transport by volume diffusion. A deviation from $n = 3$ implies the contribution of other transport mechanisms. For example, for interface transport $n = 4$ and if the interface reaction is rate controlling $n = 2$ ([5,13]). After each anneal at a particular temperature at least seven measurements were made on each grain boundary. The number of boundaries mea-

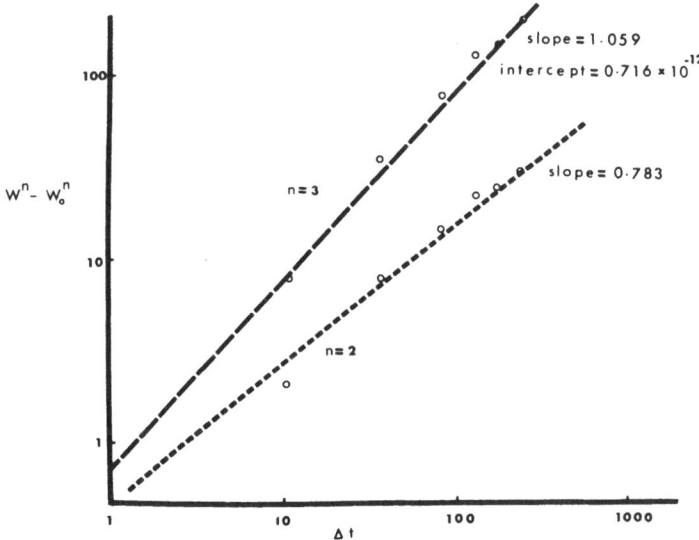

Fig. 6. Grain boundary groove width as function of time.

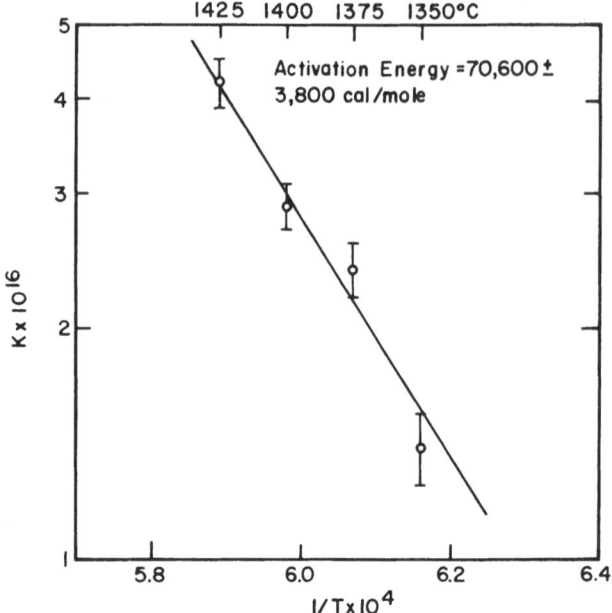

Fig. 7. Logarithm of kinetics constant versus $1/T$.

sured for each temperature was about five. The data in Fig. 6 represent a typical set of values of $W^n - W_0^n$ from the width measurements. At 1425° and 1400°C a best fit of all the data was found for $n = 3$, corresponding to volume diffusion. However, at the lower temperatures 1375°C and 1350°C plots corresponding to Fig. 6 showed significant departures from $n = 3$, indicating appreciable effects of the interfacial diffusion. For the purpose of comparison with available diffusion data in nickel we have assumed $n = 3$ at all temperatures. The average values of K are plotted versus $1/T$ in Fig. 7, yielding an activation energy for the groove development process of (70.6 ±3.8) kcal. This value is close to the activation energy for diffusion of aluminum in nickel ([14]), indicating that volume diffusion of aluminum through the metal phase is the rate-controlling mechanism. Further support of this conclusion will be shown in the next section by comparison between the calculated and the measured K.

CALCULATION OF K AND γ_i

For the evaluation of K from the analysis previously outlined published data on the diffusion of oxygen ([15]), aluminum ([14]), and nickel ([16]) in nickel were used. The interfacial tension γ_i was taken as the value established in the present experiment from the equilibrium angle measurements. The aluminum

TABLE II

Kinetic Constant for Grain Boundary Grooving*

T (°C)	n_{Al} (atoms/cm^3)	n_0 (atoms/cm^3)	$K_{meas} \times 10^{16}$	$K_{calc} \times 10^{16}$ from Eq. (3)
1350	6.0×10^{18}	2.4×10^{19}	1.6	0.19
1375	1.0×10^{19}	4.7×10^{19}	2.4	0.42
1400	2.0×10^{19}	6.0×10^{19}	2.9	1.2
1425	7.4×10^{19}	2.4×10^{20}	4.2	5.4

*Activation energy for grain boundary grooving = $(70,600 \pm 3800)$ cal/mole. Activation energy for volume diffusion of Al in Ni $\approx 64,000$ cal/mole. D_0 in nickel = $7.21 \times 10^4 \exp[-79,000/RT]$ cm^2/sec. D_{Al} in nickel = $1.87 \exp[-64,000/RT]$ cm^2/sec.

and oxygen concentration were independently measured by mass spectrographic analysis and are listed in Table II. It was found that Al and O were not present in the stoichiometric ratio for Al_2O_3. The calculated values of K are listed in Table II together with the experimental values. Considering the errors from all the experimental data involved in calculating K the agreement is good at high temperatures and indicates the general validity of the coupled diffusion analysis used. At lower temperatures volume diffusion through nickel phase is not the controlling mechanism and agreement between the calculated and the measured K is not expected.

It is possible to calculate the aluminum oxide–nickel interfacial tension for temperatures where volume diffusion is controlling by using the measured kinetics constant. For 1425°C γ_i is calculated to be 1940 ergs/cm^2 using the values of K, diffusivities, and solubilities given in Table II. The true value of γ_i probably lies between 1940 ergs/cm^2 and the value of 2500 ergs/cm^2 determined from the equilibrium angle measurements.

For the present system the magnitudes of the calculated effective diffusivity, taking account of volume and composition constraints, do not differ by large factors from the diffusivity of aluminum in nickel. In order to test the coupled diffusional analysis it would be desirable to increase the aluminum concentration sufficiently and suppress the oxygen solubility in nickel. According to theoretical predictions the effective diffusivity for the mass transport process should then approach that of oxygen in nickel. Similar experiments with Ni–Al alloys as the metal phase are reviewed elsewhere ([17]).

SUMMARY

A technique has been developed for the study of mass transport at internal interfaces in composite systems. For the process of grain boundary grooving the linear dimensions of the profile develop as $t^{1/3}$ at high temperatures, indicating that volume diffusion is the dominant transport mode. The magnitude of the kinetic constant is in good agreement with an analysis based on the coupling of the various atomic fluxes, although further experiments in different composition ranges are required for a rigorous test. The kinetics constant K experimentally established here should be extremely useful in analyzing more complicated geometrical situations.

Finally, the kinetics constant yields a value for the solid–solid interfacial tension which compares well with the valve obtained independently from the equilibrium shape measurements. In cases where diffusivities and concentration are accurately known the present technique thus affords the possibility of interfacial tension measurement from kinetic studies alone.

The value obtained for the solid–solid interfacial tension from the dihedral angle measurements is 2500 ergs/cm².

ACKNOWLEDGMENTS

The authors wish to thank Mr. B. Addis and Mr. R. Shewchuck for their assistance in setting up the equipment. Mrs. C. Newton performed the metallography necessary for the chemical analysis, which was done by Dr. R. Skogerboe of the Materials Science Center chemical analysis facility. This work was supported by the Advanced Research Projects Agency through the Materials Science Center at Cornell University.

REFERENCES

1. S. Sarian, Ph. D. Thesis, Cornell University, 1965.
2. C. Wagner, Z. Elektrochem. 65: 581 (1961).
3. I. M. Lifshitz and V. V. Slyozov, J. Phys. Chem. Solids 19: 35 (1961).
4. R. W. Heckel and R. L. DeGregorio, Trans. AIME 233: 2001 (1965).
5. W. W. Mullins, "Morphologies Governed by Capillarity," in: Metal Surfaces, American Society for Metals, Metals Park, Ohio, 1962.
6. Che-Yu Li, J. M. Blakely, and A. H. Feingold, Acta Met. 14: 1347 (1966).
7. Che-Yu Li and R. A. Oriani, in: Proceedings of the Bolton Landing Conference on Oxide Dispersion Strengthening, 1966, in press.
8. W. W. Mullins, Trans. AIME 28: 354 (1960).
9. W. H. Sutton and S. Chorne, Report No. R65SD2, General Electric Space Sciences Laboratory (1965).
10. W. D. Kingery, J. Am. Ceram. Soc. 37: 42 (1954).
11. S. P. Maiya, Ph.D. Thesis, Cornell University, 1966.
12. M. Humenik and W. D. Kingery, J. Am. Ceram. Soc. 37: 18 (1954).
13. W. W. Mullins, J. Appl. Phys. 28: 333 (1957).
14. R. A. Swalin and A. E. Martin, Phys. Rev. 96: 840 (1954).
15. C. J. Smithells and C. E. Ransley, Proc. Roy. Soc. (London) Ser. A 155: 195 (1936).
16. K. Monma, H. Suto, and H. Oikawa, J. Jap. Inst. Metals 28: 188 (1964).
17. A. H. Feingold and Che-Yu Li, Acta Met. 16: 1101 (1968).

DISCUSSION

R. S. Gordon (*University of Utah*): Is there any diffusion of Ni down the Al_2O_3 grain boundary? If so, would not the value of γ for Al_2O_3 be different from the value (Kingery) you have used?

Answer: The analysis of the mass transport rate does involve the value of the interfacial tension between Ni and Al_2O_3. In our experiment this can be derived from the Al_2O_3 vacuum values by comparison of the grain boundary groove angles or can be calculated from the measured kinetic coefficients using the literature values of the various diffusivities. The two values are in reasonable agreement considering the uncertainties in the diffusivities and experimental errors. While the effect you mention may indeed exist, it cannot easily be demonstrated due to the rather large total accumulated errors.

Alan Franklin (*ARPA*): Nickel diffusion in the Al_2O_3 grain boundary could change the grain boundary energy at the Ni end and make the use of the same value at each end somewhat doubtful. This might account for the high $Ni–Al_2O_3$ interface energy. One wouldn't really expect strain energy in Ni at 1400°C.

Answer: The validity of the assumption of a constant grain boundary tension for Al_2O_3 depends on the rate of diffusion of nickel in the boundary. There do not appear to be any relevant data.

We expect the solid $Ni–Al_2O_3$ interface to have a different value of interfacial tension from that of the liquid $Ni–Al_2O_3$ interface. Even if diffusional process can relax long-range strain fields there may nevertheless be short-range effects, perhaps due to dislocations associated with the interface.

Chapter 17

Sintering of Two-Component Oxide Systems with Compound Formation*

D. A. Venkatu† and G. C. Kuczynski

*Department of Metallurgical Engineering and
Materials Science and Radiation
Laboratory,** University of Notre Dame
Notre Dame, Indiana*

Sintering mechanisms in $MgO-Fe_2O_3$, $NiO-Fe_2O_3$, and $ZnO-Fe_2O_3$ have
been studied on systems composed of oxide spheres and plates. Except in the
very early stages, interdiffusion plays a dominant role in sintering. During
the early stages of sintering the process is motivated by the vacancy gradients
due to the sharp neck curvature, as indicated by the fifth-power law verified
for sintering in homogeneous systems. The kinetics of growth of ferrite into
the component oxides have been studied in the systems $MgO-Fe_2O_3$ and
$NiO-Fe_2O_3$. In the $ZnO-Fe_2O_3$ system sintering is dependent on the rate
of evaporation of zinc oxide above 1130°C. Below 1130°C interdiffusion in
the solid state is of greater importance.

INTRODUCTION

The processes discussed in this chapter belong to the class of solid-state reactions in which the reaction between two solid compounds results in the formation of a third. The compounds in question are divalent metal oxides reacting with ferric oxide to form ferrites. The interaction between the sintering process and that of compound formation complicates the interpretation of the experimental results.

The present study is by no means the first one dealing with the mechanisms of ferrite formation from the mixture of its component powders. Numerous excellent reports have been published on this subject. However,

*Based on part of the thesis submitted by D. A. Venkatu to the Graduate School of the
University of Notre Dame in partial fulfillment of the requirements for the degree of
Doctor of Philosophy.
†Present address: Department of Ceramic and Metallurgical Engineering, Clemson University, Clemson, South Carolina.
**The Radiation Laboratory of the University of Notre Dame is operated under contract
with the U. S. Atomic Energy Commission. This is A.E.C. document C00-38-405.

in a majority of them the emphasis was on the variation of physical properties with time and termperature of heating rather than on the process of sintering itself. The scope of this work was to investigate the effects of compound formation upon the kinetics of sintering.

EXPERIMENTAL

The ferric oxide pellets were prepared by pressing the powder of hematite at 30 tons/in.^{2} and then sintering them in air at temperatures ranging from 300° to 1300°C. The total sintering time was about 100 hr. Pellets 97.7% of theoretical density were obtained.

Zinc oxide and nickel oxide powders were pressed isostatically. Zinc oxide pellets were sintered in air at 900°C for 100 hr, resulting in compacts 85.5% theoretical density. Nickel oxide specimens were sintered in air at 1000°C for 120 hr, yielding compacts of only 78% theoretical density. The spheres of these oxides were obtained by crushing the pellets and then grinding them in a modified Jordan air-grinder ([1]). The spheres thus obtained had diameters varying from 0.5 to 0.8 mm. In addition MgO single-crystal spheres of uniform diameter of 1 mm were employed.

The sintering experiments consisted of heating the combinations of spheres and pellets at various temperatures and for various intervals of time. In order to prevent the spheres from rolling off the plates during handling small flat-bottomed cups were pressed from the oxides and then sintered. The specimens were withdrawn from the furnace after various time intervals, mounted in Lucite without pressure by polymerizing the monomer *in situ*, sectioned, and the neck diameters measured under the microscope.

RESULTS

Zinc Oxide–Ferric Oxide System. All efforts to obtain a good section displaying the neck formed between the ZnO sphere and the Fe_2O_3 plate were unsuccessful despite many precautions taken in handling the specimens. Even when the samples were removed from the furnace without dislodging the spheres they were buoyed up in liquid monomer during mounting. However, the contact areas at the necks were clearly visible, as can be ascertained from Fig. 1. They were fairly circular and contained very fine grains. The fact that the surrounding region is out of focus indicates that the necks were slightly raised above the plane of the oxide plate.

Nickel Oxide–Ferric Oxide System. The shape and size of the necks between nickel oxide spheres and ferric oxide plates obtained at various temperatures are represented in Figs. 2–4. At lower temperatures and for

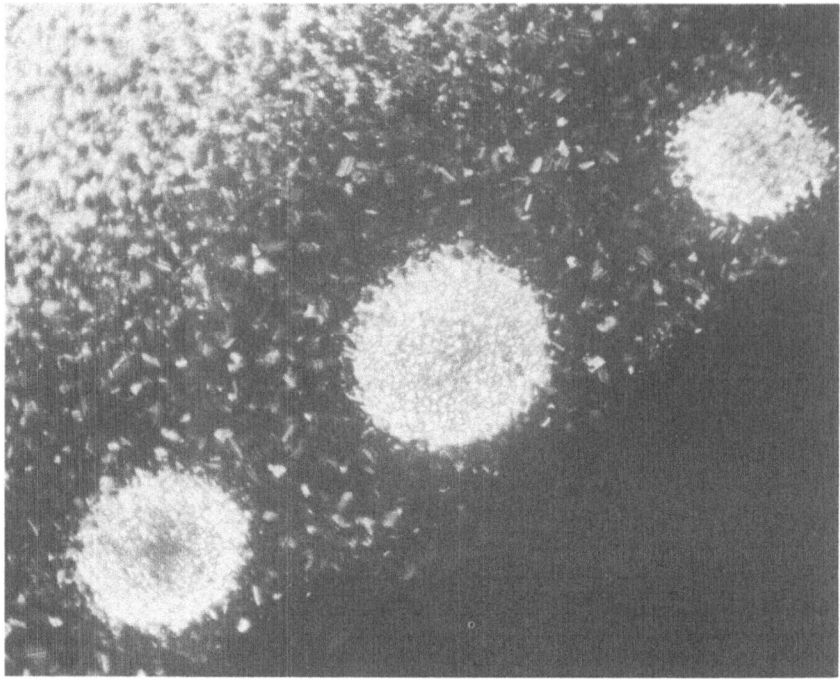

Fig. 1. Contact areas at the neck of zinc oxide spheres sintered to ferric oxide plate at 1128°C for 15.5 hr. Polarized light, 100 ×.

relatively short annealing times the neck had a familiar saddle shape with small external radius of curvature, as can be seen in Fig. 2. At higher temperatures or after long sintering times at lower temperatures the necks resembled more truncated cones or short cylinders, as depicted in Fig. 3 and 4. The development of porosity within the spinel phase on the ferric oxide side was noted.

The plot of the ratio of the neck and sphere radii, x/a, as a function of sintering time t is represented in Fig. 5. At the lowest temperature, 1224°C, for relatively short sintering times this ratio varies as $t^{0.2}$, which agrees with the well-known sintering relation for a diffusion-controlled sintering process ([2]). At higher temperatures or after longer sintering times at lower temperatures the neck radius was proportional to $t^{0.57}$.

At all sintering temperatures the neck growth was arrested for a certain interval of time when the x/a ratio approached a value between 0.4 and 0.5. With subsequent heating the neck started to grow again, with the time relation identical to that before the arrest. The time interval during which neck growth was arrested was longer the lower was the sintering temperature. The time at which the arrest of growth occurred was shorter the higher was the sintering temperature.

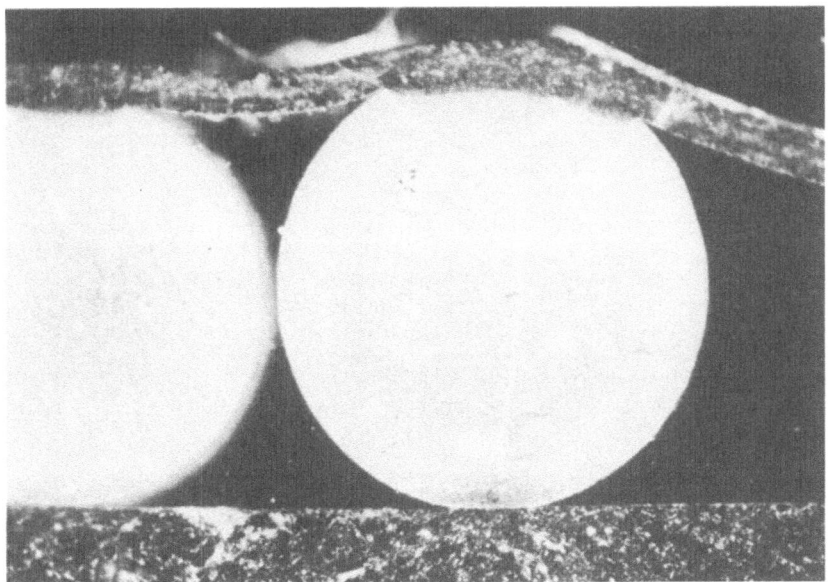

Fig. 2. Polycrystalline nickel oxide sphere sintered to polycrystalline ferric oxide sphere plate at 1160°C for 23 hr. Polarized light, 100×.

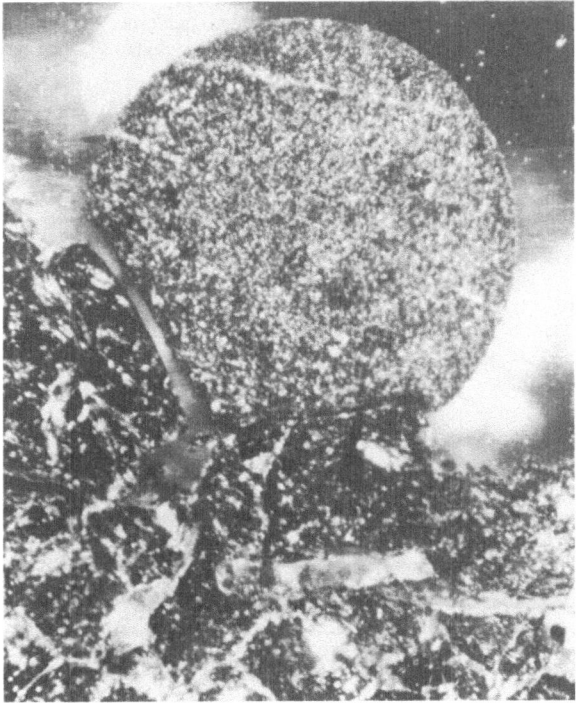

Fig. 3. Polycrystalline nickel oxide sphere sintered to polycrystalline ferric oxide plate at 1224°C for 76 hr. Polarized light, 75×.

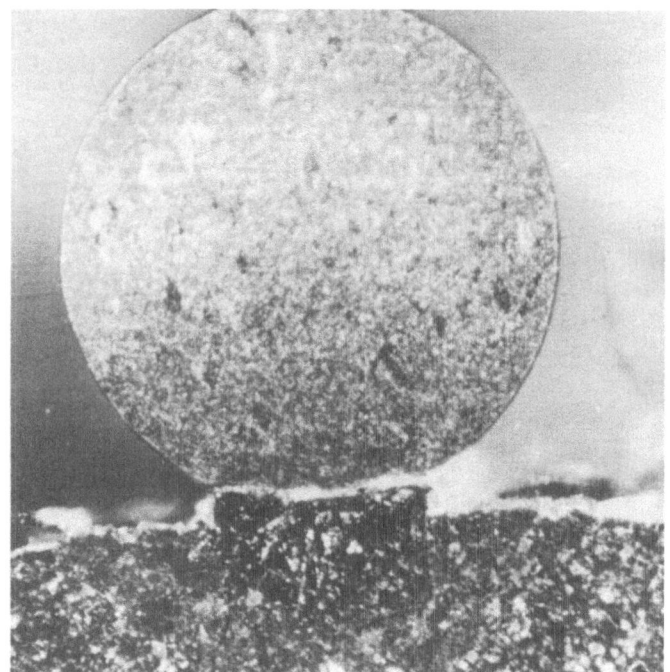

Fig. 4. Polycrystalline nickel oxide sphere sintered to polycrystalline ferric oxide plate at 1340°C for 7 hr. Polarized light, 75×.

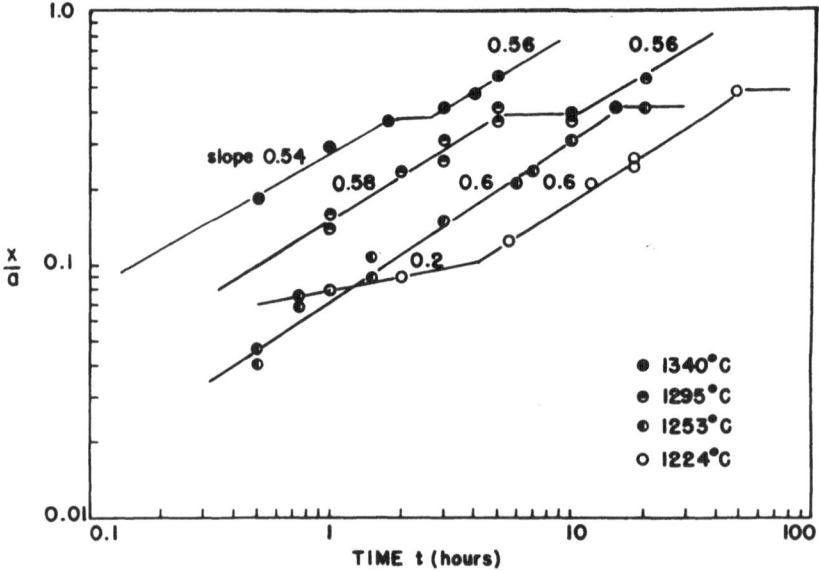

Fig. 5. Time dependence of the neck to sphere diameter ratio x/a for nickel oxide spheres sintered to ferric oxide plates at temperatures indicated.

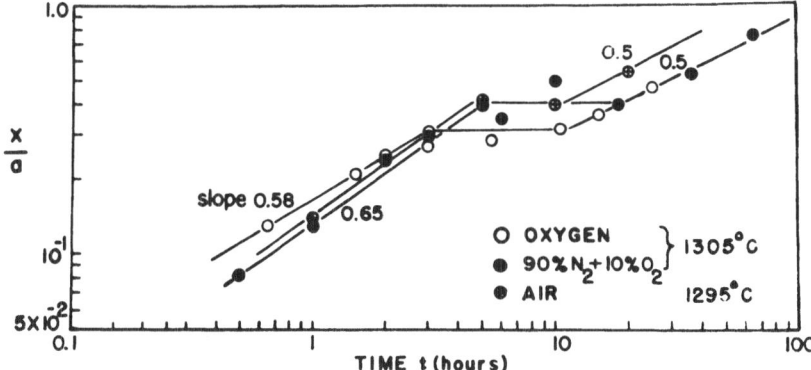

Fig. 6. Time dependence of the neck to sphere diameter ratio x/a for nickel oxide spheres sintered to ferric oxide plates at 1305°C in atmospheres indicated.

The effect of partial oxygen pressure upon the rate of sintering at 1305°C was also investigated. The specimens were heated in pure oxygen and in a 90% nitrogen, 10% oxygen mixture. The x/a versus time plots summarizing results of these experiments are given in Fig. 6, along with the values obtained for sintering in air at 1295°C. The rate of neck growth and the arrest of sintering seem to be unaffected by changes in oxygen pressure.

In the experiments with nickel oxide the sphere diameters varied between 0.45 and 1.00 mm. No effect of the sphere size upon the rate of sintering was observed.

Magnesium Oxide–Ferric Oxide System. In these experiments either magnesium oxide single-crystal spheres of 1 mm diameter were sintered to polycrystalline ferric oxide plates or ferric oxide polycrystalline spheres were sintered to single-crystal blocks of magnesium oxide. In general the results were similar to those obtained in the nickel oxide–ferric oxide system.

At low temperatures and relatively short sintering times necks with small external radius of curvature were observed, with deformation bands originating in the neck region as shown in Fig. 7. At higher temperatures these bands disappeared (Fig. 8) and the necks assumed conical shape, as shown in Figs. 9 and 10. In these last two photographs large pores and cracks in the ferrite phase may be noted. The contact areas between spheres and plates at low sintering temperatures had polyhedral shape (Fig. 11) which upon longer heating at higher temperatures slowly changed into more circular ones.

The plot of the x/a versus sintering time is given in Fig. 12. At 1180° and 1226°C this ratio is proportioned to $t^{0.2}$. Above 1226°C this ratio varies as $t^{0.65}$ except for sintering at 1302°C. At 1302°C after 12 hr of sintering the neck stopped to grow and started again after 40 hr. The same phenomenon was observed at 1340°C after 25 hr of heating.

Fig. 7. Polycrystalline ferric oxide sphere sintered to a single crystal of magnesium oxide at 1090°C for 51 hr. The sphere was held in position by nickel foil visible in the upper part of the photograph. Polarized light, 100 ×.

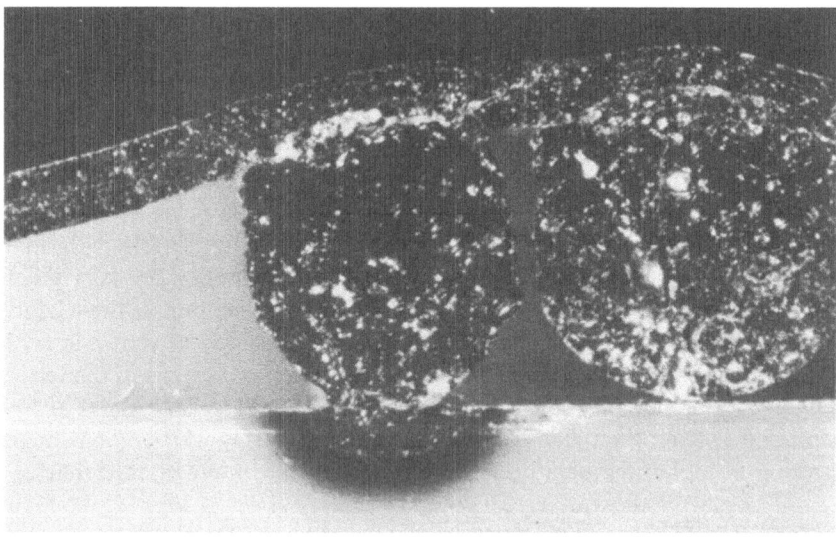

Fig. 8. Polycrystalline ferric oxide sphere sintered to single crystal of magnesium oxide at 1252°C for 59 hr. The sphere was held in position by nickel foil visible in the upper part of the photograph. Polarized light, 100 ×.

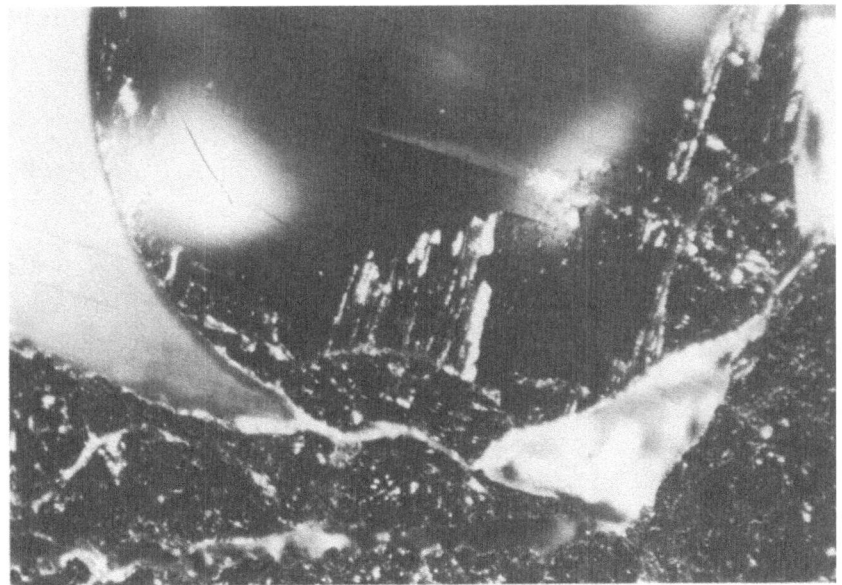

Fig. 9. Single-crystal magnesium oxide sphere sintered to polycrystalline ferric oxide plate at 1302°C for 51 hr. Polarized light, 75×.

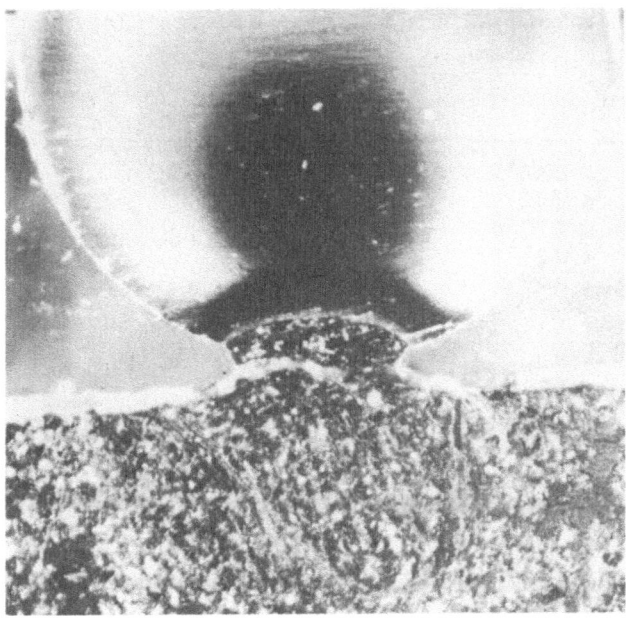

Fig. 10. Single-crystal magnesium oxide sphere sintered to poly-crystalline ferric oxide plate at 1340°C for 8.25 hr. Polarized light, 75×.

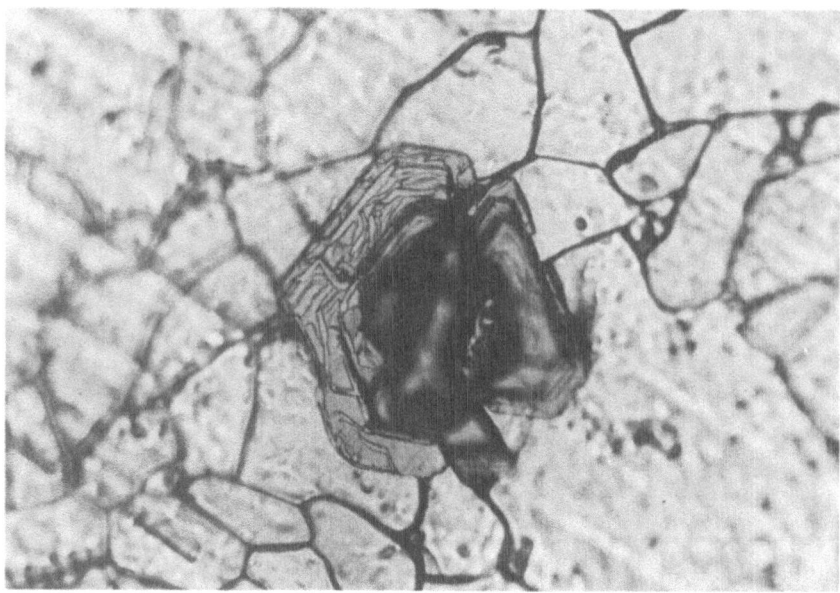

Fig. 11. Contact area between a single-crystal magnesium oxide sphere and the poly-crystalline plate of ferric oxide; sintered at 1226°C for 6 hr. Bright field illumination, 400×.

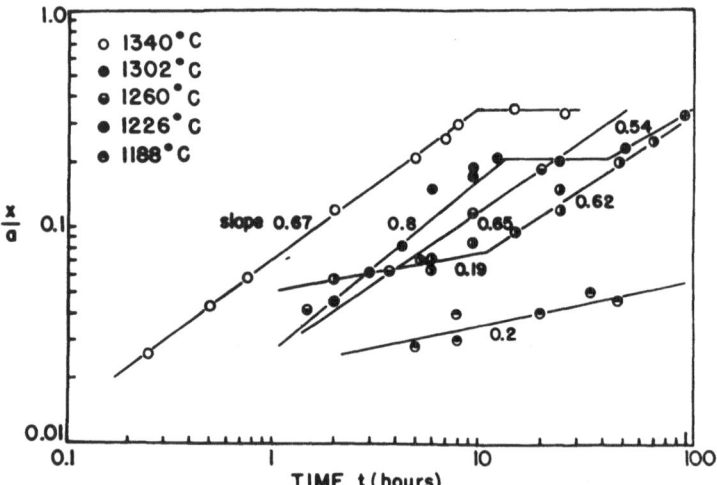

Fig. 12. Time dependence of the neck to sphere diameter ratio x/a for magnesium oxide spheres sintered to ferric oxide plates at temperatures indicated.

DISCUSSION OF THE RESULTS

The studies of sintering in multicomponent metallic systems revealed that this process is greatly affected by interdiffusion. The work of Kuczynski and Stablein [3,4] indicated that osmotic phenomena (Kirkendall effect) play a far more important role than the flow due to the capillary forces. This was to be expected, because the diffusion fluxes due to the latter gradients are much weaker than those caused by the difference in chemical potentials across the neck. The unequal diffusion fluxes cause the change of the geometry of the neck and consequently completely alter the kinetics of sintering. Kuczynski and Stablein [3] observed an arrest in the neck growth between copper and nickel and between nickel and gold wires caused by the increase of the radius of curvature of the neck resulting from vacancy precipitation at the neck base. The growth of these pores resulted in grooves of large radii of curvature, which reduced the vacancy gradient and thus slowed the neck growth. In the sintering of oxides described in the previous section the situation is further complicated by the formation of an intermediate spinel phase.

The formation of the ferrite phase is accompanied by large volume changes. In the case of $MgO-Fe_2O_3$ the linear expansion owing to the change in lattice constants in the formation of $MgFe_2O_4$ from MgO and Fe_2O_3 is 2.25 %. These strains are apparently an important factor in slowing the nucleation of the spinel phase during the early stages of sintering when the boundaries between the ferrite and component phases are expected to be coherent. The deformation bands visible in the MgO crystal in Fig. 7 are probably caused by these stresses. They were always observed at relatively low temperatures and short sintering times. The anomalous slowing down of the sintering rate at low temperatures observed by Kuczynski [5] is attributed to the same effect.

Carter [6] demonstrated by marker experiments that $MgAl_2O_4$ is formed by counter ion diffusion as first suggested by Koch and Wagner [7]. However, Kooy [8] pointed out that this mechanism cannot operate in the $MgO-Fe_2O_3$ system. Indeed, according to the available equilibrium diagram of $MgO-FeO-Fe_2O_3$ the majority of Fe_2O_3 dissolves in ferrite with the Fe^{3+}/Fe^{2+} ratio almost identical to that occurring in Fe_3O_4. Only a small amount of Fe_2O_3 enters the ferrite phase without any reduction of Fe^{3+} ions and produces cation vacancies such as occur in γ-Fe_2O_3. Thus he concludes that in view of composition at the phase boundaries the formation of $MgFe_2O_4$ takes place mainly by the counter-diffusion of Mg^{2+} and Fe^{2+} ions. Consequently at the ferrite/Fe_2O_3 boundary oxygen is given off owing to reduction of Fe^{3+} to Fe^{2+}. At the MgO/ferrite interface oxygen is taken up, converting Fe^{2+} to Fe^{3+} ions. Oxygen is transported through an external phase. The composition at the Fe_2O_3/ferrite boundary varies more rapidly with oxygen pressure than the corresponding composition of MgO/ferrite interface; hence the rate of formation of the spinel phase should increase with falling oxygen pressure.

According to Kooy the perpendicular displacement of the spheres and plates shown in the photographs reproduced in this chapter is a consequence of the Kirkendall effect. If at the respective interfaces the reduction and oxidation reactions were to take place homogeneously, the sphere and plate should move away from each other, because oxygen is transferred from a shell of larger radius to a shell of smaller radius, the ratio of hemisphere volumes adjacent to Fe_2O_3 and MgO being 2:1. Unfortunately, the reduction reaction is far from homogenous. Molecular oxygen formed at the Fe_2O_3/ferrite interface apparently collects in the voids, which under its pressure grow large and even open up cracks, as exemplified by Figs. 3 and 9. This precluded any sensible measurements of the rates of the perpendicular growth of the necks.

As mentioned above, neck growth at the beginning follows the familiar fifth-power law as in the case of monocomponent systems. However, with the increased interface area interdiffusion is enhanced, and consequently the gap between sphere and plate as well as the external radius of the neck curvature increase. Thus further extension of the neck radius due to capillary forces becomes negligible. The observed lateral growth of the almost-cylindrical neck is due to the oxidation gradient. Indeed, on the external surface one should expect the composition to be very much like that existing at MgO/ferrite interface, while in the interior this composition should be closer to that at the Fe_2O_3/ferrite boundary. Therefore Fe^{2+} ions diffuse to the surface where they become oxidized and join MgO molecules transported there by surface diffusion to form $MgFe_2O_4$. This process is not unlike that of oxidation of metals, so it is not surprising that the neck growth follows a nearly parobolic law. The swelling of the ferrite phase near the neck base visible in Figs. 2 and 10 is caused by the same mechanism.

The arrest of sintering rate observed in $NiO-Fe_2O_3$ and $MgO-Fe_2O_3$ systems is due to the formation of internal voids and cracks discussed above. They reduce the diffusion flux to the point where further neck growth is negligible. The sintering is resumed after appreciable shrinkage of the pores, and sealing of the cracks takes place. The onset of the slowing of the neck growth should depend only on the volume of Fe_2O_3 transported. Therefore in a given oxide system it should occur approximately at the same x/a ratio. As can be ascertained from Figs. 6 and 12 the neck growth in $NiO-Fe_2O_3$ is arrested at an x/a value of about 0.45 and in $MgO-Fe_2O_3$ at about 0.3, independent of sintering temperature. Activation energies of the diffusion processes controlling the neck growth were obtained by plotting the logarithm of time required to obtain a certain constant value of x/a against reciprocal absolute temperature. They were 94 kcal/mole for $MgO-Fe_2O_3$ and 120 kcal/mole for $NiO-Fe_2O_3$. It is interesting to note that the activation energy obtained from the time at which the neck growth arrest is first observed in $NiO-Fe_2O_3$ system is 122 kcal/mole, the same as that obtained for neck growth in this system. This is another indication that the formation of porosity which seems to be responsible for this arrest is controlled by the same mechanism as neck growth, in accordance with the hypothesis advanced above. These activation energies are higher than those for self-diffusion of cations in spinels. However, as was pointed out by Wagner ([9]), diffusion coefficients measured under concentration gradients, as for example in the case of oxidation of metals, may be quite different from the coefficients of self diffusion.

In the $ZnO-Fe_2O_3$ system sintering seems to proceed by two different mechanisms, depending on the temperature. The plot represented in Fig. 13 indicates that at temperatures lower than 1130°C the neck grows according to a nearly parabolic law $x^{2.1} \propto t$, while above this temperature $x^{2.8} \propto t$. The activation energies obtained for these mechanisms are 81 kcal/mole for low temperatures and 136 kcal/mole for higher temperatures. The activation energy of 81 kcal/mole is close to the activation energy of either Zn^{2+} diffusion (82 kcal/mole) or Fe^{3+} diffusion (85 kcal/mole) through zinc ferrite ([10]). The high-temperature process is most probably that of evaporation and condensation, for which the relation $x^3 \propto t$, close to the observed one, has been predicted ([2]). This hypothesis is further strengthened by the observation

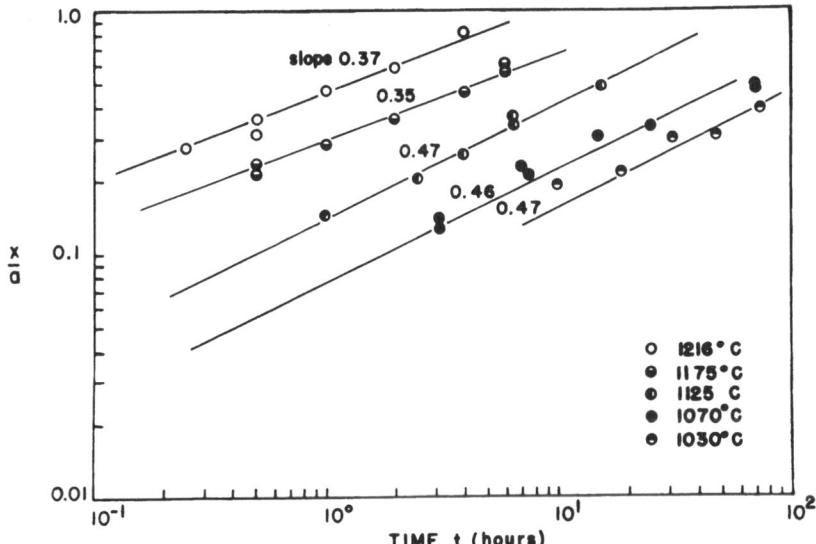

Fig. 13. Time dependence of neck to sphere diameter ratio x/a for zinc oxide spheres sintered to ferric oxide plates at indicated temperatures.

Fig. 14. Polycrystalline ferric oxide spheres sintered to polycrystalline zinc oxide plate at 1136°C for 32.5 hr. Note dark layer of ferrite formed on the surface of ferric oxide particles. Polarized light, 100×.

that above 1130°C the layer of ferrite forms around the ferric oxide sphere as shown in Fig. 14, indicating copious evaporation of zinc oxide at these temperatures. Furthermore, the value of the activation energy 136 kcal/mole obtained from sintering experiments compares favorably with the energy of vaporization of zinc oxide, 123 kcal/mole ([11]). It may be added that Norris and Parravano ([12]), in their study of sintering of zinc oxide spheres, also noticed a change in the sintering mechanism at 1170°C from bulk diffusion to that of evaporation and condensation.

ACKNOWLEDGMENTS

The initial stage of the work reported in this paper was sponsored by the Office of Naval Research and the remainder was performed under the auspices of the U. S. Atomic Energy Commission.

REFERENCES

1. P. Jordan, in: Proc. 3rd International Symposium on Reactivity of Solids, Elsevier Publishing Co., Amsterdam, 1957, p. 423.
2. G. C. Kuczynski, "Self-Diffusion in Sintering of Metallic Particles," *Trans. AIME* 185: 169 (1949).
3. G. C. Kuczynski and P. F. Stablein, Jr., "Sintering in Multicomponent Systems," Proc. 4th International Symposium on Reactivity of Solids, Elsevier Publishing Co., Amsterdam, 1960, p. 91.
4. P. F. Stablein, Jr. and G. C. Kuczynski, "Sintering in Multicomponent Metallic Systems," *Acta Met.* 11: 1327 (1963).
5. G. C. Kuczynski, "Formation of Compounds by Sintering," in: 5th International Symposium on Reactivity of Solids, Munich, 1964, p. 353–361.
6. R. E. Carter, "Mechanism of Solid State Reaction between Magnesium Oxide and Aluminum Oxide and Ferric Oxide," *J. Am. Ceram. Soc.* 44: 116 (1961).
7. E. Koch and C. Wagner, "Formation of Ag_2HgI_4 from AgI and HgI_2 by Solid State Reaction," *Z. Phys. Chem.* B34: 317 (1926).
8. C. Kooy, "Material Transport in Solid State Reactions," in: 5th International Symposium on Reactivity of Solids, Elsevier Publishing Co., Amsterdam, 1965, p. 21.
9. C. Wagner, Atom Movements, American Society for Metals, Metals Park, Ohio, 1951, p. 153.
10. O. Kubaszewski and B. E. Hopkins, Oxidation of Metals and Alloys, Butterworth and Co., London, 1962, p. 33.
11. Gmelins Handbuch der Anorganischen Chemie, Zink, 8 Ed., Verlag Chemie, Berlin, 1957, p. 803.
12. L. F. Norris and G. Parravano, "Sintering of Zinc Oxide," *J. Am. Ceram. Soc.* 46: 449 (1963).

DISCUSSION

J. E. Burke (G.E.): You say that this process involves an intermigration of divalent ions, yet you start out with Fe_2O_3, presumably trivalent ions, on the one side and a variety of divalent oxides on the other side. Then you have the same numbers of ions diffusing each way. You have MgO and Al_2O_3, so you will have three magnesiums going in for two aluminas coming out. I can see the swelling that comes about. But if indeed it is divalent ions going each way, you have approximately equal numbers for the exchange.

Answer: The ratio of the divalent cations to trivalent cations in the ferrite is always one to two. Hence the ferric oxide dissolves in it with a ratio of Fe^{2+}: Fe^{3+} which is, almost identical to that occurring with Fe_3O_4. This comes from the conversion of an equivalent number of oxygen ions into oxygen molecules, which are deposited at the ferrite–ferric oxide interface. The oxygen is recovered at the ferrite–monovalent oxide interface. I believe this is what Kooy had in mind.

J. White (University of Sheffield): The phenomenon referred to by Dr. Kuczynski is a consequence of the equilibrium relationships in the system $MgO–FeO–Fe_2O_3$, which was first investigated in detail by Woodhouse and myself[13].

Our diagram shows that at temperatures approaching 1388°C in air the spinel phase that coexists in equilibrium with Fe_2O_3 contains very little magnesia; it approaches the composition of Fe_3O_4, but with an excess of oxygen in it. At the same temperature and oxygen pressure the spinel coexisting with magnesia approaches closely to $MgO–Fe_2O_3$ in composition. Hence Mg^{2+} ions diffuse through the spinel towards the spinel–Fe_2O_3 interface while Fe^{2+} ions diffuse towards the spinel–MgO interface.

Neglecting the excess of oxygen in the spinel–Fe_2O_3 interface we can write, for the reactions occurring at that interface,

$$3Fe_2O_3 = 2Fe^{2+}Fe_2^{3+}O_4 + \tfrac{1}{2}O^2$$

followed by

$$2Fe^{2+}Fe_2^{3+}O_4 + 2Mg^{2+} = 2MgFe_2O_4 + 2Fe^{2+}$$

giving

$$3Fe_2O_3 + 2Mg^{2+} = 2MgFe_2O_4 + 2Fe^{2+} + \tfrac{1}{2}O_2$$

which is the equation given by Kooy.

The reaction at the spinel–MgO interface is then

$$3MgO + 2Fe^{2+} + \tfrac{1}{2}O_2 = MgFe_2O_4 + 2Mg^{2+}$$

This means that oxygen is lost to the gas phase at the spinel–Fe_2O_3 interface and picked up again at the spinel–MgO interface, the amount of this oxygen being $\tfrac{1}{12}$ of that in the spinel formed. Thus there is a Kirkendall shift of markers which approaches $\tfrac{1}{12}$ of the thickness of the spinel layer formed.

R. Coble (M.I.T.): I gathered from the photographs that there was a swelling outward· on the Fe_2O_3 side toward the MgO and there was a hill that built up under the sphere of MgO or the divalent iron oxide. It seems to me that Professor White is predicting a movement the other way, or do I misinterpret?

J. H. White (University of Sheffield): Kooy pointed out that when the diffusion takes place across a narrow neck, as is the case of a sphere in contact with a plane, roughly hemispherical regions of spinel will be formed on each side of the neck, the volume of the hemisphere on the Fe_2O_3 side being twice that on the MgO side. Consequently the lateral movement associated with the addition of a given number of O^{2-} ions to the spinel lattice at the spinel–MgO interface is greater than that associated with the removal of the same number of O^{2-} ions from the spinel lattice at the spinel–Fe_2O_3 interface. Hence the displacement is towards the Fe_2O_3 side.

This description is somewhat idealized since it assumes that the oxygen can escape to the gas phase at the one interface and be picked up at the other, which is not always true. In porous compacts of mixed MgO and Fe_2O_3 powders, however, transfer of oxygen can occur through the pores. Some years ago we showed that, when compacts consisting of Fe_2O_3 particles embedded in MgO particles were fired, expansions occurred and holes appeared where the Fe_2O_3 particles had occurred [14]. In this case a spherical shell of spinel would form round the Fe_2O_3 particles so that the movement had to take place outwards from the Fe_2O_3, leaving holes in the center.

ADDITIONAL REFERENCES

13. J. White and D. Woodhouse, "Phase Relationships of Iron Oxide-Containing Spinels III," *Trans. Brit. Ceram. Soc.* **54**: 333 (1955).
14. J. Goodison and J. White, in: *Agglomeration* (W. A. Knepper, ed.), Interscience, New York, 1962, p. 25.

Chapter 18

Powder Compact Studies of Initial Sintering*

D. L. Johnson

Associate Professor of Materials Science
Northwestern University

A method of analyzing initial sintering data which is insensitive to uncertainties in the time zero and initial compact length is reviewed. This method permits a considerably more reliable determination of the mechanisms of material transport in the sintering process than the usual log shrinkage–log time plots. One may with confidence distinguish among volume diffusion, grain boundary diffusion, or both acting concurrently. Contributions of surface diffusion and evaporation–condensation to total mass transport are discernible. This technique was applied to isothermal shrinkage data of compacts of alumina prepared from several different powders. It was found that compacts which deviated from the ideal compact (uniformly sized spheres heated instantaneously to the sintering temperature) followed the shrinkage models after some small amount of abnormally rapid shrinkage, resulting in an effective initial length and time zero which were different from the experimental values. While these differences were small, they had a large effect on the log shrinkage–log time plot. Alumina apparently sinters by a grain boundary diffusion mechanism, with surface diffusion important at lower temperatures.

SINTERING MODELS

Several models have been proposed relating diffusion coefficients and temperature to neck growth between two spheres or between a sphere and a plane by center-to-center approach of the spheres. There are differences among the models in boundary conditions as well as in geometry approximations. The former result in major differences in the time dependence of shrinkage and neck growth, while the latter result in minor differences. The models used in the present chapter are those which are based on the boundary conditions considered to be applicable to sintering mechanisms which produce shrinkage and which, furthermore, utilize the more accurate geometry approximations.†

*The research on which this chapter is based was supported by the Advanced Research Project Agency of the Department of Defence, through the Northwestern University Materials Research Center.
†For a review of the various models see [6].

Most sintering experiments have been interpreted assuming that a single mechanism is responsible for materials transport. The slope of plots of the logarithm of shrinkage of a compact or the neck diameter between two spheres versus time has been used as an index of the mechanism of material transport according to the models of Kuczynski ([1]), Kingery and Berg ([2]), and Coble ([3]). With few exceptions the slopes were considered to be close enough to the expected value for volume diffusion that most authors have named volume diffusion as the sintering mechanism. However, according to the models of Johnson and Cutler ([4]), Johnson and Clarke ([5]), and Berrin and Johnson ([6]), the slope of the log shrinkage–log time plot can vary from 0.32 to 0.47 depending upon the relative importance of grain boundary and volume diffusion. Furthermore, Singu ([7]) has shown that the slope of this plot is increased by concurrent surface diffusion and decreased by a vapor transport mechanism. Finally, Johnson and Cutler ([4]) showed that errors in the zero of time and/or shrinkage produce curvature and changes of slope on the plot. Thus it is concluded that the logarithm of shrinkage or neck diameter versus logarithm of time plot should be used with great caution and never as the sole plot to determine sintering mechanisms.

Johnson and Clarke ([5]) and Berrin and Johnson ([6]) have presented a method of plotting sintering data which is considerably more sensitive than the conventional log–log plot and from which, furthermore, both the volume and grain boundary self-diffusion coefficients can be calculated. The plot is based upon the following equations:

$$\left[\frac{a^3 \rho}{a + \rho}\right]\dot{y} = \frac{2\gamma\Omega D_v}{\pi k T r^3}\left[\frac{A_v}{a}\right] + \frac{4\gamma\Omega b D_B}{k T r^4} \tag{1}$$

for shrinkage and

$$\left[\frac{A_v a \rho}{a + \rho}\right]\dot{a} = \frac{4\gamma\Omega D_v}{k T r^3}\left[\frac{A_v}{a}\right] + \frac{8\pi\gamma\Omega b D_B}{k T r^4} \tag{2}$$

for neck growth, where a is the neck radius divided by r, with r the sphere radius, ρ is the minimum radius of curvature of the neck divided by r, $y = \Delta L/L_0$ is the fractional shrinkage, $\dot{y}$ is the shrinkage rate, A_v the neck surface area divided by r^2, γ the specific surface free energy, Ω the volume of diffusing species, D_v the volume self-diffusion coefficient, D_B the grain boundary self-diffusion coefficient, b the thickness of the region of enhanced diffusion at the grain boundary, k Boltzmann's constant, and T the absolute temperature.

Equation (1) can be applied to the shrinkage of compacts provided that the compact consists of uniform spheres, the number of particle-to-particle contacts is independent of shrinkage, and no extraneous effects occur during heat-up. The terms in brackets are dependent upon the shrinkage y and the relative importance of mass transport from the sphere surface to the neck. Johnson and Clarke determined the shrinkage dependence of these terms in Eq. (1) assuming no surface sources of material and calculated from the shrinkage data of compacts of spherical particles of silver both volume and grain boundary self-diffusion coefficients which are in excellent agreement with the tracer values ([5]).

Equation 2 is not useful in its present form because of the difficulty in measuring $\dot{a}$, and Berrin and Johnson have given approximate forms of it ([6]).

Unfortunately, Eq. (1) is even more sensitive to errors in the zero of shrinkage than the log–log plot ([8]), although it is independent of errors in the time zero. As a matter of fact, any equation which includes either time or shrinkage to any power other than the first power will be affected by systematic errors in time and shrinkage.

Under certain conditions either volume or grain boundary diffusion can be predominant. Grain boundary diffusion can be neglected if

$$rD_v/2\pi bD_B > 10 \tag{3a}$$

while volume diffusion can be neglected if

$$rD_v/2\pi bD_B < 0.1 \tag{3b}$$

These are somewhat arbitrary but unseable limits. If either of the conditions (3a) or (3b) obtain, Eq. (1) can be approximated and transformed into the following equations ([8,9]):

$$L = L_0[1 - \alpha(KD)^m \dot{y}^{-m}] \tag{4}$$

$$\dot{y}^{-(1+m)} = \beta(KD)^{-m} \tag{5}$$

where D is an appropriate self-diffusion coefficient, L the instantaneous length of compact (or center-to-center distance for two spheres), L_0 the initial length, and the other symbols are listed in Table I. Analysis of synthetic data indicates that plots based on these equations will yield straight lines for the first 2% and 3% shrinkage for volume diffusion and grain boundary diffusion, respectively. These equations are insensitive to systematic errors, since either L or

TABLE I

	α	β	m	K
Volume diffusion	2.56	0.80	0.97	$\gamma\Omega/kTr^3$
Grain boundary diffusion	0.84	3.64	0.48	$\gamma\Omega b/kTr^4$

t appear to the first power and $\dot{y}$ is in error only slightly for an error in L_0 (and therefore y). Any shrinkage or time corrections are immediately apparent on the plots. Furthermore, the plots can be used to determine whether both volume and grain boundary diffusion are significant or if surface diffusion or vapor transport are modifying the shrinkage behavior [8]. These equations thus provide a powerful means of elucidating sintering mechanisms.

SINTERING OF ALUMINA

Several investigators have attempted to determine the mechanism of material transport in the initial sintering of alumina. Kuczynski et al. [10] and Coble [3] measured neck growth between large single-crystal spheres, while Coble [3] and others [11-13] followed the shrinkage of powder compacts. Johnson and Cutler [11], Bagley [12], and Keski [13] applied shrinkage and time corrections using a technique which requires a certain amount of judgment and results in some uncertainty as to the mechanism of sintering. They reported a grain boundary diffusion mechanism except for alumina of less than 2 μ diameter doped with TiO^2 [12] and for one type of "pure" alumina, for which volume kinetics were reported [13]. Coble, on the other hand, did not apply corrections as such to his shrinkage data, but he took the time and length zeros as the values at which the sample reached maximum size (due to thermal expansion). Since some shrinkage occurs during heat up, he did in fact apply both shrinkage and time corrections. He ascribed the sintering to volume diffusion, although the compacts appeared to follow Kingery and Berg's model rather than his own. The shrinkage correction which results from Coble's experimental method is of such a nature as to steepen the log shrinkage–log time plots. The slopes fell between those values for volume diffusion only and grain boundary diffusion only found from the present model.

Because conflicting opinions exist as to the mechanism of sintering of alumina and since the conclusions were drawn from "corrected" data, it was felt desirable to reexamine the sintering of alumina using the equations which are insensitive to systematic time and shrinkage errors.

Data for sintering of four types of alumina which had been reported previously were analyzed using the above equations. The raw shrinkage curves were originally drawn by a strip-chart recorder activated by a sensitive linear differential transformer. The samples were disks 1.25 in. in diameter and 1/16 in. thick which were plunged into the preheated furnace after a very brief prefiring at a temperature low enough to cause no shrinkage. Heat-up times were on the order of 1 to 3 min, depending upon the temperature.

Thirty or more points were taken at constant length intervals for the first 5% linear shrinkage of each run. A computer was used to calculate the time derivative of shrinkage and the various values needed for plotting the data.

The aluminas studied were Alcoa A-14, which had been elutriated to a particle size range of 3–5 μ diameter, a high-purity powder prepared by calcining hydrated aluminum chloride which had been obtained by dissolving high-purity aluminum metal in HCl, Linde Cl.O [12], and Alcoa 2456G [13]. The latter three powders were clustered, with ultimate particles of about 0.3, 0.3, and 5 μ diameter, respectively.

RESULTS

Figures 1 and 2 show the data for the A-14 plotted according to Eq. (4) for grain boundary and volume diffusion, respectively. It is readily observed that the grain boundary plot exhibits straight lines for most temperatures after a brief curved portion, while the volume plots are all curved. There is apparently a small amount of abnormally rapid shrinkage initially, resulting in an effective L_0 about 0.003 in. less than the measured L_0. This shrinkage may be due to particle nonsphericity and size distribution. The lowest temperature runs are curved even on the grain boundary plot, indicating that surface diffusion is probably significant at these temperatures [8]. The fact that

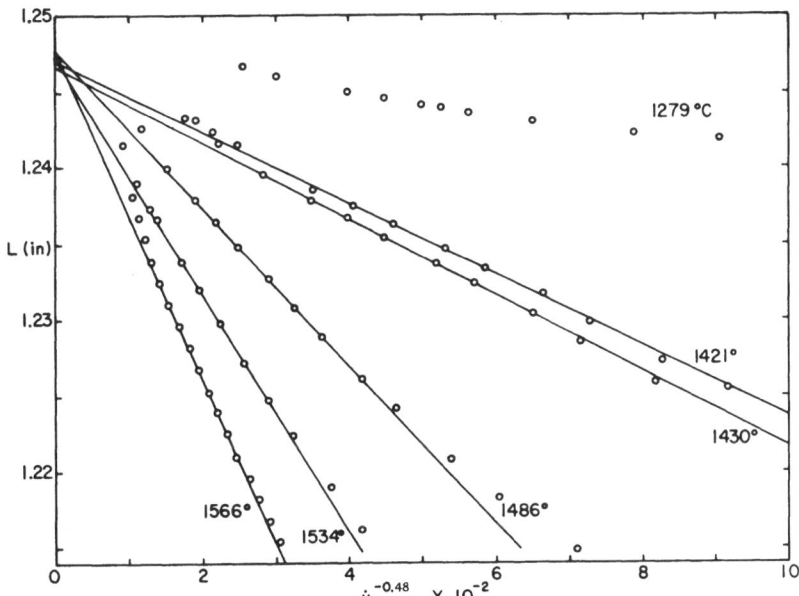

Fig. 1. Shrinkage isotherms for Alcoa A-14 plotted according to the grain boundary diffusion model.

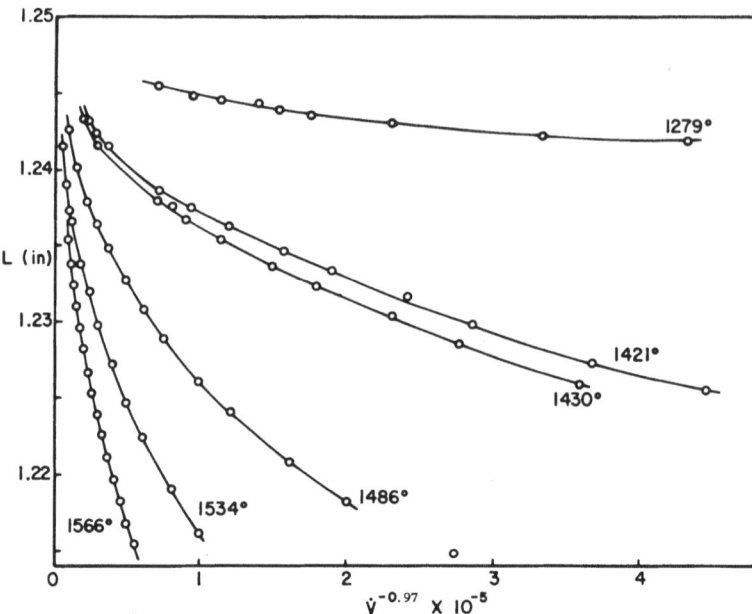

Fig. 2. Shrinkage isotherms for Alcoa A-14 plotted according to the volume diffusion model.

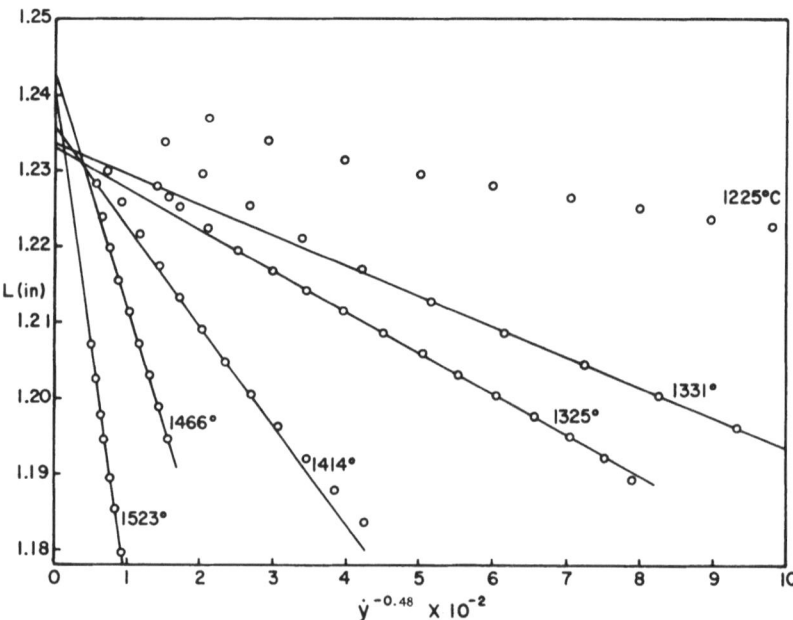

Fig. 3. Shrinkage isotherms for high-purity alumina plotted according to the grain boundary diffusion model.

the extrapolated length is independent of temperature is indicative that only grain boundary diffusion, with no significant surface diffusion or other complicating mechanisms, is operating ([8]).

Similar plots for the clustered high-purity powder also showed straight lines only on the grain boundary plot (see Fig. 3). However, the difference between the extrapolated length and the initial length is considerably greater than for the A-14 alumina. In addition, the extrapolated length is temperature sensitive, indicating that something is complicating the sintering behavior. Analysis of data for other clustered powders demonstrated that the interpretation of mechanisms involved in sintering of such powders is difficult. Johnson and Cutler ([11]) did fit such data to their grain boundary diffusion model, but the present, more objective analysis shows that this is only approximate at best. The large shrinkage correction, where straight lines are obtained on the L versus $\dot{y}^{-0.48}$ plots, together with the fact that the shrinkage rate is sensitive to the compacting pressure for clustered but not for dispersed powders, indicates that the clusters are broken somewhat on compaction and the individual cluster fragments densify rather quickly, after which they behave like single particles ([11]). The deviation from linearity, and the temperature-sensitive extrapolated length, could conceivably be the result of grain growth, and such a possibility ought to be investigated in future work.

Figures 4 and 5 show the plots for Linde Cl.O alumina from data

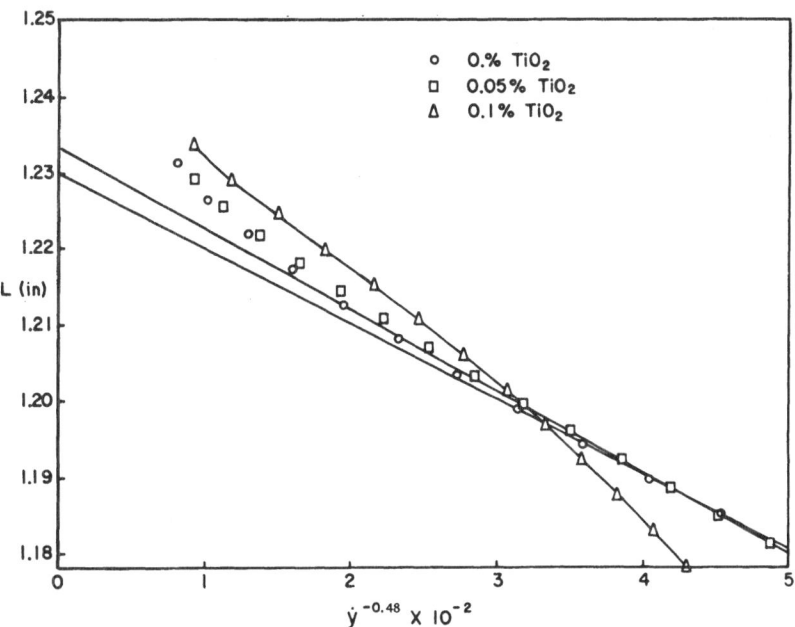

Fig. 4. Shrinkage isotherms for pure and titania-doped Linde C1.0 alumina plotted according to the grain boundary diffusion model. $T = 1300°C$. Data from Bagley ([12]).

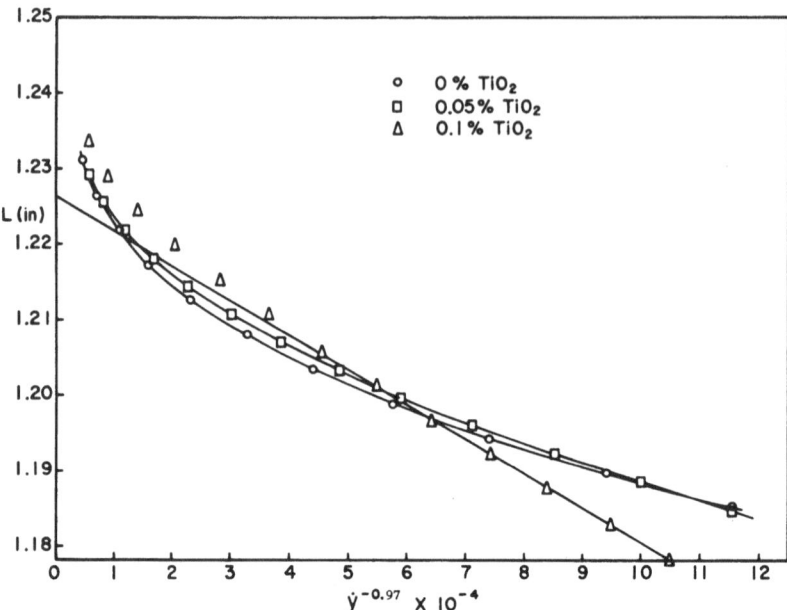

Fig. 5. Shrinkage isotherms for pure and titania-doped Linde C1.0 alumina plotted according to the volume diffusion model. $T = 1300°C$. Data from Bagley [12].

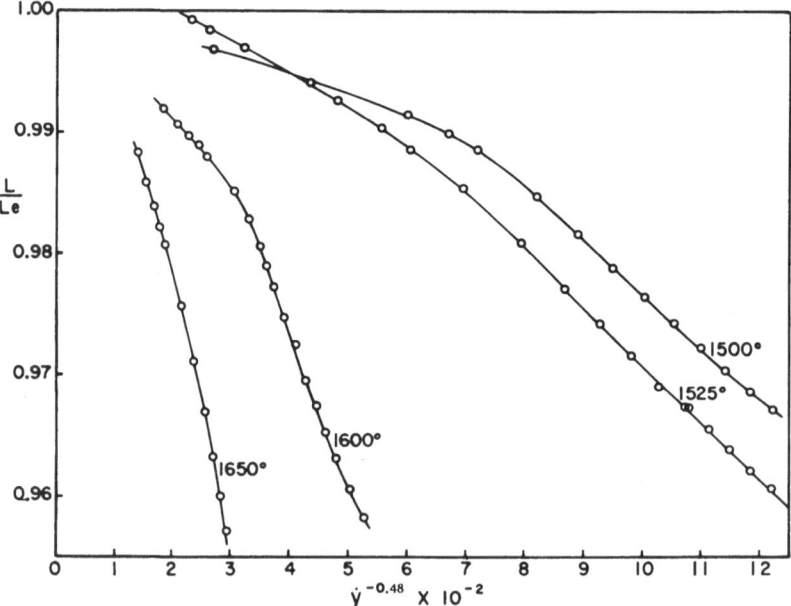

Fig. 6. Shrinkage isotherms for Alcoa 2456G alumina plotted according to the grain boundary diffusion model. Data from Keski [13].

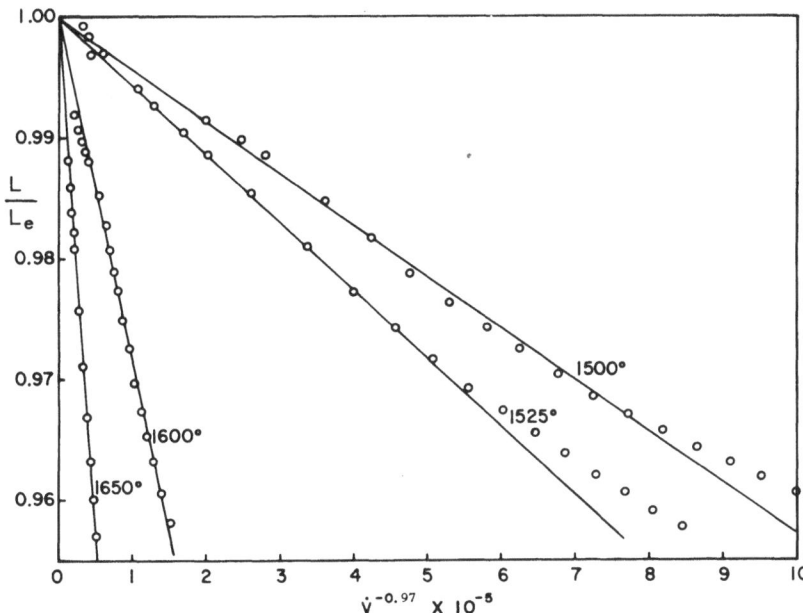

Fig. 7. Shrinkage isotherms for Alcoa 2456G alumina plotted according to the volume diffusion model. Data from Keski ([13]).

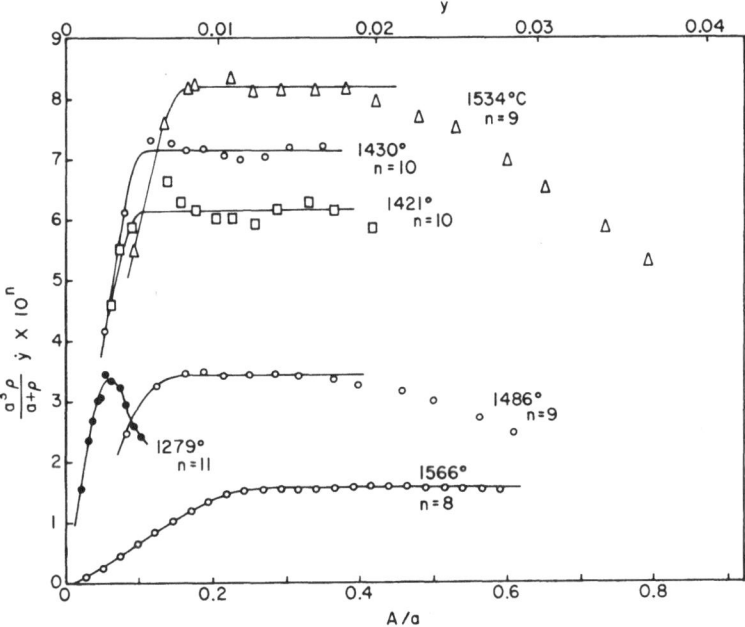

Fig. 8. Shrinkage isotherms for Alcoa A-14 plotted according to Eq. (1).

reported by Bagley. Note that doping with TiO_2 changes the apparent mechanism from grain boundary to volume diffusion.

Some of Keski's ([13]) data for shrinkage of Alcoa 2456G alumina are shown in Figs. 6 and 7. The values of L_0 were not reported, so the ordinate is normalized to the extrapolated length L_e. There is no information as to the differences of these extrapolated lengths from L_0, and this hinders the assessment of the significance of the results.

Figure 8 shows the data for the Alcoa A-14 replotted according to Eq. (1) using the corrected lengths obtained from Fig. 1. The slope and intercept on such a plot should contain the volume and grain boundary diffusion coefficients, respectively. The initial part of each curve has an approximately zero intercept, while the subsequent part has an approximately zero slope. The other aluminas which had straight lines on the grain boundary plots (Figs. 3 and 4) also produced curves similar to those shown in Fig. 8. The aluminas which had straight lines on the volume plots (Figs. 5 and 7) gave nearly straight lines through the origin when plotted according to Eq. (1).

Diffusion coefficients can be calculated from the data if values for the various parameters in Eq. (1) are known. The specific surface free energy was taken to be 950 ergs/cm^2 and Ω was 1.4×10^{-23} cm^3. The latter value assumes oxygen ion diffusion to be rate controlling. If aluminum ion diffusion is rate controlling, the resulting diffusion coefficients should be multiplied by $\frac{2}{3}$. The particle radius is the most difficult parameter to fix. The value for A-14 can be taken as 2 μ with a fair amount of confidence, since this powder was monodispersed with a relatively narrow size distribution. For the clustered powders the effective particle size is a function of pressing pressure, as indicated by the influence of pressing pressure on the sintering rate of clustered powders ([11]). Particle radii of 2.5, 0.5, and 0.25 were chosen for the 2456G, high-purity, and C1.0 aluminas, respectively. It is presumed that there may be considerable error in these values.

Apparent volume diffusion coefficients calculated for the 2456G alumina from the slopes of the curves in Fig. 7 are shown in Fig. 9 compared with the tracer diffusion coefficients reported by Oishi and Kingery ([14]) and Paladino and Kingery ([15]). Diffusion coefficients which were calculated from the initial parts of the curves in Fig. 8 and from similar plots for the other aluminas are also shown in Fig. 9. Values of bD_B were obtained from the slopes of the curves is Figs. 1, 3, and 4 and are displayed in Fig. 10. Some of the runs of the 2456G alumina showed a swing toward grain boundary kinetics at higher shrinkages, with the resulting bD_B values shown in Figure 10. It must be pointed out that the ordinates of the points for the high-purity and C1.0 powders may be in error by plus or minus an order of magnitude or more because of the uncertainty in particle radius. The points shown for the 2456G alumina probably

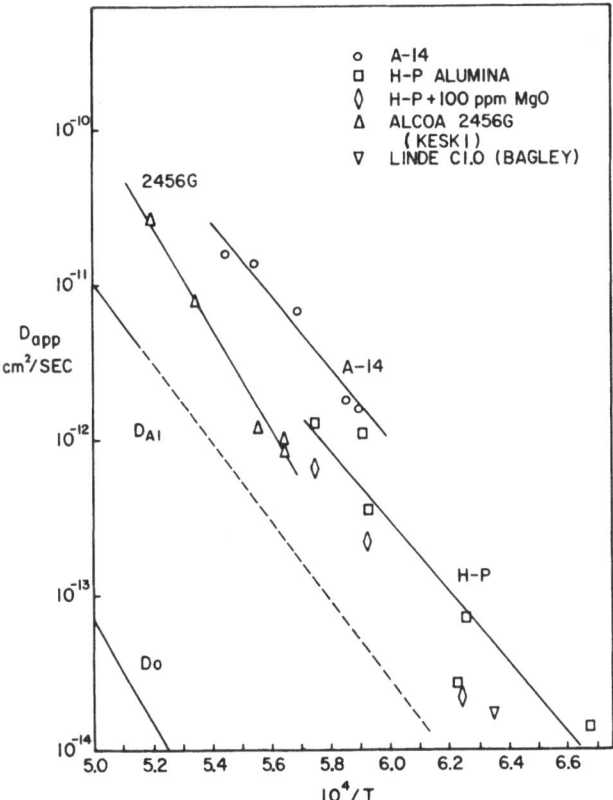

Fig. 9. Apparent diffusion coefficients calculated from the volume kinetics portions of the shrinkage isotherms compared with the reported self-diffusion coefficients for aluminum and oxygen.

represent minimum values, since the radius used was that reported by Keski to be about the mean radius of the ultimate particles in the clusters.

The least-squares lines shown in Figs. 9 and 10 result in the following diffusion coefficients:

Alcoa A-14

$$bD_B = 1.5 \exp - \frac{143,600 \pm 5900}{RT}$$

$$D_{app} = 67.3 \exp - \frac{105,300 \pm 10,600}{RT}$$

Alcoa 2456G

$$D_v = 4.8 \times 10^6 \exp - \frac{152,000 \pm 9800}{RT}$$

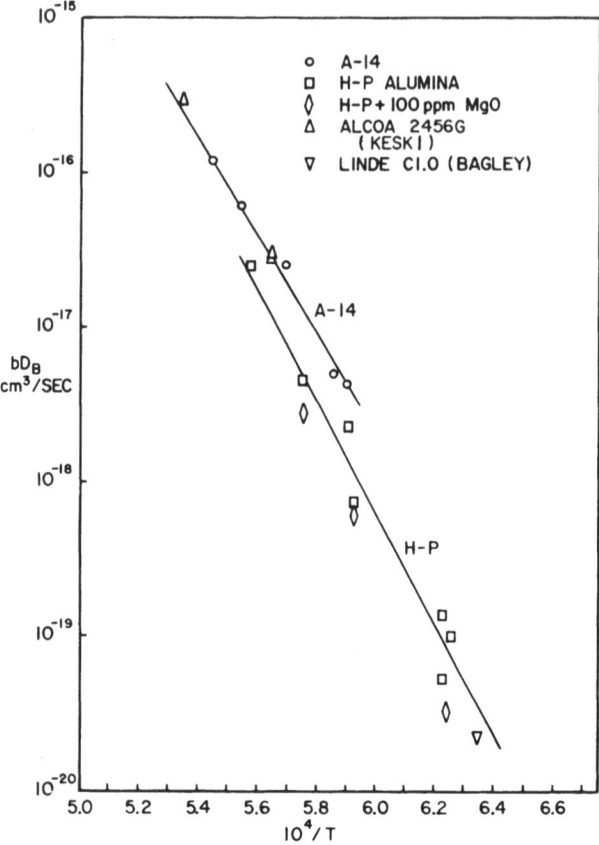

Fig. 10. Grain boundary diffusion parameters calculated from
the grain boundary kinetics portions of the shrinkage isotherms.

High purity

$$bD_B = 2 \times 10^3 \exp - \frac{163,600 \pm 13,700}{RT}$$

$$D_{app} = 10.2 \exp - \frac{103,000 \pm 21,400}{RT}$$

where D_{app} represents the diffusion coefficients calculated from the slope
of the initial part of curves like those in Fig. 8 and the indicated error is one
standard deviation.

Diffusion coefficients were calculated from the data obtained by measuring necks between spheres or between a sphere and a plate by Kuczynski
et al. ([10]) and Coble ([3]). There is too much scatter in the data points on any

given run to determine the sintering mechanism with confidence, although their results appear to lie closer to the grain boundary diffusion than to the volume diffusion model. This latter observation is particularly evident on Coble's log (a) versus log $(2r)$ plot, which has a least-squares slope of -0.63, compared with -0.66 or -0.75 for grain boundary or volume diffusion, respectively ([6]). Coble compares his results with Kingery and Berg's ([2]) model for volume diffusion. The boundary conditions they used in deriving their model are not consistent with those for volume diffusion to the grain boundary. If Coble's use of Kingery and Berg's model is justified, then different boundary conditions are appropriate for large spheres and for small particles in a compact. This is not beyond the realm of possibility.

Assuming grain boundary diffusion, Coble's data yield a bD_B ranging from 3.2×10^{-15} to 10×10^{-15} for particle diameters ranging from 40 to 3500 μ, with time and temperature constant at 100 hr and 1600°C. The extrapolated value for A-14 is 2.5×10^{-16} at this temperature. If volume diffusion according to the present model is assumed, the values of D_v vary from 4.85×10^{-11} to 47.5×10^{-11}. The volume diffusion model of Kingery and Berg ([2]) gives D_v between 2.84×10^{-11} and 1.64×10^{-11} over the same particle size range. The data of Kuczynski et al. ([10]) are essentially in agreement with Coble's neck growth data and lie a little above the extrapolation of the A-14 data.

Coble's ([3]) data for shrinkage were analyzed using Eqs. (4) and (5). The results more nearly approximated grain boundary diffusion than volume diffusion kinetics, with the extrapolated length slightly larger than L_0. It was noted previously that Coble's experimental technique resulted in a negative correction in L_0, so the above results would tend to indicate that the correction was too large. However, these data are probably not too reliable, because the powder is highly clustered. The present author was unable to rationalize data on similar powder, Linde A ([17]). It may be noted that compacts of Linde A which had been pressed at 48,900 psi ([17]) showed much higher shrinkage rates than Coble's compacts, which had been pressed at 7500 psi. Shrinkage rates comparable to Coble's were observed at about 100°C lower temperatures. This is probably a pressing-pressure effect ([11]).

DISCUSSION

It is not possible at this point to make a completely definitive statement about the sintering mechanism or mechanisms in alumina. However, most of the kinetic data seem to indicate grain boundary diffusion kinetics. The most completely characterized powder (sized Alcoa A-14) followed the grain boundary model very closely after a small, temperature-independent length correction and a short volume kinetics portion. Clustered powders either showed similar kinetics or exclusively volume kinetics, but only after substantially larger length corrections.

The length correction seems to be a function of the deviation of the real compacts from the ideal model compact. Johnson and Clarke's·[5] data on the shrinkage of compacts of closely sized spheres of silver required no correction. One of their size fractions, which had a considerably broader size distribution (a standard deviation of 13.5% of the mean radius compared with less than 5% for the other fractions), yielded anomalous results and was not included in their publication. It was recently found, however, that a length correction of about 0.4% of the initial length would bring both the form of the plot and the calculated D_v and bD_B into agreement with the other data. This indicates the importance of having a narrow size distribution in order to interpret sintering data with confidence.

The degree of linearity of all except the lowest-temperature curve of Fig. 1, the temperature insensitivity of the extrapolated length, and the small difference between the measured initial length and the extrapolated length (about 0.25% of L_0) seem to indicate that the predominant sintering mechanism for this material is grain boundary diffusion. Surface diffusion appears to be important only at the lower temperatures. The same general conclusions can be reached about the high-purity powder, but with less confidence because of the temperature dependence of the extrapolated length, the magnitude of the indicated length correction, and the fact that the powder is clustered rather than monodispersed. The temperature sensitivity of the extrapolated length for this material may be a manifestation of a surface diffusion contribution to neck growth.

The occurrence of a volume kinetics portion of the curve followed by an abrupt change to grain boundary kinetics has been discussed by Johnson and Berrin ([16]) in terms of a model of a wide region of enhanced diffusion adjacent to the grain boundary. If this is valid, then the D_{app} calculated from the initial volume kinetics portion of Fig. 8 and similar plots is identified with the grain boundary self-diffusion coefficient D_B. The width of the region of enhanced diffusion b is then on the order of a few hundred Angstroms, as calculated from the present data, and increases with increasing temperature. The extent of the volume kinetics portion calculated from Bagley's data increased with TiO_2 doping. For aluminas of less than about 1-μ radius the doped curves showed no break to grain boundary kinetics, while larger particle size powders did. The volume kinetics observed for the Alcoa 2456G may be a similar impurity effect. However, any interpretation of the latter data must be tentative, since the initial lengths are not given and therefore there is no way to determine the magnitude or possible temperature sensitivity of the length correction. It appears, however, on the basis of Figs. 9 and 10 that the behavior of this material is not too different from that of the A-14.

The neck growth data of Kuczynski *et al.* ([10]) and Coble ([3]), while yielding diffusion coefficients somewhat larger than the A-14 as analyzed above, are in order of magnitude agreement with the latter data.

Finally, a word of caution must be said about too confident an acceptance of the foregoing conclusions. Although the results seem to be relatively consistent, there is a real need for more data. There needs to be accurate simultaneous shrinkage and neck growth data on the same particles, preferably, or at least on particles from the same batch. Shingu ([7]) was able to elucidate the relative importance of surface diffusion by measuring the shrinkage of compacts of spherical particles of iron and then measuring neck growth in the same compacts. He also determined by metallographic techniques the presence or absence of grain boundaries in the necks between particles.

The following experiments are proposed to eliminate the remaining unknowns in the understanding of the sintering of alumina:

1. Simultaneous accurate measurements of neck size and shrinkage as functions of time, temperature, and particle size over ranges of each as wide as possible. This should be done with spherical particles with narrow size distributions.

2. The use of Eqs. (4) and (5) so that errors due to heat-up or nondeality of the compacts will not affect the results.

3. Simultaneous measurement of grain growth to determine if the grain boundaries migrate away from the interparticle necks.

These experiments, although they are by no means simple, should permit a definitive statement of initial sintering kinetics.

CONCLUSIONS

According to the data on hand, alumina apparently sinters by a grain boundary diffusion mechanism after a short period of volume diffusion kinetics. The activation energy for sintering of several aluminas is on the order of 150 kcal/mole. There is order-of-magnitude agreement between the shrinkage data of a sized monodispersed powder and neck growth measurements reported for larger spheres. Although surface diffusion is apparently of minor importance except at lower temperatures, there are no direct measurements to confirm this conclusion.

ACKNOWLEDGMENTS

The author is grateful to Dr. R. D. Bagley and Dr. J. R. Keski for copies of their Ph.D. theses and to Dr. H. U. Anderson for helpful discussions and for providing a copy of his paper prior to publication.

REFERENCES

1. G. C. Kuczynski, "Self Diffusion in Sintering Metal Powders," *Trans. AIME* **185**: 169 (1949).
2. W. D. Kingery and M. Berg, "Study of the Initial Stages of Sintering of Solids by Viscous Flow, Evaporation-Condensation, and Self-Diffusion," *J. Appl. Phys.* **26**: 1205 (1955).
3. R. L. Coble, "Initial Sintering of Alumina and Hematite," *J. Am. Ceram. Soc.* **41**: 55 (1958).
4. D. L. Johnson and I. B. Cutler, "Diffusion Sintering: I, Initial Stage Sintering Models and Their Application to Shrinkage of Powder Compacts," *J. Am. Ceram. Soc.* **46**: 541 (1963).
5. D. L. Johnson and T. M. Clarke, "Grain Boundary and Volume Diffusion in the Sintering of Silver," *Acta Met.* **12**: 1173 (1964).
6. L. Berrin and D. L. Johnson, "Precise Sintering Models for Initial Shrinkage and Neck Growth," in: Sintering and Related Phenomena, (G. C. Kuczynski, N. A. Hooton, and C. F. Gibbon, eds.), Gordon and Breach, N. Y. 1967 p. 445.
7. P. H. Shingu, "Effect of Competitive Mechanisms upon Densification during the Initial Stage of Sintering and Sintering Kinetics of Iron," Ph.D. Thesis, Northwestern University, 1967.
8. D. L. Johnson, "Use of Powder Compacts in Studying Initial Sintering Kinetics," *Bull. Am. Ceram. Soc.* **45**: 803 (1966); presented before the Fall Meeting, Basic Science Division, American Ceramic Society.
9. H. U. Anderson, "Initial Sintering of Rutile," *J. Am. Ceram. Soc.* **50**: 235 (1967).
0. G. C. Kuczynski, L. Abernathy, and J. Allan, "Sintering Mechanisms of Aluminum Oxide," in: *Kinetics of High Temperature Processes* (W. D. Kingery, ed.), Technology Press, Cambridge, Mass., and John Wiley and Sons, New York, 1959, p. 163.
1. D. L. Johnson and I. B. Cutler, "Diffusion Sintering: II, Initial Sintering Kinetics of Alumina," *J. Am. Ceram. Soc.* **46**: 545 (1963).
2. R. D. Bagley, "Effects of Impurities on the Sintering of Alumina," Ph.D. Thesis, University of Utah, 1964.
3. J. R. Keski, "Effects of Manganese Oxide on the Sintering of Alumina," Ph.D. Thesis, University of Utah, 1966.
4. Y. Oishi and W. D. Kingery, "Self-Diffusion of Oxygen in Single and Polycrystalline Al_2O_3, "*J. Chem. Phys.* **33**: 480 (1960).
5. A. E. Paladino and W. D. Kingery, "Aluminum Ion Diffusion in Aluminum Oxide," *J. Chem. Phys.* **37**: 957 (1962).
6. D. L. Johnson and L. Berrin, "Grain Boundary Diffusion in the Sintering of Oxides," in: Sintering and Related Phenomena, (G. C. Kuczynski, N. A. Hooton, and C. F. Gibbon, eds.), Gordon and Breach, N. Y. 1967, p. 369.
7. D. L. Johnson, "The Kinetics of the Sintering of Alumina," Ph.D. Thesis, University of Utah, 1962.

DISCUSSION

R. L. Coble (MIT): I would like to discuss the validity of the simultaneous use for transport in sintering of steady-state solutions which have been derived independently.

Johnson has shown how the steady-state solutions for lattice and boundary diffusion ontributions to shrinkage can be added, and how to conduct data analysis to extract the ndependent contributions to the overall shrinkage. Similarly, Shewmon and Wilson ssumed that the total rate of neck growth $\dot{x}_t$ was equal to the sum of the independent $\dot{x}$

contributions from lattice diffusion, boundary diffusion, surface diffusion, and evaporation condensation. In this case there are the two additional contributions to the overall rate. In both cases it has been assumed that the summation of steady-state models is valid.

For a special case in which the surface diffusion coefficient is very large and the lattice diffusion coefficient is small (and with neglect of any other modes of transport) it can be shown that the contribution of lattice diffusion to overall neck growth could be negative rather than positive as Wilson and Shewmon assumed. That being the case, we may then ask at what stage or under what conditions does the additive equation become quantitatively acceptable?

Johnson has omitted consideration of the surface diffusion effects because it does not contribute to shrinkage. If the surface diffusion coefficients measured from boundary grooving and scratch flattening data in the variety of materials of interest are correct in materials for which it is assumed that lattice or boundary diffusion causes shrinkage, then it is clear that interference of shrinkage by surface diffusion is expected in a qualitative sense. That is, it is expected to affect the time dependence. Consequently, the inference which I would draw is that the complex time dependence may result from surface diffusion interference with the lattice or boundary diffusion contribution to shrinkage rather than from simple mixing of lattice and boundary diffusion contributions.

A second complication is that the neck growth that would take place by surface diffusion prior to the observation of any shrinkage would have changed the geometry of the neck region from that which is assumed to some other, unknown, unspecified, unmeasured geometry. Consequently, it is my view that the analysis of a complex time dependence on an assumed geometric change which is probably different from that which is actually occurring during the intervals of measurement is not a satisfactory basis on which to separate the contributing mechanisms to the process.

Let us turn now to the neck growth problem to demonstrate in its simplest form the problem associated with interference by a pair of competing mechanisms. In a steady-state solution for neck growth by lattice diffusion (assuming vacancies are the predominant defects contributing to transport) at any neck size the gradient of vacancy concentration is positive from the center of the line joining a pair of spheres outward to the neck surface. That is, the concentration of vacancies increases with increasing radius from the center position to the neck. Alternatively, if neck growth takes place solely by surface diffusion (with a zero lattice diffusivity), then the gradient of defect concentrations left under the migrating neck surface decreases with increasing radii out to the neck surface. This results directly if we assume that the equilibrium concentration of defects (fixed by the surface curvature) is formed at the surface at each position and that concentration is buried beneath the surface as additional material is filled in to the neck by surface diffusion. Thus the two steady-state models considered independently give positive and negative gradients for the vacancies under the neck surface, respectively. If for the second case (when surface diffusion is predominant) we now assume that the lattice diffusivity is not zero but just finite, the migration of the defects under the influence of the negative gradient will contribute to neck growth in a negative manner. Thus in this case our intuition about the possible additivity of the simple rate equations is clearly invalid. Qualitatively a wrong effect would be predicted. The next question, not now answerable, is over what range of different values of the ratios of surface to lattice diffusivities the quantitative effects are anywhere near satisfactory? This we can not now answer and I think that it is unnecessarily speculative to do so.

What is needed is a solution to the moving boundary problem with simultaneous satisfaction of the surface and bulk LaPlacian's everywhere for all of the species which can migrate.

Answer: Shingu has synthesized sintering data, including a surface diffusion contribution, and plotted it according to Eqs. (4) and (5). These plots become nonlinear if the surface diffusion contribution is appreciable, and the extrapolated length is temperature sensitive. Thus we conclude that grain boundary diffusion is apparently not too significant except at lower temperatures for the A-14 alumina.

Most recently we have devised a method of taking surface diffusion into account explicitly and obtaining the surface diffusion coefficient from the data. Thus we will no longer need to rely on the plots used here to infer the presence or absence of significant surface diffusion.

Your comment about the frozen-in vacancy concentration gradient is interesting and could perhaps occur under the conditions you outline. However, I am quite certain that this situation would never obtain under conditions where measurable shrinkage is occurring. The reason is simply that a large number of atoms or ions must be moved to provide the shrinkage. For instance, for particles of 1-μ radius on the order of tens of atom layers must be transported from the grain boundaries to the neck surface to result in 1% linear shrinkage. No frozen-in vacancy concentration gradients could be found under conditions where this many atoms are moving.

ADDITIONAL REFERENCE

18. D. L. Johnson, "New Method of Determining Volume, Grain Boundary, and Surface Diffusion Coefficients from Sintering Data," *J, Appl. Phys.*, Jan., 1969.

Chapter 19

Densification Kinetics During Nonisothermal Sintering of Oxides*

C. S. Morgan

Metals and Ceramics Division
Oak Ridge National Laboratory
Oak Ridge, Tennessee

The intial densification kinetics in the sintering of powder compacts have been examined for five oxides: ZrO_2–CaO, MgO, CaO, CeO, and SiO_2. Study of the initial sintering, the densification which occurs as the temperature of the specimen is increased, permits the observation of densification kinetics which may not be diffusion controlled. The densification kinetics demonstrate several features which are not compatible with a diffusion-controlled process. For example, there is a disproportionately large increase in the densification rate when the temperature is increased compared to that expected from a volume diffusion process. Other tests indicate that isostatic pressing of partially sintered powder compacts causes densification on subsequent heat treatment. An explanation of these effects by a diffusion sintering process, either volume or grain boundary diffusion, is not evident; however, a qualitative explanation based on dislocation transport of material can be proposed. This explanation rests on the premise that as the temperature is increased, the stress required for motion and generation of dislocations decreases.

INTRODUCTION

Students of sintering are agreed that there are three distinct processes that can contribute to densification of a crystalline powder compact: dislocation movement, volume diffusion, and grain boundary diffusion Under "steady-state," isothermal sintering all of these processes will be diffusion controlled and follow diffusion kinetics Such a value as the activation energy can only distinguish between grain boundary diffusion on the one hand and volume diffusion or plastic flow on the other. (In this chapter plastic flow refers to transport of material by dislocation movement.) The reason for this is that dislocation motion during isothermal sintering is subject, because of the necessity for such processes as climb, to volume diffusion kinetics.

*Research sponsored by the U. S. Atomic Energy Commission under contract with the Union Carbide Corporation.

Under nonisothermal conditions, however, densification kinetics may not be diffusion controlled. If the temperature is being raised, the stress to move (and to generate) dislocations will steadily decrease. Material transport by dislocation motion will occur, and the extent of this transport may be dependent primarily on the final temperature rather than on the time of the heat treatment. In other words, as the temperature is increased, a steady procession of dislocations which were previously anchored will move. The facilitation of motion of dislocations by diffusion processes may be overshadowed by the contribution of the increasing temperature.

Densification kinetics not controlled by diffusion can be seen only in powders where there is extensive sintering at temperatures equivalent to only a fraction of the melting point. Such a condition is most easily obtainable with a high-melting refractory oxide such as thoria. It has been found that under some conditions small-particle-size ThO_2 powders and ThO_2–CaO powders do sinter with nondiffusion-controlled densification kinetics [1,2]. In this chapter results of densification with other oxides—ZrO_2–CaO, MgO, CeO_2, CaO, and SiO_2—are reported along with the effect on sintering of powder compacts of an intermediate isostatic pressing step.

There are three distinctive effects seen in the densification kinetics of ThO_2 and ThO_2–CaO powders and in the sintering of oxides reported here: (1) Regardless of the heating rate, the density obtained varied only slightly for specimens brought to temperature but not held at the temperature for a measurable time. This will be called a temperature-dependent densification end point. (2) A disproportionately large increase in rate of densification with increase in temperature is observed compared to the increase expected using a volume diffusion model of sintering. (3) A maximum and a minimum occur in the densification rate when the temperature of a powder compact is raised at a uniform rate. This characteristic is very sensitive to impurity content [1].

It is not evident that these effects can be explained on the basis of a densification process dependent on volume diffusion, grain boundary diffusion, or on dislocation motion which is diffusion controlled. However, they are readily explainable qualitatively on the basis of the properties of dislocations, particularly the sensitivity of dislocation motion to temperature and stress [1].

EXPERIMENTAL

The calcia-stabilized ZrO_2 (14.2 mol % CaO) was obtained from the General Electric Research Laboratory, where sintering and diffusion studies had been made on it by Jorgenson [3] and by Rhodes and Carter [4]. The ZrO_2–CaO was ground at ORNL in an alumina mortar and pestle to obtain a powder suitable for sintering. This powder contained 0.78 wt.% Al_2O_3 after the grinding, an amount that is sufficient to cause grain boundary melting.

However, there was no evidence of grain boundary melting in specimens taken only to 1450°C, nor in those taken to 1600°C for short times.

The MgO and CaO powders were "Baker analyzed," the CeO_2 powder was from Fisher Scientific, and the SiO_2 was a submicron powder from Vitro Laboratories. Powders were prepressed, broken up with a mortar and pestle, and repressed into small pellets at approximately 15,300 psi.

Powder compacts were carefully dried and sintered in air. During fast-heating-rate tests the thermocouple was embedded in powder (the same as that used to make the compact being tested), and the compact was placed over it. Specimens were not held a measurable time at temperature except when the time at temperature is specifically stated. The densities after sintering were obtained either by immersion techniques if the specimen was warped or by geometric measurements.

The determination of the effect of isostatic pressure on densification during subsequent heat treatment involved an initial heat treatment to partially sinter the pressed-powder compacts followed by a pressing step and a final heat treatment. This was accomplished in one of two ways. In the first method compacts were brought to a set temperature at a heating rate of $3°C \sec^{-1}$ with no hold at the final temperature. These partially sintered compacts were then isostatically pressed at room temperature in rubber bags immersed in the oil cylinder of a hydraulic press. Isostatic pressures were maintained for about 15 sec. The isostatically pressed compacts were measured again to determine the extent of shrinkage during pressing and then given another heat treatment in which they were brought to a temperature either 150° or 200°C below the first temperature. The extent of further shrinkage was then determined In a second procedure, used only with ThO_2, the compacts were held for 1 hr at 1250°C in the initial heat treatment. For the heat treatment following the isostatic pressing they were brought to 1250°C at $3°C \sec^{-1}$.

RESULTS AND DISCUSSION

Calcia-Stabilized Zirconia

The densities of calcia-stabilized zirconia powder compacts brought to temperature at different heating rates are shown in Fig. 1. For a final temperature of 1450°C there is a temperature-dependent end point for densification, although at heating rates below $6°C \sec^{-1}$ the "end point" vanishes. For specimens taken to 1600°C there is no temperature-dependent densification end point, or at least it is less evident. Apparently, when the geometric change accompanying desification is carried into the range obtained at 1600°C further densification requires diffusion support. The densification due to a

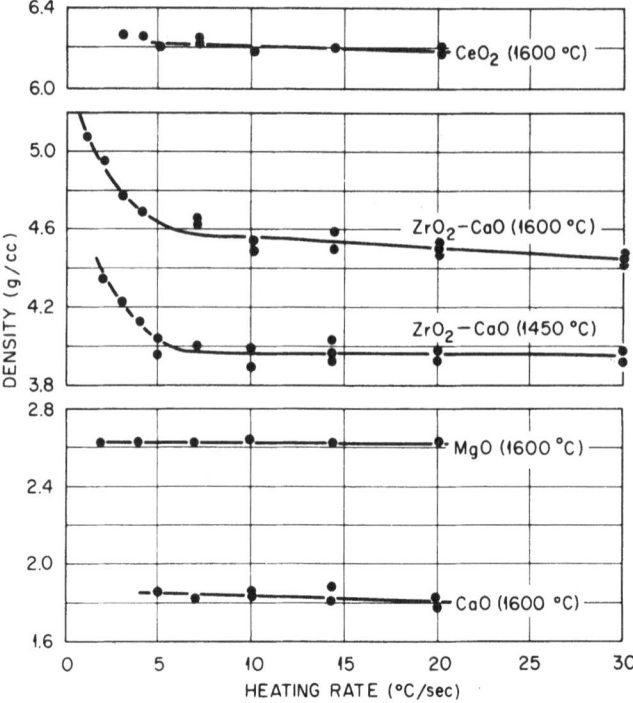

Fig. 1. Density versus heating rate for powder compacts brought to temperature but not held at temperature. (The points for MgO are the average of several specimens.)

rush of dislocation motion as the temperature is being raised does not predominate over the densification dependent on diffusion processes. Of course, the diffusion-controlled densification is always present, but in the 1450°C case and with many highly sinterable oxides it is much less noticeable.

Tests in which the incremental sintering of ZrO_2–CaO compacts is noted in additional heating time at the same temperature and at a higher temperature after a hold at a selected temperature and is compared with the results expected from volume diffusion sintering are given in Table I. In these tests powder compacts were held 10 min at a temperature T_1; then some were held at T_1 for an additional 10 min; and others were taken to a temperature 150°C higher (T_2) and held for a total of 1 min (including temperature increase time)

The last column of Table I gives the factor by which the observed densification ratio exceeds the one obtained from Kingery and Berg's equation. If one assumes that the additional increase in densification rates during the higher temperature excursion is due to dislocation motion without much

TABLE I

Densification Rate Ratios χ for ZrO_2–CaO Powder Compacts

T_1 (°C)	T_2 (°C)	Diffusion Coef. T_2/T_1 *	χ_{KB} †	χ_{obs}	χ_{obs}/χ_{KB}
1200	1350	18.3	9.5	44	4.6
1300	1450	13.2	7.3	27	3.7
1450	1600	8.8	5.3	10	1.9

*Rhodes and Carter ([4]).

†χ_{KB} is the ratio of the densification rate at T_2 to the densification rate at T_1 obtained by substituting the temperature and diffusion coefficients in the equation used by Kingery and Berg ([6]) for sintering by volume diffusion.

or any support by diffusion processes, it is evident that the amount of material transport by dislocation motion (not made possible by diffusion processes) declines relatively as the geometry of the powder compact changes. This is in agreement with the intuitive assumption that as the powder compact becomes less sinterable because of the sintering that has already taken place, the amount of dislocation motion that takes place rapidly while the temperature increases is decreasing.

The isothermal sintering of ZrO_2–CaO specimens at 1350°C yielded a straight line plot for $\log L/L_0$ versus log time. The diffusion coefficient calculated from the volume diffusion equation of Johnson and Cutter ([5]) was 7×10^{-13} cm²/sec (using an estimated value of 0.5μ for the particle radius). This compared with 1.3×10^{-14} for the zirconium diffusion coefficient obtained by extrapolating the tracer data of Rhodes and Carter ([4]).

Magnesium Oxide

Examination of the initial densification kinetics of MgO powder compacts proved difficult because of the tendency of compacts to bulge when subjected to rapid heating. Substantial water evolution occurs because highly sinterable powders must have large surface areas and because the surface of MgO is hygroscopic. Several MgO powder preparations were made and tested, and it was found that twice-prepressed "Baker analyzed" MgO powder made the most satisfactory compacts. Some of the the other MgO powders tested had higher sinterability but did not give as consistent results because of greater tendency to swell.

The densities of MgO compacts brought to 1600°C at different heating rates are shown in Fig. 1. The small decline in density with increasing heating rate shows that MgO exhibits a rather definite temperature-dependent densification end point under these conditions. Data for the density of MgO plotted against temperature for specimens brought to temperature at a heating rate of 2°C sec⁻¹ made a smooth curve within experimental error, i.e., the curve had no inflection such as seen in Fig. 2 for CaO.

When the temperature was raised MgO compacts also exhibited a disproportionately large increase in densification rate compared with that expected from the increase in the volume diffusion coefficient. The observed ratio in the 1450°–1600°C range was 40, compared to 4.1 calculated by substituting the diffusion coefficients ([7]) and temperatures in the Kingery and Berg equation for volume diffusion sintering.

Silicon Dioxide

The extremely small silica particles made compacts which showed a sharp increase in densification rate with temperature increase in the range 1100°–1250°C. The rate increased 73 times with the 150°C increase in temperature. Diffusion coefficients were not available for a calculation of the rates expected from volume diffusion.

Calcium Oxide

The curve for density versus heating rate for CaO compacts is shown in Fig. 1 and suggests a temperature-dependent densification end point. The density versus temperature relation for CaO compacts brought to temperature is shown in Fig. 2. This curve shows a sharp inflection that indicates a definite maximum and minimum in the densification rate as the temperature is being

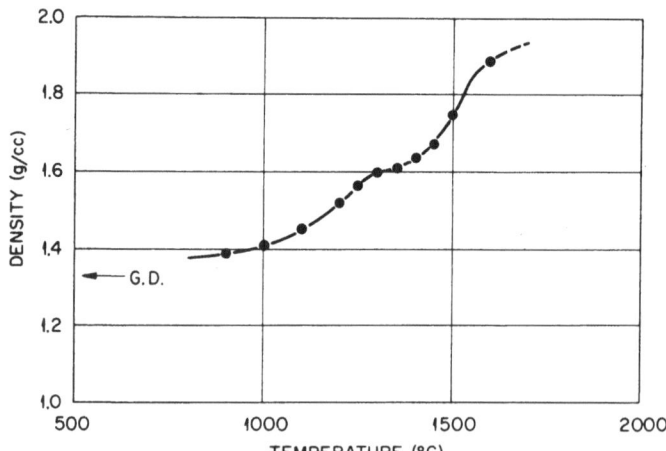

Fig. 2. Density versus temperature for CaO powder compacts heated at a uniform rate of 3°C/sec⁻¹.

TABLE II

Densification Rate Ratios χ for CaO Powder Compacts

T_1 (°C)	T_2 (°C)	Diffusion Coef. T_2/T_1 *	χ_{KB}	χ_{obs}	χ_{obs}/χ_{KB}
1300	1450	9.5	5.6	19	3.4
1450	1600	6.7	4.3	10.5	2.4

*Lindner ([8]).

raised at a uniform rate. Similar maxima and minima were seen during the densification of some thoria compacts. Calcia responded to temperature increase as shown in Table II.

Ceric Oxide

A curve for CeO_2 compact density versus heating rate is included in Fig. 1 and indicates a temperature-dependent densification end point. For a series of CeO_2 compacts subjected to uniformly increasing temperature the densification rate has been demonstrated to show a maximum and a minimum ([1]). However, the CeO_2 used in these tests showed no maximum and minimum, again suggesting the sensitivity of the initial densification versus temperature curve to the purity of the material being tested. The ratio of the rate of densification during heating to and at 1600°C compared to the rate during a longer hold at 1450°C was 22. While diffusion coefficients were not available for comparison with the rate expected from volume diffusion sintering, it seems probably that there was a substantial increase over that expected from a volume diffusion sintering process.

Isostatic Pressing Results

Sintering tests coupled with isostatic pressing for ThO_2 and ThO_2–CaO compacts are shown in Table III. There is a small but definite density increase, which is attributable to the isostatic pressing. The reality of this small increase was verified by hundreds of tests. The effect of isostatic pressing is more clearly evident in Fig. 3. Here the increase in extent of sintering as a result of isostatic pressing is shown by plotting the compact density as a function of the number of times it is heated to the same temperature. During the heating after isostatic pressing there is a jump in the density.

Partially sintered compacts prepared from MgO and from CaO tended to densify during isostatic pressing, but showed no further densification during subsequent heatings to a lower temperature. The addition of silica sol (equivalent to 10–15% SiO_2) to an aqueous slurry of MgO caused compacts

TABLE III

Densification Response of ThO₂ and ThO₂-CaO Powder Compacts to Isostatic Pressing

Compact density after initial firing	Isostatically pressed 45,000 psi	To 1250°C at 3° sec⁻¹	Increase (from highest previous density)
1 hr 1250°C	ThO₂ density (g/cm³)		
6.31	6.32	6.34	0.02
6.30	6.30	6.31	0.01
6.31	6.31	6.33	0.02
To 1450°C at 3° sec⁻¹			
8.11	8.10	8.13	0.02
8.16	8.16	8.19	0.03
	ThO-1.4 wt.% CaO		
7.00	7.00	7.06	0.06
6.91	6.92	6.95	0.03
6.84	6.84	6.87	0.03

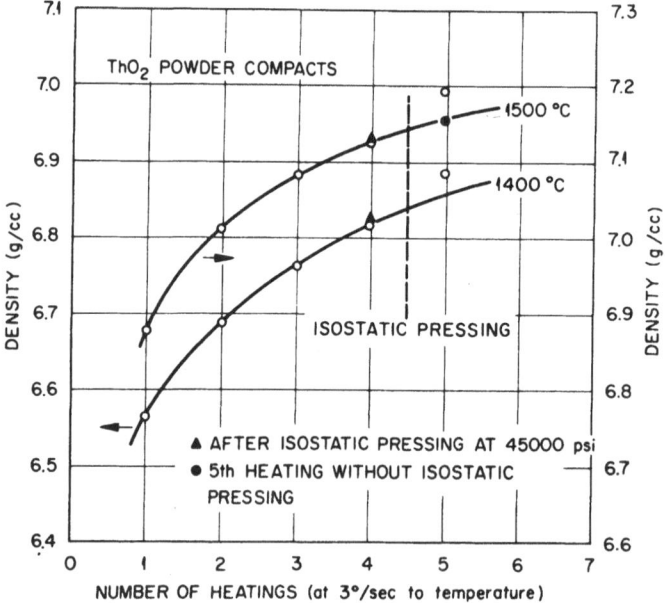

Fig. 3. Density versus number of heatings to temperature at 3°C/sec⁻¹ (no hold at temperature) for ThO₂ powder compacts. Between the fourth and fifth heating the partially sintered compacts were isostatically pressed. Each point is the average of two specimens.

prepared from the dried slurry to show an enhanced sinterability as a result of isostatic pressing, while the amount of further densification during isostatic pressing was greatly reduced. MgO compacts taken to 1450°C, isostatically pressed at 20,000 psi, and reheated to 1300°C had an average density increase of 0.024 g/cm³, whereas a maximum increase of 0.011 g/cm³ was seen on reheating without the isostatic pressing step. The cause of this effect will be discussed later.

Observing an isostatic pressing effect on the densification of calcium oxide proved more difficult. Addition to the powder slurry of small ThO_2 or SiO_2 particles in the form of sols had no effect. Homogenizing the particle-doped CaO powder in a Waring blender had no effect. It was found, however, that if CaO was prepared from calcium oxalate which had been precipitated with concurrent addition of ThO_2 sol, an effect could be observed. CaO–6% ThO_2 prepared in this manner and brought to 1400°C at 3°C sec⁻¹ densified only slightly on isostatic pressing at 20,000 psi. Specimens to be used in isostatic pressing tests were repeatedly heated to 1200°C until the density increase with each cycle had fallen to 0.01 g/cm³. This was necessary because CaO–6% ThO_2 tended to densify continually on reheating to 1200°C. Specimens brought to 1400°C, "exhausted" at 1200°C, and then isostatically pressed at 20,000 psi showed a density increase of 0.015–0.022 g/cm³ on bringing to 1200°C. After the CaO–6% ThO_2 specimens had been put through the 1400°C treatment and a series of 1200°C cycles solution in nitric acid indicated that the ThO_2 was still present as discrete particles.

DISCUSSION

The small but definite densification enhancement resulting from isostatic pressing is consistent with material transport by dislocation motion. The pressing enhancement could result as follows. Dislocations are put into the structure of the partially sintered powder compacts by isostatic pressing at room temperature. On subsequent heating the dislocations move and affect the material transport. In oxides in which the dislocations move easily at room temperature, e.g., MgO and CaO, the effect on subsequent thermal densification appeared to be reduced or absent. Addition of particles presumed to act as pinning points resulted in densification enhanced by isostatic pressing.

These conclusions are compatible with Radcliff's [9] observation of what is thought to be a similar phenomenon with respect to isostatic pressure generation of dislocations which move on subsequent activation. He found that if iron–carbon alloys were isostatically pressed, discontinuous yielding was eliminated. Bullen et al. [10] observed a similar effect in iron specimens containing carbon and were also able to obtain up to 60% plastic deformation in chromium specimens after isostatic pressing at 10^4 atms ([11]). Without the isostatic pressing step the chromium specimens were essentially brittle.

It was assumed by both investigating groups that dislocations were put in the specimens during isostatic pressing and that these dislocations moved during subsequent stress–strain tests. The elimination of the discontinuous yield point was less evident in pure iron specimens, suggesting that dislocations were around Fe_3C particles in the iron carbon alloys. Radcliffe and Warlimont ([12]) used electron transmission microscopy to verify that dislocations were put in the specimens during isostatic pressing and that they were concentrated at Fe_3C particles.

Some or all of the three distinctive effects previously reported in the densification kinetics of thoria ([1]) occur in each of the five oxides studied here. These effects as well as the isostatic pressing effect can be explained on the basis of the properties of dislocations. The question can be asked whether there is an alternative explanation or whether it is necessary to assume a large dislocation motion contribution which is substantially free of support by diffusion processes. A satisfactory alternative is not evident. Any alternative involving classical sintering concepts must be a diffusion-controlled process, but the densification kinetics do not fit the kinetics of either volume diffusion or grain boundary diffusion. After the early heating brings the specimen to a temperature where significant particle–particle bonding occurs particle rearrangement is thought to be quite small or absent. It also seems that the compact topology would be such that only at a few specialized points would there be stresses to cause grain boundary sliding. A substantial dislocation motion contribution to the densification therefore appears to be present in the initial stage sintering. The isostatic pressing effect corroborates the significance of dislocations.

The extent of the dislocation contribution indicated by the densification kinetics appears to vary with the sinterability of the compacts, which for a given material is dependent primarily on the shape and size of the particles involved. The results for progressive sintering of the ZrO_2–CaO compacts demonstrate this. On sudden increase of the temperature from 1300°C the densification kinetics suggest a very large dislocation motion contribution without diffusion control. But after further sintering, which takes place when higher temperatures are used, there is less evidence of dislocation motion without diffusion control.

These results agree with the concept that initially the geometry of the powder compact is such that there are many particle contacts which have small radius of curvature, and therefore high stress exists to move material into the neck area. When the temperature is raised dislocations and dislocation sources which were inactive at the lower temperature become operational, with the result that there is a burst of densification. As the geometry of the compact changes, the sintering stresses are reduced and either the number of moving dislocations or their velocity quickly decreases. Each subsequent temperature increase will produce a smaller effect.

Undoubtedly, dislocation motion would continue under isothermal conditions, but it would be diffusion controlled. It seems likely that as sintering continues, the ratio of dislocation transport to diffusion transport would decline; however, studies of the type reported here give information only on the initial or nonisothermal stage of densification.

CONCLUSIONS

Although these densification studies do not delineate the relative contribution of various material transport processes, they do suggest that the characteristics of material transport by dislocation motion should be seriously considered in evaluating potential sintering procedures. Thus the influence on dislocation motion of additives, heating cycle, compact topology, and stress should be considered. With small particle powders the massive densification occurring as the temperature is raised must be considered. Sintering models which completely disregard dislocation motion would seem suspect, except possibly during very late stages of sintering. Ordinarily, models are applied to the early stages of sintering.

Giving proper credit to dislocation motion may explain the abnormally high diffusion coefficients often obtained by application of diffusion models of sintering to the observed densification or neck growth data.

REFERENCES

1. C. S. Morgan and C. S. Yust, *J. Nucl. Mater.* **10**: 182 (1963).
2. C. S. Morgan, C. J. McHargue, and C. S. Yust, *Proc. Brit. Ceram. Soc.* No. 3, pp. 177–84 (1965).
3. P. J. Jorgenson, "Diffusion Controlled Sintering in Oxides," General Electric Research Laboratory Report No. 65–RL–3994M, 1965.
4. W. H. Rhodes and R. E. Carter, *J. Am. Ceram. Soc.* **49**(5): 244–49 (1966).
5. D. L. Johnson and J. B. Cutter, *J. Am. Ceram. Soc.* **46**(11): 541–45 (1963).
6. W. D. Kingery and M. Berg, *J. Appl. Phys.* **26**: 1205 (1955).
7. R. Lindner and C. D. Parfitt, *J. Chem. Phys.* **26**: 182–85 (1957).
8. R. Lindner, *Acta Chem. Scand.* **6**, 468–74 (1952).
9. S. V. Radcliffe, in: *Symposium on the Irreversible Effects of High Pressure and High Temperature on the Properties of Materials*, Philadelphia, 1964. pp. 141–162.
10. F. P. Bullen, F. Henderson, M. M. Hutchinson, and H. L. Wain, *Phil. Mag.* **9**: 285–97 (1964).
11. F. P. Bullen, F. Henderson and H. L. Wain, *Phil. Mag.* **9**: 803–15 (1964).
12. S. V. Radcliffe and H. Warlimont, *Phys. Status Solidi* **7**(2): K67–69 (1964).

DISCUSSION

D. L. Johnson (Northwestern University): You have reported heating rates as high as 150°C sec^{-1}. How are you sure the compact itself is actually at the temperature indicated? It seems that even for thin compacts there would be a large lag in temperature at the center of the compact. Perhaps at the higher rates the center of the compact was much cooler and the overall densification was limited by this cool core.

Answer: This is a good question and we have considered it with the problems you present in mind. At the higher heating rates there may be a lag in the core temperature. To determine the effect of this we carried out tests under three different conditions: (1) Where the temperature of the specimen exceeded that of the thermocouple, (2) where the temperature of the specimen lagged behind the thermocouple, and (3) where as near as we could achieve it the two temperatures were the same. The three cases resulted in different densities being achieved, but in each case the density achieved varied only slightly with heating rate.

F. V. Lenel, (Rensselaer Polytechnic Institute): The densification of compacts which have been presintered, isostatically pressed, and resintered seems to be an effect similar to one observed in metal compacts. Copper compacts will shrink a small amount (small compared to the shrinkage during normal sintering at temperatures near the melting point) in the temperature range where the internal stresses introduced during compacting are relieved. The driving force for this shrinkage is internal stress rather than stress due to surface tension forces. It may be possible to speak about internal stresses (a dislocation structure in the very fine particles) produced in the powder which are relieved during heating to the sintering temperature. In this case the unusual densification kinetics of the fine powders may also be attributed to these internal stresses.

Answer: We do not think that appreciable internal stresses are introduced into the partially sintered compacts during the isostatic pressing. There would be elastic strains in the compact and dislocations would be created at high stress points but would not move. An analogous situation might be the pressing together of two perfectly flat blocks with a micron-size particle between them. The particle would press into one of the blocks, creating dislocations but not measurably deforming the block. In the isostatically pressed compacts dislocations have been introduced but no significant stresses have been added other than the stress fields around the dislocations because there is little plastic deformation at room temperature. When the temperature is raised the dislocations move because of the stresses created by the surface tension.

M. R. Montierth (Corning Glass Works): The interpretation of the dislocation movement contribution during reheating following isostatic pressing was made on the basis of sintering data only. Could not these data also be interpreted to mean that the isostatic pressing resulted in a breakage within the compact such that the initial rapid sintering conditions were again approached?

Answer: That is a good point to consider. We cannot say with absolute certainty that there was no effect of that nature, but we do not believe that it is a significant factor. In calcia and magnesia compacts where we observed appreciable densification during the isostatic pressing we did not detect densification that could be attributed to the isostatic pressing during the subsequent reheating. The isostatic pressing effect on subsequent densification was seen only in compacts that had little or no measurable densification during isostatic pressing. Therefore we do not believe that there was much or any effect from shattering of the particles.

Alan Franklin (ARPA): The experiments in which a sudden jump in temperature produced an increase in sintering too large to be accounted for on the basis of the temperature dependence of volume diffusion coefficients might still be explained on the basis of volume diffusion if it is assumed that insufficient time is allowed at the higher temperature to reestablish the lower gradient in vacancy concentration appropriate to the higher temperature. Hence diffusion at the higher temperature is abnormally high because the vacancy gradient is abnormally high.

Answer: This is a point which we have not considered, and I am not certain what the effect would be. However, the vacancy gradient causing material transport effecting densification would be created by the surface tension stresses causing the sintering. The vacancy demand in the interior of the particles resulting from the temperature increase would create a vacancy gradient which would not contribute to densification. This gradient would move cacancies from all surfaces to the interior. It seems likely that the movement of vacancies to effect densification—from the weld neck to a suitable surface—would actually be supressed until the vacancy equilibrium concentration had built up, because some vacancies from the neck would move to the particle interior. In addition, the concentration of vacancies in the gradients causing densification would require time to build up after the temperature increase before the increase would have full effect on sintering by volume diffusion.

R. S. Gordon (University of Utah): Would not you have to know something about the dislocation yield stresses in your material and some mechanical qualities before you could disprove that you have dislocation motion?

Answer: While an analysis of the actual stresses and dislocation movement in the complicated powder compact is not practical, we do know that in powder compacts of the particle size we have the stresses created by the surface tension can easily exceed the macroscopic yield stress of the material as the temperature is raised. There is a large literature (for example, the work by Gilman and Johnston) showing that the movement of dislocations is extremely sensitive to stress and temperature. The densification kinetics which we observe are in qualitative agreement with the kinetics of dislocation motion.

J. E. Burke (G.E.): I believe I understood you to say that you think the sintering rate is too rapid to be accounted for on the basis of a diffusion-controlled process when you jump the temperature and that you think of this higher temperature as increasing the mobility of the dislocations. I think in general the activation energy for the diffusion processes is considerably higher than the activation energy for dislocation mobility, so that it seems to me that it would in general work the other way.

Answer: The extensive studies of Gilman and Johnston on LiF show a tremendously high activation energy for dislocation movement. Other results have suggested the same thing for many materials. We know from studies at ORNL that ThO_2 creep is very temperature sensitive. On the other hand, there are a few deformation studies in metals where the yield stress showed little temperature dependence, usually over a relatively narrow temperature range. We believe the dislocation motion activation energy in our specimens in the temperature range used is much higher than the diffusion activation energy.

Chapter 20

The Behavior of Oxygen in Oxides During Sintering*

R. H. Condit

Chemistry Department
Lawrence Radiation Laboratory, University of California
Livermore, California

The need for studying details of the transport mechanisms during sintering is outlined. A technique for doing this is proposed which makes use of the nuclear activation of the rare isotope of oxygen O^{17} so that it can be used as if it were in effect a radioactive tracer. A test of the feasibility of the use of this as a marker in the region of neck growth between grains of ceramic material is then reported.

INTRODUCTION

Oxygen is an important constituent in ceramic materials, and it would seem that measurements of its behavior during the sintering process in ceramics should contribute to our understanding of that phenomenon. In many cases it appears that oxygen may be the slower diffusing species, and therefore its migration may then be the rate-limiting step. In other cases it has been proposed that oxygen diffusion is relatively rapid within the interface region between two adjoining grains and that cation migration may then be rate limiting [1].

In many studies information about sintering has been available only in terms of the shrinkage of a mass of fine-grained material at some temperature as a function of time. The assumption is often made that such measurements could be used in a straightforward manner to indicate diffusion by bulk, surface, or grain boundary mechanisms or to indicate material transport by plastic deformation or as a vapor phase. Since different laboratories have reached different conclusions about the rates for a given material and about the plausibility of almost every proposed mechanism, the field of sintering research has become one of the more controversial within the science of ceramics.

A likely cause for experimental difficulties is the importance of factors

---

* Work performed under the auspices of the U.S. Atomic Energy Commission.

which are not easily susceptible to mathematical analysis. These generally try to relate rates of neck growth between particles or shrinkage of particle compacts to such more or less physically homogeneous phenomena as diffusion, plastic deformation, or vapor transport. Impurity effects are often cited as sources of trouble. These may serve to either enhance or suppress the diffusion of the slower moving species, and differences in impurities between materials used by different laboratories may be a source of reported discrepancies. What might be called anomalous impurity effects should also be considered, however. Small amounts of water in the gas environment may strongly influence the tendency for grains to become faceted at the same time that they are joining ([2]). The formation of whisker bridges between particles has also been observed ([3]), and this seems to be a function of gas composition. Within the solid itself a redistribution of impurities might be expected to occur concurrently with sintering. Foreign material on the surface of the grains may enter into solution, or some constituents in solution may tend to precipitate if the sintering anneal is carried out at a lower temperature than that employed for the original material preparation. If insoluble microinclusions are present, they may provide sources or sinks for vacancies, may punch out dislocation networks on thermal cycling, or otherwise modify the ideal shrinkage process

The substructure of the material being studied may also be of considerable importance. Dislocation pipe diffusion of oxygen within magnesium oxide has been observed, for example ([4]), and it is possible that the presence of dislocations in the high concentrations introduced by grinding a coarse material into fine grains before the anneal may well influence sintering rates insofar as they depend on oxygen transport.

To return to the mathematical theories which have been advanced, it should be mentioned that the basic assumptions and boundary conditions have been questioned, i.e., the actual distribution of stresses within two particles as they knit together may be significantly different from those which have been commonly assumed ([5]), and this could affect the rate of shrinkage as a function of time. Moreover, it has been difficult to set up a mathematical treatment which provides a consistent description of the sintering process as it proceeds from the initial stages to later stages of pore isolation, pore shrinkage, and ultimate densification even though the basic physicochemical phenomena may be the same throughout.

In view of the manifold difficulties with simple, parametric approaches it would seem that students of sintering might be well advised to become much more concerned with the elucidation of the details of mechanisms, and this should be done by looking at the processes which take place within individual grains as they join together. Various experimental tools come to mind which can have satisfactory resolving powder for seeing into the small regions within a neck. It will be the purpose of this chapter to describe one of these, a means for following the behavior of oxygen tracers. This investigation is in its early stages and it is appropriate that attention be directed toward establishing the value of the method, because it is important that experiments in the field of sintering be capable of generating meaningful results if they are to be considered worth carrying out at all.

PROPOSED EXPERIMENTS

The use of inert markers has played an important role in helping us gain an understanding of solid-state kinetics, and they have been used in sintering studies ([6]). However, it is possible that markers can interfere with the processes being studied, and if work on a very fine scale is intended, this becomes a major objection. The use of radioactive tracers permits the location of very small markers, but the ultimate in compatibility between marker and matrix is achieved if the marker is chemically identical with the matrix. In other words, a tracer for one of the atomic species naturally present is best, and in the present case it would appear desirable to have a tracer for oxygen, since, as mentioned above, it is a major component in most ceramic materials. Oxygen tracers have value in cases where they are equivalent to inert markers. On the other hand, observation of the motion of the tracer by diffusion when it participates in material transport should be instructive.

Unfortunately, there are no convenient radioactive isotopes of oxygen. The stable isotopes are O^{16} (natural abundance being 99.76%), O^{17} (0.04%), and O^{18} (0.20%). At present the rarer isotopes can be purchased commercially in enrichments up to 99% O^{18} and about 10% O^{17}, factors of 500 and 250, respectively, above their natural concentrations, and it has been shown that their use together with nuclear activations which make them radioactive can provide for their detection as if they were radiotracers ([7]). The only difference is that they are made radioactive after a diffusion or reaction experiment rather than having been radioactive from the beginning.

The availability of an oxygen tracer permits us to consider several types of marker experiments. One of these which we are undertaking can be understood by reference to Fig. 1. Two spheres can be prepared, one having its surface layer labeled with tracer, and the other without tracer. The distribution of the tracer within the neck area after a sinter anneal as compared with the initial distribution should provide a clue about the mechanism of neck formation. The following should be mentioned:

1. If the joint results by plastic deformation of the two spheres, there should be negligible transport of material away from the interface region, and the amount of tracer there should be the same as on the original surface of the labeled sphere.

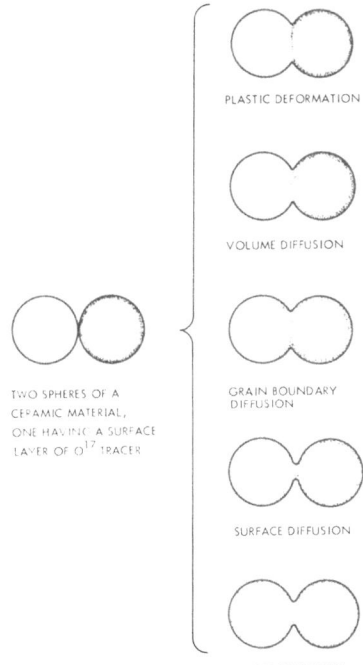

Fig. 1. Ways in which oxygen might migrate during sintering.

2. If sintering has occurred by volume diffusion with oxygen diffusion limiting the rate, there should be appreciable penetration of the tracer into the bulk of the labeled sphere, and there should also be some loss of material from the interface region. The perturbation of sintering upon a simple diffusion penetration profile distribution has not been analyzed in the literature as far as we have been able to discover.

3. If the neck has grown as a result of grain boundary transport of material away from the interface between the spheres, the tracer should be depleted in the center of the neck and piled up around its rim.

4. If surface diffusion has been important, the migration of the tracer from the surface of the labeled onto the unlabeled sphere should be apparent and the concentration of tracer should decrease with distance from the original labeled grain. In this case there should be no shrinkage of the distance between sphere centers.

5. If vapor transport has been important, the redistribution of tracer might be expected to give a more or less homogeneous layer over the two spheres with extra buildup in the neck region. As with surface diffusion no shrinkage would pe expected. Small temperature gradients in the sinter assembly should lead to a characteristic buildup of tracer on the low-temperature side of each cavity.

The way in which such redistributions of tracer can be observed by activation and subsequent autoradiography will now be discussed.

Ion Bombardment Activation

The use of the O^{18} tracer in ceramics studies has been previously described ([7]). In this former investigation the activation $O^{18}(p, n)F^{18}$ was carried out with bombardments from a cyclotron to give 2.8-MeV protons. These convert some of the tracer to the radioactive isotope fluorine-18 which has a half-life of 110 min and decays with the emission of a 0.6-MeV positron. This half-life is long enough to allow such manipulations as autoradiography to be carried out on the specimen. In the present study an activation technique developed by Holt ([8]) has been used which employs O^{17}. This is activated by the reaction $O^{17}(d, n)F^{18}$ using deuterons having energies of about 1.0 MeV, which gives the F^{18} as before. Bombardment energies up to 1.5 MeV are preferable but not essential. The bombardment energies required for this reaction are less than when using protons and the cost of a suitable van de Graaff thereby becomes no more than that of a modestly priced electron microscope. The cross section for this reaction is slightly less than for the protons and the concentrations of O^{17} are normally smaller, but since the beam voltage is lower, the bombardment currents can be increased without heating damage to the sample. As with proton bombardment, the activation is fairly specific for oxygen, and only a few undesired, interfering activities can be produced. Among these are B, Na, and to a lesser extent Si.

The availability of this activation reaction permits us to carry out a marker experiment of the type outlined above. The pair of spheres can be imbedded in a suitable matrix and sectioned so that a surface is presented which cuts through the two spheres, revealing them as a dumbbell pair (see Fig. 2a). This surface can be activated by ion bombardment. The resultant F^{18} activity distribution is a representation of the distribution of the oxygen tracer. The nuclear reaction is exothermic by 3.4 MeV and this energy, together with the 1.0 MeV carried in by the deuteron, is partitioned between the products of reaction in inverse proportion to their masses. The share going to the F^{18}, about 220 keV, is not sufficient to drive it more than about 0.1 μ through most solids ([9]). In addition, the bombardment energy put into our specimens has not heated them above 100°C, and it does not seem likely that appreciable diffusion of either the O^{17} or the F^{18} can occur at this temperature.

The amount of activation produced by a typical bombardment with 20 μA deuteron current spread over a square centimeter would produce 10^{11} atoms of F^{18} during a 1-hr irradiation if the target surface were 10% O^{17}. This would still represent less than 1 ppm of F^{18} within the region of

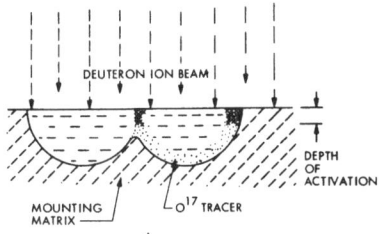

Fig. 2. Examination of sintered spheres. A
pair of sintered spheres is first imbedded in a
matrix and then sectioned. (*A*) The exposed
surface is irradiated with deuterons. (*B*) The
specimen is then covered with a photographic
film and the darkening of the film can be taken
as an indication of the location of activity.

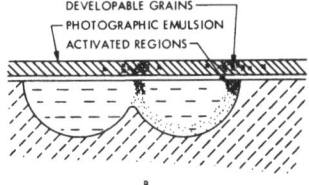

activation, and the chemical and physical disruption of the specimen is
negligible.

A particularly important feature of ion bombardment is that it produces
activity within a very shallow layer at the surface. In the case of 1-MeV
deuteron bombardments the density of resultant activity falls off by a factor
of two within 3 μ and by 10 within 10 μ from the surface. For purposes of
achieving high autoradiographic resolution this shallow activation is essential,
and it can be made shallower still if the irradiation beam impinges at a grazing
angle. A depth of activation of about 1 μ is attainable in this way.

The distribution of activation can be determined by the technique of
autoradiography, where a photographic film is placed over the surface of the
activated specimen (Fig. 2*b*). It is interesting to note that the resolution
which may be needed for studying the details of sintering cannot be achieved
by using most radioactive tracers in the usual way because "cross-fire"
subsurface activity would then obscure features of the distribution at the
surface.

Autoradiography

The positrons emitted by the fluorine-18 have the same properties as
beta rays (high-energy electrons) insofar as the response of a photographic
film is concerned. They are eventually annihilated, with the production of
gamma rays, but the response of most films to gamma radiation of the energies involved here is roughly 0.1 % of their response to the positrons. The
mean range of the positrons is about 2000 μ in water, 1300 μ in silica, 700 μ
in alumina, and 450 μ in lead. In all of these materials the amount of absorp-

tion which the positron flux can suffer while traveling distances of less than 10 μ can be assumed to be less than 2%. Therefore from the standpoint of absorption of the radiation in the sample, its mounting, or the films normally used no absorption correction is necessary. However, back reflection of that portion of the positrons which initially proceed from the surface down into the specimen can be a source of difficulty in the quantitative analysis of auto-radiographs. This back reflection will give an image roughly as though there were a source of activity similar in configuration to the surface layer but buried at a depth which equals the mean range. Consequently, a point source will have a blurred region about its autoradiographic image. The total flux of the back-reflected radiation from the virtual "source" may be as high as 70% of the forward-emitted flux [10]; the amount of back reflection goes up with the density of the material. Consequently, it is possible that a point of activity at the edge of a specimen of silica mounted in lead may appear to be stronger than an identical one in the center of the silica surface. One can make correction for such effects during the analysis of film density profiles if one wishes to employ computer techniques.

There have been numerous reviews of autoradiography which include discussions of photographic films [11,12]. As a rule of thumb, beta or positron fluxes of 10^7 to 10^9 give film densities which are convenient to measure. The yield of developable grains seems to run between 0.1 and 1.0 per positron, and the density of a film will also depend on its thickness. With our typical targets, which have been irradiated for $\frac{1}{2}$ hr, sufficient activity is obtained to give a good exposure with Kodak Autoradiograph Plates Type A in a few minutes. On the other hand, Kodak stripping emulsion AR-10 will require 1 hr for similar darkening. In general, the resolution of a film goes up with decreased grain size, while its sensitivity goes down, and that is the difference between these two films. The Type A plates have a grain size of about 0.65 μ and an emulsion thickness of 25 μ, with a 1 μ overcoat. The active specimen is normally placed on top of a plate. In this procedure small departures from perfect flatness of the specimen surface cause imperfect contact between specimen and plate. Therefore the resolution to be expected with the plates is probably no better than 30 μ at best, but these plates are convenient for a quick examination of the activation produced.

The AR-10 film has an emulsion thickness of 5 μ, no overcoat, and a grain size of about 0.15 μ. This emulsion is supplied on glass plates from which it can be stripped and transferred onto the surface of the specimen. This is done by first floating the emulsion on top of water and then bringing the specimen up to it from underneath the water, lifting it out, and allowing the wet film which now clings to the specimen to dry. The intimate contact permits resolutions of 5 to 10 μ. The use of these films has been fully described elsewhere [13].

A further improvement in resolution is possible with the use of melted emulsions. These are painted onto a surface, and when solidified can have a thickness of less than 1 μ ([14]). In a study of tritium distribution in steel resolutions of a few tenths of a micron have been achieved ([15]). It should be understood that "resolution" in such studies as these has a slightly different meaning from that which is commonly used. The measurement of such resolutions is only possible by looking at the emulsion in the electron microscope and counting the distribution of individual, developed silver grains. The correlations between the positions of these grains and the features of interest on the specimen can then be determined. The use of such techniques would probably be necessary if pipe diffusion were to be studied, but they do not seem to be required for initial studies of sintering.

Resolution and sensitivity are not completely separable autoradiographic variables, particularly when quantitative measurements on a micro scale are being attempted. With large images it is a relatively simple matter to expose the film under standard conditions of temperature, humidity, and source-film geometry and to develop the film by standard processing. The film density as a function of exposure can then be calibrated by carrying out a series of exposures to find the characteristic response curve of the film. The activity of a source can be measured within a per cent or two in this way. On the other hand, measurements on a micro scale require the same precautions, but the random nature of the radioactive decay and the fact that the photographic image is made up of individual grains introduce statistical fluctuations in grain densities. Thus if an exposure has resulted from a flux of 10^8 positrons per cm^2 and each one has produced a developed grain of silver, this is an average density of 100 in a 10-μ square. The mean fluctuations in a number N which counts a random process is $N^{1/2}$, and in the present example this means that within the square a 10% noise factor must be allowed for. The distribution of grains within the emulsion is also subject to these same statistical laws but the density is higher, about 15/μ^2 in AR-10, and the fractional variations from this source are negligible.

As a result of the random nature of radioactive decay, it appears that a sort of indeterminacy principle operates in quantitative microautoradiography. One can measure the amount of flux through large areas rather precisely or one can measure the location of a point source within a few microns, but one cannot locate and measure both quantities with precision. For sintering studies, however, this does not detract from the value of the technique. If diffusion of the oxygen tracer over appreciable distances is important, the measurement of concentration as a function of distance can be carried out satisfactorily. Conversely, if the tracer has not spread out, its position can be found.

In summary, the autoradiographic technique can give resolutions of

5 to 10 μ when standard, commercial stripping films are used and the source of activity is sufficiently strong. Higher resolutions may be possible, down to less than a micron, but the measurement and data analyses become considerably more difficult.

EXPERIMENTAL PROCEDURES

Some of the incidental experimental procedures which have been developed in the course of this investigation should be mentioned. The initial implantation of the tracer has been carried out by a diffusion anneal in an apparatus of the same type as used for the measurement of oxygen diffusion coefficients ([16]). Spheres of MgO or Al_2O_3 of 1 mm have been placed in a platinum crucible, enclosed within a bulb containing the enriched oxygen-17 gas, and annealed at 1650° by induction heating of the crucible Activation of these spheres followed by autoradiography reveals that much of the activity on the surface is confined to a few points. More uniform layers are obtained in SiO_2 slabs heated at these same temperatures. Since the MgO and Al_2O_3 spheres were made by tumbling abrasion, the presence of microcracks in the surface is probable. This by itself raises some question about measurements of oxygen diffusion which depend entirely on a monitoring of the change in isotopic content of the gas in which the specimen is annealed without looking at the solid afterward.

After the oxygen impregnation diffusion anneal the ceramic is frequently flecked with tightly adhering crystals of platinum. If the platinum is not removed, it is possible to obtain false indications of oxygen exchange with the ceramic upon activation because the platinum will contain large amounts of oxygen. The removal of platinum cannot be done mechanically, and the use of aqua regia is dangerous because it tends to dissolve small amounts of the ceramic also. It has been found that gallium is a good solvet for platinum when heated to about 800°C. The specimens after their O^{17} anneal, together with 10 g of gallium, are placed in a quarty bulb and heated gently while a vacuum is pulled on the bulb to avoid oxidation of the metal. Considerable outgassing and frothing of the gallium occurs, but after $\frac{1}{2}$ hr the sample can be cooled and the gallium brushed off the ceramic, taking the platinum with it.

The sinter anneals have been carried out in rhenium foil supports within a tungsten crucible. Rhenium is used in contact which ceramics because it has been found to be less reactive than tungsten. The actual arrangement of ceramic materials need not adhere to the model indicated in Fig. 1, since other geometries are also likely to be instructive. Specifically, plates having a surface layer loaded with O^{17} tracer in contact which unlabeled spheres should have some advantages.

Following the sintering anneal the specimen is carefully transferred to

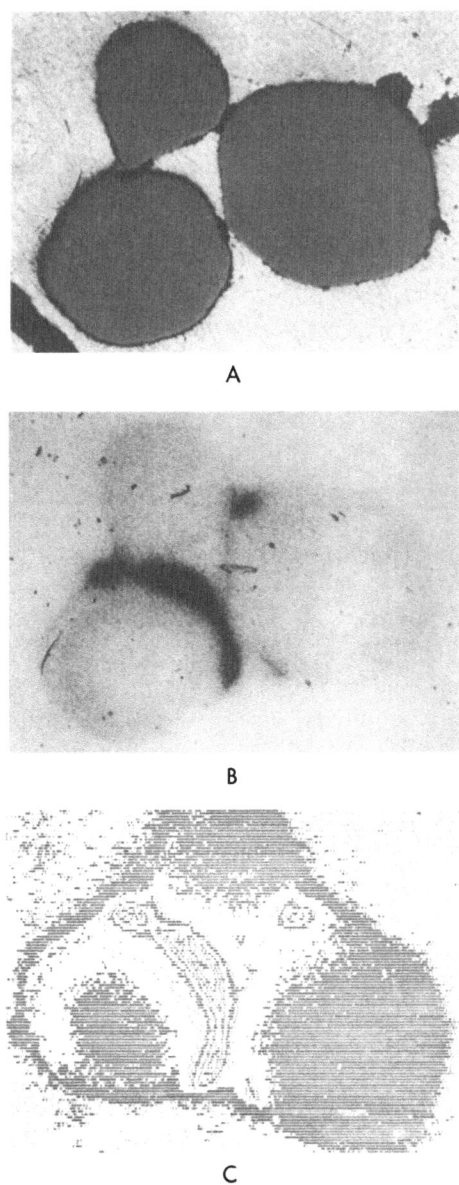

Fig. 3. (A) Mounted specimen, (B) autoradiograph, and (C) isodensity trace, 20×, reduced
by 25 per cent for reproduction. Three spheres
of MgO mounted in Woods metal are shown.

an aluminum metal cup 1 cm in diameter and 3 mm high. Wood's metal is then cast around the specimen and the surface is ground away, as indicated in Fig. 2. To achieve penetration of the Wood's metal into all cavities between the spheres, the sintered assembly is contained in a small evacuated tube, the Wood's metal is melted over it, and air is admitted and drives the molten metal into the specimen, where it solidifies. It is then sectioned. One may examine a series of sections through a specimen if desired by repeating the sectioning-irradiation-autoradiography cycle.

For the deuteron ion bombardment the specimen in its casting is attached to the flat surface of a steel cold finger with liquid nitrogen. An adhesive of silicone grease is suitable. This is pasty at room temperature, but is a strong cement at low temperature and serves to conduct heat away from the specimen during irradiation. The Wood's metal has not been found to melt until currents of 30 μA at 1.0 MeV (i.e., 30 W) are exceeded.

Following irradiation the specimen is highly active due to small amounts of carbon contamination from diffusion pump oil in the ion beam pipe. This gives a 10-min half-life product as a result of the reaction $C^{12}(d, n)N^{13}$, which must be allowed to decay before the activity due to the F^{18} can be measured.

An assembly of three irradiated MgO spheres is shown in Fig. 3A. In this experiment the sample was annealed at 1700°C for 21 hr in air. All three spheres were initially annealed in O^{17}, and the redistribution of tracer indicated by the autoradiograph, Fig. 3B, suggests that there has probably been vapor transport under a temperature gradient. While this is not a marker experiment of the type initially discussed, it serves to illustrate the fundamentals of the technique.

The autoradiograph was obtained on a Kodak Type A plate. This was scanned with a densitometer, which records the film optical density. The scan is presented in Fig. 3C in the form of isodensity contours, which can be converted to intensity of flux measurement by use of a film response calibration. Since the densitometer detector which provides the signal that controls the recording pen can also feed a signal onto recording tape, the data can be processed by computer operations. Conversion from film density to radiation flux, corrections for background, corrections for back reflection, averaging of values from neighboring traces to suppress noise, and integrations to find the total amount of tracer within various areas can thereby be performed automatically. Therefore given a good autoradiograph its information content data can be translated into measurements of material transport of the tracer during sintering.

CONCLUSION

At the present time the activation and autoradiographic technique is being applied to the study of oxygen tracer migration during sintering.

Spheres can be produced conveniently and sintered successfully up to about 1 mm in diameter. If resolutions of 10 μ are obtained, this amounts to about 1% of the overall dimensions of the specimens. Processes taking place on a smaller scale should be detectable with the use of more refined procedures, resolutions of less than 1 μ being possible. Thus it appears that the application of these techniques should make new information available about sintering phenomena.

ACKNOWLEDGMENTS

This investigation has benefitted from discussions with a number of people. Of particular help have been J. B. Holt and L. Himmel at this Laboratory, D. L. Johnson of Northwestern University, and D. Lazarus of the University of Illinois. The deuteron activations were carried out with the help of the staff of the 2-MeV van de Graaff accelerator and the densitometer traces were prepared by C. H. Dittmore at this Laboratory.

REFERENCES

1. A. E. Paladino and R. L. Coble, "Effect of Grain Boundaries on Diffusion-Controlled Processes in Aluminum Oxide," *J. Am. Ceram. Soc.* **46**: 133–36 (1963), 46: 460 (1963).
2. G. C. Kuczynski, L. Abernathy, and J. Allen, "Sintering Mechanisms of Aluminum Oxide," in: *Kinetics of High Temperature Processes* (W. D. Kingery, ed.), John Wiley, New York, 1959, pp. 163–172.
3. L. F. Norris and G. Parravano, "Sintering of Zinc Oxide," *J. Am. Ceram. Soc.* **46**: 449–52 (1963).
4. J. B. Holt and R. H. Condit, "Oxygen-18 Diffusion in Surface Defects on MgO as Revealed by Proton Activation," in: *Materials Science Research*, Vol. 3, (W. W. Kriegel and H. Palmour III, eds.), Plenum Press, New York, 1966, pp. 13–29.
5. D. L. Johnson, to be published.
6. J. Brett and L. Seigle, "The Role of Diffusion Versus Plastic Flow in the Sintering of Model Compacts," *Acta Met.* **14**: 575–82 (1966).
7. R. H. Condit and J. B. Holt, "A Technique for Studying Oxygen Diffusion and Locating Oxide Inclusions in Metals by Using the Proton Radioactivation of Oxygen-18," *J. Electrochem. Soc.* **111**: 1192–94 (1964).
8. J. B. Holt, private communication.
9. B. G. Harvey, "Recoil Techniques in Nuclear Reaction and Fission Studies," *Ann. Rev. Nucl. Sci.* **10**: 235–58 (1960).
10. G. Friedlander and J. W. Kennedy, *Introduction to Radiochemistry* John Wiley, New York, 1949, p. 165.
11. R. G. Ward, "Microradiography and Autoradiography," in: *The Physical Examination of Metals*, 2nd ed. (B. Chalmers and A. G. Quarrell, eds.), Edward Arnold, London, 1960, pp. 825–53.
12. R. H. Condit, "Autoradiographic Techniques in Metallurgical Research," in: *Techniques in Metals Research*, Vol. 2, part 4 (R. F. Bunshah, ed.), John Wiley, New York, in press.

13. W. E. Lotz and P. M. Johnson, "Preparation of Microautoradiographs with the Use of Stripping Film," *Nucleonics* **11**(3): 54 (March, 1953).
14. D. M. Prescott, "Autoradiography with Liquid Emulsion Methods," in: *Methods in Cell Physiology I* (D. M. Prescott, ed.), Academic Press, New York, 1964, pp. 365–70.
15. C. B. Gilpin, D. H. Paul, S. K. Asunmaa, and N. A. Tiner, "Electron Microautoradiography and its Application to the Study of Hydrogen Distribution in Steel," in: *Advances in Electron Metallography*, Special Technical Publication No. 396, Am. Soc. Testing Materials, 1966, pp. 7–20.
16. J. B. Holt, "Self-Diffusion of Oxygen in Single-Crystal Beryllium Oxide," *J. Nucl. Mater.* **11**: 107–10 (1964).

DISCUSSION

C. H. Greene (Alfred University): I have a question on your assumption about the tabulation of the energy of the deuteron particle as it goes into the bulk. If the energy of one of these particles loses only 25%, it would seem to me you still could get the activation of the oxygen isotope just below the surface. If this is true, then how do you know that what you are seeing in your radiograph is not the fact that you are also attenuating the positron as it is emitted from the same activated molecule?

Answer: The cross section for the reaction $O^{17}(d, n)F^{18}$ seems to drop to zero at an energy of about 0.4 MeV. Therefore as the deuteron beam passes into the solid, it can produce the reaction product to a depth of only about 5–10 μ even though the final stopping point for the deuterons may be 20–30 μ below the surface. To a casual glance the curve showing the yield as a function of depth might appear to be an exponential decay type of junction. The range of the positrons through the solid and the autoradiograph film would be many hundreds of microns, and so there is clearly not much loss from internal absorption within the specimen. We have actually sectioned irradiated specimens to determine the profile of resultant activation, and we have satisfied ourselves that activation occurred only in the surface region.

H. M. Davis (Army Research Office—Durham): Although you mentioned the possibility of diffusing O^{18}, you seem to have concentrated on the use of O^{17}. Why do you do this in preference to O^{18}, which is more abundant?

Answer: The cost of an ion accelerator to generate 1-MeV particles is much less than for 3-MeV particles. A 1-MeV machine such as a van de Graaff is relatively simple to operate, and one can start it up easily and carry out an irradiation in the time it might take to get a more complicated machine into operation. These are our reasons for making use of the O^{17} activation rather than the O^{18} activation with protons, which does require the higher energy. The cost of the O^{17} for an equivalent enrichment is considerably greater, to be sure, but the capital costs of equipment are much less, and the difficulty of carrying out the experiment is less.

Chapter 21

The Role of Plastic Flow by Dislocation Motion in the Sintering of Calcium Fluoride*

A. R. Hingorany, F. V. Lenel, and G. S. Ansell

Department of Materials Engineering
Rensselaer Polytechnic Institute
Troy, New York

The kinetics and the mechanism of material transport in sintering calcium fluoride was investigated. For this purpose spheres of hot-pressed calcium fluoride containing a fine dispersion of stabilized zirconia as markers were sintered for various times and at various temperatures to plates of the same composition. Both conventional sintering without application of an external load and sintering with an external load on the sphere were studied. The rate of growth of the neck between sphere and plate was determined by sectioning, polishing, and photographing the sintered samples. The photographs of the neck regions were carefully examined in order to determine whether the marker particles extended throughout the diameter of the neck or whether a marker-free zone existed in the outer region of the neck. Only in samples sintered for long times (400–1000 hr) did the markers not extend through the entire neck regions, but left a marker-free zone near the ends of the neck. The extent of the marker-containing central region strongly depended upon the presence of an external load. With load the marker-containing region occupied a much larger percentage of the entire neck. From the strong influence of the external load upon the marker containing region of the neck it is concluded that material transport in sintering of calcium fluoride takes place by diffusional flow only in the later stages of sintering, where the stress due to surface tension forces becomes small. In the early stages of sintering, where the stress due to surface tension forces is higher, and particularly when an additional stress is supplied by an external load, material transport takes place by slip, i.e., dislocation motion. The range of stress which separates the two mechanisms is estimated to be in the neighborhood of 10^5 dyn/cm^2 (1.4 psi).

INTRODUCTION

The mechanism of material transport in the sintering of powder compacts has been the subject of considerable controversy in the last few years. In

---

*This paper is based on a thesis submitted by A. R. Hingorany in partial fulfillment of the requirements for the Degree of Doctor of Philosophy to the Department of Materials Engineering, Rensselaer Polytechnic Institute.

particular, for solids of low volatility the controversy has centered upon the relative roles played by slip, i.e., by the movement of dislocations on the one hand and on the other by diffusional flow in the sintering process. Lenel and Ansell ([1]) have argued that material transport by slip is responsible for sintering when the stresses due to either surface tension forces or to an externally applied load are high; when the stresses are low material transport by diffusional flow occurs. This shift in transport mechanism with stress arises from the different rate sensitivities to stress of these two transport mechanisms. The dimensional changes observed in sintering, be they shrinkage in actual compacts or neck growth in model experiments, are proportional to the strain rate observed in conventional plastic deformation. This strain rate in turn is proportional to the stress for diffusional flow, but is proportional to the 4.5th power of the stress in the particular case of slip controlled by dislocation climb in pure metals and in general to a relatively high power of the stress in most processes of slip by dislocation motion. Because of this very different relationship between strain rate and stress for the two transport mechanisms one would expect that the transition range between the two mechanisms would be relatively narrow.

In a recent model experiment Burr et al. ([2]) attempted to determine the transition stress range. They sintered a silver sphere to a silver plate under an external stress sufficiently large so that stresses due to surface tension forces could be neglected. The change in the rate of growth of the neck between the sphere and the plate was observed when the load applied to the sphere was suddenly changed. At an average stress of 7×10^5 dyn/cm^2 on the silver sphere it was found that the rate of neck growth, and therefore the strain rate, was proportional to the 1.8th power of stress, indicating that the shift from a mechanism by diffusional flow to one by slip takes place near this stress. At higher average stresses of $8\text{--}10 \times 10^6$ dyn/cm^2 the expected relationship of strain rate as proportional to the 4.5th power of the stress was confirmed.

Another experimental approach to the study of the material transport mechanism in sintering has been taken by Brett and Seigle ([3]). Seigle et al. had suggested previously ([4]) that mass transport by slip can be critically distinguished from that by diffusion by the study of the movement of inert fiducial markers. Inert markers remain unmoved during material transport by diffusional flow, but are swept along during transport by slip, i.e., dislocation motion. Accordingly, Brett and Seigle produced nickel and copper wires consisting of a fine dispersion of alumina in nickel and of alumina and zirconia, respectively, in copper. They sintered aggregates consisting of three of these wires twisted together and observed the appearance of the neck area formed between the wires. The experiments on copper wires did not lend themselves to conclusions on the mechanism of material transport. In those on aggregates from alumina-containing nickel wires Brett and Seigle observed that "marker movement could not be discriminated with certainty at neck diameters less than approximately 20% of the wire diameter."

For neck growth up to this size Brett and Seigle left the question of transport mechanism open. However, when the necks grew to larger sizes they were definitely free of marker particles, indicating that in the formation of these larger necks transport took place by diffusional flow.

Brett and Seigle's experimental approach appears to be particularly suitable for studying the effect of stress upon the mechanism of material transport. As the neck grows, the stresses exerted by surface tension forces steadily decrease. When the stress becomes sufficiently low transport by slip is no longer possible and no further transport of particles into the neck can take place. Approximate calculations [5] of the stresses due to surface tension on the basis of the concept of virtual work have been attempted in order to find a value of stress for an array of wires of a certain diameter for any given ratio of neck diameter to wire diameter.

Barron [5], in an experiment similar to the one by Seigle and Brett, produced wires consisting of a fine dispersion of aluminum oxide in a silver matrix. Arrays of two or three wires were sintered together and the width of the neck region containing oxide marker particles was determined. From this data the maximum stress for which transport by diffusional flow is possible was calculated. This stress is in reasonable agreement with the value of the transition stress obtained by Burr et al.

While the studies discussed so far were concerned with the sintering of metals, the question of material transport in sintering is as pertinent for ionic solids as it is for metals. At the temperatures where sintering is important extensive plastic deformation by dislocation motion is possible in both single crystals and polycrystalline aggregates of ionic solids. It was therefore decided to adapt the Brett and Seigle technique of sintering with fiducial markers to an ionic crystalline solid, calcium fluoride.

Calcium fluoride can be plastically deformed at elevated temperatures, as was shown by Phillips [6] and by Burn and Murray [7]. The vapor pressure near its melting point as measured by Schulz and Searcy [8] and by Blue et al. [9] is small enough to preclude material transport by evaporation and condensation. Data for the surface tension (specific surface energy) of various crystal faces of calcium fluoride has been obtained by Benson and Claxton [10] and by Gilman [11]. The sintering of calcium fluoride was studied by Allison and Murray [12]. Their dilatometric measurements on powder compacts did not permit any clear-cut conclusions with regard to sintering mechanism. Yust and Morgan [13], on the other hand, who sintered calcium fluoride crystals together and observed the presence of dislocations in the neck region using etch pit techniques, came to the conclusion that slip is involved in the process.

In the sintering experiments described in this chapter the markers consisted of stablized zirconia. Calcium fluoride spheres containing a fine dispersion of this marker material were sintered to plates also containing zirconia markers. In sintering runs at a given sintering temperature and for a given sintering time assemblies of spheres and plates were sintered with and without an external load. The sintered aggregates of sphere and plate were sectioned and the position of the markers in the neck region observed. By comparing the width of the neck region containing markers in samples sintered without and with an external load, the effect of stress upon the material transport mechanism was demonstrated. An attempt was made to determine the stress level down to which transport by slip and below which transport by diffusional flow takes place in the sintering of calcium fluoride.

MATERIALS

The calcium fluoride was obtained as CP grade with a purity of 99.975 % from Fisher Scientific Company. Its average particle size was 0.3 μ. Stabilized zirconia, a cubic form of zirconia containing 3 to 7 % CaO, which does not have a phase transformation, was available in the form of hollow spheres. It was crushed and ground. The particle size fraction finer than 5 μ was separated out and used as marker material.

The calcium fluoride and stabilized zirconia were blended in a cone blender for 12 hr and then ball milled for 24 hr in a cemented carbide mill with steel balls. Most zirconia particles were well distributed, although agglomeration could be observed in a few places, as will be seen in the photomicrographs.

Either the powder mixture or the pure calcium fluoride was hot pressed into plates $\frac{1}{2}$ in. in diameter and either $\frac{1}{8}$ in. or $\frac{1}{4}$ in. thick in a graphite die at 1250°C under a pressure of 3000 psi. The minimum density of the hot-pressed material was 96 % of calculated theoretical density; however, most of the plates were 99 % dense.

The hot-pressed $\frac{1}{8}$-in. thick plates were polished on a series of emery papers with paraffin dissolved in kerosene as a lubricant. Both sides of the plates were polished parallel to within ± 0.001 in. to a final thickness of approximately 0.1 in. After polishing the plates were washed in benzene, then in acetone, and then allowed to dry.

The hot-pressed $\frac{1}{4}$-in. thick plates were broken into three or four pieces. Using a method suggested by Carter et al. ([14]), these pieces were ground into spheres approximately 0.1 in. in diameter by letting them bounce on the surface of a rotating horizontal grinding wheel while being confined in a hollow tube. By using successively finer grinding wheels and finishing on 300 grit silicon carbide paper pasted on a wooden lap wheel, spheres spherical to within ± 0.001 in. with a diameter controlled to ± 0.001 in. were obtained.

EXPERIMENTAL PROCEDURE

The two methods of assembling the marker-containing calcium fluoride spheres and calcium fluoride plates which were used for the sintering experiments are illustrated schematically in Figs. 1(a) and 1(b). It is seen that in the first method, Fig. 1(a), the calcium fluoride spheres with markers were sintered to the calcium fluoride plate with markers without any external load applied to the spheres. The calcium fluoride pieces on the edge of the plate acted to keep the spheres from rolling off the plate. The assembly was contained in an enclosure formed by the two platinum cups. The calcium fluoride powder outside the enclosure served as getter. In the second method three calcium fluoride spheres with markers were positioned firmly in three blind holes approximately 0.01 in. deep and forming an equilateral triangle. The holes were drilled into a calcium fluoride plate without markers. The load applied to the spheres consisted of a polished calcium fluoride plate with markers, the platinum cup, and an external load on top of the cup. The space between the inner and outer platinum cups was filled with calcium fluoride powder as a getter. Care was taken that a minimum of powder landed on top of the platinum cup, because such powder would add an unknown amount

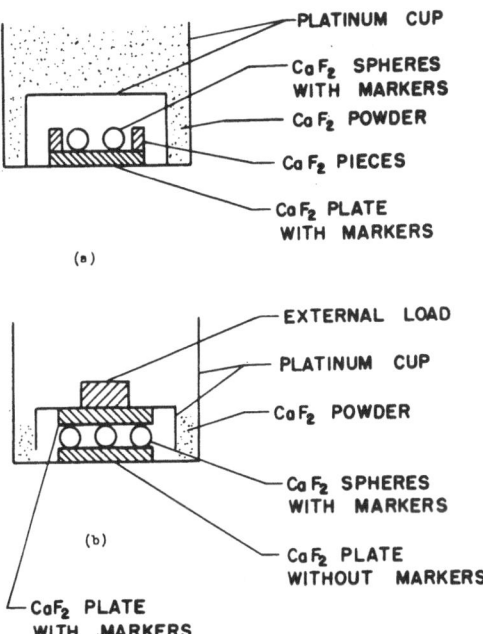

Fig. 1. Schematic view of assembly of samples (a) without an external load and (b) with an external load.

of additional weight to the known weight of plate, platinum cup, and external weight resting on the spheres.

In the assemblies sintered with an external load the necks formed between the spheres and the top plate, both of which contained markers, were used for observation and measurement. The only purpose of the marker-free lower plates in the assembly was as a control. In the final polishing of the samples for microscopic examination diamond powder was used, which could possibly become embedded in the calcium fluoride and give the impression that markers were present. No such effect was observed.

The use of calcium fluoride powder as a getter during the sintering operation was required since it had been found that in sintering spheres to plates in air or even in dried nitrogen without the use of the getter the surface of the calcium fluoride would react and form a porous layer of a material other than calcium fluoride. Based on the findings of Phillips ([6]) on calcium fluoride single crystals and of Meisser ([15]) on magnesium fluoride, it is believed that the porous layer was calcium oxide formed by reaction of calcium fluoride with water vapor, with intermediate formation of an oxyfluoride. The use of calcium fluoride powder as a getter completely prevented this undesirable reaction.

For the sintering experiment the assemblies shown in Figs. 1(a) and (b) were placed on a platform which was a 2-in. long flat-ground portion at one end of an alumina rod 12 in. long and 1 in. in diameter. The other end of the rod was clamped in a horizontal position. A horizontal tube furnace 18-in. long, of 2 in. inside diameter, and having three split windings individually controlled by rheostats was mounted on casters so that it could be moved over the samples with a minimum of vibration. One end of the furnace tube was closed except for a gas inlet tube through which dry nitrogen was fed into the furnace. The temperature of the furnace was controlled using an on–off controller and a thermocouple near the winding of the furnace. The temperature was measured with a thermocouple near the sample and found to vary $\pm 2°C$ for the duration of runs at 1260° and 1280°C and $\pm 4°C$ for runs above 1300°C. The temperature variation along the length of the zone in which the assemblies were located was less than $\pm\frac{1}{2}°C$.

In order to make a sintering run, the furnace was brought up to temperature and then wheeled over the samples a short distance at a time over a total period of about 15 min in order to prevent thermal shock. After the furnace was in position loose Fiberfrax high-temperature insulation was fitted between the alumina rod and the inside of the furnace to improve temperature control and to minimize back diffusion of water vapor.

The time necessary to bring the samples to the sintering temperature was less than an hour. Since most of the runs lasted hundreds of hours and the sintering rate depends upon temperature exponentially, the heat-up time was neglected and the count of sintering time was started when the temperature reached within 10°C of the desired sintering temperature.

After each sintering run the sintered samples were mounted in epoxy resin while they were still inside the platinum cups in order to prevent any movement of the spheres relative to the plates. After the epoxy resin had set the samples were cut with a diamond wheel just inside the inverted platinum cup. The position of the sphere on the plate could then be seen. This made it possible to produce successive sections of the sample by polishing off layers parallel to the axis running through the center of the sphere and of the neck. The samples were rough polished on a flat polishing stone used for polishing natural rock samples. Increasingly finer silicon carbide abrasive powder from 200 to 600 grit was used suspended in water. For examination under the microscope the samples were fine polished on nylon cloth with 3-μ diamond paste as abrasive and diamond lapping oil as lubricant. As successive parallel sections are taken, the apparent neck diameter observed will increase, reach a maximum which is the actual diameter at the point where the plane of polish goes through the centers of sphere and neck, and then decrease.

By alternately polishing and measuring the apparent neck diameter with a filar eyepiece under the microscope, the actual neck diameter was determined. As a check, the sphere diameter was measured simultaneously with the neck diameter. The maximum apparent sphere diameter was found to equal the sphere diameter measured before the sintering experiment. In a few cases neck diameters were observed and measured by lifting the spheres off the plate after sintering. These cross sections were in all cases very nearly circular. This observation justified the assumption of circular cross section made when neck diameters are determined by the method of polishing off successive layers.

By examining the as-polished structure under the microscope at a magnification of 300×, the marker particles could be clearly resolved and their presence or absence in the outer neck region determined. In those samples in which no marker particles were present in the outer neck region the maximum neck diameter up to which markers were visible was measured on photomicrographs. Since at the magnification needed to resolve the marker the entire neck cross sections could not be accommodated in one photograph, sectional photographs were taken, assembled in composite photographs, and rephotographed.

The neck was not always photographed when the apparent neck diameter was at its maximum, i.e., when the actual neck diameter showed. However, as seen in Fig. 2, the actual diameter of the region with markers GH can be constructed from the apparent diameter of the region with markers EF which is measured on the micrographs. On a circle with diameter AB,

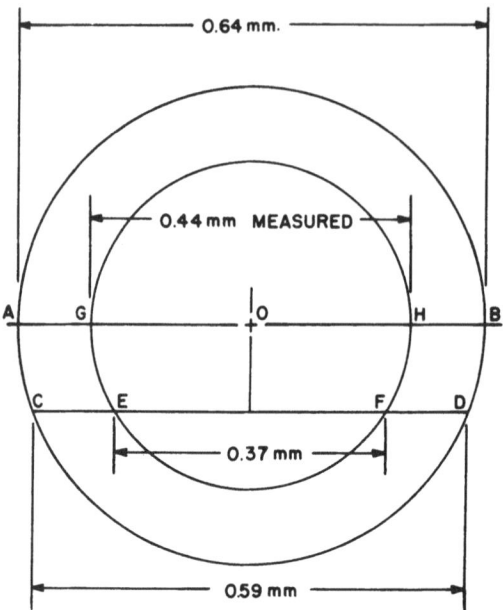

Fig. 2. Geometrical construction to measure the actual width of the marker-containing region.

which is the actual neck diameter as determined by repeated polishing and measuring, a chord CD is drawn equal to the neck diameter, as seen on the photomicrograph. On this chord the distance EF equal to the apparent diameter of the region with markers on the photomicrograph is plotted. A circle drawn with radius OE around O at the center will intersect the line AOB at points G and H.

RESULTS AND DISCUSSION

As seen in Table I, five experimental runs, I to V, were made. For each run the sintering temperature and time are given in columns 2 and 3. In each run assemblies were sintered both with and without an external load. In each assembly there were up to four samples, but only those samples for which data on neck diameter were determined were given designations (column 4). Column 5 gives the weight of the external load resting on each of the three spheres, which is equal to one-third the total load on the assembly, making the reasonable assumption that the load is distributed evenly. The samples in assembly 8 of run IV, sintered without an external load, varied in sphere diameter. Two of them were 2.54 and two of them 1.63 mm in diameter,

TABLE I

Experiments Sintering Calcium Fluoride Spheres to Calcium Fluoride Plates

Run	Sintering temp. (°C)	Sintering time (hr)	Sample number	External load (g)	Sphere diameter a (mm)	Neck diameter x (mm)	Ratio (x/a)
I	1259	74	1A	3.10	2.54	0.77*	0.30
			1B	3.10	2.54	0.79*	0.31
			1C	3.10	2.54	0.80*	0.32
			2A	None	2.54	0.40	0.16
II	1278	219	3A	3.27	2.54	0.90*	0.35
			3B	3.27	2.54	0.91*	0.36
			4A	None	2.54	0.54	0.21
III	1322	466	5A	3.35	2.54	1.14	0.45
			5B	3.35	2.54	1.14	0.45
			6A	None	2.54	0.64	0.25
			6B	None	2.54	0.80	0.32
IV	1330	947	7A	1.33	2.54	1.44	0.57
			8A	None	2.54	0.96	0.38
			8B	None	2.54	0.96	0.38
			8C	None	1.63	0.76	0.47
			8D	None	1.63	0.76	0.47
V	1342	950	9A	0.62	2.54	1.66	0.65
			10A	None	2.54	1.33	0.52
			10B	None	2.54	1.28	0.50
			10C	None	2.54	1.44	0.57

*Neck diameter measured by lifting sphere from plate after sintering.

while the samples in all other assemblies were 2.54 mm in diameter (Column 6). The neck diameters are shown in column 7 and the ratios of neck to sphere diameter in column 8.

No attempt was made to develop a quantitative relationship between the neck diameter and the sintering time or temperature. In general, the neck diameter of spheres of a given size, 2.54 mm, increased with increasing sintering time and temperature. The maximum variation in neck diameter for samples sintered under identical conditions was 20%, but for most samples the agreement was much better.

The ratio of neck diameter to sphere diameter for samples 8C and 8D, with a sphere diameter of 1.63 mm, is greater than for samples 8A and 8B with a sphere diameter of 2.54 mm. This would be expected because of the higher surface tension stress in the samples with the smaller sphere diameter.

By comparing the samples in the same run, it can be seen that those with an external load grew a larger neck than those sintered without an external load. The increase in neck area due to an externally applied stress is most strikingly demonstrated in run IV near the melting point of calcium fluoride, in which sample 7A, sintered with the small external load of 1.33 g, had a neck diameter 50% greater than samples 8A and 8B, sintered without a load.

The compressive stress due to the applied external load can be readily computed by dividing the force in dynes due to the load by the cross-sectional area of the neck. From the effect of the applied load on sintering rate one would estimate that the stress due to this load is of the same order of magnitude as that due to surface tension forces.

A calculation of the compressive stress due to surface tension forces acting between the centers of the spheres in a two-sphere model was made by Leary ([16]). He calculated the total surface free energy of the model by multiplying the surface area by the specific surface free energy (surface tension). He then assumed that the decrease in surface area of the two-sphere model was caused by material in the polar caps of the spheres being transported into the neck. By taking the derivative of the decrease in surface energy due to the decrease in surface area with respect to the change in distance between the sphere centers, he calculated a compressive force and from this force a stress by dividing the force by the area of the neck. The use of this calculation for a sphere-on-plate rather than a sphere-on-sphere model should introduce only a moderate error. However, when stresses were calculated using this model, choosing a value for specific surface free energy of 500 ergs/cm^2 based on the data of Benson and Claxton ([10]) and of Gilman ([11]), it became apparent that the computation overestimates the stresses due to surface tension. The reason for this, as will be shown later, is the assumption of center-to-center motion of the spheres (motion of the center of the sphere toward the plate in the sphere-on-plate model) made in the calculation. Such motion was not observed in these experiments and the model is therefore not applicable. At present no calculation of stress due to surface tension is available which would be more satisfactory.

The most important data in this investigation were those on the extent to which the markers are spread through the neck region. These data were obtained on photomicrographs such as those in Figs. 3–6, which show cross sections of necks. In polishing the samples, severe pullout of grains was observed. This was most pronounced in areas around the edges and the neck region. The dark areas in the photomicrographs are the regions from which grains have been pulled out during polishing. In spite of the pullouts, the ends of the necks stand out clearly because they are filled with resin, which appears light grey in the micrographs.

In some cases, when the samples were polished for considerable depth

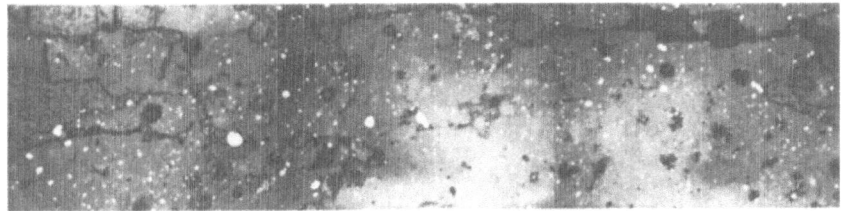

Fig. 3. Cross section of neck of sample 5B from run III, sintered 466 hr at 1322°C with an external load of 3.35 g.

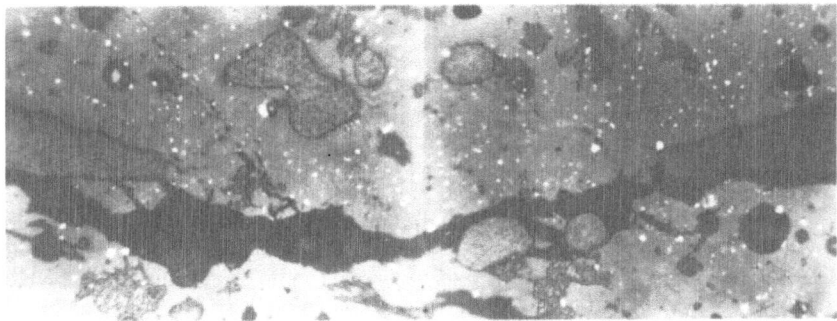

Fig. 4. Cross section of neck of sample 6A from run III, sintered 466 hr at 1322°C without an external load.

Fig. 5. Cross section of neck of sample 9A from run V, sintered 950 hr at 1342°C with an external load of 0.62 g.

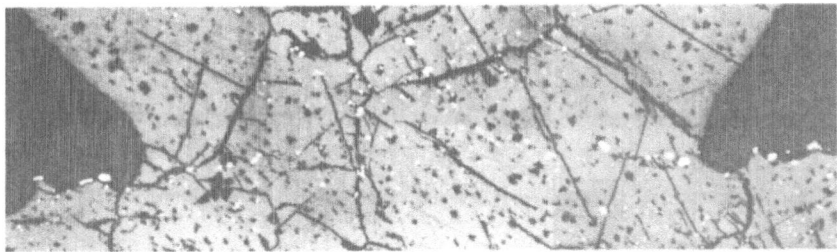

Fig. 6. Cross section of neck of sample 10C from run V, sintered 950 hr at 1342°C without an external load.

the dimensions of the samples became too small to handle comfortably and the samples were remounted in epoxy resin. Here some of the pullout regions were filled with epoxy resin. Typical examples of pullout regions filled with resin are seen in Figs. 5 and 6. The presence of these areas is the result of pullout during polishing and does not mean that the initial spheres and plates had high porosity.

The zirconia marker particles appear white in the micrographs. It is evident that in Figs. 4, 5, and 6 there are marker-free regions near the end of the neck, while in Fig. 3 there is no such marker-free region. Data on the distribution of the markers in the neck are tabulated in Table II for all those samples which were examined under the microscope for the presence or absence of a marker-free region at the ends of the neck. The presence of such a region is indicated in column 2. The neck diameter x and the width of the region y up to which markers were visible are given in columns 3 and 4. The ratios of neck diameter to sphere diameter a, of diameter y to sphere diameter, and of diameter y to neck diameter are given in columns 5, 6, and 7. Column 8 shows the external load in grams for those samples where such a load was applied, and column 9 shows the stress S_L due to the external load when the neck diameter reached width y of the marker-containing region. The table indicates that for runs I and II, which were sintered at relatively low temperatures (below 1300°C) and short times (less than 400 hr) the two samples which were examined showed markers extending to the very end of the neck even though these samples were sintered without an external load. In run IV samples 8A and 8C were both sintered without a load under identical conditions. The relative width of the marker-containing zone compared to the entire neck in sample 8A was 46%, in accordance with the lower stress due to surface tension for a sphere 2.54 mm in diameter, while this relative width was 84% for sample 8C, in which the stress due to surface tension was higher because the sphere diameter was only 1.63 mm.

Those parts of the neck region in Figs. 3–6 where there are markers should be carefully scrutinized. All of the photomicrographs show the markers well distributed, and no agglomeration of markers at the boundary between sphere and plate can be detected. If markers were left in the neck area due to a diffusional flow mechanism in which vacancies are transported from the surface of the neck to the grain boundary or grain boundaries in the neck region, i.e., by movement of the sphere toward the plate, one would expect to see a high concentration line of markers along the neck. Since no such line is observed, it is concluded that the initial marker-containing zone in the neck is produced by transport of material by slip. Once the stress has decreased below the transition value and diffusional flow becomes the controlling transport mechanism, most of the material transport is by vacancy diffusion from the surface of the neck, which has a small radius of curvature

TABLE II

Data on Distribution of Markers in the Necks

Run and sample number		Marker distribution*	x (mm)	y (mm)	x/a	Ratio y/a	Ratio y/x	Load L (g)	$S_L = \dfrac{980L}{\pi y^2/4}$ (×10⁵ dyn/cm²)
I	2A		0.40	0.40	0.16	0.16	1	None	—
II	4A	E	0.54	0.54	0.21	0.21	1	None	—
III	5A	E	1.14	1.14	0.45	0.45	1	3.35	—
	5B	E	1.14	1.14	0.45	0.45	1	3.35	—
	6A	M	0.64	0.44	0.25	0.18	0.69	None	None
IV	7A	E	1.44	1.44	0.57	0.57	1	1.33	0.8
	8A	M	0.96	0.44	0.38	0.18	0.46	None	None
	8C	M	0.76	0.64	0.47	0.39	0.84	None	None
V	9A	M	1.66	1.21	0.65	0.48	0.73	0.62	0.5
	10A	M	1.33	0.48	0.52	0.19	0.37	None	None
	10B	M	1.28	0.47	0.50	0.19	0.37	None	None
	10C	M	1.44	0.45	0.57	0.18	0.31	None	None

*E—To end of neck. M—Marker-free region.

toward the adjacent surface regions of plate or sphere, which have larger radii of curvature.

In Figs. 5 and 6 it will be noted that the distribution of the markers around the neck is unsymmetrical. In Fig. 5, a specimen sintered under load, the marker-free region of the neck is confined to the plate resting on top of the sphere. In Fig. 6, a specimen sintered without load, the marker-free region is reversed, being confined to the sphere. No explanation for this curious phenomenon is available at present.

Comparisons of Fig. 3 with Fig. 4 and of Fig. 5 with Fig. 6 illustrate the effect of an external load upon the width of the neck and upon the distribution of the markers. Sample 5B, shown in Fig. 3, was sintered with an external load of 3.35 g and has a considerably greater neck diameter than sample 6A, shown in Fig. 4, which was sintered without an external load. In spite of this larger neck diameter, the markers in sample 6A (Fig. 4) extend to the very end of the neck, while in sample 5B (Fig. 3) there is a small but definite marker-free zone near the ends of the neck. The increase in the width of the marker-containing zone due to the imposition of an external load is even more clearly demonstrated in Figs. 5 and 6, representing samples 9A and 10C, respectively. Both samples were sintered in run V for 950 hr at 1342°C. In sample 10C (Fig. 6), in which no external load was applied, the neck diameter is 57% of the sphere diameter. Only the region of the neck near its center, amounting to 31% of the entire neck, contains markers, while the regions near the ends of the neck are marker-free. The application of a small external load of 0.62 g caused the neck diameter to increase to 65% of the sphere diameter, but the marker-containing region of the neck is considerably wider than for sample 11C, taking up 73% of the entire neck. The stress due to the external load at the time when the neck diameter reached 1.21 mm, i.e., the width of the marker-containing zone, was 0.5×10^5 dyn/cm².

It is evident that with the external load the markers reached to the end of the neck or, at any rate, occupied a much larger proportion of the neck, than when the samples were sintered without external load. It is inconceivable that the origin of stress, i.e., whether it is due to surface tension forces or to an external load, should have an influence upon the material transport mechanism. The mechanism should be determined only by the magnitude of the stress and not by its origin. The increase in the width of the inner marker-containing zone of the neck by the imposition of the external load, i.e., by an increase of the total stress acting in the neck, strongly supports the idea that the marker-containing zone was formed through material transport by slip rather than by diffusional flow. Only an approximate estimate is possible for the transition stress range, above which material transport by slip and below which material transport by diffusional flow take place. Since in sample

9A the stress due to the external load (0.5×10^5 dyn/cm^2) must be about the same as that due to the surface tension force, which is the only stress acting in samples 10A, 10B, and 10C, the total stress at which the transition from slip to diffusional flow takes place should be in the neighborhood of 10^5 dyn/cm^2 or 1.4 psi for calcium fluoride of the grain size used in the experiment, i.e., 40-μ average grain diameter.

CONCLUSIONS

The results of this investigation indicate that material transport during sintering of polycrystalline calcium fluoride takes place by slip due to dislocation motion when the stress exerted by surface tension forces or by a combination of surface tension forces and an external force is high. When the stress becomes smaller than a critical stress, which is estimated to be in the neighborhood of 10^5 dyn/cm^2 (1.4 psi) for calcium fluoride with an average grain size of 40 μ, material transport takes place by diffusional flow.

ACKNOWLEDGMENTS

The support of this research by the Research Division of the United States Atomic Energy Commission is gratefully acknowledged.

REFERENCES

1. F. V. Lenel and G. S. Ansell, "Creep Mechanisms and their Role in the Sintering of Metal Powders," in: *Modern Developments in Powder Metallurgy* Vol. 1 (H. H. Hausner, ed.), Plenum Publishing, New York, 1966, p. 281.
2. M. F. Burr, F. V. Lenel and G. S. Ansell, "Influence of Pressure upon the Sintering Kinetics of Silver," *Trans. Met. Soc. AIME* **239**: 557 (1967).
3. J. Brett and L. Seigle, "The Role of Diffusion Versus Plastic Flow in the Sintering of Model Compacts", *Acta Met.* **14**: 575–582 (1966).
4. R. Baluffi, F. Rosi, and L. Seigle, "Self Diffusion of Metals and Allied Phenomena," Progress Report IX, AEC Contract AT(30-1)GEN 367, April 5, 1954.
5. D. Douglas Barron, "The Role of Plastic Deformation in the Sintering of Silver Wires Containing Inert Marker Particles," Master of Engineering Thesis, Rensselaer Polytechnic Institute, January 1966; to be published.
6. W. L. Phillips, Jr., "Deformation and Fracture Process in Calcium Fluoride Single Crystals," *J. Amer. Cer. Soc.* **44**: 499 (1961).
7. G. Burn and G. T. Murray, "Plasticity and Dislocation Etch Pits in CaF$_2$," *J. Amer. Ceram. Soc.* **45**: 251 (1962).
8. D. A. Schulz and A. W. Searcy, "Vapor Pressure and Heat of Sublimation of Calcium Fluoride," *J. Phys. Chem.* **67**: 103 (1963).
9. G. D. Blue, J. W. Green, R. G. Bautista, and J. L. Margrave, "The Sublimation

Pressure of Calcium (II) Fluoride and the Dissociation Energy of Calcium (I) Fluoride," *J. Phys. Chem.* **67**: 877 (1963).

10. G. C. Benson, and T. A. Claxton, "Calculation of the Surface Energy of the 110 Face of Some Crystals Possessing the Fluorite Structure," *Can. J. of Phys.* **41**: 1287 (1963).

11. J. L. Gilman, "Surface Energies of Crystals," *J. Appl. Phys.* **31**: 2213 (1960).

12. E. B. Allison, and P. Murray, "A Fundamental Investigation of the Mechanism of Sintering," *Acta Met.* **2**: 487 (1954).

13. C. S. Yust and C. S. Morgan, "Dislocation Movement in Sintering of Calcium Fluoride," paper presented at 65th Annual American Ceramic Society Symposium, May 1963.

14. J. L. Carter, E. V. Edwards, and I. Reingold, "Ferrite Sphere Grinding Technique," *Rev. Sci. Instr.* **3**: 946 (1959).

15. D. R. Meisser, "Kinetics of High-Temperature Hydrolysis of Magnesium Fluoride: I, Evaluation of Reaction Mechanism," *J. Am. Ceram. Soc.* **48**: 452, (1965).

16. E. A. Leary, Rensselaer Polytechnic Institute, private communication.

DISCUSSION

C. S. Morgan (Oak Ridge National Laboratory): Was there any way you could tell how much of the material in the marker area had arrived there by volume diffusion and how much had arrived by surface diffusion? If you noted the movement of the center of the spheres together, you might have been able to tell.

Answer: Unfortunately, we do not have any data on center-to-center motion, which would give an answer to the question of where the material in the growing neck came from, i.e., whether it came from the free surface of the sphere adjacent to the neck or whether it came from the interior of the neck volume.

Alan Franklin (ARPA): Your problem of hydrolysis or oxidation of calcium fluoride is a serious one. You can get appreciable oxidation sufficient, for instance, to control the effect of the mobility of the fluoride even though you cannot see any damage to the surface. If one looks through an ultraviolet microscope in the region of 200 mμ, which is very sensitive to the oxygen, you can more sensitively detect whether or not there is oxygen present.

Answer: Thank you very much for this comment. We have not looked at the surface with ultraviolet light, but have only relied on examination in the visible range.

J. E. Burke (General Electric): There has been a lot of discussion about two mechanisms which involve dislocation movement and diffusion. Grain boundary sliding must operate under some circumstances. You have looked at many specimens very carefully; have you seen any evidence of this occurring? I know it is difficult in the oxide case because the closed structure varies so much in something like alumina. But in a metal such as copper at what particle size would you expect to see this dislocation movement mechanism important in sintering?

Answer: No, we have not seen any evidence of grain boundary sliding. In powder metallurgy we are generally concerned with particles smaller than 150 μ. For sintering of particles of this size I believe that dislocation motion is important at least in the early stages of sintering. The final elimination of the pores may be due to diffusional flow processes.

D. L. Johnson (Northwestern University): We have proposed a different interpretation for our silver results, for which the particle diameter was in this range. We got almost exact agreement with the volume and boundary diffusion coefficients. Could you explain our result?

Answer: I am aware of the results you obtained on the shrinkage of silver compacts and your calculations, which indicate that the results can be interpreted by material transport through grain boundary and volume diffusional flow. At present I cannot tell where the discrepancy between our differing interpretations comes in.

John H. Hensler (University of Melbourne): One point which crops up in the interpretation of the creep experiments by dislocation movement is the incredibly small number of dislocations which had to move to account for the results. I think it was a strain rate of a 0.1 % per hr in 100-μ material, which in the number of dislocations moving is about one every several hours, or something rather impossible like that. The matter of the stress component in the dislocation system in Dr. Lenel's work, which was fairly close to the melting point of the material (whereas in Dr. Morgan's work the temperature was about half the melting temperature), is important with respect to the theories of sintering from diffusion of dislocations, particularly with the diffused ones pinned. The stress is extremely large and the temperatures are near the melting point. In situations where the operative temperatures are moderately high the stresses are usually not large and the approximation of Herring's derivation breaks down. As to how one comes to an approximation for an exponential relationship, it is not really surprising in stress exponents that one is not found. In metals stress experiments they have been found, and so presumably this could be interpreted as meaning that something else entirely is happening. When attempting to fit models to experiments performed at temperatures at or close to the melting point, I think that high-elasticity experiments tend to indicate stresses which are structure sensitive rather than stresses in which there is diffusion or dislocation movement. Going back go Dr. Lenel's comments on grain boundary sliding, for the deformation of magnesium oxide which we have been studying in Melbourne we feel that shear at grain boundaries is a predominant mechanism, for which stress exponents have been determined under some conditions at approximately 1, and under other conditions at approximately 4. Larger grains of magnesium oxide remain despite extensive stress deformation. Perhaps this is partial substantiation for Dr. Lenel's comment that the grain growth can in part occur by a dislocation movement mechanism. Dr. Lenel obtained evidence that suggests that the introduction of the markers has an effect on the ultimate behavior of the material. I think that the introduction of markers in creep experiments also interfered somewhat with the normal behavior of the sintering.

Answer: I cannot say categorically that the markers have not interfered with the sintering process. However, the markers were relatively large, 2-5 μ in diameter, and therefore should not have much effect in strengthening the calcium fluoride by dispersion strengthening.

Chapter 22

A Reconsideration of Stress and Other Factors in the Kinetics of Densification

H. Palmour, III, R. A. Bradley, and D. R. Johnson

North Carolina State University at Raleigh
Raleigh, North Carolina

Experimental factors which are important in describing the kinetics of densification during hot pressing or sintering are considered, and evidence from the literature is presented to demonstrate that strain energy introduced into a material by premilling or explosively shocking the material has a significant effect on densification kinetics. An empirical model similar in form to the Zener–Holloman kinetic expression for deformation by creep is proposed to describe the rate of densification in terms of temperature, remnant porosity, remnant surface area, applied stress, and the excess internal energy available from annealable defects, and an experimental approach for testing the validity of the model is described.

INTRODUCTION

Narrow channels of meaningful communication about densification (and the closely related process of deformation) appear to exist between those with scientific interests on the one hand and those concerned with practical engineering objectives on the other. Unfortunately, scientists usually find it necessary to simplify their kinetic models and to establish rather restrictive boundary conditions in order to make them mathematically tractable. Many of the terms employed in such models convey well-defined conceptual quantities, but are not amenable to direct experimental measurement. Consequently, it is not surprising that such models, though valid scientifically, are not always readily adaptable to practical densification problems. The practicing engineer is more likely to rely on personal judgment developed by cut-and-try methods with real materials than he is to accept predictions based on an obviously oversimplified model which is predicated upon an unrealistically idealized material or process.

Despite all their know-how, it is unfortunately true that many production-oriented technologists find it difficult to generalize their experiences, and can extend this know-how to include new processes or materials only after a reasonably long period of cut-and-try. Usually, these practical-minded people are neither interested in, nor willing to investigate, densification behavior outside the narrow range of acceptable production practice. Consequently, their experimental results, though often voluminous, are from the viewpoint of the scientists almost single-point values, and hence of no particular assistance to one who is attempting to identify a general densification process.

It is deplorable, but increasingly obvious, that the simplifying assumptions which made possible real scientific advances in understanding the nature of densification during the last two decades now may be overly restrictive in terms of complete experimental descriptions which are important to further advances in technology and in science. We now sense a new era in the study of densification phenomena, an era in which new tools for the engineer—in particular the computer—will be bringing about the need for, and the means of obtaining, much more general and far more accurate empirical descriptions of materials behavior. We can expect these findings to both assist and challenge the scientific community as new knowledge evolves.

IDENTIFICATION OF VARIABLES

In this section experimental variables appropriate to a description of densification are identified on the basis of logical inferences from experimental findings. Table I summarizes the densification data obtained by Pearson et al. ([1]) in their study of sintering behavior and abrasion resistance in grinding balls produced from alumina subjected to various amounts of pregrinding (dry ball milling). The tabulation also includes pertinent data on the surface area of the ground particulates ([2]).

These data suggest that *temperature*, *time at temperature*, and *grinding time* are experimentally significant variables. The effect of grinding time is much larger at lower temperatures than at higher ones; obviously, some interaction exists between them. The effects of increases in temperature and/or grinding time are more pronounced when the porosity is high; therefore *porosity* may be a significant variable. In addition, the *initial porosity* may be a factor.

It is also evident from the table that the influence of grinding time can be attributed at least in part to increases in *specific surface area*. However, surface area alone may not account for the effect; other investigators have demonstrated that dry grinding of alumina also introduces *strain energy* detectable by X-ray line-broadening. As Fig. 1 illustrates, Lewis and Lindley ([3]) found that premilling caused significant enhancement of sintering, which they evaluated in terms of increased compressive strength. The strength increase correlated much better with strain energy resulting from premilling than it

TABLE I

Effect of Dry Grinding Time Upon Initial Surface Area, Initial Density, and Final Density for Alumina Balls Sintered at Various Temperatures*

Grinding time (hr)	Initial surface area (m²/g)	Initial fractional density†	Fractional density† of specimens fired 1 hr at					
			1395°C	1460°C	1520°C	1550°C	1600°C	1650°
0	7.5	N.A.	N.A.	N.A.	N.A.	N.A.	N.A.	N.A.
4	10.0(0%)	0.484(0%)	—	0.826(0%) [70.8%]	0.896(0%) [85.1%]	0.919(0%) [89.9%]	0.954(0%) [97.2%]	0.970(0%) [100.5%]
8	11.2(12%)	0.509(5.2%)	—	0.913(10.4%) [79.4%]	0.964(7.6%) [89.4%]	0.977(6.2%) [91.8%]	0.987(3.4%) [93.8%]	0.990(2.1%) [94.5%]
17	13.0(30%)	0.519(7.3%)	0.892 [71.8%]	0.951(15.0%) [83.2%]	0.981(9.5%) [89.0%]	0.984(7.0%) [89.6%]	0.990(3.8%) [90.7%]	0.990(2.0%) [90.6%]

*After Pearson et al. ([1]). The alumina balls were isostatically pressed balls initially 1.2 in. in diameter produced from 99.8% purity Alcoa low-soda calcined alumina containing 0.05% added MgO. Parentheses indicate increases attributable to grinding time longer than 4 hr. Brackets indicate increases over initial density attributable to firing conditions for a given grinding time.
†Theoretical density considered to be 3.987 g/cm³.

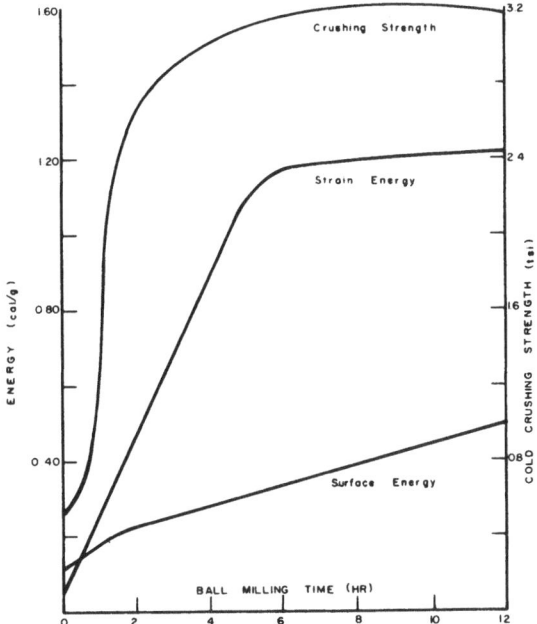

Fig. 1. Crushing strength for α-Al_2O_3 of different strain and surface energies after sintering for 2.5 hr at 1300°C. After Lewis and Lindley ([3]).

did with the more modest increases in total surface energy. Lewis and Lindley point out that

> "Active powders sinter more readily and at lower temperatures than nonactive powders because of greater surface area and distortion of the crystal lattice. The driving energy for sintering is usually assumed to be the excess free energy of the active state above that of the normal equilibrium state."

They conclude that large amounts of strain energy can be introduced in ceramics ($\leqslant 7$ cal/g), which should be considered in characterizations of the sintering behavior of ceramic powders.

In explosively shocked alumina Bergmann and Barrington ([4]) found at least a 350% increase in line broadening, but only a modest 29% increase in the measured surface area which was reported as a consequence of shocking. In material sintered at 1600°C the shocked alumina was about 32% more dense than unshocked control specimens. (At lower sintering temperatures the effect of shocking presumably would have been even greater.) It was reported that

> "The particle size of the shocked alumina was only moderately smaller than for the untreated Al_2O_3.... This difference in particle size was not sufficient to account for the large difference in response to sintering observed between the two aluminas."

They conclude that

> "The introduction of large numbers of defects (strain) into the crystal by the explosive shock treatment may correspond to cold working in metals. If it is assumed that such cold worked ceramics behave similarly to cold worked metals on subsequent heat treatment ... such materials should display the phenomenon of primary recrystallization when heated to some minimum temperature, similar to observations made by Klein et al. ([5])."

Other striking examples of the introduction and retention of strain energy in ceramic particulates (in this case in partially sintered compacts) and of its subsequent contribution to enhanced densification is afforded by the *partial sinter–isostatic press–resinter* experiments reported by Morgan ([6]). Discontinuous jumps in densification upon reheating were observed in various materials. These effects could not be reconciled with diffusional sintering processes, but were attributed instead to material transport by dislocation motion. Morgan concludes that

> "A substantial dislocation motion contribution to the densification therefore appears to be present in the initial stage sintering. The isostatic pressing effect corroborates the significance of dislocations."

From the foregoing evidence it is apparent that dry milling time or grams of explosive are not the really significant canonical variables in densification, whereas the *excess surface energy* and *internal strain energy* induced in particulate materials by such treatments are variables which can be measured experimentally and which cannot safely be ignored in densification processes. Since other forms of energetic pretreatment, e.g., neutron irradiation, are also capable of activating materials by creating lattice defects and distortions, it seems reasonable to categorize the variables associated with any preconditioning of particulates as *excess surface energy* and *annealable excess internal energy*.

The temperature–time dependence of the influence of these two types of excess energy upon densification can be expected to take different forms. The driving force for densification attributable to excess surface energy (proportional to the the product of remnant surface area and unit surface energy) is likely to remain in effect over the whole range of conditions so long as surface area remains, and will tend to operate by any available mechanism to achieve the more stable final state of densification. Annealable excess energy, on the other hand, is "frozen in" and remains essentially unavailable until the temperature is high enough to develop mobility sufficient to permit annealing. Once the defects have been dissipated—generally a rapid process—this source of excess energy becomes exhausted. Therefore such effects are most likely to be observed at moderately low temperatures and in the first relatively brief periods of time after the annealing temperature is reached.

Figure 2 illustrates one possible sighting of such an effect in pure alumina premilled for 17 hr in experiments by Pearson ([2]). These data (single-point values for which no replicates are available) seem to suggest that at times

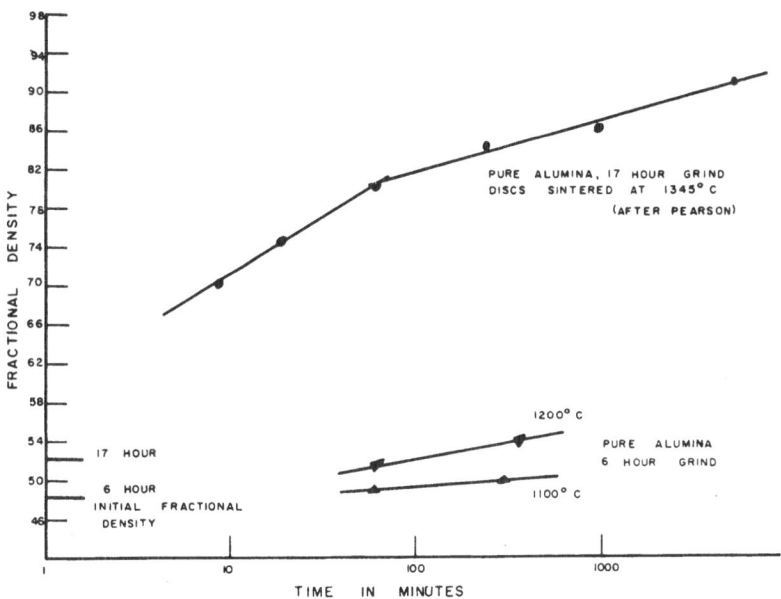

Fig. 2. Time dependence of densification for premilled alumina. After Pearson ([2]).

shorter than about 60 min the densification rate is appreciably higher than is the case at longer times. Significantly, the increase above green density is greater during the first 9 min than it is during the next 900+ minutes. Even though some shrinkage probably occurred prior to reaching 1345°C, the 1100° and 1200°C data for a comparable alumina premilled 6 hr suggest that almost all the shrinkage occurred at temperatures above 1200°C. On this basis and at the heating rates employed the fractional density increased from 0.52 to approximately 0.70 during the final 6 min of heating plus the first (unrecorded) 9 min after temperature was attained. It is interesting to note that the onset of rapid densification seems to occur above 1300°C, where experimental evidence characteristic of plastic flow phenomena has been observed in alumina by other investigators ([7,8]).

Figure 3 schematically summarizes the principal factors which, on the basis of the foregoing arguments, must be considered as variables in any densification process. The basic driving force for densification derives from the *total remnant surface energy*. An apparent thermal activation energy, Q, must be overcome in going from the less dense to the more dense state, so *temperature* is obviously important. As the caption implies, the kinetic relationships will also be determined in part by the *remnant porosity* and by *externally applied stresses*, if any. An important additional factor has been introduced here and is denoted as the *annealable excess energy*, q, which pro-

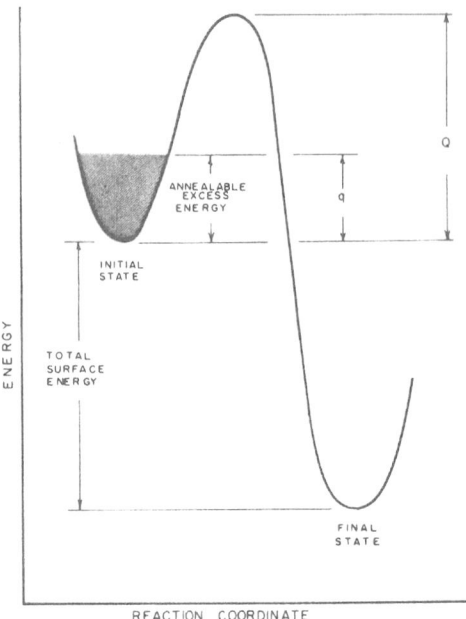

Fig. 3. Schematic representation of energy relationships influencing densification kinetics at any
given porosity and stress.

vides for the possibility of kinetic contribution from nonthermodynamic
defects introduced in the material by prestrain, irradiation damage, or other
energetic treatments which could occur prior to (or even during) the process
of densification.

DENSIFICATION AS A CREEP PROCESS

In this section an experimentally tractable creep rate equation, long
used to described deformation in metals, is suitably modified to permit empirical description of the densification kinetics of ceramics and ceramic-like
materials. This particular approach is predicated upon a few simple assumptions:

1. Densification is at least in part a physicochemical process; clearly,
it is necessary to achieve an effective phase separation in the gas–solid system
before the solid can be fully densified.

2. All the pertinent independent variables (whatever their number) are
well behaved and can be experimentally identified and appropriately scaled
so that their combined effect describes the process.

3. Densification processes involve large populations of randomized particles in randomized local environments, and are thus statistical in character; hence statistical methods of experiment design and analysis, together with computer processing of data, are likely to be appropriate and desirable.

4. The kinetic model employed will resolve mathematically to a form in which the effect of the individual variables (including higher order terms if present) can be treated in an additive fashion, as in an expanded Taylor's series model; such forms are well suited to statistical analyses of the relative significance of the additive components.

5. The forms in which the principal variables are treated will be compatible with those of existing kinetic models, thus facilitating scientific correlation and interpretations.

The first of these assumptions implies a special concern with the particulate properties and purity of starting materials and has unusual pertinence to the concept of rate-controlled densification discussed in a later section. The second assumption in effect restricts the variables to those over which the experimenter can exercise deliberate control, either by preselection (discrete variables) or during the course of densification itself (experimental variables). The third assumption recognizes the scatter inherent in such experimentation and acknowledges that the rules of evidence about the significance of the findings must be considered. The fourth assumption is based in large part upon practical considerations of data processing, but it also provides for the rational evolution of a descriptive model on the basis of acceptance or rejection of each term upon merit as attested by statistical significance. The final assumption almost goes without saying; what is intended is a relationship which can improve communication between technologists and scientists.

Development of Creep Equations for Densification

In this section it is demonstrated that the densification of particulate materials during hot pressing may be adequately and conveniently described by a modified form of the creep rate equation developed by Zener and Hollomon ([9]):

$$\dot{\varepsilon} = A\sigma^n e^{-Q/RT} \tag{1}$$

where $\dot{\varepsilon}$ is the creep rate [$\dot{\varepsilon} = dl/l\,dt$, where l is length], A is a proportionality constant, σ is the applied stress in psi, T the temperature in degrees absolute, Q the apparent activation energy for mass transport, and R the universal gas constant.

The stress exponent, n, should be constant for a single mass-transport mechanism and typically assumes a value of unity for volume-diffusion-controlled viscous flow processes ([10]) or a value of three to four for plastic-flow-controlled processes ([11]).

At first inspection Eq. 1 seems well suited to densification by hot pressing, in which case the creep strain rate would be replaced by the rate of densification. However, it leaves something to be desired in that the effect of porosity on the rate of densification is ignored; in addition, this equation would predict a finite shrinkage rate at the end point condition of zero porosity, where shrinkage must obviously cease.

Kriegel et al. ([12]) modified Eq. (1) by inserting the porosity variable as a power function P^m. Rummler and Palmour ([13]) demonstrated the validity of the porosity term as an independent variable and found two ranges of porosity dependence. Above 12–15% porosity m was about 4, but at lower porosities m was approximately 1.1.

Rummler ([14]) proposed the use of MacClelland's ([15]) corrected stress term, $\sigma/(1 - P^{2/3})$, to account for the effect of porosity on the applied stress and demonstrated that P^m remained a statistically significant variable over and above its passive role as a stress modifier. With these two modifications Eq. (1) becomes

$$\dot{\varepsilon} = AP^m\sigma_e^n e^{-Q/RT} \tag{2}$$

where σ_e is the effective applied stress, $\sigma/(1 - P^{2/3})$. Rummler found that Eq. (2) reliably describes the densification kinetics of vacuum hot-pressed $MgAl_2O_4$ (~98.5% pure) over the range 1260–1390°C, with $m \approx 1.10$, $n \approx 1.06$, and $Q \approx 87.2$ kcal/mole ([13]).

In the final stages of hot pressing, i.e., at low porosity values (< 0.15) Eq. (2) described over 95% of the total variation in Rummler's observed strain rate data. This is remarkably good correlation; nonetheless, Eq. (2) is deficient as a densification model, and becomes increasingly so at higher levels of porosity. It is deficient in that it predicts zero strain rate at zero applied stress, i.e., stresses due to the surface tension of the material are ignored, and densification by sintering is in effect disallowed. In reality, during the initial stages of hot pressing of a very fine particulate material the stresses arising from surface tension may grossly outweigh the applied stress. Very fine $MgAl_2O_4$ derived from coprecipitates was hot pressed under rate control at North Carolina State University by Barnes and Palmour ([16]) and by Choi ([17]) under an initial applied stress of only 200–400 psi, whereas the calculated stress arising from surface tension at the onset of densification was approximately 9000–11,000 psi.

The basic concepts embodied in describing densification in terms of creep kinetics were adapted to the case of pure sintering by Palmour and Johnson ([18]). They proposed substituting the product of surface energy per unit area, γ, and the surface area per unit volume, ϕ, for σ as the driving force in Eq. (2)

and replacing the linear strain rate, $\dot{\varepsilon}$, with a volume shrinkage rate as the dependent variable.

Palmour et al. [19] assumed the forces for densification arising from surface tension and applied stress to be additive; they modified Eq. (2) to

$$\dot{\varepsilon} = AP^m(\sigma_s + \sigma_e)^n e^{-Q/RT} \tag{3}$$

where $0 < \dot{\varepsilon} < \dot{\varepsilon}_{max}$, σ_s is the stress attributable to surface forces $= k\gamma\phi$ $= k\gamma S\rho_{th}D$, with k a factor to convert dynes/cm^2 to psi, γ is the net surface energy per unit area in ergs/m^2, ϕ the surface area per unit volume in m^2/cm^3, and S is the specific surface area in m^2/g.

This model applies, in principle, to all phases of stress-augmented sintering; i.e., from free sintering ($\sigma_e = 0$) to low values of applied stress ($\sigma_s > \sigma_e$) to conventional ($\sigma_s \leqslant \sigma_e$) and ultrapressure hot pressing ($\sigma_s \ll \sigma_e$). Equation (3) has another important feature, in that it is intended that the densification process be subjected to rate control [12,16,18,19]. Once having reached a "safe" maximum densification rate $\dot{\varepsilon}_{max}$, the rate should be controlled (and even systematically decreased) within acceptable limits by careful manipulation of temperature and/or stress in order to prevent trapped porosity containing gases which are difficult to remove, since such entrapments prevent complete densification. Barnes and Palmour [12,16] empirically developed rate control methods in working with very reactive $MgAl_2O_4$ having high specific surface area (> 25 m^2/g), and later demonstrated that the densification kinetics could effectively be described in a relationship similar to Eq. 2 [16].

If the evidence introduced earlier about the role of annealable excess internal energy in densification is taken into account, the kinetic relationship is considered to take the form

$$dD/D\, dt = A\,(1-D)^m(\sigma_s + \sigma_e)^n \exp[-(Q-q)/RT \tag{4}$$

$$0 < (dD/D\, dt) < (dD/D\, dt_{max})$$

where $dD/D\, dt$ is the rate of densification*; $dD/D\, dt_{max}$ the maximum safe rate of densification; D the fractional density $= 1 - P = \rho/\rho_{th}$; P the fractional porosity $= 1 - D$; ρ the density in g/cm^3; ρ_{th} the theoretical density; Q the apparent activation energy for mass transport; and q the apparent excess internal energy associated with annealable defects $\approx k_2\,\Delta c\,\psi$, with k_2 a proportionality constant, Δc the change in concentration of annealable defects, and ψ the apparent excess energy per defect.

---

*$dD/D\, dt = \dot{s}_v = dv/v\, dt$; in a hot pressing die with constant cross-sectional area $dD/D\, dt$ $= \dot{\varepsilon} = dl/l\, dt$.

Equation (4) is a very general relationship. In contrast to most existing models for sintering and hot pressing kinetics, it does not exclude major variables and does not require restrictive assumptions. It is probable that the number of significant terms, as well as the values of the materials constants, will change over different ranges of densification and under differing experimental conditions. In fact, the way such changes ocur can be expected to assist significantly in quantitative characterizations of the densification behavior of the material in question.

In selected situations this broad relationship can be shown to be equivalent to other kinetic models already familiar in the literature. If σ_s and q are known to be insignificant factors, and if $m \approx 1.1$ and $n \approx 1$, then it has been demonstrated ([13]) that the reduced form of Eq. (4) resembles the diffusional creep expression proposed for hot pressing by Rossi and Fulrath ([20]). However, as Amato et al. ([21]) have pointed out, the Rossi and Fulrath treatment could easily have erred in interpreting atomistic mechanisms in terms of stress dependency, since it did not take surface tension stresses into account.

Amato et al. sensed the significance of surface forces in hot pressing, but did not have any kinetic model at their disposal with which they could account for them; Eq. (4) does afford such a possibility. Similarly, Lewis and Lindley ([3]) and Bergmann and Barrington ([4]) sensed the importance of strain energy in sintering, but had no suitable kinetic model in which to incorporate their findings. Equation (4) does offer such an option.

An Experimental Approach

If one's objective is to produce a material with near theoretical density, the experimental variable which must be controlled is the densification rate ([12,16,17,18,19]). Engineers who have contended with sintered products filled with trapped gas and porosity when sintering rates were too high have known this intuitively for a long time. Whereas stress and temperature can be controlled directly during the densification process, densification rate can be controlled only indirectly by appropriate modifications of temperature and stress level. The other variables important to the process—porosity, surface area, and defect energy—are affected initially by the methods of powder preparation and consolidation, but during the densification process they are beyond the direct control of the engineer. Clearly, however, they must decrease as densification progresses, and experimental documentation of such relationships is much needed.

In order to anticipate and control densification rates, one must understand—or at least sense—the interplay between densification rate, applied stress, temperature, porosity, remnant surface area, and defect energy throughout the densification process. Research on the kinetics of densification is now in progress in our laboratory in an attempt to better define this relationship. High-purity cocrystallized and calcined magnesium aluminate is vacuum hot-pressed in graphite dies in an Instron testing machine equipped with a Brew high-temperature furnace.

The initial weight, height, and diameter of the prepressed compacts are measured and recorded along with the zero position of the Instron crosshead as indicated by a precision dial gauge which measures crosshead motion. Shrinkage rate is a function of the rate of movement of the Instron crosshead and can be independently varied over wide ranges by the investigator. By monitoring crosshead motion from the onset of shrinkage ($\rho \approx 0.2\rho_{th}$) to essentially full density ($\rho \approx \rho_{th}$) and applying corrections to account for changes in length on the load column due to changes in temperature and stress, it is possible to calculate Δh, the change in height of the compact. The height of the compact at any instant can then be determined by subtracting Δh from the initial height and can be used to calculate instantaneous porosity and strain rate.

Surface area, ϕ, is determined empirically from an auxiliary study of surface area versus shrinkage. The net unit surface energy, γ, is assumed to be essentially constant with a value on the order of 10^3 ergs/cm^2.* The calculated surface stress, $k\gamma\phi$, is added to the effective applied stress, σ_e, to obtain a total stress which takes into account the significant driving forces due to remnant surface area.

As long as they remain within acceptable limits all the variables are allowed to change freely. Approximately 150 separate observations of strain rate, stress, temperature, and displacement are made during the consolidation of each hot-pressing compact. Eight runs have been made to date, yielding a total of approximately 1200 observations. The processing of this data is awaiting experimental determination of the relationship between remnant surface area and remnant porosity in partially densified compacts.

The hot-pressing technique developed in this investigation and the insight gained in the kinetic studies have been used to hot press small compacts to very near theoretical density, as indicated by their high degree of translucency. They were outgassed during a temperature arrest prior to the onset of shrinkage and were formed under a modified version of rate control ($\dot{\epsilon}_{max} < 0.01$ min^{-1}) at applied stresses $\leqslant 14{,}000$ psi and temperatures $\leqslant 1500°C$.†

---

*The surface free energy has been calculated from thermodynamic data by Bruce ([22]). The value for spinel is appreciably higher than most oxides, $\gamma_{s-v} \approx 1975$ ergs/cm^2 at 1450°C. The net surface energy is the difference between γ_{s-v} and the boundary surface energy γ_{s-s}. It is the *net* surface energy which we estimate to have a value $\approx 10^3$ ergs/cm^2.

†Rate-controlled sintering has been demonstrated to be effective for spinel in preliminary experiments and also has been successfully employed in sintering conventional polyphase ceramic compositions containing liquid phases.

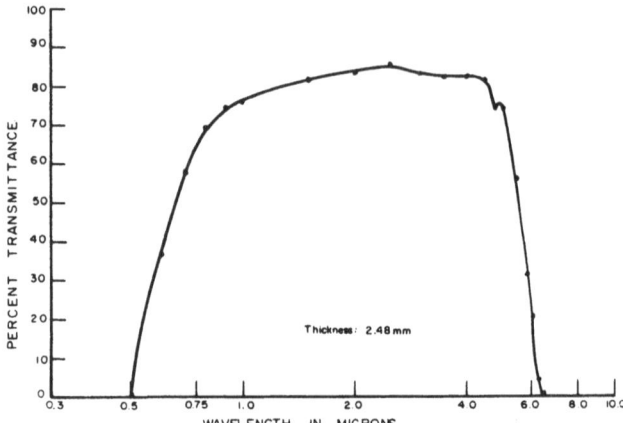

Fig. 4. Optical transmittance of hot-pressed spinel disk. After Carnall ([23]).

When vacuum hot pressed in metal dies at higher pressures and lower temperatures this same spinel material has been densified to an essentially pore-free, fully transparent condition ([23]). Figure 4 reproduces optical transmittance data from such an undoped $MgAl_2O_4$ disk 1 in. in diameter and 2.48 mm thick (after optical polishing).

These spinel materials are densified at stresses and temperatures at which plastic deformation by multiple slip processes is readily induced in densified polycrystalline spinel ([17,24]) and there is no reason to think that the material is behaving differently in a hot-pressing environment. Not only is a proper kinetic description of the process likely to require a nonlinear stress dependency ($n \approx 3$–4) of the sort provided by Eq. (4), but work hardening and subsequent strain-anneal processes are also known to be likely consequences of dislocation interactions. Hence q, the excess energy term in Eq. (4), may also be required to account adequately for this effect.

CONCLUSION

Experimental evidence obtained from the literature and from current research has been cited to demonstrate that densification processes are responsive to at least five independent experimental variables, *temperature, remnant porosity, remnant surface area, applied stress*, and the *concentration of nonthermodynamic defects*, which represent annealable excess internal energy. An empirical model for densification kinetics incorporating these variables has been proposed which provides for nonlinear dependences of densification rate upon porosity and total stress. The proposed relationship is compatible with existing (but more restrictive) kinetic models now in the scientific literature. It is also well suited to data reduction by statistical methods and to process characterization and control by computer techniques.

ACKNOWLEDGEMENTS

Support of this research by the U. S. Army Research Office, Durham, N. C., is gratefully acknowledged.

REFERENCES

1. A. Pearson, J. E. Marhanka, G. MacZura, and L. D. Hart, "Dense, Abrasion-Resistant, 99.8% Alumina Ceramic," *Bull. Am. Ceram. Sec.* **47**(7): 654–658 (1968).
2. A. Pearson, Aluminum Company of America, personal communication, June 1967.
3. D. Lewis and M. W. Lindley, "Enhanced Activity and the Characterization of Ball-Milled Alumina," *J. Am. Ceram. Soc.* **49**(1): 49 (1966).
4. Oswald R. Bergmann and Jonathan Barrington, "Effect of Explosive Shock Waves on Ceramic Powders," *J. Am. Ceram. Soc.* **49**(9): 502 (1966).
5. M. J. Klein, F. A. Rough, and C. C. Simons, "Structure and Annealing Behavior of Explosively Shocked magnesia Single Crystals," *J. Am. Ceram. Soc.*, **46**(7): 356–58 (1963).
6. C. S. Morgan, this volume, Chapter 19.
7. G. E. Gross and Paul L. Gutshall, A Study of the Physical Basis of the Mechanical Properties of Ceramics, Contract AF–33–(615)–2669, Bimonthly Progress Report No. 7, 15 July–14 September 1966.
8. E. Passmore, A. Mochetti, and T. Vasilos, "The brittle–Ductile Transition in Polycrystalline Aluminum Oxide," *Phil. Mag.* **13**(126): 1157 (1966).
9. C. Zener and J. H. Hollomon, "Plastic Flow and Rupture of Metals," *Trans. Am. Soc. Metals* **33**: 163–235 (1944).
10. F. R. N. Nabarro, "Deformation of Crystals by Motion of Single Ions," Report on a Conference on Strength of Solids, Physical Society of London, pp. 75–90, 1948; C. Herring, "Diffusional viscosity of a Polycrystalline Solid," *J. Appl. Phys.* **21**: 423 (1950).
11. J. R. Weertman, "Theory of Steady-State Creep Based on Dislocation Climb," *J. Appl. Phys.* **26**: 1213–1217 (1955).
12. W. W. Kriegel, H. Palmour III, and D. M. Choi, "Preparation and Mechanical Properties of Spinel," in: *Proceedings of a Symposium, British Ceramamic Research Association*, Stoke-on-Trent, July 1964 (P. Popper, ed.) Academic Press, London, 1965.
13. D. R. Rummler and H. Palmour III, "Vacuum Hot-Pressing of Magnesium Aluminate Spinel," *J. Am. Ceram. Soc.* **51**(6): 320–326 (1968).
14. D. R. Rummler, "Hot-Pressing Kinetics of Magnesium Aluminate in Vaccum," Unpublished M. S. Thesis, Department of Mineral Industries, North Carolina State University, 1966.
15. J. D. McClelland, "Kinetics of Hot-Pressing," in: *Powder Metallurgy—Proceedings of an International Conference*, New York, 1960, (W. Leszynski, ed.), Interscience Publishers, New York, 1961, pp. 157–171.

16. L. D. Barnes and H. Palmour III, "Rate-Controlled Hot Pressing of Spinel," paper presented at the Fall Meeting of the Basic Science Division of the American Ceramic Society, Pennsylvania State University, October, 1966. [In preparation for Publication.]

17. D. M. Choi, "Flow and Fracture of Hot Pressed Polycrystalline Spinel at Elevated Temperatures," Unpublished Doctoral Dissertation, North Carolina State University, Raleigh, 1965. (Available through University Microfilms, Ann Arbor, Michigan.)

18. H. Palmour III and D. R. Johnson, "Phenomenological Model for Rate-Controlled Sintering," in: *Sintering and Related Phenomena* (G. C. Kuczynski *et al.*, eds.), Gordon and Breach, New York, 1967, pp. 779–791.

19. H. Palmour III, D. R. Johnson, and D. R. Rummler, "General Phenomenological Model for Stress Augmented Sintering," paper presented before the Basic Science Division, 68th Annual Meeting of the American Ceramic Society, Washington, D. C., May 7–12, 1966. [In preparation for publication.]

20. R. C. Rossi and R. M. Fulrath, "Final State Densification in Vacuum Hot Pressing of Alumina," *J. Am. Ceram. Soc.* **48**(11): 558–64 (1965).

21. I. Amato, R. L. Colombo, and A. M. Petruccioli Balzari, "Hot Pressing of Uranium Oxide," *J. Nucl. Mater.* **20**: 210–214 (1966).

22. R. H. Bruce, "Aspects of the Surface Energy of Ceramics: I—Calculation of Surface Free Energies," in: *Science of Ceramics*, Vol. 2 (G. H. Stewart, ed.), Academic Press, New York, 1965, Chapter 24, pp. 359–367.

23. E. Carnall, Jr., Eastman Kodak Company, personal communication, June 1967.

24. H. Palmour III, "Multiple Slip Processes in Magnesium Aluminate at High Temperatures," *Proc. Brit. Ceram. Soc.* **6**: 209–224 (June 1966).

DISCUSSION

R. L. Coble (M.I.T.): The final equation given for densification during hot pressing predicts that the densification rate would be zero as the density becomes equal to the theoretical value; thus in integrated form an infinite amount of time is required to reach density.

Having worked both on the mechanisms of transport during densification and as a technologist, I've also thought about the problem of connecting the two, that is, of applying the information gained in model systems to technological problems. One of the questions raised from an engineer's point of view is with respect to the *reproducibility* of fabrication with any given material. The other important aspect of the question is with respect to *predictability* for either changing variables or for changing systems entirely. The equation presented here includes many of the operational variables pertinent for hot pressing; a number of additional material and operational variables which have been shown to effect sinterability are not included. Therefore the control of reproducibility using this equation would require that the atmosphere, particle shape, purity contents, etc. be held constant. While that may be satisfactory for maintaining reproducibility in an operational sense, it's clearly unsatisfactory for the problems associated with predictability when changing variables.

Another way of organizing the variables for consideration of control is simply to list them in order of importance without specifying anything about their mutual relationships. For sintering I have listed the variables as follows:

1. Composition.
   (a) General.
   (b) Impurities.
2. Temperature.
3. Time.
   Heating rate (not separable from time–temperature).

4. Particle size.
5. Bulk density.
6. Particle size distribution.
7. Density distribution.
8. Atmosphere.
9. Surface energy and boundary energy.

For application to hot pressing the applied pressure is approximately equivalent in importance to the particle size. If the raw materials and compacts formed from them are characterized completely, it seems to me that reproducibility in behavior is reasonable to expect, but predictability is lacking because the precise interrelationships among the variables are lacking with the multiplicity of possible material transport mechanisms taking place.

Answer: We have given a lot of thought to the implications of this model [Eq. (4)] and its relationship to other variables by which materials commonly have been characterized. We feel that most of those you have listed really have been adequately covered by the independent variables in our model. By working in terms of the fractional porosity (or fractional density, whichever you prefer) and the remnant surface area, we can take into account the effective contribution of all the particulate properties, including particle size, particle size distribution, and even particle and pore shapes. We are developing the capability for experimentally measuring remnant surface areas in partially densified compacts, as well as the initial surface area of the starting material. In our very fine spinel materials (typically 0.03 μ in diameter) particle size measurements *per se* become extremely difficult and very unreliable, whereas the accuracy of surface area measurement gets better as the particles get smaller. I would like to point out that initial density is reflected in our model; the remnant porosity term is defined as the initial porosity minus that which has been removed by shrinkage. The initial surface area also is included, since remnant surface area is defined as initial area minus that which has been traded off for shrinkage.

Atmosphere is certainly important, but we feel that it is best treated—experimentally and mathematically—as a discrete variable.

Coble: Where would you say atmosphere is implicit in that formula?

Answer: It would be discrete; in any given solution the constants A, m, n, Q, etc. would be determined for a given atmosphere and a given starting material. Changes in those constants as functions of discrete changes in composition, partial pressure, etc. presumably would provide a quantitative measure of the influence of that environment (or purity) on the densification process, and, in addition, might indicate something about the nature of the mechanism by identifying which of the independent variables were affected.

NOTE ADDED IN PROOF

Heckel and Youngblood([25]) recently reported on structural effects measured by X-ray line-broadening in α-Al_2O_3 and MgO powders following explosive shock treatments such as those described in Refs. 4 and 5. The X-ray particle size of shocked powders was below the actual particle size: X-ray size measurements were considered to sense the spacing of dislocation arrays, as in cold-worked metals. The elastic strain energy was calculated to be 0.51 and 0.37 cal/g for the MgO and Al_2O_3 respectively, i.e., the excess internal energy was about equal to the surface energy of the powders.

ADDITIONAL REFERENCE

25. R. W. Heckel and J. L. Youngblood, "X-ray line broadening of explosively shocked MgO and α-Al_2O_3 powders," *J. Am. Ceram. Soc.* **51**(7): 398–401 (1968).

Chapter 23

Glow Discharge Sintering of Alumina*

C. E. G. Bennett and N. A. McKinnon

CSIRO Division of Applied Mineralogy
Melbourne, Australia

Experimental work is reported on the rate of sintering of alumina compacts sintered in a microwave-induced plasma. It is shown that the sintering rate is very much greater for plasma-sintered specimens than for specimens sintered by conventional means. Evidence indicates the generation of plasma inside the pores of the compact, and the increased rate of densification is related to an increased rate of diffusion.

This note essentially concerns the observation that the rate of sintering of alumina in a microwave-stimulated glow discharge (plasma) is much greater than that obtained with normal heating techniques ([1]). Associated with this phenomenon is the fact that the grain size in the final high-density product is much finer than in conventionally sintered specimens at the same density, and consequently the mechanical strength is appreciably higher ([2]).

The experimental apparatus consists of a silica or alumina sintering tube inserted through a waveguide which is fed by coaxial line from a 1-kW, 2450-MHz magnetron. The wave guide is tuned to produce a peak of electric field intensity in the region of the sintering tube. With a low pressure (1–50 torr) of a suitable atmosphere inside the tube a plasma is created by the electric field. Because of recombination phenomena on surfaces immersed in the plasma a pressed compact situated in the center of the sintering tube and waveguide may be rapidly heated to a temperature which is controlled by the power input, the pressure of the gas, and, as far as the peak temperature is concerned, by the nature of the gas atmosphere.

Comparative densification curves for pressed compacts of Linde A alumina are given in Fig. 1. It will be seen that the rates of densification of the specimens sintered in an air plasma are very much greater than the rates for identical specimens sintered in air in a conventional furnace.

Figure 2 shows the maintenance of strength levels at high densities for plasma-sintered specimens, whereas the strength of conventionally

---

*Paper presented by L. S. Williams.

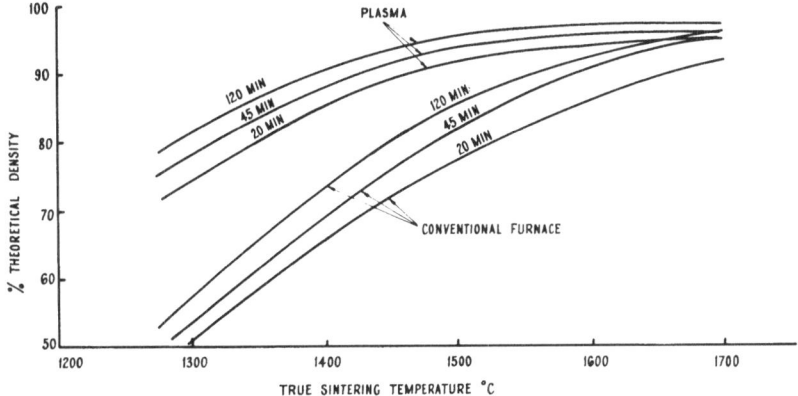

Fig. 1. Sintering of Linde A alumina in air. Green density ≈44% theoretical.

sintered specimens falls drastically. The difference in behavior is consistent with the grain sizes of the specimens concerned—averaging 4–10 μ and 50–150 μ for plasma and conventional methods, respectively. Figure 3 shows the two fracture surfaces. (The test bars used were machined to a standard square form from isostatically pressed rods after sintering and were tested in four-point bending.)

A critical factor in the assessment of the experimental results is the reliability of temperature measurements. Because of direct heating of the

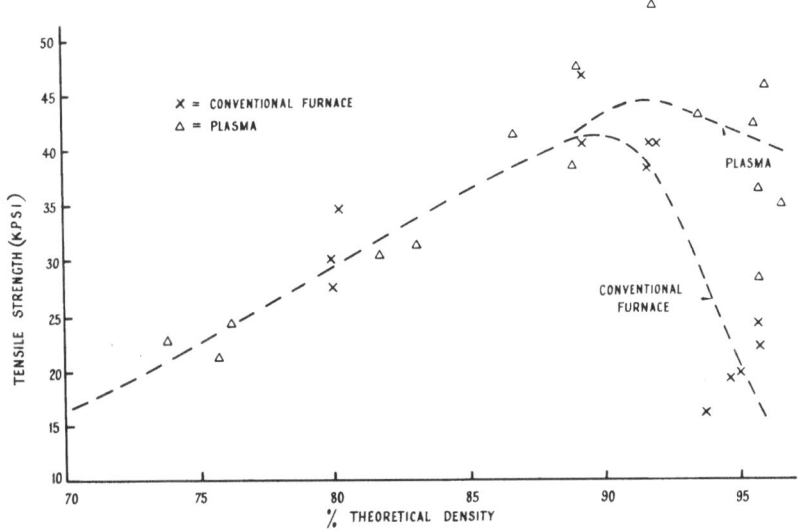

Fig. 2. Bend bar data on Linde A alumina.

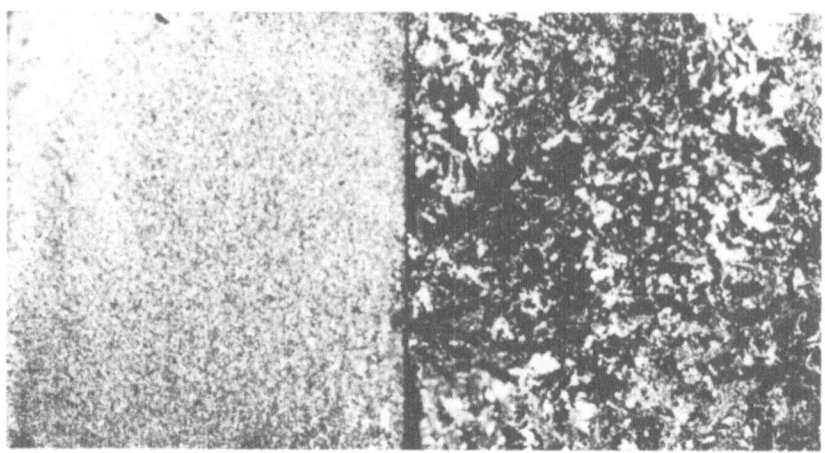

Fig. 3. Fracture surfaces of fine-grained plasma-sintered and coarse-grained conventionally-sintered alumina bars. 200 Å of aluminum vacuum-deposited to increase reflectivity. Magnification ×20.

metallic components thermocouples cannot be used. With optical and radiation pyrometers there is significant interference from the discharge at temperatures below about 1500°C, while with the sintering tube having "cold walls" the low emissivity of the pressed compact produces a marked departure from black body conditions. An apparent versus true temperature correction curve was established by "push-out" experiments from near-blackbody conditions measured by reference thermocouple and optical pyrometer to simulated cold wall conditions, where the optical pyrometer and a quick-reading radiation pyrometer were used to establish apparent temperatures. In practice, the interference from the discharge to the optical pyrometer readings is overcome by measuring during a brief interruption to the high-tension supply to the magnetron.

One of the initial hypotheses advanced concerning the difference between glow discharge and conventional sintering was based on the expected release of adsorbed gases and precursor residues through ionic and electron bombardment. The effective surface energy of the oxide powder would thus be increased so that more rapid sintering might be expected, at least in the initial stages. However, on a simple surface-cleaning hypothesis it would seem that once the major portion of the adsorbed gases is removed and a rapid sintering rate is established the material should continue to sinter comparably in a conventional furnace. This is not so. For example, a specimen sintered for 20 min in the discharge at 1300°C reached 74.4% of theoretical density. In a subsequent 100 min at 1300°C in a conventional furnace the density only moved slightly to 74.7%. Returned to the plasma for a further

100 min at 1300°C a density of 83.1%, just slightly in excess of that expected in 120 min of plasma sintering alone at 1300°C, was obtained. The same type of result occurred in similar experiments at 1535°C and 1650°C. Thus in 100 min at 1650°C in a conventional furnace a specimen initially sintered for 20 min in the plasma moved to a density equivalent to about 45 min total in the plasma. It is thus clear that there are positive effects on the rate of sintering throughout the whole of the sintering range in the plasma method.

Heating in the plasma comes from the energy of recombination of the various ionized species originally produced by the effects of the electric field. The contribution from dielectric heating due to the field itself is minimal in the case of alumina. Studies of the sintering of aluminas both finer and coarser than Linde A show that the gap between the two families of densification curves is wider, the finer the particle size of the starting material. Such evidence points to the generation of plasma inside the pores of the compact, and so it may be postulated that the increased rate of densification is related to an increased rate of diffusion due to (1) physical damage to the lattice by creation of point defects by bombardment and/or (2) local increases in temperature inside the pores, perhaps primarily in the regions adjacent to interfaces between particles.

Further studies are in progress on the kinetics of the plasma sintering process as observed in Linde A and other aluminas. Preliminary work on other oxides such as beryllia, titania, magnesia, zirconia, hafnia, and thoria shows positive effects in some cases but not in others. The reasons for this behavior are still conjectural, but may be connected with particle size and distribution, with the presence of strongly chemisorbed species, or with the type of crystal structure of the oxide concerned. Of special interest is the mechanism which prevents or at least minimizes the rate of grain growth in plasma-sintered material.

ACKNOWLEDGMENTS

The authors would like to thank their colleague, Dr. L. S. Williams, who first suggested this use of the plasma technique for sintering, for useful discussions. They are also grateful to Dr. F. K. McTaggart of the Division of Mineral Chemistry, CSIRO, on whose parallel application of plasma methods to the stimulation of chemical reactions they have drawn heavily for technical guidance.

REFERENCES

1. R. A. Dugdale, "The Application of the Glow Discharge to Material Processing," *J. Mat. Sci.* **1**: 160–169 (1966).

2. C. E. G. Bennett, N. A. McKinnon, and L. S. Williams, "Sintering in Gas Discharges," *Nature* **217**: 1287–1288 (1968).

DISCUSSION

J. E. Burke (G.E.): Is the heating done by heat transfer from the plasma? In addition, have you done any microscopic studies?

Answer: Yes, it is direct heat transfer. Microscopic studies are in progress. We are impregnating these porous bodies—we are initially more interested in the porous range—with furfuryl alcohol, producing carbon decomposition, and then examining these carbon-filled oxide bodies optically. We are also doing conventional open and closed porosity measurements of course, and some ancillary electron microscope studies are in progress.

Burke: The usual custom that we have used has been to impregnate specimens with a plastic, but it is essentially impossible to distinguish microscopically between the two phases. Carbonization sounds like a very neat little trick.

Answer: We took our lead from G. M. Fryer and J. P. Roberts, "Some Techniques for Microscopical Examination of Ceramic Materials," Trans. Brit. Ceram. Soc. 62: 537 (1963).

Dr. Hayne Palmour (North Carolina State): Did you monitor your pressure in your vacuum chamber or do a gas analysis on what was being pumped out during the time you began to sinter?

Answer: The pressure is regulated and observed fairly frequently throughout the run.

Palmour: Did you see the pressure burst through at the time you really began to get densification?

Answer: That's a very good point. The pressure in the sintering tube is dependent on the balance between a throttled vacuum pump and a controlled leak. It is also affected by the increasing temperature of the specimen, so that very careful experiments would be needed to distinguish such a rise in pressure. Perhaps I should explain that so far we have concentrated on the overall phenomenology in order to complete a patent specification—we are now able to go back and look at the more scientific aspects.

Chapter 24

Vaporization of Cuprous Oxide and Other Dissociating Oxides

R. H. Campbell and M. O'Keeffe

Chemistry Department
Arizona State University
Tempe, Arizona

The behavior of dissociating oxides at high temperatures is briefly reviewed. At low pressures the composition of the oxide will correspond to an equilibrium oxygen pressure quite different from the ambient oxygen pressure. Evaporation rate studies on cuprous oxide in various atmospheres show that this oxide dissociates according to Cu_2O (solid) $\longrightarrow$ $2Cu$ (gas) $+ \frac{1}{2}O_2$ (gas).

Many high-temperature processes involving oxides are studied under conditions such that one or more components have appreciable volatility. Apart from its intrinsic interest as a high-temperature reaction, evaporation has some interest in that an evaporating material is not in equilibrium with its surroundings. For a dissociating oxide of variable stoichiometry this observation has some interesting consequences, which we explore in this chapter.

We shall be concerned with oxides that dissociate in the vapor phase according to the overall reaction

$$MeO_x \text{ (solid)} \longrightarrow MeO_{x-n} \text{ (gas)} + (n/2)O_2 \text{ (gas)} \qquad (1)$$

Some well-known examples include SnO_2^1, for which $x = 2$ and $n = 1$, ZnO^2 and NiO^3, for which $x = n = 1$, and Cu_2O^4, for which $x = \frac{1}{2}$ and $n = \frac{1}{2}$.

THEORY

A number of different pressures appear in the theoretical expressions and it is convenient at the outset to define the following sysmbols, P_{O_2} is the equilibrium oxygen pressure of the sample under consideration, $P_{O_2}^{ext}$ the oxygen pressure in the gas phase surrounding the sample, P_M the pressure of gaseous MeO_{x-n} in equilibrium with the solid (the subscript M is used for MeO_{x-n}

throughout), $P_0 = [P_{0}^{n/2}(P_{0} - P_{0}^{ext})]^{2/(n+2)}$, the significance of which appears below, and $\bar{P}$ the total pressure in the gas phase.

We proceed now to review the possible experimental conditions under which vaporization might be observed.

Close to Equilibrium

This is the situation in, for example, a Knudsen cell in which the size of the orifice is very small compared to the area of the sample. In this experiment the true vapor pressure of the sample is measured and, as the amount of gas escaping from the cell is very small, the composition of the sample is fixed. It is also usually assumed that the solid is in equilibrium with its surroundings in a flow system when the rate of flow of gas is small.

In a Rapid Flow of Inert Gas or In a Vacuum

Under these conditions evaporating material is lost irrevocably at a rate proportional to the partial pressure of the component in question. The rate of loss of component i in units of moles per unit time is given by the well-known expression ([5]):

$$N_i = AP_i\alpha_i(2\pi M_i RT)^{-1/2} \tag{2}$$

Here M_i is the molecular weight of species i, A is the area of the sample, and α_i is a numerical factor less than or equal to unity and known as the evaporation coefficient. The magnitude of α_i will depend on the mechanism of evaporation, but for simple atoms and molecules it is normally of the order of unity ([6]).

Equation (2) allows us to derive a simple criterion for distinguishing between a "rapid" flow of gas and removal of vapor at the equilibrium pressure. In a slowly flowing gas stream of pressure $\bar{P}$ moving over the sample at a rate of S moles/sec the equilibrium rate of removal of material is

$$N_i = SP_i/\bar{P} \tag{3}$$

For the gas to be in equilibrium with the solid, the rate as given by Eq. (3) must be much greater than that given in Eq. (2), i.e.,

$$A\alpha_i(2\pi M_i RT)^{-1/2} \gg S/\bar{P} \tag{4}$$

or

$$\frac{A\alpha_i \bar{P}}{S(MT)^{1/2}} \gg 2.3 \times 10^{-9} \tag{5}$$

where A is in cm², $\bar{P}$ is atmospheres, S in moles/sec. Under laboratory conditions the inequality ([5]) normally holds for gases at 1 atm, but at low pressures the nonequilibrium condition will obtain.

For an arbitrary oxide N_{O_2} and N_M will in general be very different, so that as evaporation proceeds, the composition will change, until at some final composition congruent evaporation is occurring. If the final composition is MO_x, then

$$(n/2)N_M = N_{O_2} \tag{6}$$

or, from Eq. (2)

$$\gamma \frac{2}{n} \equiv \left(\frac{P_{O_2}}{P_M}\right)\frac{\alpha_{O_2}}{\alpha_M}\left(\frac{M_M}{M_{O_2}}\right)^{1/2} = 1 \tag{7}$$

In Fig. 1(a) we show an isotherm for a hypothetical Me–O₂ system, and in Fig. 1(b) γ [as defined in Eq. (7)] is shown as a function of x for the same system calculated on the assumption that the only gaseous species are Me and O_2. It may be seen that there is only one composition at which $\gamma = 1$ and that this composition may involve one or two solid phases ([7]).

In a Rapid Flow of Gas with Constant $P_{O_2}^{ext}$. The situation here is the same as above with the steady-state condition again given by Eq. (6). The difference is that now the composition will depend on $P_{O_2}^{ext}$.

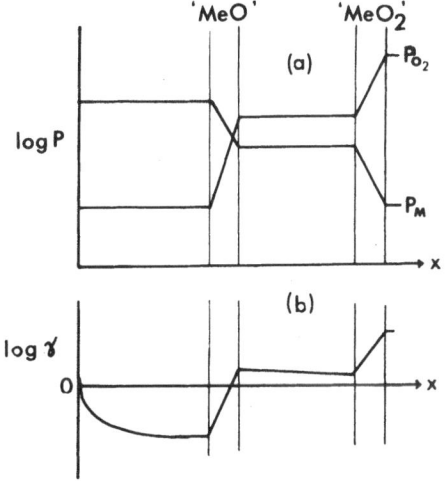

Fig. 1. (a) Isotherms for a hypothetical Me–O₂ system. (b) γ [Eq. (7)] as a function of x for the same system, showing the composition of congruent evaporation.

The rate of loss of MeO_{x-n} will be

$$N_M = P_M \alpha_M (2\pi M_M RT)^{-1/2} \tag{8}$$

The rate of loss of oxygen will be the rate of evaporation minus the rate of condensation:

$$N_{O_2} = (P_{O_2} - P_{O_2}^{ext}) \alpha_{O_2} (2\pi M_M RT)^{-1/2} \tag{9}$$

It is worth noting at this stage that Eqs. (6), (8), and (9) require that $P_{O_2} \neq P_{O_2}^{ext}$. The difference between these two quanities is obtained as follows. The equilibrium constant for reaction (1) is

$$P_{O_2}^{n/2} P_M = \exp(-\Delta G^0 / RT) \tag{10}$$

where ΔG^0 is the standard free energy change for the reaction as written. Combining Eq. (10) with Eqs. (6), (8), and (9) yields

$$P_{O_2}^{n/2}(P_{O_2} - P_{O_2}^{ext}) = \frac{n\alpha_M}{2\alpha_{O_2}} \frac{M_{O_2}}{M_M} \exp\left(\frac{-\Delta G^0}{RT}\right) = P_0^{(n+2)/2} \tag{11}$$

In Fig. 2 we have plotted (in units of P_0) P_{O_2} as a function of $P_{O_2}^{ext}$ for the case $n = 1$. It may be seen that for $P_{O_2}^{ext} > 10P_0$, $P_{O_2} \approx P_{O_2}^{ext}$; however, for $P_{O_2}^{ext} < 10P_0$, $P_{O_2} > P_{O_2}^{ext}$, and in fact P_0 is the lowest possible value of P_{O_2} attainable.

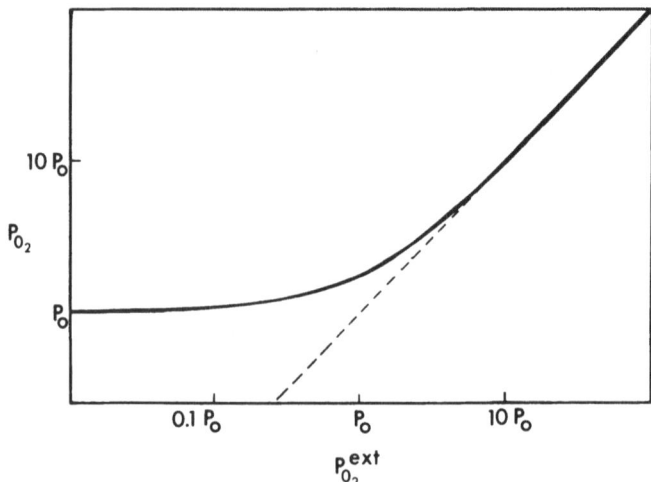

Fig. 2. The variation of the equilibrium oxygen pressure P_{O_2} with external pressure $P_{O_2}^{ext}$ for a rapidly evaporating oxide MeO.

To take an example, for ZnO at 1500°K, $P_0 = 2 \times 10^{-6}$ atm (assuming $\alpha_{O_2} \approx \alpha_M$). If some property of ZnO (such as electrical conductivity) that depends on P_{O_2} is to be measured under these conditions it should only be done with $P_{O_2}^{ext} > 2 \times 10^{-5}$ atm.

Static System

High temperature experiments are very frequently carried out in a closed system with the oxygen pressure, $P_{O_2}^{ext}$, maintained at a constant pressure, but with P_M very close to zero as a consequence of part of the walls of the apparatus being cold. When the total pressure is low the situation is exactly as described above for case 2b, and the same care is required in interpreting low pressure data. As an example of the difference in behavior of an oxide at low pressures as opposed to in a flow system we might cite the observations of Zirin and Trivich[8] who found that Cu_2O had an apparently anomalous thermoelectric power at low total pressures but behaved normally in a flow system with a total pressure of 1 atm and the same partial pressure of oxygen.

As the total pressure is increased in a static system, a point is reached when the rate of loss of material as given by eqs. 8 and 9 is reduced by a factor β as the evaporating molecules have to diffuse through the surrounding gas: β will be a function of the total pressure $\bar{P}$. A knowledge of $\beta(\bar{P})$ is essential to a correct interpretation of the effect on the evaporation rate of varying $P_{O_2}^{ext}$. This point is illustrated in the next section.

STUDIES ON CUPROUS OXIDE

High-temperature work, particularly gravimetric studies ([9]), on cuprous oxide are hampered by volatilization of the oxide. We have studied this process as a function of ambient pressure using a vacuum microbalance in a static system. Two series of experiments have been performed: (1) in pure oxygen and (2) in nitrogen with a constant partial pressure of oxygen determined by the presence in the system of metallic copper at the same temperature as the oxide.

Figure 3 shows the steady-state rate of weight loss as a function of total pressure in these experiments at two temperatures. From the observation that the conductivity rapidly reaches an equilibrium value under these conditions it is assumed that evaporation is proceeding congruently. For dissociative evaporation according to

$$\tfrac{1}{2}Cu_2O \longrightarrow Cu \text{ (gas)} + \tfrac{1}{4}O_2 \text{ (gas)}$$

the rate of weight loss in grams per unit area per unit time is

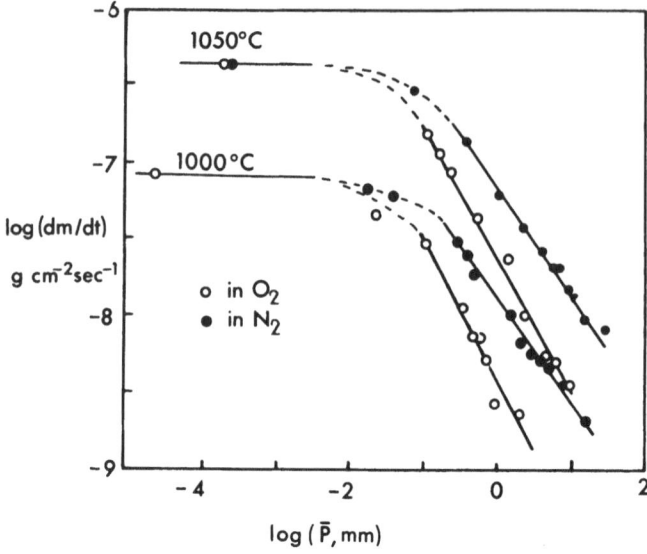

Fig. 3. Rate of weight loss of Cu_2O in N_2 and O_2 as a function of total pressure.

$$
\begin{aligned}
dm/dt &= M_{Cu}N_{Cu} + M_{O_2}N_{O_2} \\
&= (4M_{Cu} + M_{O_2})N_{O_2} \\
&= (4M_{Cu} + M_{O_2})(P_{O_2} - P_{O_2}^{ext})2\pi M_{O_2}RT)^{-1/2}\alpha_{O_2}\beta(\bar{P})
\end{aligned} \quad (12)
$$

For the experiments in nitrogen the only variable on the right-hand side of Eq. (12) is $\beta(\bar{P})$. Accordingly, if it is assumed that $\beta(\bar{P})$ is unity at very low

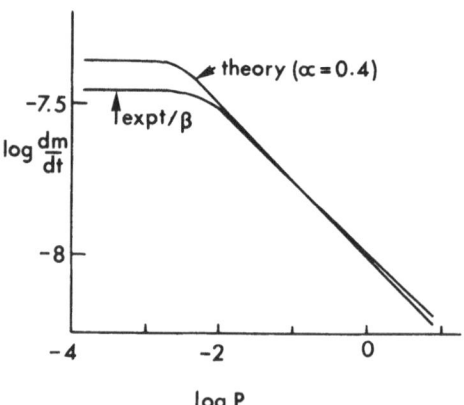

Fig. 4. Theoretical and corrected rate of weight loss of Cu_2O as a function of oxygen pressure.

pressures, β can be calculated for any other value of $\bar{P}$. The same value of β can be used for evaporation in oxygen due to the close similarity in physical properties of O_2 and N_2. In this way we calculate the rate of evaporation in oxygen corrected for the diffusion factor β. The resulting calculated rate of weight loss is shown in Fig. 4. Also shown in Fig. 4 is the theoretical rate of weight loss as given by Eq. (12) when P_{O_2} is calculated from the theoretical expression Eq. (11), viz.,

$$P_{O_2}^{1/4}(P_{O_2} - P_{O_2}^{ext}) = \tfrac{1}{4}(32/63.54)\exp(-\Delta G^0/RT) \qquad (11')$$

with ΔG^0 taken from standard thermodynamic tables ([10]). In comparing the theoretical and experimental results we have fitted the curves at one point by taking $\alpha = 0.4$. This is close to the theoretical value $\alpha = \tfrac{1}{3}$ ([11]). More importantly, the agreement of the theoretical and experimental pressure dependences of the rate of evaporation is valuable evidence that we are indeed dealing with a dissociation reaction in this case.

ACKNOWLEDGEMENT

This work was sponsored by the Air Force Office of Scientific Research, Office of Aerospace Research, United States Air Force, under AFOSR grant Nr 719–65.

REFERENCES

1. C. L. Hoenig and A. W. Searcy, *Am. Ceram. Soc.* **49**: 128 (1966).
2. C. L. Hoenig, Ph.D. Thesis, University of California, UCRL-7521 (1964).
3. R. T. Grimley, R. P. Burns, and M. G. Inghram, *J. Chem. Phys.* **34**: 664 (1961).
4. R. H. Campbell, M. S. Thesis, Arizona State University, 1966.
5. M. Knudsen, *Ann. Phys.* **47**: 697 (1915).
6. R. P. Burns, *J. Chem. Phys.* **44**: 3307 (1966).
7. G. N. Lewis and M. Randall, *Thermodynamics* (ref. ed., K. S. Pitzer and L. Brewer) McGraw-Hill, New York 1961, Chapter 33.
8. M. H. Zirin and D. Trivich, *J. Chem. Phys.* **39**: 870 (1963).
9. M. O'Keeffe and W. J. Moore, *J. Chem. Phys.* **36**: 3009 (1962).
10. U. S. Bureau of Mines, Bulletins 584 (1960), 592 (1961), and 605 (1963).
11. J. P. Hirth and G. M. Pound, *J. Chem. Phys.* **26**: 1216 (1957).

DISCUSSION

J. A. Pask (*University of California at Berkeley*): What are the vapor species above the cuprous oxide in this case? Do you get molecules of Cu_2O?

Answer: The vapor species in this case are copper and oxygen. It is also possible to get

Cu_2O, but it must be a minority species, because if the reaction were mainly solid going to molecular vapor, the behavior in oxygen would be exactly the same as in nitrogen. The fact that oxygen suppresses the rate of vaporization shows that one is getting molecular oxygen in the gas phase. It is possible, but rather improbable, that a lower oxide of copper is produced. The observed pressure dependence of evaporation supports the proposed reaction of dissociation into copper atoms and molecular oxygen.

A. K. Kuriakose (Norton Company): Has anyone observed the vapor species by mass spectrometer?

Answer: Not in the case of cuprous oxide. There has been surprisingly little done with the simpler transition metal oxides, but those that have been studied have largely been found to dissociate. Oxides such as zinc oxide and stannic oxide dissociate on evaporation to give oxygen, and in the case of stannic oxide a lower oxide of tin. It should be remarked that we were particularly interested in the steady-state evaporation in different ambient atmospheres and this is rather difficult to study mass-spectrometrically.

W. M. Robertson (North American Aviation Science Center): Do you see copper residing on the surface of the cuprous oxide?

Answer: No, it evaporates. You find it only in the cold parts of the apparatus. We find the copper condensed on the colder parts of the balance hang-down tube.

Robertson: The reason I was asking is that I have had occasion to do a small amount of work with this and ended up with particles or pieces of copper sitting on the surface of the oxide; this leads to my position that not all the copper goes away.

Answer: You had a closed system all at one temperature, presumably. However, in this case the copper that evaporates is trapped on the cold walls, where in fact it reoxidizes. Microscopic examination has never revealed metallic copper on the crystal surface.

H. B. Johnson (Pittsburgh Corning Corporation): Why would you expect an α of 0.3 in a rather complex evaporation process like this?

Answer: I am not entirely sure I can answer that question. I expect it is because it is frequently observed. In other words, I am not alarmed that it is 0.3 or close to 0.3. In the case of Al_2O_3 and In_2O_3, which likewise dissociate upon evaporation, Burns ([12]) found that α was a function of temperature but was close to 0.3 for all vapor species near the melting point of the oxide. Judging from the correlation of the available experimental data given by Burns (his Fig. 6) we would anticipate in our experiments, carried out at about 0.85–$0.9\ T_m$, that $0.1 < \alpha < 0.3$.

P. W. M. Jacobs (University of Western Ontario): The theory of Hirth and Pound which you are using was developed for metals, and you only expect α to be 0.3 for surface diffusion when the limiting step is surface diffusion between surface steps, which are generated from the edges of the evaporating powder. Now in actual fact α is frequently found to be much less than 0.3 in dissociating solids, so I am also surprised you get such a high value. Second, a relatively minor point; you assume your value for β to be the same in oxygen and nitrogen. This is theoretically unsound—each of them depends on the diffusion coefficient in the gas phase, which again would depend on the reduced mass of the diffusion species, and therefore on the mass in the gas phase. Now in comparing oxygen and nitrogen you are obviously making this error.

Answer: In answer to the second question—I agree entirely but the error should be rather small. This is precisely why we used nitrogen. Let me emphasize that we are concerned with diffusion of the same species (copper and oxygen) in the gas phase. The reason that we thought that diffusion in nitrogen and in oxygen would be very similar is that the collision cross section of the molecules should be rather similar in the two cases, since they are both diatomic molecules of similar size and mass.

ADDITIONAL REFERENCE

12. R. P. Burns, *J. Chem. Phys.* **44**: 3307 (1966).

Chapter 25

Kinetics and Mechanism of the Reaction Between TiO₂ and SrCO₃*

S. F. Hulbert and M. J. Popowich†

Department of Ceramic Engineering
Clemson University
Clemson, South Carolina

The kinetics of the reaction between strontium carbonate and titanium dioxide were studied in the temperature range 800° to 975°K in order to test the validity of existing solid-state reaction models and to determine the effect of deviations from stoichiometry on solid-state reactions. Anatase, rutile, and nonstoichiometric rutile were isothermally reacted with strontium carbonate, and reaction rate data were obtained using thermogravimetric analysis. The kinetics of reaction between strontium carbonate and stoichiometric anatase were found to be described by the Ginstling–Brounshtein equation. The activation energy for this reaction is 66.6 kcal/mole and the frequency factor is 4.5×10^9 min^{-1}. The kinetics of the reaction between strontium carbonate and both stoichiometric and nonstoichiometric rutile were found to be described by a nuclei growth rate equation based on the transport-controlled growth of spheroids of any shape (plates, spheres, etc.) from a constant number of nuclei in a matrix of composition different from that of the product. The activation energy for the reaction between strontium carbonate and stoichiometric rutile is 97.8 kcal/more and the frequency factor is 3.2×10^{15} min^{-1}. The fact that the solid-state reaction model applicable to the reaction between strontium carbonate and titania is dependent on the structure of titania is explained by considering the process of epitaxial growth. Defects were introduced into the rutile lattice by heating under vacuum and in a carbon monoxide atmosphere. As the O:Ti ratio of rutile decreased from 2.00 to 1.97, a decrease in the activation energy and frequency factor was observed. Both methods of defect preparation resulted in an equivalent effect on the kinetics of the reaction.

INTRODUCTION

The solid-state reactions between SrCO₃ and TiO₂ (anatase), TiO₂ (rutile), and nonstoichiometric TiO₂ (rutile) were examined in an attempt to gain a

---

*This work was supported by the National Aeronautics and Space Administration and is based on a thesis submitted by M. J. Popowich in partial fulfillment of the requirements for the degree of Master of Science in the Department of Ceramic Engineering, Clemson University, Clemson, South Carolina.

†Presently with E. I. DuPont de Nemours and Co. at Niagara Falls, New York.

better understanding of solid-state reactions in general. To accomplish this objective the kinetics of these reactions were studied with emphasis on:

1. The rate equation which best fits the experimental data, i.e., the fraction of reaction completed versus time.

2. The mechanism of the reaction: (a) whether the reaction is phase-boundary- or transport-controlled, (b) whether the transport mechanism is unidirectional solid-state diffusion [Wagner ([1]) diffusion, Hauffe ([2]) diffusion], vapor, etc., and (c) whether the geometry of the reaction is nucleus growth or diffusion through a continuous product layer.

3. What effect changes in the defect nature of one of the reactants has on the overall reaction.

There are many obstacles involved in making a kinetic study of even the simplest of solid-state reactions ([3]). In most solid-state reactions the impenetrable barrier to obtaining satisfactory rate data is the analysis of the products. This difficulty arises because conventional solvent and chemical separation techniques are not applicable. Another experimental difficulty arises from the fact that knowledge of the defect type and concentration must be known before any quantitative interpretation may be developed as to how the defect state affects solid-state reactions. Care must be taken at all times to keep the entire investigation within the boundary conditions set up by the methods used to investigate the data.

The specific reaction studied was chosen with these experimental difficulties in mind. The problem of determining the fraction of reaction completed was overcome by choosing a reaction in which a gaseous product is evolved. Thus by measuring the change in weight due to gas evolution, a continuous record of the per cent of reaction may be obtained as a function of time. The change in weight of the reaction sample was determined with the aid of a thermogravimetric apparatus.

KINETIC ANALYSIS OF SOLID-STATE INTERACTION IN POWDERED COMPACTS

The rate of a solid-state reaction is controlled either by the chemical combination at the reaction interface or by the transport of reactants to the reaction zone. In diffusion-controlled reactions, in systems including only plane surfaces, unidirectional diffusion processes, and constant diffusion coefficients the thickness of the product layer y is related to reaction time t by the well known parabolic rate law

$$y^2 = kDt \tag{1}$$

where k is a proportionality constant and D is the diffusion coefficient of the migrating species.

Most ceramic processes are carried out by intimately mixing fine powders. In 1927 Jander ([4]) applied the parabolic rate law developed for planer interface reactions to powdered compacts. Jander's model is based on the following assumptions:

1. The reaction under consideration can be classified as an additive reaction.

2. Nucleation, followed by surface diffusion, occurs at a temperature below that needed for bulk diffusion, so that a coherent product layer is present when bulk diffusion does occur.

3. The chemical reaction at the phase boundary is considerably faster than the transport process, and thus the solid-state reaction is bulk-diffusion controlled.

4. Bulk diffusion is unidirectional.

5. The product is not miscible with any of the reactants.

6. The reacting particles are all spheres of uniform radii.

7. The ratio of the volume of the product layer to the volume of the materials reacted is unity.

8. The increase in the thickness of the product layer follows the parabolic rate law [Eq. (1)].

9. The diffusion coefficient of the species being transported is not a function of time.

10. The activity of the reactants remains constant on both sides of the reaction interface.

Let V denote the volume of material still unreacted at time t; then

$$V = 4/3\pi(r - y)^3 \tag{2}$$

where r is the initial radius of the reacting particles. Letting x be the fraction reaction completed at time t, the volume of material unreacted is also given by

$$V = 4/3\pi r^3(1 - x) \tag{3}$$

Equations (2) and (3) can be equated to yield

$$y = r[1 - (1 - x)^{1/3}] \tag{4}$$

Combining Eq. (4) with Eq. (1) and rearranging yields

$$k_j t = \frac{2kDt}{r^2}[1 - (1 - x)^{1/3}]^2 \tag{5}$$

Equation (5) is the well-known Jander equation relating the fraction of reaction completed to time, where k_J is the rate constant. In order to determine the rate constant for an isothermal solid-state reaction, the fraction of material reacted must be determined as a function of time, and then, according to the Jander model, a plot is made of $[1 - (1 - x)^{1/3}]^2$ versus time. This plot, often referred to as a Jander analysis, should give a straight line whose slope is the rate constant k_J. If the Jander model applies to the system being studied, the rate constant should not drift as the reaction proceeds. If the rate constant does drift with time, another model must be sought. The empirical equation developed from the Jander physical model for solid-state reactions states that the rate constant is proportional to the diffusion coefficient of the species being transported and inversely proportional to the square of the radius.

It is often found that Jander's equation does not represent solid-solid reaction data, indicating that a more complicated situation actually exists.

Kroger and Ziegler ([5,6]) indicated that Jander's assumption of a constant diffusion coefficient was not applicable to all solid systems, particularly during the early stage of a reaction. Kroger and Ziegler used Jander's geometry (Jander's assumptions 1–7) and assumed that the diffusion coefficient of the transported species was inversely proportional to time. The Kroger–Ziegler equation is

$$k_{\text{KZ}} \ln t = (2k/r^2) \ln t = [1 - (1 - x)^{1/3}]^2 \tag{6}$$

Zhuravlev et al. ([7]) modified the Jander relation by assuming that the activity of the reacting substances was proportional to the fraction of unreacted material $(1 - x)$. Their relationship between fraction of reaction completed and time is

$$k_{\text{ZLT}}t = \left[\left(\frac{1}{1 - x}\right)^{1/3} - 1\right]^2 \tag{7}$$

Ginstling and Brounshtein ([8]) arrived at a model using Jander's assumptions with the exception of the parabolic rate law. They indicated that the parabolic rate law asserted that the reaction surface area remained constant; however, when they considered spherical particles this surface actually decreased in area as the reaction proceeded. They discarded the parabolic rate law in favor of an equation relating the growth of the product layer to Barrer's ([9]) equation for steady-state heat transfer through a spherical shell. The Ginstling–Brounshtein equation is

$$k_{\text{GB}}t = 2kDt/r^2 = 1 - \tfrac{2}{3}x - (1 - x)^{2/3} \tag{8}$$

Carter ([10,11]) further improved the Ginstling–Brounshtein model by accounting for differences in the volume of the product layer with respect to that of the reactants. Carter also used Barrer's equation to represent the rate of product formation and included a Z term to account for the change in volume, where Z represents the volume of the reaction product formed per unit volume of the reactant consumed:

$$k_{cv}t = \frac{2ktD}{r^2} = \frac{Z - [1 + (Z - 1)x]^{2/3} - (Z - 1)(1 - x)^{2/3}}{Z - 1} \qquad (9)$$

Valensi ([12]) earlier developed the same solid-state reaction model mathematically from a different starting point. Thus Eq. (9) is referred to as the Valensi–Carter equation.

Dunwald and Wagner ([13]) derived an equation for solid-state reaction analysis based on a solution to Fick's second law for diffusion into or out of a sphere. Serin and Ellickson ([14]) expressed the Dunwald–Wagner equation in terms of the fractional completion of the process:

$$k_{DW}t = \pi^2 Dt/r^2 = \ln[6/\pi^2(1 - x)] \qquad (10)$$

Although all the models discussed thus far are limited by the fact that they are based on the reaction of spherical particles of the same radius, they have been shown to represent many actual solid-state reactions ([15,16]). There have been attempts to introduce particle size gradation into a workable model; however, these have resulted in models which involve complicated mathematics and contain parameters that are difficult to measure. Models including particle size gradation have been advanced by Miyagi ([17]) (based on Jander's assumptions), Sasaki ([18]) (based on Carter's assumptions), and Gallagher ([19]) (based on the Dunwald–Wagner assumptions).

In the case where the reaction starts only at the contact zones between particles and the reaction proceeds by diffusion through the contact zones Jander's assumption that the surface of one component is completely and continuously covered with particles of the other component is obviously not valid. To take into account the effect of the number of contact points Komatsu ([20]) introduced into the Jander equation the mixing ratio of the two components, the ratio of the radius of the two components, and a parameter which describes the packing state of the powders.

The solid-state reaction models for powdered compacts thus far discussed have been based on the assumption that initially surface diffusion rapidly coats the surface of the reacting particle with a continuous product layer. The subsequent rate of reaction is taken to be the rate of diffusional growth of the product blanket. There is, however, another way of looking at the initial product formation and subsequent growth. This approach considers the nucleation of products at active sites and the rate at which the nucleated particles grow. According to Welch [21], such a mechanism is possible whenever the product phase is partially miscible in one of the reactants.

There is increasing interest in nucleus growth mechanisms and many mathematical models have been advanced relating nucleation and nuclei growth rates to the kinetics of solid-state reactions [22-28]. The general form of the kinetic equations for nuclei growth models is

$$\ln(1 - x) = -kt^m \tag{11}$$

where m is a parameter which is a function of reaction mechanism, the number of nuclei present, the composition of parent and product phase, and the geometry of the nuclei. Christian [28] has summarized the values of m which may be obtained for various boundary conditions. If a solid-state reaction can be represented by a nuclei growth model, according to Eq. (11) a plot of $\ln\{\ln[1/(1 - x)]\}$ versus $\ln t$ (nuclei growth analysis) should yield a straight line with slope m and intercept $m \ln k$.

When diffusion through the product layer is so rapid that the reactants cannot combine fast enough at the reaction interface to establish equilibrium the solid-state reaction is phase-boundary controlled. The product layer is discontinuous when the molar volume of the product phase is considerably different from that of the reactant upon which it is growing. According to Laidler [29], when a discontinuous product phase occurs the rate-determining step may be the chemical process occurring at the phase boundary. Under these circumstances the rate is determined by the available interface area, and such processes are referred to as topochemical.

Equations relating x and t have been derived for simple geometrical systems assuming (1) the reaction rate is phase-boundary controlled, (2) the reaction rate is proportional to the surface area of the fraction of unreacted material, and (3) the nucleation step occurs virtually instantaneously so that the surface of each particle is covered with a layer of product. The models developed from the foregoing boundary conditions are termed phase boundary or contracting volume kinetic models. For a sphere reacting from the surface inward [29] the relationship between x and t is

$$k_{PB}t = (u/r)^t = 1 - (1 - x)^{1/3} \tag{12}$$

and for a circular disk reacting from the edge inward or for a cylinder [30]

$$k_{PB}t = (u/r)^t = 1 - (1 - x)^{1/2} \tag{13}$$

where u is the velocity at which the interface advances into the particle. Note that rate constants for phase boundary kinetic models are inversely proportional to the radius, whereas for transport-controlled kinetic models they are inversely porportional to the square of the radius.

Equations analogous to classical rate equations have often been applied to solid-state reactions. The integrated form of the general kinetic equation based on the concept of an order of reaction is

$$kt = \frac{1}{n-1}\left[\frac{1}{(1-x)^{n-1}} - 1\right] \tag{14}$$

where n is the so-called order of the reaction. For certain values of n Eq. (14) leads to some of the equations based on physical models. When $n = \frac{2}{3}$ Eq. (14) is identical to Eq. (12). When $n = \frac{1}{2}$ Eq. (14) is identical to Eq. (13). When the rate-determining step is the nucleation process and when there is equal probability of nucleation at each active site one obtains by analogy with radioactive decay a kinetic equation of the first order ([31]). Whenever the rate of reaction is proportional to the volume of unreacted material present it is, according to the nomenclature of classical reaction kinetics, a first-order reaction. At present values of n other than $\frac{1}{2}$, $\frac{2}{3}$, and 1 lead to equations with no obvious physical significance.

EXPERIMENTAL PROCEDURES

The chemical reactions listed in Table I were studied isothermally in air using thermogravimetric analysis to monitor the per cent reacted versus time. The thermogravimetric analysis apparatus consisted of a Kanthal furnace, a temperature controller capable of maintaining a constant temperature $\pm 3°C$, a temperature recorder, and an Ainsworth recording balance. From the recorded weight loss curves the fraction of reaction completed (x for $SrCO_3 + TiO_2 \rightarrow SrTiO_3 + CO_2 \uparrow$) was calculated as the ratio of the weight loss at time t to the theoretical total weight loss.

After determining the weight-loss curves with the thermobalance the phases formed were determined by conventional X-ray diffraction techniques using Cu $K\alpha$ radiation.

The raw materials used were Baker-analyzed $SrCO_3$ and TiO_2 (anatase). The particle size range of $SrCO_3$ was 1–60 μ with an average size of 15 μ and the particle size range of TiO_2 was 5–90 μ with an average size of 20 μ.

Anatase was converted to rutile by heating at 1200°C in a platinum crucible for 2 hr ([32]). X-ray analysis showed complete conversion to rutile. The rutile was ground in a mortar and pestle to obtain the same average particle size as the original material.

Nonstoichiometric rutile was prepared by heating rutile in a platinum crucible at a pressure of 5×10^{-5} mmHg at temperatures ranging from

1000°C to 1100°C and times ranging from 1 to 10 hr. The foregoing heat treatments resulted in oxygen to titanium ratios ranging from $TiO_{1.995}$ to $TiO_{1.970}$. Nonstoichiometric rutile was also prepared by heating rutile in a carbon monoxide atmosphere at 1000°C. Several nonstoichoimetric rutile samples were heated in air at 1000°C for 2 hr. No weight gain was observed, indicating that the rutile does not reoxidize in air at this temperature. Each nonstoichiometric rutile was ground in a mortar and pestle to obtain the same average particle size as the original material.

Equimolecular mixtures of the reactants were throughly mixed in a mortar and pestle. The mixture was then cold pressed into 0.7 pellets using a $\frac{3}{8}$in. die and 5000 psi pressure.

PRESENTATION AND DISCUSSION OF RESULTS

X-ray patterns of the products formed by the reaction between strontium carbonate and anatase, rutile, and nonstoichiometric rutile each showed $SrTiO_3$ to be the only phase present. X-ray patterns of partially reacted compacts revealed the presence of strontium titanate, strontium carbonate, and titania. The absence of strontium oxide indicates that the carbon dioxide evolution was associated with the reaction between $SrCO_3$ and TiO_2 to form a product phase and not with the decomposition of $SrCO_3$ to SrO. The data of Wanmaker and Radielovic ([33]) for the rate of dissociation of strontium carbonate indicates that interference from dissociation in measuring the kinetics of interaction between $SrCO_3$ and TiO_2 is negligible below 900°C. Above this temperature a correction may be necessary for the dissociation of $SrCO_3$. However the fact that the same kinetic model can be used to represent the data above and below this temperature suggests that the correction term is small.

The fraction reaction completed versus time curves at 900°C for the first two reactions listed in Table I are given in Fig. 1. Fraction of reaction completed versus time curves for the reactions between strontium carbonate and nonstoichiometric titania are not included because the distribution of these curves with respect to the strontium carbonate–stoichiometric rutile reaction curve is highly dependent on the temperature chosen.

In order to determine whether one of the nuclei growth equations fit the data, a plot of $\ln\{\ln[1/(1 - x)]\}$ versus $\ln t$ was made for each set of data. In every case the nucleus growth analysis produced a nonlinear relationship for the reaction between strontium carbonate and anatase (see Fig. 2), indicating that nuclei growth models are invalid for analyzing this solid-state reaction. The nuclei growth analysis for the reaction between rutile and strontium carbonate shows that this reaction can be broken into three distinct regions (see Fig. 3), with each region represented by a different nuclei

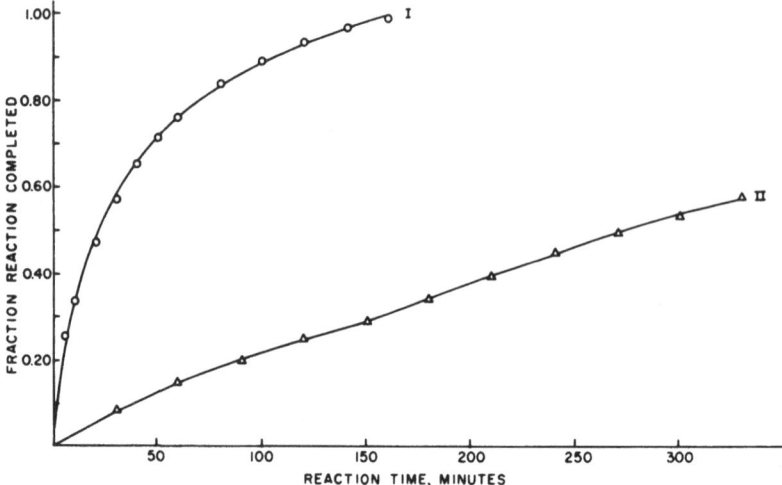

Fig. 1. Isothermal reaction curves (900°C). I. The reaction between anatase and strontium carbonate. II. The reaction between rutile and strontium carbonate.

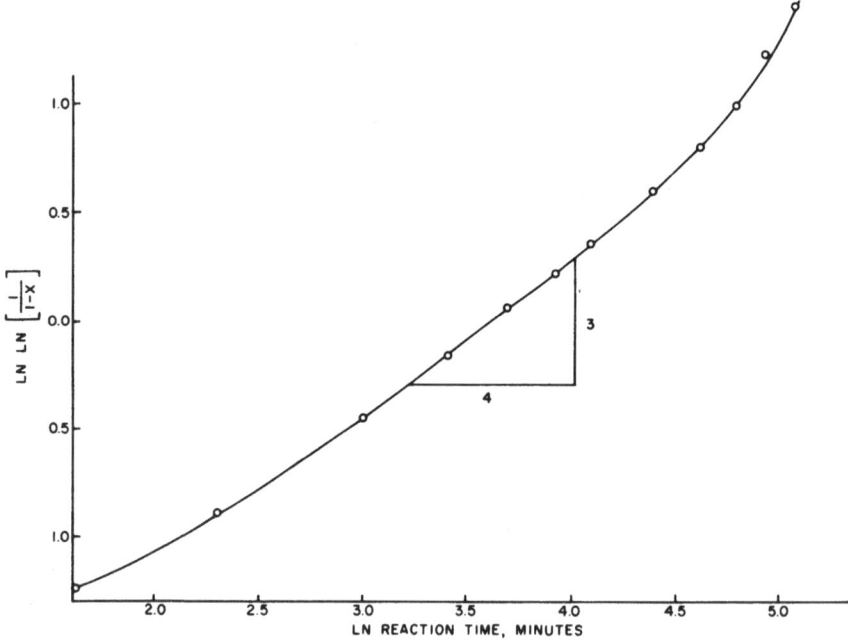

Fig. 2. Nucleus growth analysis of the 900°C isothermal reaction between anatase and strontium carbonate.

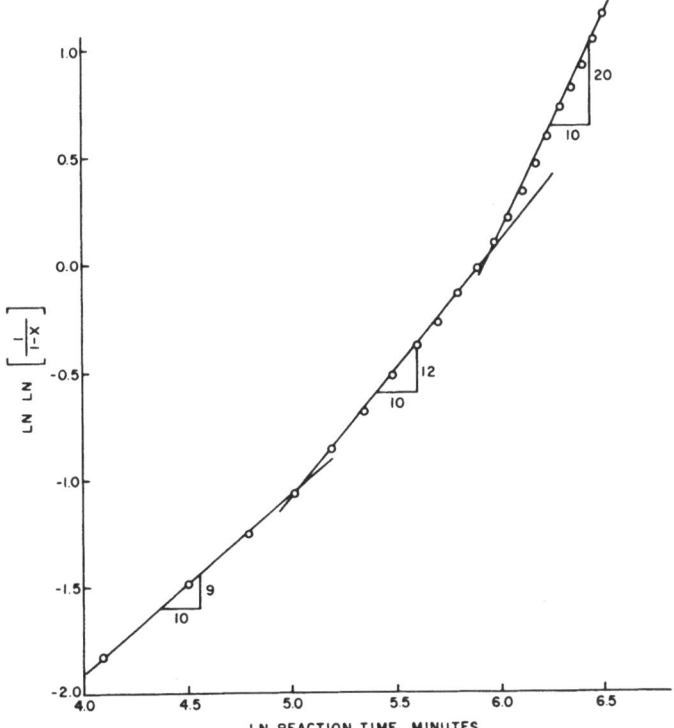

Fig. 3. Nucleus growth analysis of the 900°C isothermal reaction between rutile and strontium carbonate.

growth model. The slope of the nucleus growth analysis plots has an average value of approximately 1.0 in region one, 1.25 in region two, and 2.0 in region three. The first region represents from 0 to 20–30% of completion, the second region from 20–30 to 70–80% of completion, and the third region from 70–80 to 100% of completion.

The kinetic data for each set of isothermal reactions listed in Table I were substituted into the Jander, Kroger–Ziegler, Zhuravlev–Leskhin–Tempel'man, Ginstling–Brounshtein, Dunwald–Wagner, contracting-sphere, contracting-cylinder, nucleus-growth (with $m = \frac{3}{4}, \frac{5}{4}$, and $\frac{3}{2}$), and classical first-order reaction rate equations and rate constants were calculated. The rate constants were then plotted as functions of isothermal reaction time (see Figs. 4–7). In order for a rate equation to represent a reaction, the reaction rate constant should be independent of the reaction time under isothermal conditions.

The Carter–Valensi equation was not used in analyzing the data, due to lack of high-temperature density for computing the Z value. Giess [15] has shown that the Carter–Valensi correction becomes significant only when the ratio of product volume to reactant volume exceeds a value of 2. In all of the reactions listed in Table I using room temperature density data a ratio of less than two is calculated.

The Miyagi, Sasaki, and Gallagher models were not considered, due to the difficulty of applying them to experimental data. Attempts were made to control the particle size to conform to the boundary conditions of the simpler diffusion models. Rate constants for the reaction between anatase

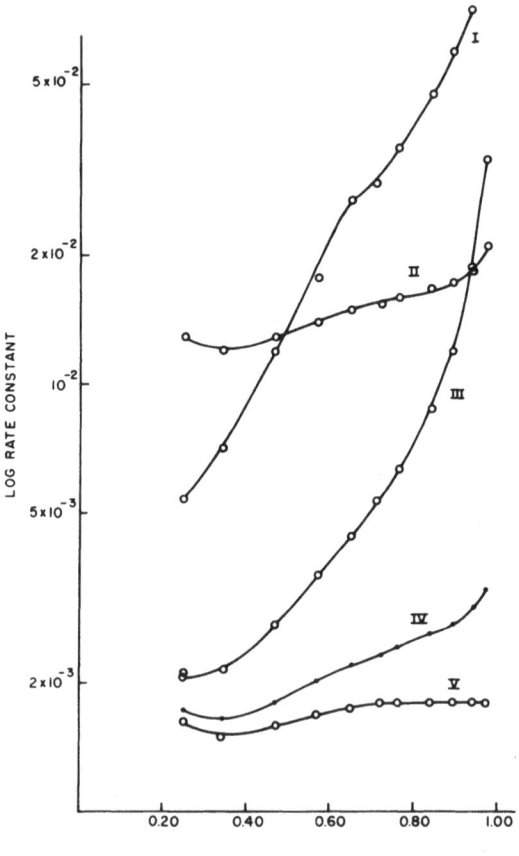

FRACTION REACTION COMPLETED

Fig. 4. Analysis of rate constants for the 900°C isothermal reaction between anatase and strontium carbonate. I. Kroger–Ziegler; II. Dunwald–Wagner; III. Zhuravlev–Lesokhin–Tempel'man; IV. Jander; V. Ginstling–Brounshtein.

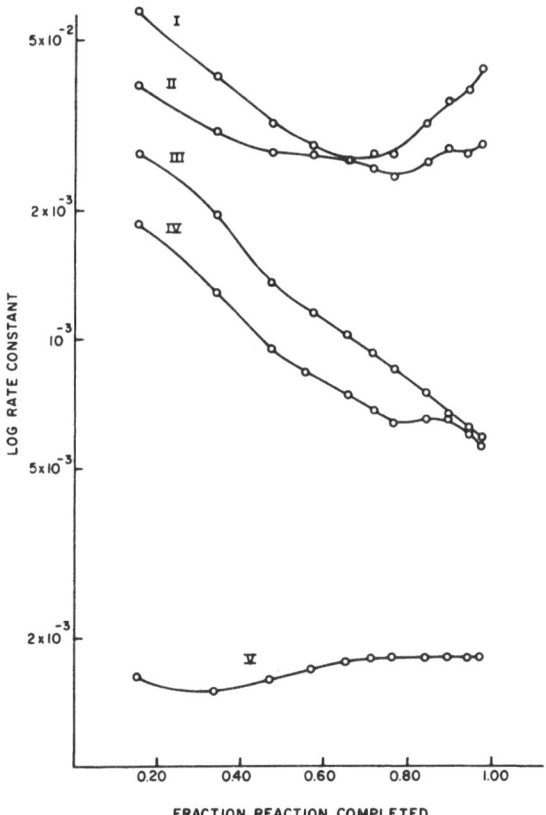

Fig. 5. Analysis of rate constants for the 900°C isothermal reaction between anatase and strontium carbonate. I. Classical first order; II. nucleus growth $kt = \{\ln [1/(1 - x)]\}^{4/3}$; III. contracting cylinder; IV. contracting sphere; V. Ginstling–Brounshtein.

and strontium carbonate calculated using the Ginstling–Brounshtein equation yielded an approximately horizontal plot of rate constant versus time. Rate constants calculated using the other models showed a pronounced drift with time (see Figs. 4 and 5).

The applicability of the Ginstling–Brounshtein rate equation to this reaction indicates that the mechanism of the reaction consists of the formation of a coherent product layer around one of the reactants with unidirectional diffusion of the other reactant as the rate-controlling step. To determine the mobile species in the reaction an inert marker study was made ([34]). Disks of the reactant materials were pressed and platinum markers placed in the anatase so that they were flush with the surface of the disk. Strontium

carbonate and anatase were placed face to face and reacted. The product
was observed to form only on the anatase side of the markers. This indicates
that the formation of the product layer is around the anatase particles and
that the reaction takes place by the unidirectional motion of SrO. No direct
experimental evidence is available to account for the manner in which the
SrO is transported during the reaction. Due to the relative ionic size, the
diffusion of oxygen rather than strontium is considered the rate-controlling
step.

Rate constants for the reaction between rutile and strontium carbonate
calculated using both the contracting cylinder equation and the nucleus
growth rate equation with $m = \frac{5}{4}$ yielded nearly horizontal plots of rate

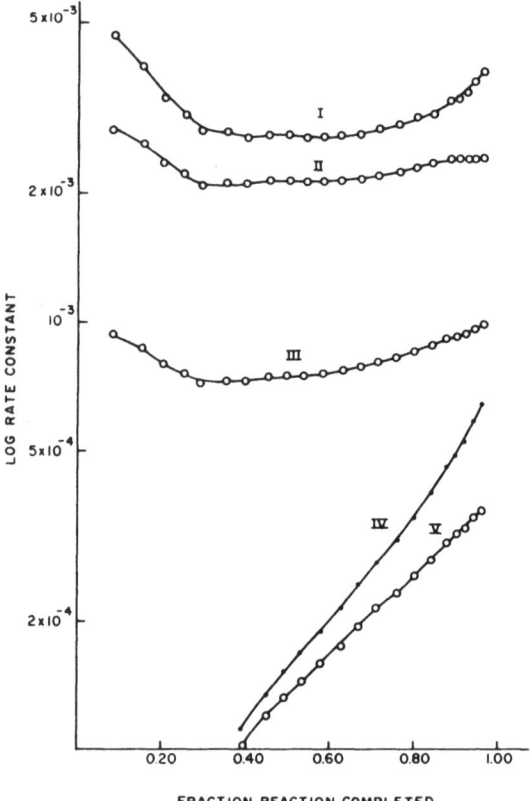

Fig. 6. Analysis of rate constants for the 900°C isothermal reaction between rutile and strontium carbonate. I. Nucleus growth $kt = \{\ln [1/(1 - x)]\}^{4/5}$; II.
contracting cylinder; III. contracting sphere; IV: Jander;
V. Ginstling–Brounshtein.

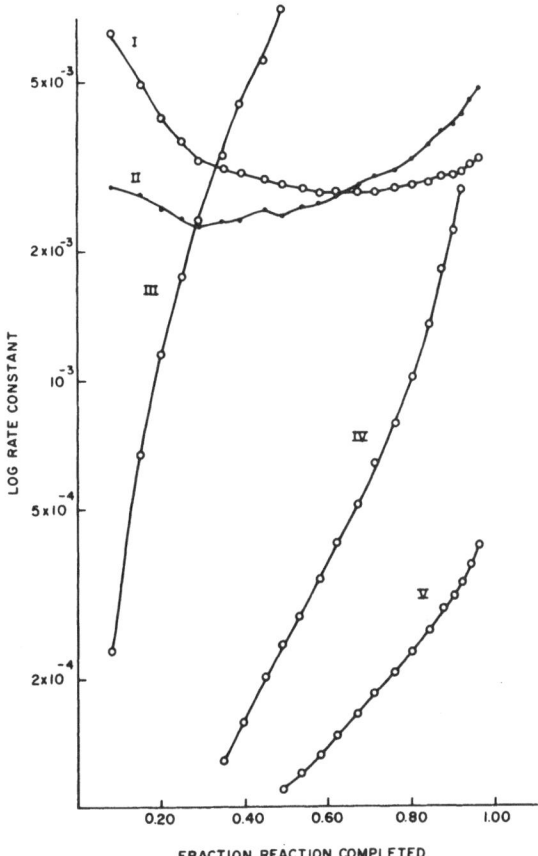

Fig. 7. Analysis of rate constants for the 900°C isothermal reaction between rutile and strontium carbonate. I. Nucleus growth $kt = \{\ln [1/(1 - x)]\}^{2/3}$; II.
classical first order; III. Kroger–Ziegler; IV. Zhuravlev–
Lesokhin–Tempel'man; V. Dunwald–Wagner.

constant versus time. Rate constants calculated using the other models
showed a pronounced drift with time (see Figs. 6 and 7). Inert marker experiments using rutile and strontium carbonate again showed that products
formed only on the titania side of the markers, showing that the reaction
takes place by unidirectional motion of SrO.

The Ginstling–Brounshtein model was used in constructing an Arrhenius
plot for the reaction between strontium carbonate and anatase. Arrhenius
plots were constructed for the reaction between strontium carbonate and
rutile (see Fig. 8) using both the contracting cylinder model and the nucleus
growth model with $m = \frac{5}{4}$. Activation energies calculated from the two reac-

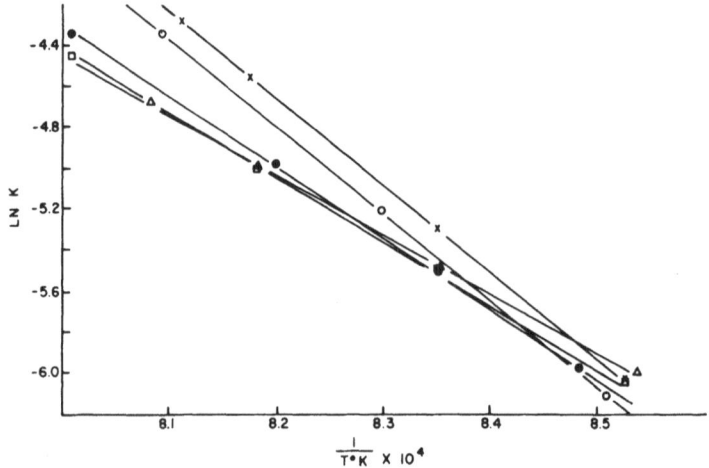

Fig. 8. Arrhenius analysis of nucleus growth rate constants ($m = \frac{5}{4}$) for reaction between rutile and strontium carbonate. $\times$: $TiO_2 + SrCO_3$; $\bigcirc$: $TiO_{1.995} + SrCO_3$; $\bullet$: $TiO_{1.983} + SrCO_3$; $\square$: $TiO_{1.975} + SrCO_3$; $\triangle$: $TiO_{1.970} + SrCO_3$.

tion rate equations were within experimental error. The experimental activation energies and frequency factors for all the reactions studied are listed in Table I.

The observed change in activation energy with stoichiometry suggests that the phase boundary model is untenable for the reaction between strontium carbonate and rutile. If the reaction were phase-boundary controlled in the case of stoichiometric rutile, then when the activation energy for the

TABLE I

Reaction	Activation energy (kcal/mole)	Frequency factor (min⁻¹)
$SrCO_3 + TiO_2$ *	66.6	4.5×10^9
$SrCO_3 + TiO_2$ †	97.8	3.2×10^{15}
$SrCO_3 + TiO_{1.995}$ **	86.1	2.1×10^{13}
$SrCO_3 + TiO_{1.983}$ **	72.1	5.8×10^{10}
$SrCO_3 + TiO_{1.975}$ **	67.2	7.3×10^9
$SrCO_3 + TiO_{1.970}$ **	61.4	6.8×10^8
$SrCO_3 + TiO_{1.975}$ ‡	65.5	6.01×10^9

*Anatase.
†Rutile.
**Rutile reduced under vacuum.
‡Rutile reduced in CO.

chemical reaction becomes lower than that for diffusion through the product layer one would expect in the case of $TiO_{1.970}$, that a rate equation based on a diffusion mechanism would apply. However, the reaction between $SrCO_3$ and $TiO_{1.970}$ could not be represented by any of the solid-state reaction models based on bulk diffusion through a continuous product layer. Only the contracting cylinder model and the nucleus growth model with $m = \frac{5}{4}$ were found appropriate for expressing and correlating the data for all the reactions studied between strontium carbonate and nonstoichiometric rutile.

A nuclei growth equation with m equal to 1.25 is based on the transport-controlled growth of spheriods of any shape (plates, needles, spheres) from a constant number of nuclei in a matrix of different composition ([28]). The assumption of a constant number of nuclei or, alternatively, that all nuclei are present at the beginning of the reaction, is not as restrictive as it first appears. Since the volume nucleation rate I is so sensitive to the degree of supersaturation, a very small amount of continuous precipitation will change the mean concentration of solute in the untransformed matrix by an amount sufficient to decrease I by one or more orders of magnitude. Results obtained for the kinetics of continuous precipitation can generally be interpreted only by assuming that all nuclei are present at the beginning of the reaction. Examination of the boundary conditions used in deriving the nucleus growth equation ($m = 1.25$) as well as the observed activation energies and frequency factors gives no evidence which would restrict the use of this equation in representing the reaction between strontium carbonate and rutile.

It has been seen that two different reaction models are applicable to the reaction between strontium carbonate and titania, depending on whether titania is present in the form of anatase or rutile. The change in mechanism may be explained by considering the process of epitaxial growth. The strontium titanate product has a cubic structure with $a = 3.92$ Å. The anatase structure is tetragonal with $a = 3.73$ Å and $c = 9.37$ Å. Rutile is tetragonal with $a = 4.58$ Å and $c = 2.95$ Å. It can be seen that in anatase the conditions are favorable for epitaxial growth, since its (001) plane has dimensions similar to the (001) plane of strontium titanate. In this case the product layer may grow on the "seed" crystals of anatase and bulk diffusion will control the rate of the reaction. In the case of rutile apparently none of the crystal planes have dimensions similar to a corresponding plane in strontium titanate. The product phase must grow from individual nuclei and a nuclei growth model will be applicable.

It is seen in Table I that the effects of deviation from stoichiometry on the kinetics of the reaction between rutile and strontium carbonate are very pronounced. On the basis of the applicability of the nucleus growth models it appears that increasing the defect concentration lowers the frequency factor as well as the activation energy for diffusion of the rate-controlling species in rutile. There are several possible explanations for these effects. A high vibrational frequency is associated with larger effective force constants between atoms in the lattice. A defect causes a local disturbance and this will have an effect on the vibrational frequencies V of atoms in the vicinity of the defect and in general will result in a reduction of V. One could thus expect a slight lowering of the frequency factor with increased defect concentration. However, the magnitude of the effect of defect concentration on the frequency factor suggests that the opportunity for diffusion of the rate-controlling species is being drastically curtailed. The presence of titanium interstitials would lower the frequency factor for the diffusion of strontium ions or oxygen ions through a "structure stuffing" effect. It is difficult to account for the lowering of the frequency factor on the basis that the imperfections in nonstoichiometric rutile are oxygen vacancies.

The assumption of either titanium interstitials or oxygen vacancies offers an explanation for the lowering of activation energy for diffusion of strontium ions because both would reduce the number of Sr-O bonds which must be broken. If one assumes the rate-controlling step is the diffusion of oxygen ions, then the presence of titanium interstitials should increase the activation energy, while oxygen vacancies should reduce the activation energy. Thus the rate-controlling step in the reaction between rutile and strontium carbonate is thought to be the diffusion of strontium ions in rutile.

Shannon and Pask ([32,35]) have suggested that two types of defects may predominate in nonstoichiometric titania, depending on the method of preparation. They propose that titania reduced in a vacuum contains interstitial Ti^{+3} ions, while titania reduced in a hydrogen atmosphere contains oxygen vacancies. A nonstoichiometric rutile sample was prepared in a CO atmosphere. The defect rutile was reacted with strontium carbonate and the activation energy and frequency factor were calculated. This sample had an activation energy of 65.5 kcal/mole and a frequency factor of 6.0×10^9, as compared with values of 67.2 kcal/mole and 7.2×10^9 for defect rutile of the same stoichiometry prepared under vacuum. These values are within experimental error. This would indicate either that both types of defect show an equal effect on the kinetics of the reaction or else that the same type of defect is created in both types of preparation of the defect rutile.

The results of this investigation support data obtained by internal friction ([36,37]) and oxidation studies ([38]) indicating that the formation of titanium interstitials is the mechanism through which nonstoichiometric rutile is produced.

CONCLUSIONS

1. The kinetics of the reaction between strontium carbonate and anatase were found to be described by the Ginstling–Brounshtein equation.

The activation energy for the reaction in the temperature range 800–1000°C is 66.6 kcal/mole and the frequency factor is 4.5×10^9 min^{-1}.

2. The kinetics of the reaction between strontium carbonate and stoichiometric rutile were found to be described by the $kt = [-\ln(1 - x)]^{4/5}$ nuclei growth equation. The activation energy for this reaction is 97.8 kcal/mole and the frequency factor is 3.2×10^{15}.

3. The fact that the solid-state reaction model applicable to the reaction between strontium carbonate and titania is dependent on the structure of titania was explained using considerations based on the process of epitaxial growth.

4. The introduction of defects in the form of deviations from stoichiometry in the rutile structure results in a decrease of activation energy and frequency factor of the reaction.

5. The activation energy and frequency factor of the reaction between $SrCO_3$ and nonstoichiometric rutile prepared under vacuum are approximately equal to the activation energy and frequency factor of the reaction between $SrCO_3$ and nonstoichiometric rutile prepared by heating in a carbon monoxide atmosphere. This indicates that either both methods of preparation resulted in the same type of defects, either interstitial cations or anion vacancies, or that both types of defects show an equal effect on the kinetics of the reaction.

6. The results of this investigation suggests that the formation of titanium interstitials is the mechanism through which nonstoichiometric rutile is produced.

ACKNOWLEDGMENTS

The writers thank G. D. Mackenzie of the General Refractories Research Laboratories for the particle size determination.

REFERENCES

1. C. Wagner, *Z. Physik. Chem.* **B24**: 309 (1936).
2. K. Hauffe, *Reaktionen in und an fester Stoffen*, Springer Verlag, Berlin, 1955, p. 582.
3. W. E. Brownell, "Kinetics of Reactions between Oxides in the Solid State," Ph.D. Dissertation, Pennsylvania State University Pennsylvania, 1953.
4. W. Jander, "Reactions in Solid State at High Temperatures, I," *Z. Anorg. Allgem. Chem.* **163**(1–2): 1–30 (1927).
5. C. Kroger and G. Ziegler, "Reaction Rates of Glass Batch Melting, II," *Glastech. Ber.* **26**(11): 346–53 (1953).
6. C. Kroger and G. Ziegler, "Reaction Rates of Glass Batch Melting, III," *Glastech. Ber.* **27**(6): 199–212 (1954).
7. V. F. Zhuravlev, I. G. Lesokhin, and R. G. Tempel'man, "Kinetics of the Reactions

for the Formation of Aluminates and the Role of Mineralizers in the Process," *J. Appl. Chem. USSR* **21**(9): 887–902 (1948).

8. A. M. Ginstling and B. I. Brounshtein, "Concerning the Diffusion Kinetics of Reactions in Spherical Particles," *J. Appl. Chem. USSR* **23**: 1327–38 (1950).

9. R. M. Barrer, "Diffusion in Spherical Shells, and a New Method of Measuring the Thermal Diffusivity Constant," *Phil. Mag.* **35**(12):802–11 (1944).

10. R. E. Carter, "Kinetic Model for Solid State Reactions," *J. Chem. Phys.* **34**(6): 2010–15 (1961).

11. R. E. Carter, "Addendum: Kinetic Model for Solid State Reactions," *J. Chem. Phys.* **35**(3): 1137–38 (1961).

12. G. Valensi, "Kinetics of the Oxidation of Metallic Spherules and Powders," *Compt. Rend.* **202**(4): 309–12 (1936).

13. H. Dunwald and C. Wagner, "Measurement of Diffusion Rate in the Process of Dissolving Gases in Solid Phases," *Z. Physik. Chem.* (*Leipzig*) **B24**(1): 53–58 (1934).

14. B. Serin and R. T. Ellickson, "Determination of Diffusion Coefficients," *J. Chem. Phys.* **9**: 742–47 (1941).

15. E. A. Giess, "Equations and Tables for Analyzing Solid State Reaction Kinetics," *J. Am. Ceram. Soc.* **46**(8): 374–76 (1963).

16. D. L. Branson, "Kinetics of Zinc Aluminate Spinel Formation," Ph.D. Dissertation, Ohio State University, Columbus, Ohio, 1964.

17. S. Miyagi, "A Criticism on Jander's Equation of Reaction Rate Considering the Statistical Distribution of Particle Size of Reacting Substance," *J. Japan. Ceram. Soc.* **59**: 132–35 (1951).

18. H. Sasaki, "Introduction of Particle Size Distribution into Kinetics of Solid State Reactions," *J. Am. Ceram. Soc.* **47**(10): 512–16 (1964).

19. K. J. Gallagher, "The Effect of Particle Size Distribution on the Kinetics of Diffusion Reactions in Powders," in: *Reactivity of Solids*, (G. M. Schwab, ed.) Am. Elsevier, New York, 1965, pp. 192–203.

20. W. Komatsu, "The Kinetic Equation of the Solid State Reaction: The Effect of Particle Size and Mixing Ratio on the Reaction Rate in a Mixed Powder System," in: *Reactivity of Solids* (G. M. Schwab, ed.), Am. Elsevier, New York, 1965, pp. 182–191.

21. A. F. E. Welch, "Solid–Solid Reactions," in: *Solid State Chemistry* (W. E. Garner, ed.), Butterworth and Co., London, 1953, pp. 297–310.

22. P. W. M. Jacobs and F. C. Tompkins, "Classifications and Theory of Solid Reactions," in: *Solid State Chemistry* (W. E. Garner, ed.), Butterworth and Co., London, 1953, pp. 297–310.

23. M. E. Fine, *Phase Transformations in Condensed Systems*, Macmillan Co., New York, 1964, pp. 47–78.

24. M. Avrami, "Kinetics of Phase Change," *J. Chem. Phys.* **7**: 1103–12 (1939).

25. M. Avrami, *J. Chem. Phys.* **8**: 212–24 (1940).

26. M. Avrami, *J. Chem. Phys.* **9**: 177–84 (1941).

27. B. V. Erofe'ev, "Generalized Equations of Chemical Kinetics and Its Application in Reactions Involving Solids," *Compt. Rend. Acad. Sci. URSS* **52**: 511–14 (1946).

28. J. W. Christian, *The Theory of Transformations in Metals and Alloys*, Pergamon Press, New York, 1965, pp. 471–495.

29. K. J. Laidler, *Chemical Kinetics*, McGraw-Hill, New York, 1965, pp: 316–318.

30. J. H. Sharp, G. W. Brindley, and B. N. Narahari Achar, "Numerical Data for Some Commonly Used Solid State Reaction Equations," *J. Am. Ceram. Soc.* **49**(7): 379–82 (1966).

31. A. Bielanski, J. Nedoma, W. Turowa, "Studies on the Polymorphic Transformations

of Sodium Fluoroberyllate," in: *Reactivity of Solids*, (G. M. Schwab, ed.) Am. Elsevier, New York, 1965, pp. 90–99.

32. R. D. Shannon and J. A. Pask, "Kinetics of the Anatase–Rutile Transformation," *J. Am. Ceram. Soc.* **48**(8): 391–97 (1965).

33. W. L. Wanmaker and R. Radielovic, "The Dependence of the Rate of the Dissociation of Strontium Carbonate in Some Mixtures on Particle Size and Composition," in: *Reactivity of Solids*, (G. M. Schwab, ed.), Am. Elsevier, New York, 1965, pp. 529–539.

34. L. C. Correa da Silva and R. F. Mehl, "Interface and Marker Movements in Diffusion in Solid Solutions of Metals," *Trans. AIME* **191**(2): 155–73 (1951).

35. R. D. Shannon, "Phase Transformation Studies in TiO_2 Supporting Different Defect Mechanisms in Vacuum-Reduced and Hydrogen-Reduced Rutile," *J. Appl. Phys.* **35**(11): 3414–16 (1964).

36. R. D. Carahan and J. O. Brittain, "Point-Defect Relaxation in Rutile Single Crystals," *J. Appl. Phys.* **34**(10): 3095–3104 (1963).

37. J. B. Wachtman, Jr. and L. R. Doyle, "Internal Friction in Rutile Containing Point Defects," *Phys. Rev.* **135**: A276 (1964).

38. P. Kofstad, K. Hauffe, and H. Kjollesdal, "Investigation on the Oxidation Mechanism of Titanium," *Acta. Chem. Scand.* **12**: 239–266 (1958).

DISCUSSION

J. E. Burke (G. E. Research Laboratory): You mentioned in the beginning that you tested the ability of TiO_2 to oxidize in air at a variety of temperatures. Did you imply by that you then ran the strontium carbonate–TiO_2 reactions in air? Is it then assumed that the stoichiometry is unchanged during the reaction? It seems surprising to me that you have enough mobility to cause a reaction or diffusion of components and still not to change the stoichiometry of the material.

Answer: The kinetics of the reaction between strontium carbonate and titanium dioxide were observed in air in the temperature range 800–975°C. The assumption was made that the stoichiometry of the unreacted fraction of titania remained constant during the course of the reaction. Inert marker studies showed that the product phase formed on the titania side of the markers. This indicates that reactions take place by the unidirectional motion of strontium ions. The defect character appeared to be quenched in the nonstoichrometric rutile samples. It was observed that reheating a nonstoichiometric rutile sample which had been quenched to room temperature and exposed to the atmosphere to the temperature where it was originally reduced caused very little change in weight. To cause a detectable change in defect character it was necessary to heat at a temperature several hundred degrees higher than the original reducing temperature. The work of Haul and Dumbgen ([39]) and Dominik and MacCrone ([40]) may offer a possible explanation. Haul and Dumbgen have shown from self-diffusion coefficient measurements that the oxygen vacancy concentration in their specimen was constant and independent of the oxygen partial pressure at temperatures $\approx 1000°C$, i.e., independent of the state of reduction, which occurred presumably by the formation of Ti^{+3} interstitials. Dominik and MacCrone, explaining the decrease in peak height (low-temperature dielectric loss tangent of reduced rutile) with increasing impurity concentration, assumed that as the concentration of trivalent impurities is increased, the oxygen vacancies become progressively more and more associated with two trivalent impurities. It appears that the titania we were working with had sufficient impurities to stabilize the oxygen vacancies at 1000°C.

M. R. Montierth (Corning Glass Works): You quoted Komatsu in your paper and I think it is interesting to note that he gave three basic criteria for a meaningful powder compact analysis. The first is a small-sized fraction for a given particle, the second is that you have a dilute solution with respect to one of the reactants, and the third is that you get a continuous contact in cells with one reactant around large particles of the other reactant. [I notice in your work that you mixed equal ratios of $SrCO_3$ and TiO_2 and that the difference in particle size was only a factor of 2.] I was wondering whether you would like to comment on the validity of your results in view of this.

Answer: Strontium titanate is usually prepared industrially by reacting equimolecular mixtures of fine powders of $SrCO_3$ and TiO_2, and therefore this seemed a reasonable place to begin the investigation on the kinetics of interaction between $SrCO_3$ and TiO_2. Komatsu [20] pointed out that all models based on the Jander geometry assumed that the surface of the component in which the reaction takes place (Phase A) is completely and continuously covered with particles of the other component (Phase B), as though the former particles were immersed in a melt of the latter. This assumption is approximately true when the radius of A/B is very large and the amount of component B is greatly in excess of component A. Thus one would not expect an exact fit between our data and the Ginstling–Brounshtein model in the case of the reaction between strontium carbonate and anatase (see Figs. 4 and 5). However, in the case of the reaction between strontium carbonate and rutile the Ginstling–Brounshtein rate constant drifts over an order of magnitude, indicating that this model is not appropriate for expressing and correlating the data. I doubt in this case that changing the boundary conditions to more closely correspond to the Jander geometry would alter the situation significantly. Dr. Montierth's point is an important one, and in future kinetic studies, after using equimolecular mixtures to obtain a preliminary indication of the mechanism of the reaction we will use boundary conditions which more closely represent the Jander geometry in evaluating solid-state reaction models based on diffusion through a coherent product layer.

R. S. Gordon (University of Utah): How reproducible were your results? The reason that I ask is that especially in the actual stages of the reaction the number of point contacts between reactants might be quite important. But when you have equal-size particles, or roughly so, and an equimolecular mixture you are going to have a number of light particle contacts. No matter how you try to reproduce your conditions, by compression or any other means, I don't think you will be able to say that you actually have this. Only if you go to a situation where you are depending on having very large particles in a very small matrix can you reach the proper approximation.

Answer: Reactions of the type Solid A + Solid B ⟶ Solid C + Gas are highly dependent on processing parameters. In order to obtain reproducible results, all samples had to have the same mass, geometrical configuration-and processing pressure. I agree that in solid-state kinetic studies involving powdered compacts that one should use the criteria set down by Komatsu [20].

J. B. Wachtman, Jr. (National Bureau of Standards): Single crystals of TiO_2 and $SrTiO_3$ are frequently oxidized in flowing oxygen at $800°C$, so that it is surprising to me as well that you were not able to reoxidize your powder.

Have you checked to see whether you are within the one-phase region for all the compositions? It appears you might be outside the remnant as determined by Blumenthal and Whitmore. This would have a bearing on your extreme reactions in the region of the annealing stage. The second question is—how did you determine the stiochiometry of the original material? Did you depend on trying to get TiO_2 and then measuring the deviations from stoichiometry?

Answer: I don't mean to say that nonstoichiometric rutile will not oxidize in air at $1000°C$. However, for the material employed in our investigation of the rate of oxidation of nonstoichiometric rutile is several orders of magnitude lower than the rate of interaction of nonstoichiometric rutile with strontium carbonate. The reported stoichiometry of rutile after reduction in a vacuum was based on the assumption that the original material was TiO_2.

P. G. Morgan (Franklin Institute): There seemed to be a temperature rise where your strontium carbonate itself decomposes. In other words, you superimposed some of the complexities of decomposition of strontium carbonate onto further reaction with titanium dioxide, and your supply of strontium ions and oxygen ions is continuously changing as decomposition proceeds. I think you could do one of two things to simplify this. You could either carry out the reaction between the carbonate and oxide in a temperature regime where the carbonate itself would not otherwise decompose, or you could just start with strontium oxide and titanium oxide to simplify things.

Answer: In most solid-state reactions the nearly impenetrable barrier to obtaining satisfactory rate data is the quantitative analysis of the product phase. The kinetics of a solid-state reaction which results in the formation of a gaseous product can easily be studied by monitoring the rate of evolution of the gaseous product using thermogravimetric analysis. However, in indirect measurements of the rate of formation of the product phase, as by change in weight due to gas evolution, there always exists the question of whether or not the measurement actually represents the rate of the reaction being studied or if it is rather the rate of decomposition of one substance in the presence of a second phase. X-ray patterns of the product from the reaction between $SrCO_3$ and TiO_2 showed $SrTiO_3$ to be the only phase present. X-ray patterns of partially reacted compacts revealed the presence of $SrTiO_3$, TiO_2, and $SrCO_3$. Furthermore, a sudden decrease in furnace temperature of a partially reacted compact would lead to weight gain if decomposition had occurred. In the temperature range for which the kinetic data is reported, 800–975°C, no weight gain was observed with sudden reduction in furnace temperature. Reactions at temperatures deliberately high enough to independently decompose strontium carbonate (1150°C) gave weight gains with sudden temperature reduction into the stability range of strontium carbonate. Thus over the temperature interval studied it is believed that the evolution of CO_2 is associated with the reaction between $SrCO_3$ and TiO_2 to form $SrTiO_3$ and not with the decomposition of $SrCO_3$ to yield SrO.

Joseph A. Pask (University of California): The temperature range in which you were working is also the range in which the anatase–rutile transformation occurs. I wonder whether your increased reaction with the use of the anatase perhaps would give the explanation.

Answer: When studying the reaction between strontium carbonate and anatase there is a definite possibility that the anatase–rutile transformation is taking place concurrently with the reaction between strontium carbonate and titania. However, X-ray analysis of partially reacted samples failed to show the presence of rutile. Comparison of our data with the work of Shannon and Pask ([32]) on the kinetics of the anatase–rutile transformation indicates that over the temperature interval studied the strontium carbonate–titania reaction is considerably faster than the phase transformation.

H. Palmour (North Carolina State): Would you comment on how you determined the reaction temperature?

Answer: During an actual firing schedule the rate constant k is a function of time. What this means is that all of the solid-state rate equations (Jander, Ginstling–Brounshtein, etc.) which are normally applied for constant k can be used, but the product kt must be replaced by an average product designated $\overline{kt}$ and given by the equation

$$\overline{kt} = \int_0^t k(t)\, dt$$

This equation can be used to correct for the reaction that occurs during the heating and

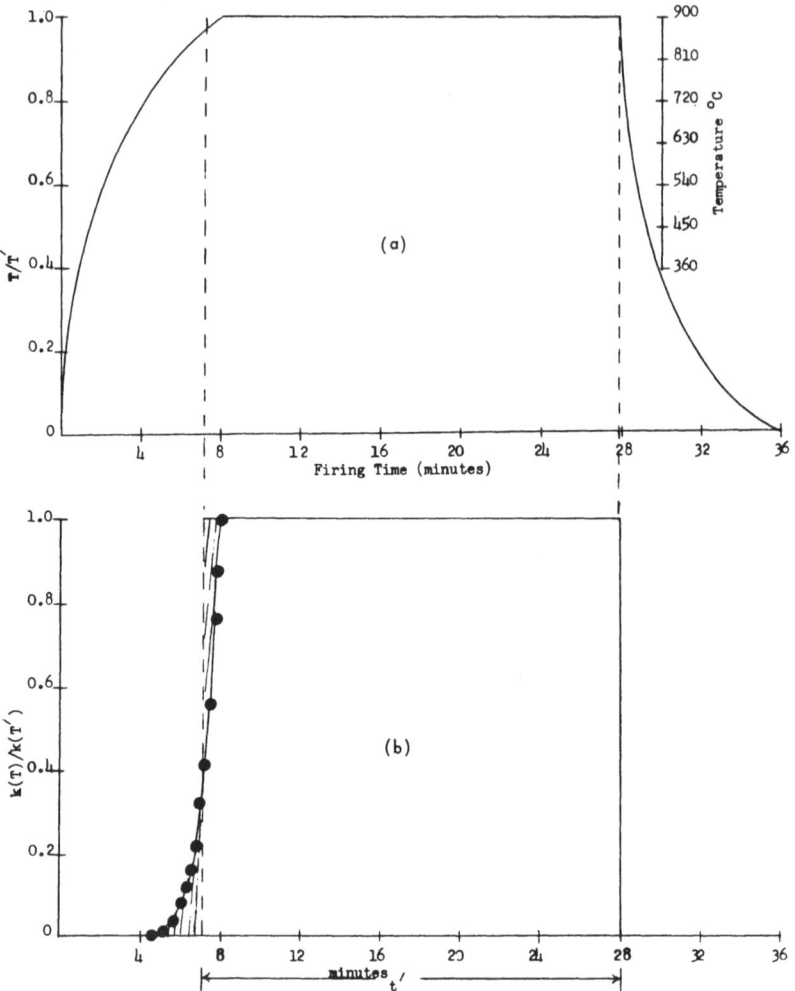

Fig. 9. Calculation of fraction of reaction completed, where k is a function of time. (a) Firing schedule. (b) Reaction schedule.

cooling of a reaction system which has been soaked at some fixed temperature. As an example of the application of the foregoing equation, consider a reaction system that has the firing schedule shown in the Fig. 9(a). The problem is to determine the time t' at temperature T' which would have produced the same amount of reaction as actually occurred during the heating, soaking, and cooling. This can be determined graphically once the T versus t data are transformed into a plot of k versus t. This has been done in Fig. 9(b) using an Arrhenius equation of the form

$$k(t) = A \exp(-Q/RT)$$

Preliminary data are used to estimate the parameter in the above equation. For the boundary condition employed in constructing Fig. 9 one observes that even though the actual soaking time was 20 min, the effective reaction time at the soaking temperature was 21 min and that 21 min, not 20 min, should be used in calculating the rate constant.

ADDITIONAL REFERENCES

39. R. Haul and B. Dumbgen, "Sauerstoff-Selbstdiffusion in Rutilkristallen," *J. Phys. Chem. Solids* **26**: 1–10 (1965).
40. L. A. K. Dominik and R. K. MacCrone, "Dielectric Relaxation of Hopping Electrons in Reduced Rutile, TiO_2," *Rhys. Rev.* **156**(3): 910–913 (1966).

Chapter 26

Oxidation Kinetics

T. J. Gray

*State University of New York College of
Ceramics at Alfred University
Alfred, New York*

M. J. Pryor

*Metals Research Laboratories
Olin Mathieson Chemical Corporation
New Haven, Connecticut*

The kinetics of oxidation are reviewed on the basis of a model which takes into account the defect constitution of metal and oxide, the formation of amorphous and crystalline films, and general morphological characteristics. Specific attention is drawn to the oxidation of aluminum and its alloys.

INTRODUCTION

The kinetics of the oxidation of metals have been studied in a wide variety of ways over more than four decades since the early pioneering work of Tammann ([1]) and Pilling and Bedworth ([2]). The classical theories of Wagner ([3]) and of Mott and Cabrera ([4]) have stimulated considerable interest, as has the absolute reaction rate theory advocated by Gulbransen ([5]). These theories suffer in common from the assumption of an oversimplified model for what is in fact a very complicated combination of kinetic processes. Evans ([6]), in numerous publications, drew attention to the cracking and cavity formation in growing oxide layers and developed an interesting model for coverage by expanding islands of oxide. Gray ([7]) and Gulbransen and Ruka ([8]) have drawn attention to the significance, in certain circumstances, of the initial gas adsorption process. In the case of superpure single crystals or thin metal films, the dissociative adsorption is frequently rate controlling, as is readily demonstrated by the enormous increase obtained through the use of atomic oxygen. In all but the simplest cases, coexistence of more than one oxide phase complicates the transport mechanism, and there is adequate evidence to suggest that, contrary to earlier views, the oxygen ion under certain specific conditions may become a significant diffusing species together with the more usual diffusion of cations.

It is, therefore, appropriate to consider in some detail a number of possible simultaneous or consecutive kinetic processes which may individually or collectively contribute significantly to the overall observed rate of oxidation. These may be summarized as follows:

1. Adsorption of gas with or without dissociation.
2. Primary formation of amorphous oxide.
3. Migration of ionic species through the amorphous oxide contributing to its thickening. This may be uniform or regionalized and may be caused by high-field condition after the Mott–Cabrera model or more generally by a concentration gradient.
4. Nucleation of a crystalline oxide phase.
5. Growth of the crystalline oxide phase.
6. Modified migration of ionic species through the polycrystalline oxide. In these circumstances the differential mobility between grain boundaries and bulk becomes a significant factor.
7. Subsequent formation of secondary phases.

The significance of the defect constitution of both metals and nonmetals has received widespread attention, particularly as an influence on semiconductivity and associated phenomena. However, it is remarkable that many of the important implications of the relevant features of these phenomena have not been considered sufficiently in correlation with corrosion process. Within the past decade ion field emission microscopy has demonstrated the localized character of the initial oxidation process originating from unique discontinuities, often dislocations. At the same time this important technique demonstrates the differences in significance among various crystallographic phases. Accurate diffusion measurements on alkali halides as well as on oxides have demonstrated the profound effect of minor impurities on transport processes, while significant oxygen migration has been recently established as an important factor often previously overlooked. The extent and character of anion and cation vacancies in correlation with oxidation kinetics has largely been ignored, save in the work initiated by McMullen and Pryor ([9]). Lattice vacancies play a crucial role in diffusion in metals and ionic solids and must be considered critically. In some instances impurities and vacancies form complexes which may be mobile (e.g., magnesium in aluminum) or immobile (e.g., silver in gold). The metallurgical significance of this complex formation has been studied by several authors ([10-12]).

It becomes immediately obvious that the morphology of oxide growth has a very profound significance, particularly in thicker films, and the electron microscopy of Pryor and co-workers [e.g., ([13])] lends much to the development of a satisfactory kinetic model, at least for the oxidation of aluminum.

Except in the circumstances of superpure and scrupulously clean surfaces of single crystals of metal and sometimes thin metal films, the adsorption process can normally be regarded as rapid, and hence will not contribute significantly to the overall kinetics. For thin films of oxide of amorphous character our understanding of the kinetics owes much to Mott and co-workers [e.g., ([4])]. However, even in the thin-film range and progressively as the film thickens, the significance of defect constitution becomes pronounced, and this significance logically can be extrapolated back to the period of the initial film formation. The work of Pryor and co-workers [e.g., ([9])] has emphasized these aspects and established a sensitive technique for their investigation which depends on correlation between dielectric loss, defect constitution, and oxidation kinetics. Morphological studies of oxide films on aluminum by Thomas and Roberts ([14]), Dignam ([15]), Doherty and Davis ([16]), and Randall and Bernard ([17]) have established conclusively the primary formation of an amorphous film followed by the nucleation and growth of a crystalline oxide located at the metal–metal oxide interface and growing into the metal. These results have been confirmed by Beck et al. ([13]), who have demonstrated that the peripheral growth of the crystals accounts satisfactorily for the oxidation kinetics except in the early and late stages, where the effect of the growth of the amorphous film dominates.

Much work has been performed on the nucleation and crystal growth in oxide films; we note particularly the work of Bernard and co-workers ([18]), Gulbransen and co-workers ([19]), Hoar ([20]), and others on copper, iron, iron alloys, and Ni–Cr, including both oxide and sulfide formation. Phenomenologically the processes are similar in the initial development of an amorphous film, nucleation, and subsequent growth of a fixed number of nuclei. It has been postulated by Bernard that once nucleation has occurred, migration into the nuclei produces an impoverished environment in which further nuclei cannot form. The number of nuclei formed may be as high as the dislocation density in the substrate metal, as suggested by Doherty and Davis ([16]) in the case of aluminum, or may be controlled by experimental variables such as oxygen pressure, temperature, and rate of surface migration. Bernard has suggested that the source of nuclei are the disordered sites generated by stresses in the oxide layer when it reaches a critical thickness.

It is illuminating to consider the oxidation of aluminum on this basis assuming a pure (99.997 %), but not superpure, metal in the form of annealed and chemically prepared polycrystalline sheet; here the initial adsorption rate for oxygen is so rapid that it cannot influence the overall kinetics. The initial stages of the oxidation, with the formation of an amorphous oxide, occur with cation transport through the oxide from the metal interface under the influence of the electric field generated by the adsorbed oxygen ions ([4]).

The field-induced oxidation process for thin films at low temperatures is shown to be governed by a rate law

$$dx/dt = n\gamma \exp[-(W + U)/kT] \, x \exp(\alpha e V/kT)$$

In this expression n is the number of metal ions crossing a unit area per unit time and may well be very considerably less than the actual number of possible available sites, perhaps 10^2 to 10^3 smaller. The first term of this expression expresses the jump probability between initial and final interstitial sites with an energy difference W separated by a saddle of activation energy U in a lattice of natural frequency γ, while the second exponential term takes into account the reduction in "saddle" height resulting from the field V across the film, with α the average distance from an interstitial site to a saddle point. From this rate law a limiting thickness of oxide film can be deduced.

For thick films and high temperatures this growth law is modified, since the ionic transport current becomes proportional to the field and a parabolic law is realized. Using reasonable values, the limiting temperature for the low-temperature process on aluminum can be estimated as 300°C, above which the parabolic rate law is indeed found to hold.

For the anodic growth of films under high-field conditions in an electrolyte such as purified boric acid ([21]) logarithmic kinetics common to those for dry oxidation in the temperature range below 300°C are observed. At temperatures between 300° and 425°C thermal oxidation of aluminum obeys classic parabolic-diffusion-controlled kinetics:

$$W_a = \sqrt{kt + C}$$

where W_a is the weight gain of amorphous oxide during time t, suggesting that although the mechanism may not be identical, diffusion of the cations remains rate controlling. At higher temperatures of 425°C and above the oxidation kinetics of aluminum become more complex, with the weight gain–time curves assuming an approximately sigmoidal shape. In the region of more complex kinetics crystalline γ-Al_2O_3 also appears as a reaction product. The crystals are known to nucleate at the amorphous oxide–metal interface and to grow very rapidly into the metal to a constant and only slightly temperature-dependent depth. Throughout their existence they are overlayed by the amorphous oxide film and do not appear to consume it.

Since the number of dislocations in the aluminum substrate which may reasonably be expected to contribute crystal nucleation sites is large, as it is likely to be in the case of rolled, polycrystalline sheets, the nucleation rate would be expected to be very high. This corresponds to instantaneous random nucleation and eliminates any significant contribution deriving from nucleation kinetics. Subsequently crystals grow laterally at a constant, temperature-dependent rate around each randomly distributed nucleus in essentially circular geometry leading, as discussed by Evans ([6]), to a kinetic relationship of the form

$$W_c = \rho\delta[1 - \exp(-\pi v^2 \Omega t^2)]$$

where W_c is the weight of crystalline oxide, ρ the density of the γ-Al$_2$O$_3$, δ the thickness of the crystals, v the radical velocity of growth of crystals, and Ω the density of nuclei for time t. The crystals continue to expand as rather flat cylinders until they touch each other, after which no further crystal growth can be observed.

There is no real doubt that the appearance of crystalline γ-Al$_2$O$_3$ at the amorphous oxide–metal interface is the result of inward migration of oxygen through the overlying amorphous film. This rate of inward migration increases rapidly with increasing temperature. It is probable that this occurs through anion vacancies in the overlying amorphous layer.

Dielectric loss measurements have shown that the crystalline γ-Al$_2$O$_3$ component of the duplex film makes essentially no contribution to the total film resistance. It is even more remarkable that the overlying amorphous aluminum oxide film continues to grow with accurately parabolic kinetics notwithstanding even the presence of an essentially complete crystalline layer situated between it and the metal. Growth of the amorphous film is likely the result of *outward* diffusion of aluminum ions, and so it appears that this outward aluminum ion diffusion (leading to amorphous γ-Al$_2$O$_3$ film formation) and the inward oxygen diffusion (leading to crystalline γ-Al$_2$O$_3$ formation) can take place without mutual hindrance and interaction.

In this particular instance the essential independence of the two growth patterns allows summation of the growth rates to give a net expression

$$W_a + W_c = \rho\delta[1 - \exp(-\pi v^2 r t^2)] + \sqrt{kt + C}$$

This analysis has been employed advantageously by Beck et al. ([13]) to establish more rigorous kinetics for the oxidation of aluminum than heretofore available. These results are illustrated in Figs. 1 and 2. Examination of the data establishes a discontinuity in the growth kinetics of the amorphous film at about 525°C, above and below which the apparent activation energies are identical. This may reasonably be associated with a change in mechanism, with a greatly reduced aluminum ion diffusion rate through the amorphous film in this temperature range without any accompanying discontinuous change in the oxygen ion diffusion rate and therefore without any discontinuity in the growth kinetics of the underlying crystalline film. These and other

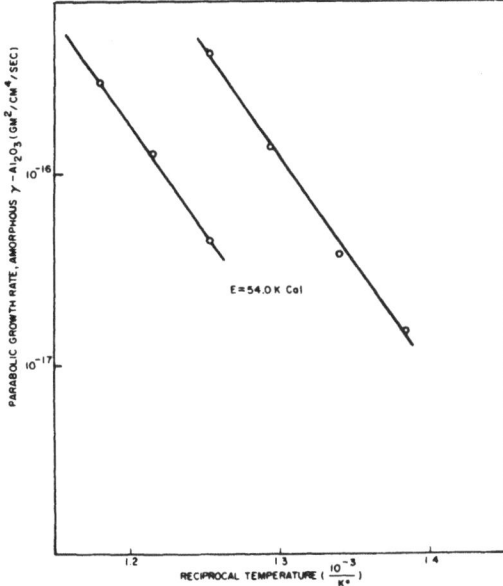

Fig. 1. Parabolic growth rate of alumina showing a discontinuity at about 525°C but exhibiting essentially identical opponent activation energies in each range.

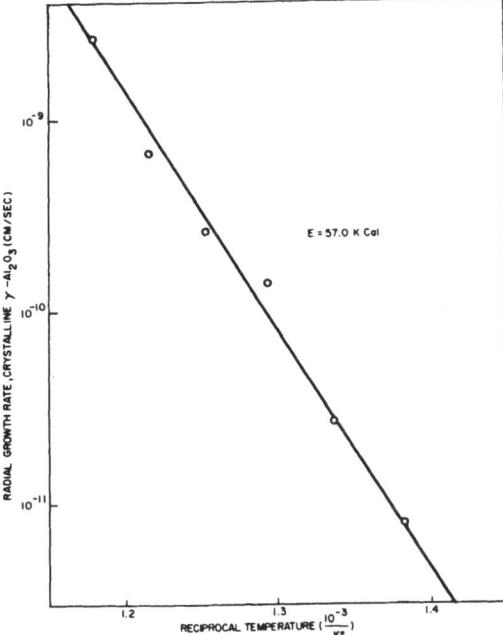

Fig. 2. Radial growth rate of crystalline alumina at the oxide–metal interface.

considerations have been discussed in detail by Beck *et al.* ([13]). However, it should be mentioned that the crystalline phase does not always develop at the metal–metal oxide interface. For passive films formed on aluminum in chromate solutions the crystalline phase develops at the outer surface, as demonstrated by Pryor ([22]) and Heine and Pryor ([23]).

EXPERIMENTAL

It is not our intention here to review all the varied techniques that have from time to time been applied to the study of the oxidation of metals. Attention will be directed to two techniques, one direct and the other indirect, both of which would normally be applied in association with replica and transmission electron microscopy, electron diffraction, and X-ray examination to characterize the oxide layer.

One of the most important methods of investigating oxidation processes is also the most direct and least ambiguous—that of microbalance gravimetric determination of oxidation kinetics. Only an automatically recording quartz beam microbalance in a greaseless high vacuum system capable of thorough degassing and operation for very long periods under static conditions with sustained high precision can be regarded as satisfactory. This excludes balances of metallic construction, particularly those including any magnetic materials and electronic components within the vacuum system, because these can never be satisfactorily degassed. Corrections for thermomolecular flow or convection currents, depending on the pressure range, and for all other pertinent factors, must be applied, after which a precision of ± 1 μg on gross weights of 0.5–1 g can be sustained over periods of 100 hr or more.

The indirect technique derives from a study of dielectric properties adopted after the study by Breckenridge ([24]) of the defect constitution in alkali halide crystals. When the dielectric properties of an oxide film are examined critically with respect to the dielectric constant ϵ, conductivity σ, resistivity ρ, and loss tangent tan δ as a function of frequency, then results resembling the idealized curves of Fig. 3 can be anticipated. The detailed analysis of these curves has been considered by many authors and recently summarized by Volger ([25]) and Gray ([26]). The analyses can be modified for a variety of conditions with or without the coexistence of interfacial polarization. This technique has been extensively applied by Pryor and co-workers ([27,28]), and the technique is rapidly finding new adherents ([29,30]).

Since there is no single AC bridge capable of making satisfactory measurements over the entire range of frequencies required, compromise measurements on several different bridges are usually necessary. This difficulty is further aggravated by the very high capacitance and low resistances frequently

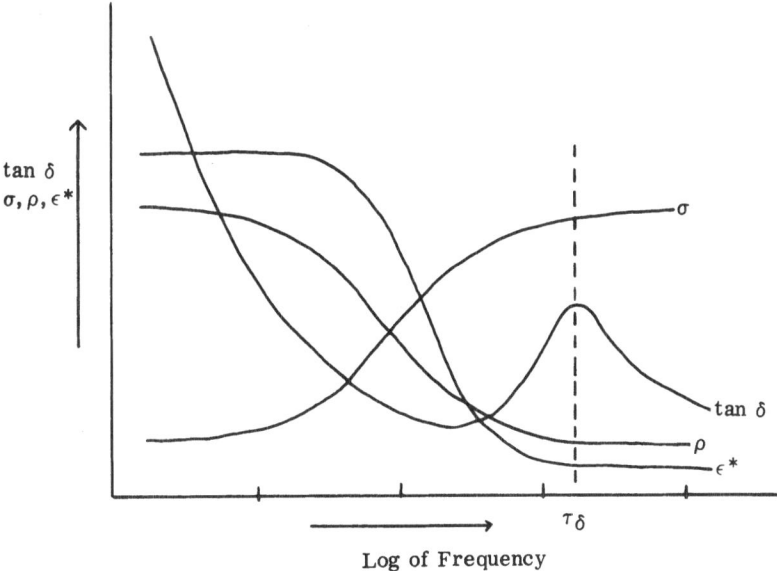

Fig. 3. Idealized curves for the variations in dielectric properties for a two-phase complex dielectric medium.

observed. Only laboratory-constructed bridges are satisfactory in the very low frequency range from 0.5 cps or lower up to 200 cps, which is often a range of considerable importance. The most suitable types have been reviewed by Gray ([26]). The range from 200 cps to 1 Mc/sec can be satisfactorily covered by a number of commercially available bridges based either on the Schering or Mayo designs. These should be used under suitable guard circuit conditions to eliminate or minimize spurious effects, and additionally should be operated under conditions permitting the superposition of a polarizing potential. Operation with small input signals (50 mV p to p, or less) is imperative, and a tuned high-gain preamplifier coupled to an oscilloscope provides the best detector system.

An alternate method using square-wave pulses has been largely used by Leach ([31]) and co-workers, following the original design of Denholm ([32]). The method, which depends on oscilloscope comparison of pulses passed through the specimen cell with those derived from an analog using standard resistors and capacitors, has the merit of continuous rapid measurement, but suffers from the basic difficulty involved in the accurate analysis of square-wave functions in a complex model. Under certain simplified limiting conditions it does provide good agreement with the more accurate bridge measurements.

Typical measurements may be performed in any standard manner adapted for dry or wet measurements which a provision for adequate guard and shielding circuits, counter electrodes, and with the normal requirements for precise thermostatic control. Because of the wide range of capacitance and frequency to be covered, separate assemblies are usually necessary at the range extremes. This is particularly the case in wet corrosion studies, where the apparent capacitance may be many microfarads per square centimeter of surface. A typical cell is illustrated by Beck et al. ([27]).

The model used in the analysis of results can be varied to suit specific cases, but the most successful assumes a series combination of oxide layer capacitance and double layer capacitance, each with its parallel leakage resistance. The resistance of the solution determined independently is normally

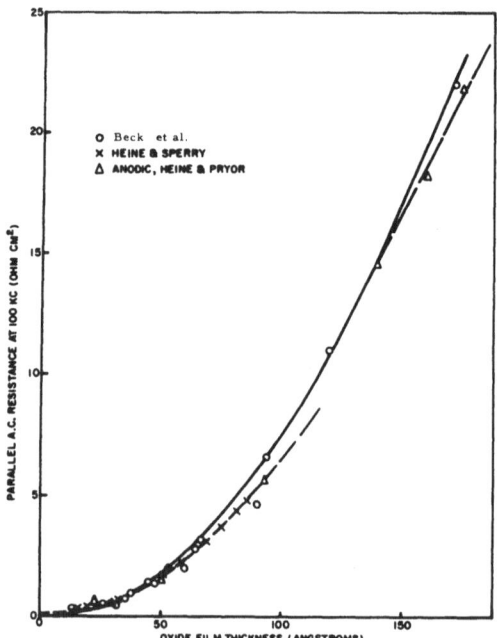

Fig. 4. Relationship of parallel AC resistance (at 100 kc/sec) to oxide film thickness for both thermal and anodic films.

very small part (less than 2%) of the total resistance, but can be allowed for where significant. Data calculated on this basis are exemplified in measurements on thermally and anodically prepared aluminum oxide films (Fig. 4). The choice of electrolyte is of considerable importance, and for the specific case of aluminum, chromate solutions commend themselves; they can be adjusted to a pH of 6, where they do not thin aluminum oxide films significantly or introduce new defects and so can be used to derive the defect constitution of the initial film; chromates can be used at higher pH up to 9.0, in which case they thin aluminum oxide films to around 20 Å without introducing defects. By this technique the distribution of defects through the thickness of the oxide can be accurately studied, as in the work of Heine and Sperry ([33]). Wood and Pearson ([29]) have extended this approach with other electrolytes to determine the distribution of defects through oxide films on niobium.

These results essentially emphasize the contribution of both ionic and electronic mobilities to the conduction process in an essentially nonstoichiometric oxide containing a wide variety of lattice defects. Following the original observation of Schilling ([34]) that a metal excess exists in the immediate vicinity of the metal–metal oxide interface for thin oxide layers on copper, Grunberg and Wright ([35]) have demonstrated similar features for thin oxide films on aluminum, magnesium, and zinc. From photo-induced electron emission it was concluded that the most probable defect was of the F^1-center type situated at an oxygen vacancy. Heine and Sperry ([33]) have been successful in establishing the distribution of such F^1 centers throughout the first 80–100 Å of oxide films on aluminum.

DISCUSSION

The kinetics of oxidation and corrosion are of immense technological importance as well as being of profound scientific significance. While laboratory investigations have related in most cases to relatively simple studies of pure metals and a few alloys, industrial interest has been directed in very sophisticated research toward reduction in the rate of oxidation. However, it should be appreciated that on occasion, such as in the development of sacrificial anodes, electrodes for primary batteries, and in the potential development of consumable metal electrode fuel cells, an enhancement in the rate of oxidation is sought.

In endeavoring to reduce the rate of oxidation many approaches have been studied in considerable detail, but none more than that of alloy selection. The corrosion rates for a typical series of commercial aluminum alloys are illustrated in Fig. 5, which demonstrates the correlation between the defect constitution of the oxide film as reflected in the dielectric loss tangent and the weight loss under standard conditions in a fixed period of time ([36]). These

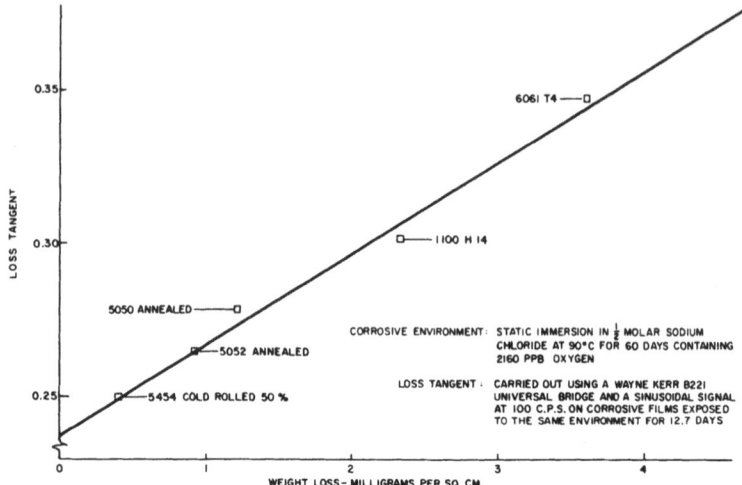

Fig. 5. Corrosion rates of commercial aluminum alloys correlated with loss tangents of corrosion films.

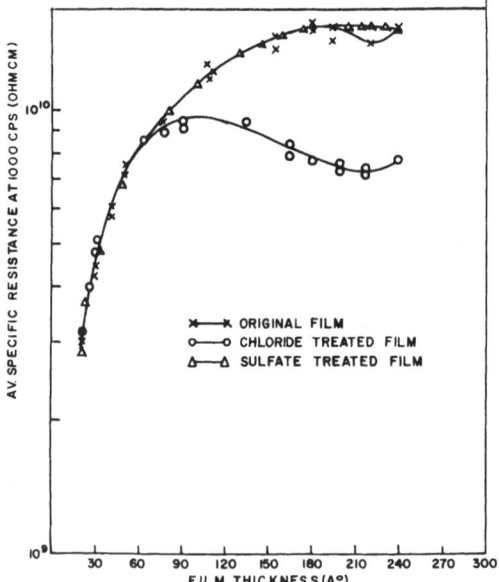

Fig. 6. Relationship between average specific resistance at 1000 cps and thickness of untreated γ-Al_2O_3 films 240-Å thick and similar films which have been treated by 16-hr immersion in either 1.0 N sodium chloride or sulfate solution at pH 6 and at 25°C: ×, original film; ○, chloride-treated film; △, sulfate-treated film.

data are in accord with independent determinations on corrosion rates with these alloys.

However, the same measurements of dielectric loss tangent observed over a range of frequencies can also serve to illustrate the inhomogeneity in an oxide film. In Fig. 6 the parallel resistance of a 240-Å thick alumina film formed anodically at 20 V in neutral ammonium tartrate solution shows a nonuniformity of loss characteristics which has been analyzed by Heine et al. [37]. A relatively low resistivity region exists 60–80 Å thick adjacent to the metal–metal oxide interface varying from about 5×10^9 ohms/cm at 25 Å to about 2×10^{10} ohms/cm at 80 Å. Thereafter the overlying oxide has a relatively constant resistivity of about 2×10^{10} ohms/cm out to 180 Å, superimposed on which there is a lower resistivity layer with a final high resistance layer in the immediate vicinity of the outer surface. These findings are generally compatible with the views of Belova et al. [38] that in the vicinity of the metal–metal oxide interface the oxide is semiconducting by virtue of excess metal (probably with an additional F^1-center component), succeeded by a relatively uniform "insulating" layer and an oxygen-excess, p-type, relatively conductive layer at the oxide–electrolyte interface.

Not noted in the work of Belova et al. [38] is the existence of a fourth low-resistivity region studied between the p-type outer layer and the uniform (insulating) layer. According to Heine et al. [37], this layer is an n-type layer containing substituted hydroxyl ions from the electrolyte. This contention has been amply proved by the recent work of Brock and Wood [30], who showed that this layer was absent when anodizing was conducted in a nonaqueous electrolyte.

Heine et al. [37] also studied the changes in resistance of the films described above on subsequent immersion in sodium sulfate and sodium chloride solutions at pH 6. Immersion conditions were selected such that there was no film thinning, crystal structure change, or detectable corrosion. The results obtained are also summarized in Fig. 6. Here it can be seen that the effect of immersion in sulfate solution is minor and of doubtful significance. Immersion in chloride, on the other hand, produces a major decrease in parallel resistance of the oxide film, as shown in Figs. 6 and 7. Analysis of the effect of chloride shows that little substitution of chloride ions occurs in the outer high-resistance p-type layer and that the majority of the conductivity change and chloride substitution occurs in the underlying n-type layer, which contains substituted hydroxyl ions.

Other complexities can be studied, as in the case of magnesium-containing alloys of aluminum. Oxidation at temperatures up to 500°C demonstrates that amorphous alumina is always the first film to form on the surface, but subsequently MgO and occasionally magnesia spinel, $MgAl_2O_4$, is observed. In cases such as these the kinetic study is only meaningful gravimetric studies

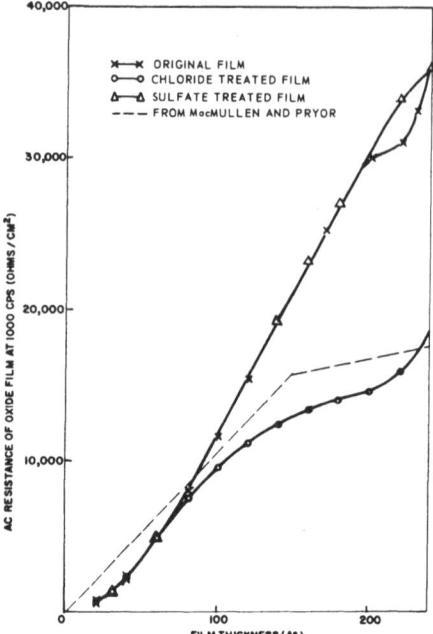

Fig. 7. Relationship between AC resistance at
1000 cps and thickness of untreated γ-Al$_2$O$_3$
films 240-Å thick and similar films which have
been treated by 16-hr immersion in either 1.0 N
or sodium chloride or sulfate solution at pH 6
and at 25°C.

are combined with morphological investigation and dielectric loss characterization. Heine [40] has observed that the apparance of MgO occurs earlier
and at lower temperatures as the magnesium content increases, and it is
observed that the film which develops before the appearance of MgO is
more protective against further oxidation than a comparable film on high-
purity alumina, but this is reversed immediately when MgO appears. The
spinel only forms at low magnesium concentration.

Essentially, these investigations establish a correlation between the
dielectric loss characteristics of a corrosion layer, more particularly, the
ionic conduction, and the rate of oxidation or corrosion. The significance of
anion substitution, accompanied in many instances by corresponding cation
vacancies, is clearly demonstrated by the profound effect of halogens. However, while this correlation is of great significance, it is also shown that in
certain cases morphology is of equal if not greater significance.

While these various investigations relate primarily to the potentially
corrosion resistant aspects, Keir *et al.* [39] have also investigated the condi-

tions under which oxidation or corrosion can be deliberately enhanced. It could be anticipated that the substitution of a higher-valent ion into a lattice site normally occupied by an aluminum ion would induce a corresponding increase in cation vacancies and in ionic conduction. This was indeed found to be the case with Sn^{4+} ions, but attempts to introduce Si, Ti, Zr, or Ge were believed to involve a precursor to compound formation without the corresponding increase in ionic conduction, so that the corrosion rate was not significantly affected. This emphasizes the importance of considering the possibility of strong chemical interaction when investigating doping to effect valence control to modify the defect constitution.

REFERENCES

1. G. Tammann, *Z. Anorg. Allgem. Chem.* **111**: 78 (1920); **39**: 869 (1926).
2. N. B. Pilling and R. E. Bedworth, *J. Inst. Metals* **29**: 529 (1923).
3. C. Wagner, *Z. Physik. Chem. (Liepzig)* **B21**: 25 (1933); **B32**: 447 (1936).
4. N. F. Mott and N. Cabrera, *Trans. Faraday Soc.* **43**: 429 (1947).
5. E. A. Gulbransen, *Trans. Electrochem. Soc.* **83**: 301 (1943).
6. U. R. Evans, *Trans. Faraday Soc.* **41**: 365 (1945).
7. T. J. Gray *et al.* (eds.), *Defect Solid State*, Interscience, New York, 1957.
8. E. A. Gulbransen and R. Ruka, *J. Metals* **188**: 1500 (1950).
9. J. J. McMullen and M. J. Pryor, in: *First International Congress on Metallic Corrosion*, Butterworth and Co., London, 1961, p. 52.
10. G. J. Dienes and A. C. Damask, in: *Reactivity of Solids*, (G. M. Schwab, ed.), Am. Elsevier, New York, 1965.
11. T. Kino, S. Kabemoto, and A. Yoshida, *J. Phys. Soc.* Japan **18**: 78 (1963).
12. R. M. J. Cotterill and R. L. Segall, *Phil. Mag.* **8**: 1105 (1963).
13. A. F. Beck, M. A. Heine, E. J. Caule, and M. J. Pryor, *Corrosion Science* **7**: 1 (1967).
14. K. Thomas and M. W. Roberts, *J. Appl. Phys.* **32**: 70 (1961).
15. M. J. Dignam, *J. Electrochem. Soc.* **109**: 184, 192 (1962).
16. P. E. Doherty and R. S. Davis, *J. Appl. Phys.* **34**: 619 (1963).
17. J. J. Randall and W. J. Bernard, *J. Appl. Phys.* **35**: 1317 (1964).
18. J. Bernard and J. Talbot, *Compt. Rend.* **225**: 411 (1947); J. Bernard and J. Bardolle, *Rev. Met.* **49**: 613 (1952); *L'Oxydation des Metaux*, Gauthier-Villars, Paris, 1962.
19. E. A. Gulbransen, W. R. McMillan, and K. F. Andrew, *J. Metals* **6**: 1027 (1954); E. A. Gulbransen and W. R. McMillan, *J. Chim. Phys.* **53**: 643 (1956); E. A. Gulbransen and K. E. Andrew, *J. Electrochem. Soc.* **106**: 511 (1959).
20. T. P. Hoar and A. J. Tucker, *J. Inst. Metals* **81**: 665 (1952-3).
21. A. Guntherschulze and H. Betz, *Z. Physik* **71**: 106 (1932).
22. M. J. Pryor, *Z. Electrochem.* **62**: 782 (1958).
23. M. A. Heine and M. J. Pryor, *J. Electrochem. Soc.* **114**: 1001 (1967).
24. R. G. Breckenridge, *J. Phys. Chem.* **18**: 913 (1950).
25. J. Volger, in: *Progress in Semiconductors*, Vol. 4, (A. F. Gibson *et al.*, eds.) Heywood, London, 1959.
26. T. J. Gray, in: *Experimental Methods in Catalytic Research* (R. B. Anderson, ed.), Academic Press, New York, 1968, pp. 293–320.
27. A. F. Beck, M. A. Heine, D. Van Rooyen, and M. J. Pryor, *Corrosion Science* **2**: 133 (1962).

28. M. A. Heine and M. J. Pryor, *J. Electrochem. Soc.* **110**: 1205 (1963).
29. G. C. Wood and C. Pearson, *Nature* **208**: 547 (1965).
30. G. C. Wood and A. J. Brock, *Nature* **209**: 773 (196); G. C. Wood, C. Pearson, A. J. Brock, and S. W. Khoo, *J. Electrochem. Soc.* **114**: 145 (1967).
31. J. S. L. Leach, NONR Contract N 62558(24)1177, 1st Biennial Report, Imperial College, London, 1958.
32. W. T. Denholm, Ph.D. Thesis, U. of Adelaide, South Australia.
33. M. A. Heine and P. R. Sperry, *J. Electrochem. Soc.* **112**: 359 (1965).
34. H. Schilling, *A. Watenforsch.* **7a**: 211 (1952).
35. L. Grunberg and K. H. R. Wright, *Proc. Roy. Soc. (London), Ser. A* **232**: 403 (1955).
36. W. H. Anthony, *J. Electrochem. Soc.*, in press.
37. M. A. Heine, D. S. Keir, and M. J. Pryor, *J. Electrochem. Soc.* **112**: 24 (1965).
38. A. P. Belova, L. G. Gorskaya, and L. N. Zakgeim, *Fiz. Tver. Tela* **3**: 1851 (1961).
39. D. S. Keir, M. J. Pryor, and P. R. Sperry, *J. Electrochem. Soc* **114**: 777 (1967).
40. M. A. Heine, unpublished work.

DISCUSSION

J. E. Burke (G. E.): I'm intrigued by your chloride argument. I presume you claim the chlorine replaces the oxygen. What is the conduction mechanism: is it electronic under those circumstances?

Answer: High-frequency (100-kc/sec) measurements of alumina films substituted with chloride ions show that the conductivity is not affected ([37]). We interpret this as meaning that chloride ion substitution does not modify the electronic resistance of the oxide film. This is also supported by cathodic polarization studies in neutral chlorides, since the form of the cathodic polarization trace in nearly neutral electrolytes is highly sensitive to the electronic resistance of the oxide film ([41]). However, pronounced changes in conductivity occasioned by introduction of the chloride into the film are observed at frequencies below 10 kc/sec ([37]). We believe that chloride substitution affects the ionic conducitivity of the oxide only, and this is supported by anodic polarization studies in chloride solutions and by previous measurements of growth kinetics of alumina films in these media ([22]). These results imply that chloride substitution results in the creation of additional cation vacancies. The only feasible mechanism appears to be the replacement of O^{--} ions by Cl^{-} ions, which results in the creation of positive charge which is neutralized by a few aluminum ions passing from the oxide into solution. The degree of chloride substitution of about 1 part per 10^3 has been estimated.

H. J. Oel (Max Planck Institute): You note that oxygen mobility should be taken into account. Would you agree then that results from isotope cation-substitution investigations will not be precisely the same as far as these data are concerned?

Answer: The whole point is that at the metal–metal oxide interface there is an excess of metal and the indirect photoexcited electron emission indicates very strongly, although not absolutely conclusively, that this is in fact an F-center mechanism. There is absolute no question of the diffusion of oxygen, for this has been measured by O^{18} diffusion. You have to take into account both anion and cation migration. I think just accepting one and ignoring the other is not a valid approach.

ADDITIONAL REFERENCE

41. D. S. Keir and M. J. Pryor, *J. Electrochem. Soc.* **102**: 605 (1955).

Chapter 27

Considerations on the Atomistics of Oxidation*

J. B. Lightstone†

Division of Engineering and Applied Physics
Harvard University, Cambridge, Massachusetts and
Ledgemont Laboratory, Kennecott Copper Corporation
Lexington, Mass.

J. P. Pemsler

Ledgemont Laboratory, Kennecott Copper Corporation
Lexington, Mass.

Experiments were performed to determine the effect of structure on the mechanism of zirconium oxidation. It is shown that there is a high volume fraction of short circuit diffusion paths in the oxide formed on Zr in the temperature range 500–1000°C. The results of tracer experiments are in agreement with a short circuit diffusion mechanism. The assumption of line diffusion is shown to imply intuitively reasonable values of both the grain boundary thickness and the ratio of the line diffusion coefficient to the coefficient for lattice diffusion.

INTRODUCTION

The oxide formed on a metal during high-temperature oxidation is usually considered to be polycrystalline. This introduces the possibility that the detailed mechanism of the oxidation may be sensitive to the microstructure ([1]). In this work the grain size of ZrO_2 formed on Zr has been determined and the observed microstructure related to the diffusion mechanism through the oxide and to the crystal structure.

---

*This work has been partially supported by the Advanced Research Projects Agency under Contract ARPA SD–88, and was submitted by J. B. L. in partial fulfillment of the requirements for the Ph.D. degree, Division of Engineering and Applied Physics, Harvard University.
†Presently with Union Carbide Corp., Linde Division, Tonawanda, New York.

MICROSTRUCTURE

Theory

The observed X-ray (hkl) reflection from crystallites will be broadened if their thickness in a direction normal to the reflecting planes, $2a_{hkl}$, is less than ~ 2000 Å ([2]). The observed integral breadth B has contributions from the instrumental broadening b and the particle size broadening β, where

$$B^2 = \beta^2 + b^2 \tag{1}$$

and

$$2a_{hkl} = 1.05\lambda/(\beta \cos \theta) \tag{2}$$

The wavelength of the radiation is λ, and θ the Bragg angle.

It has previously been suggested ([3]) that the grain size of the oxide may be ~ 100 Å, and an X-ray line-broadening technique was chosen to study particle size.

Experimental Work

Zirconium disks were oxidized at 490°, 749°, and 975°C. After cooling to room temperature the monoclinic ZrO_2 (111) reflection was examined in $CuK\alpha$ radiation. The peak, with a maximum at $2\theta = 28.17°$, was stepscanned at 0.01° intervals, and the time for 10^4 counts was recorded at each interval. The (111) peak was examined because of its angular displace-

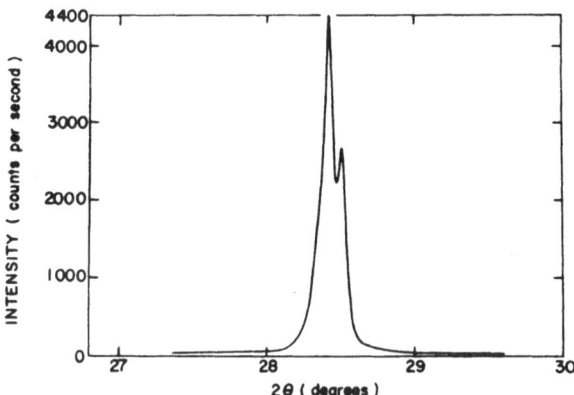

Fig. 1. The (111) peak of silicon standard at $2\theta = 28.42°$. The complete angular range was used in the calculation of the integral breadth.

TABLE I

Line Broadening Data for (111) Lines of Monoclinic ZrO_2 and Silicon

Specimen	Oxidation temperature (°C)	Peak intensity (counts/sec)	Estimated background (counts/sec)	Angular range of 2θ used in calculation of integral breadth (deg)	Observed breadth (deg)	Particle size broadening (deg)	Particle size (Å)
Si	—	4405.3	36.3	27.34–29.63	0.156	—	—
ZrO_2	490	239.2	11.2	27.35–29.34	0.449	0.422	225
ZrO_2	749	1030.9	17.5	27.49–29.25	0.358	0.323	286
ZrO_2	975	4065.0	13.2	26.97–29.39	0.311	0.269	353

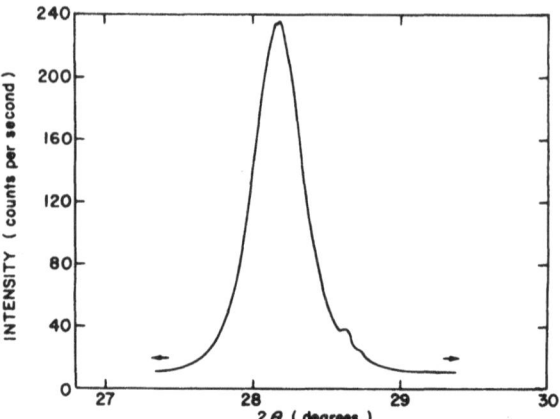

Fig. 2. The (111) peak of monoclinic ZrO_2 at $2\theta = 28.17°$ formed by oxidizing at 490°C. The nearest peaks are the monoclinic ZrO_2 (110) reflection at $2\theta = 24.48°$ and the tetragonal ZrO_2 (111) reflection at $2\theta = 30.15°$. The arrows indicate the angular range used in the calculation of the integral breadth.

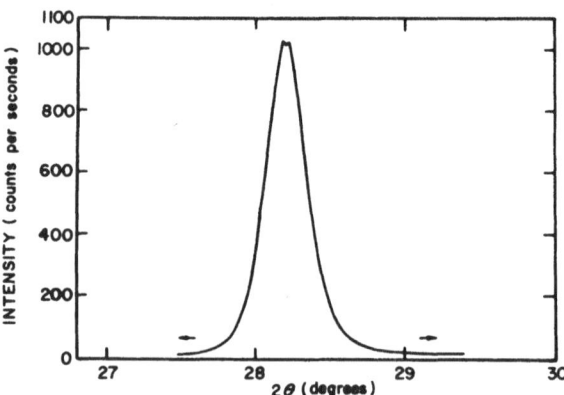

Fig. 3. The (111) peak of monoclinic ZrO_2 at $2\theta = 28.17°$ formed by oxidizing Zr at 749°C. The nearest peaks are the monoclinic ZrO_2 (110) reflection at $2\theta = 24.48°$ and the tetragonal ZrO_2 (111) reflection at $2\theta = 30.15°$. The arrows indicate the angular range used in the calculation of the integral breadth.

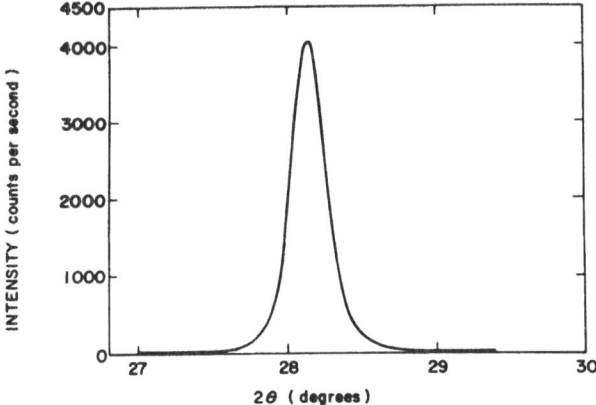

Fig. 4. The (111) peak of monoclinic ZrO_2 at $2\theta = 28.17°$ formed by oxidizing Zr at 975°C. The nearest peaks are the monoclinic ZrO_2 (110) reflection at $2\theta = 24.48°$ and the tetragonal ZrO_2 (111) reflection at $2\theta = 30.15°$. The complete angular range was used in the calculation of the integral breadth.

ment from neighboring peaks. The nearest ZrO_2 peaks are the monoclinic (110) reflection at $2\theta = 24.48°$, and the tetragonal (111) reflection at $2\theta = 0.15°$. The lowest angle Zr reflection is at $2\theta = 31.96°$ and was shown by the specimen oxidized at 490°C. A monochromator was used between sample and detector and the power stabilized.   The (111) reflection of a silicon standard at $2\theta = 28.42°$ was also stepscanned to determine the instrumental broadening b. The integral breadths B were computed (Table I) and each peak was drawn (Figs. 1–4).

Results

The grain sizes derived from the integral breadths by the use of Eqs. (1) and (2) are shown in Fig. 5. There is a gradual increase in grain size with temperature.

It has been tacitly assumed that the observed broadening has no significant contribution from strains in the oxide produced either during oxidation or during cooling. However, if the strain had been produced on cooling, the strain, and hence the line broadening, would have been expected to increase as the oxidation temperature increased, and this is not observed. A line profile analysis of the data is also being undertaken. A further assumption in the interpretation of the data is that the grain thickness normal to the surface of the crystallites having a certain orientation is representative of an average grain size.

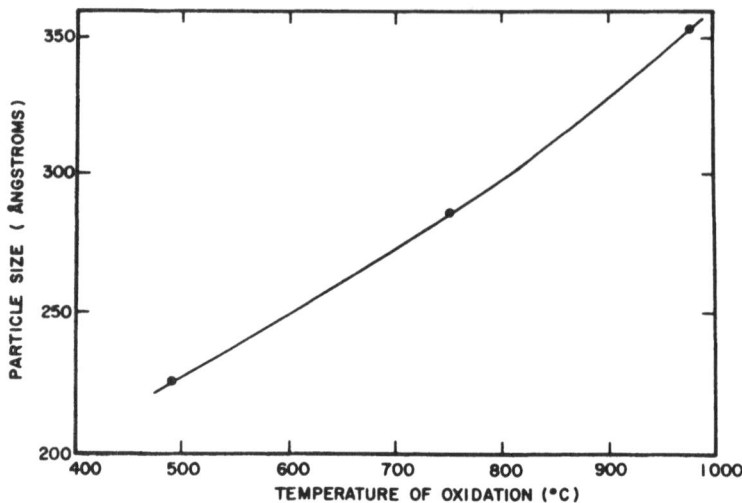

Fig. 5. Particle size of ZrO_2 formed on Zr by oxidation at various temperatures.

TRANSPORT MECHANISM

The oxidation of zirconium proceeds by the movement of oxygen ions into the oxide. With the exception of Cox and Roy ([4]), this movement has been assumed to be a lattice diffusion. However, the fact of a very fine grain size in the oxide formed by oxidation emphasizes the importance of attempting to distinguish between diffusion in the lattice and grain boundary diffusion. This can only be accomplished by directly measuring the movement of the oxygen in the ZrO_2 during oxidation.

Experimental Work

Reactor-grade sponge zirconium specimens were oxidized in oxygen of normal isotopic content (99.76% O^{16}, 0.20% O^{18}) at temperatures in the range 500°–1100°C. After a known time the normal isotopic oxygen, "O_2^{16}," was pumped out of the system and, with the sample still at temperature, oxygen enriched to 90% O^{18} was admitted to the reaction tube. After rapid cooling to room temperature the samples were examined in an Ion Microprobe Mass Spectrometer,* in which the surface of the sample is bombarded by a beam of high-energy inert-gas ions produced by a low-voltage arc discharge. The beam hits the target at an oblique angle of incidence, and succes-

---

*G. C. A. Technology Division, Bedford, Mass.

sive monolayers are sputtered, and a fraction of the material is ionized. These secondary ions are focused electrostatically in a double-focusing mass spectrometer. Sputtering rates used were in the range 1–10 μ/hr, or 3–30 monolayers per second. By determining the O^{18}/O^{16} ratio as a function of sputtering time, or depth, the O^{18} profile is also known, because the total oxygen concentration remains sensibly constant in ZrO_2.

Results

A typical plot, using a sputtering rate of 10 μ/hr, is shown in Fig. 6. The Zr sample was oxidized at 975°C for 64 hr in O_2^{16} and then for 4 hr in O_2^{18}. The ordinate represents the fraction of O^{18} exchanged, normalized to compensate for the fact that the gas is only 90% O^{18}. The $O^{18}/(O^{16} + O^{18})$ ratio of the oxide at the oxide–metal interface is only about one half of that in the gas phase. At long times the experimental points may be in error due to the fact that sputtering does not give rise to a planar interface as it goes into the oxide. There is a cratering because of the energy distribution across the beam. Various specimens were cored so that they would occupy only the central portion of the beam, where the ion energies are more nearly equal.

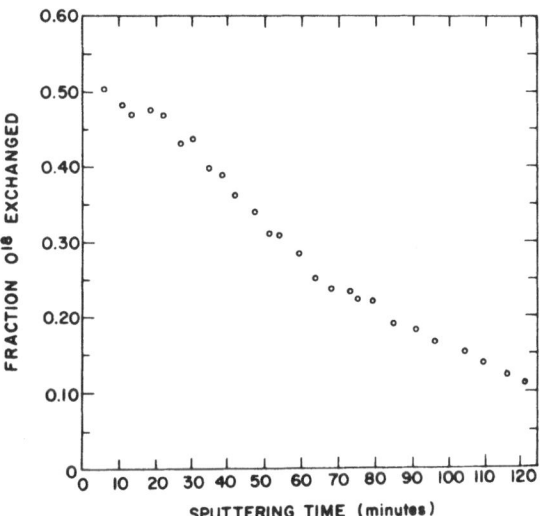

Fig. 6. The O^{18} concentration profile in ZrO_2 after oxidation of Zr at 975°C for 64 hr in O_2^{16} and 4 hr in O_2^{18}. Sputtering rate is 10 μ/hr.

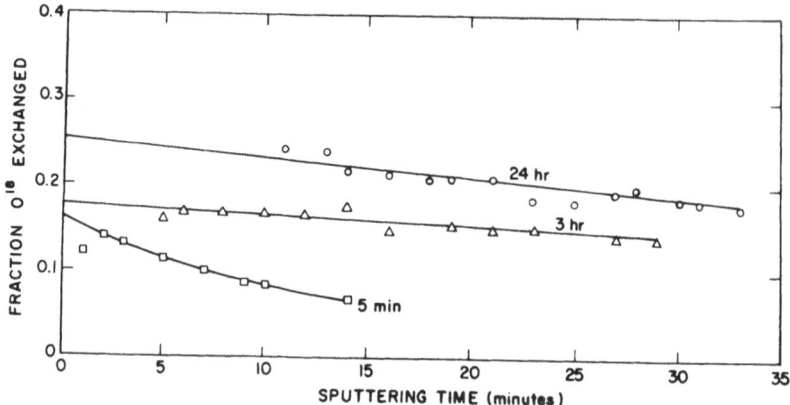

Fig. 7. The O^{18} concentration profile in ZrO_2 after oxidation of Zr at various temperatures. The figures in parentheses are the times in O_2^{16} and O_2^{18} (in hours unless otherwise stated). The sputtering rate is 1 μ/hr.

Figure 7 shows other results using sputtering rates of about 1 μ/hr for samples oxidized at various temperatures. The numbers in parenthesis give the time (in hours unless otherwise stated) in O_2^{16} and then in O_2^{18}. The mole fraction O^{18} exchanged at the surface varies markedly with temperature.

Experiments were also performed at 600° and 910°C, in which the time in O_2^{18} was varied for the same initial time in O_2^{16} (Figs. 8 and 9). At 600°C the surface fraction exchanged varied from 0.17 to 0.25. With a short O_2^{18} exposure the slow sputtering rate of 1 μ/hr gives a pronounced decrease in concentration with sputtering time or depth. At longer times the curves become more nearly horizontal. After exposures of 4 and 16 hr to O_2^{18} at 910°C the lines become horizontal for the sputtering times used. Extrapolation of

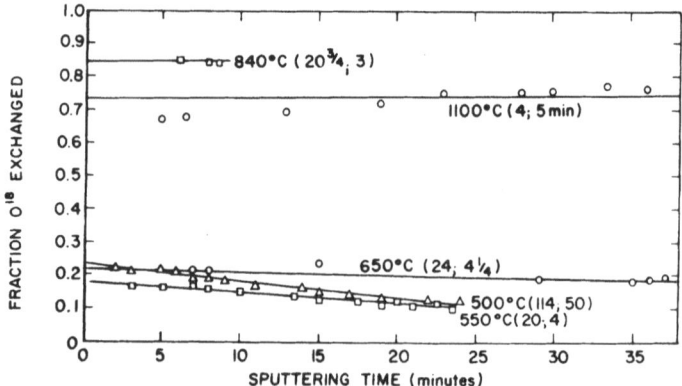

Fig. 8. The O^{18} concentration profile in ZrO_2 after oxidation of Zr at 600°C for 20 hr in O_2^{16} and in O_2^{18} for the various times shown. The sputtering rate is 1 μ/hr.

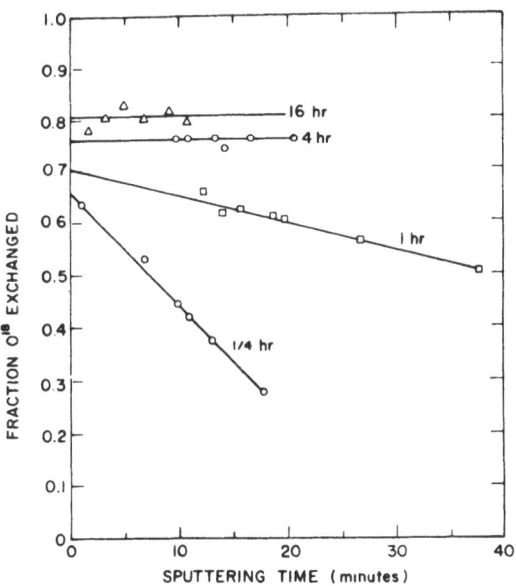

Fig. 9. The O^{18} concentration profile in ZrO_2 formed on Zr by oxidation at 910°C for 20 hr in O_2^{16} and in O_2^{18} for the various times shown. The sputtering rate is 1 μ/hr.

these curves to zero sputtering time gives the fraction exchanged in a volume at the oxide–metal interface. The experiments at 600° and 910°C will be used later in the analysis.

Theory

A value of unity for the fraction exchanged in a volume at the metal–oxide interface would indicate that the oxidation is proceeding by lattice diffusion. The fact that the fraction exchanged is much less than unity may be taken to suggest that many of the oxygen sites are not taking part in the diffusion process and that transport is occurring through easy diffusion paths in the oxide which we shall consider to be grain boundaries. We shall further assume a model in which the effective volume fraction of the grain boundaries is equal to the fraction O^{18} exchanged in a volume at the oxide–gas interface after infinitesimally short times in O_2^{18}. At short times in O_2^{18} the oxidation takes place primarily through the diffusion of oxygen in these easy diffusion paths.

However, after longer exposures to O_2^{18} there is a slow diffusion of O^{18} from sites in the grain boundaries into the body of the grains, and the

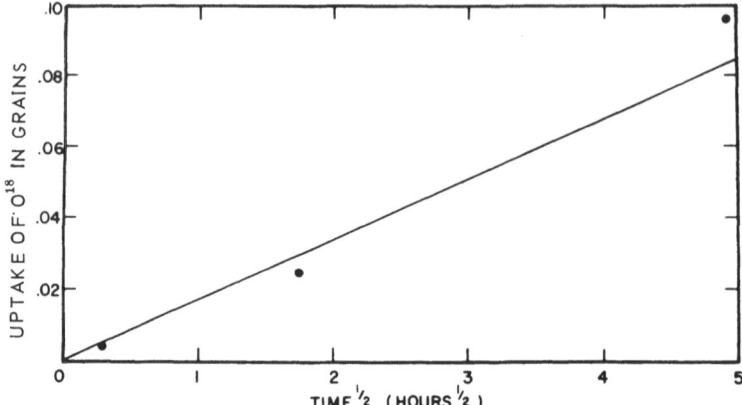

Fig. 10. The O^{18} uptake in the oxide grains at 600°C as a function of the square root of the exposure time to O_2^{18}.

experimental technique measures the average O^{18} content of the oxide being sputtered. The variation of the average O^{18} content at the oxide–gas interface with sputtering time is shown in Figs. 10 and 11. The information is normalized to represent the amount taken up by the body of the grains, considered as a fraction of the total amount taken up by the grains after infinitely long diffusion times. The terms $M_0, M_t,$ and M_∞ are the amounts of O^{18} taken up by a volume of material very near the oxide–gas interface after a very short time in O_2^{18}, after time t in O_2^{18}, and after infinite time, respectively. Then, according to our model, M_0/M_∞ is the volume fraction of easy diffusion paths.

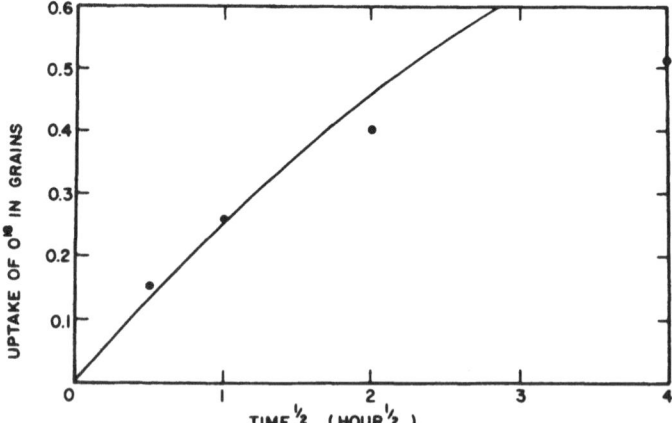

Fig. 11. The O^{18} uptake in the oxide grains at 910°C as a function of the square root of the exposure time to O_2^{18}.

Lattice Diffusion

The data at 600°C fit well on a parabolic plot characteristic of the shape of the curve for small uptake by a sphere, which may be expressed as a function of $(D_v t)^{1/2}/a$ [5], where D_v is the diffusion coefficient for oxygen in the body of the grains, which are assumed spherical, with a radius a. By substituting the value of $a = 125$ Å from Fig. 5 into the known relation [5] $(M_t - M_0)/(M_\infty - M_0) = f(D_v t/a^2)$ and using the results in Fig. 10, we find that $D_v(600°C) = 1.1 \times 10^{-20}$ cm^2 sec^{-1}.

At 910°C the experimental points deviate appreciably from the theoretical curve. This may be due to the fact that the crystallites do not approach a spherical geometry, so that the presence of sharp corners, for example, will imply a high initial rate of exchange. This is equivalent to suggesting that there may be a distribution of particle sizes rather than a unique particle size a. Substitution of $a = 166$ Å from Fig. 5 yields the value $D_v(910°C) = 4.2 \times 10^{-18}$ cm^2 sec^{-1}.

We may correlate our knowledge of the grain size from X-ray line-broadening with our estimate of the effective grain boundary volume. Let δ be the grain boundary thickness. Then the volume fraction of grain boundaries $V_f \approx 1 - (2a/2a + \delta)^3$, which implies a grain boundary thickness δ of 14 Å at 600°C and 118 Å at 910°C. If the latter value is real, a matter open to question, then it may reflect the presence of dislocation networks which have arisen during a martensitic inversion (see later section).

Line Diffusion

We may now consider the conditions for the O^{18} exposure to be short enough that the O^{18} profile represent line diffusion. This has been called type C diffusion by Harrison [6], and will obtain when

$$(D_v t)^{1/2} < \delta \tag{3}$$

This is satisfied for both $t = 5$ min at 600°C and $t = 15$ min at 910°C.

To derive the line diffusion coefficient D_l from the concentration profile of O^{18} we change the scale of the ordinate so that the fraction exchanged at the oxide-oxygen interface reads unity, and fit the curve to an error function plot [5]. It is found that $D_l(600°C) = 9.0 \times 10^{-13}$ and $D_l(910°C) = 7.7 \times 10^{-13}$ cm^2 sec^{-1}.

We may also derive a lower limit to the diffusion coefficient at 975°C by assuming that condition (3) is obeyed for oxidation for 4 hr in O$_2^{18}$ (Fig. 6), which is $D_l(975°C) > 9 \times 10^{-11}$ cm^2 sec^{-1}.

Finally, a lower limit on the diffusion coefficient at 1100°C may be obtained from the fact that at the end of 5 min (Fig. 7)

$$(D_l 300)^{1/2} > 10^{-4}/2 \tag{4}$$

which implies that $D_l(1100°C) > 8 \times 10^{-12} \, cm^2 \, sec^{-1}$.

Discussion

The analysis of the diffusion data may be criticized because of the implicit assumption that the diffusion coefficients are not functions of depth into the oxide. However, because oxidation is occurring there must be a gradient in oxygen potential, with an implied gradient in oxygen vacancy concentration, so that the diffusion coefficient D may be written $\sim D_0(1 + \alpha x)$, where x is the distance into the oxide. This is formally equivalent to considering diffusion into a cylinder rather than diffusion into a semi-infinite plane, and will be discussed in a later work.

Our treatment has implicitly assumed that the line diffusion will approximate a three-dimensional random walk. We might intuitively expect this to obtain if $(D_l t)^{\frac{1}{2}} > 2a$, so that a diffusing atom will assay a three-dimentional network of boundaries. This condition is well met for the shortest times in O_2^{18} at each temperature.

INVERSION

ZrO_2 may exist in a monoclinic form below about 1100°C and in a tetragonal form above 1100°C. The inversion between the two forms is diffusionless ([7]).

Small disks of Zr were spot-welded to a platinum ribbon resistance heater in an MRC high temperature camera. Filtered $CuK\alpha$ radiation was used. Two Pt/Pt-10%Rh thermocouples were also spot-welded to opposite sides of the disk. The camera was evacuated to 10^{-5} Torr at room temperature and subsequently pressured with a few psi of helium which had been gettered by chips of a 50/50Ti–Zr alloy. The specimen was taken up to various temperatures in the range 500–1000°C. At the desired temperature the system was pressurized with oxygen. The detector was set to oscillate over the range $2\theta = 27.4$–30.8°, which includes the (111) reflection of monoclinic ZrO_2 at 28.17° and the (111) reflection from tetragonal ZrO_2 at 30.15°. The scanning speed was 2° min^{-1}. As soon as oxygen was admitted, but not before, monoclinic and tetragonal peaks were observed at all temperatures in the range considered.

The linear absorption coefficient for ZrO_2 is $\sim 15 \mu$ for $CuK\alpha$ radiation, so that we are assaying an external layer about $7\frac{1}{2} \mu$ in thickness.

At 920°C the monoclinic and tetragonal peaks have equal intensity and continue to have approximately equal intensity after long times of oxidation, which would indicate that the tetragonal phase once formed does not transform back to monoclinic.

At 500°C the tetragonal to monoclinic intensity ratio reaches a maximum of about $\frac{1}{8}$ after about 20 min, when the oxide thickness is theoretically about $\frac{3}{4}\,\mu$.

At 920°C we could expect any tetragonal oxide formed initially at the oxide–metal interface to remain tetragonal after it moves away from the interface, because of the nature of the hysteresis in the transformation ([7]). However, this does not explain the existence of tetragonal oxide at 500° and 720°C.

In view of the very fine grain size observed the existence of the tetragonal phase at the two low temperatures may be correlated with the observation ([8]) that very small particles of ZrO_2, ~ 250 Å, formed by decomposition of various zirconium salts are stable to room temperature.

CONCLUSION

An attempt has been made to study the detailed mechanism and structure that obtains during the oxidation of zirconium. It has been shown that there is a high volume fraction of short-circuit diffusion paths in the oxide formed on Zr in the temperature range 500–1000°C. The results of tracer experiments are in agreement with a short-circuit diffusion mechanism. The assumption of line diffusion is shown to imply intuitively reasonable values of both the grain boundary thickness and the ratio of the line diffusion coefficient to the coefficient for lattice diffusion.

ACKNOWLEDGMENTS

One of us (J.B.L.) would like to offer his sincere thanks to his advisor, Prof. Arthur Bienenstock, and to Prof. David Turnbull for their constructive help, their encouragement, and their example.

REFERENCES

1. J. Bénard, paper presented at the Fall Meeting of the Electrochemical Society, Philadelphia, 1966.
2. H. P. Klug and L. A. Alexander, *X-Ray Diffraction Procedures for Polycrystalline and Amorphous Materials*, John Wiley, New York, 1954, Chapter 9.
3. J. Chute, Atomic Energy of Canada Report, AECL—1999, Aug. 1964.
4. B. Cox and C. Roy, *Electrochem. Tech.*, **4**: 121 (1966).
5. J. Crank, *Mathematics of Diffusion*, Oxford University Press, London, 1957.
6. L. G. Harrison, *Trans. Faraday Soc.* **57**: 1191, (1961).
7. B. Ya. Sukharevskii, B. G. Alapin, and A. M. Gavoish, *Dokl. Akad. Nauk SSSR* **156**: 677 (1964); G. M. Wolten, Aerospace Corporation Report ATN–63(9213)–2, 21 March, 1963.
8. R. C. Garvie, paper presented at the Fall Meeting, American Ceramic Society,

Washington, 1963; K. S. Mazdiyasni, C. T. Lynch, and J. S. Smith, *J. Am. Ceram. Soc.* **48**: 372 (1965).

DISCUSSION

John H. Hensler (*University of Melbourne*): I was particularly heartened by the ratios of lattice and boundary coefficients. A similar ratio is observed in metal systems. The ratios for boundary diffusion in magnesium oxide are also of interest. If the boundary coefficient were about 10^5 times the lattice coefficient, this would facilitate the interpretation of the same sort of results that we have obtained. In the X-ray detection work, was there any evidence of a preferred orientation relationship between the two phases?

Answer: I had done some previous work using single-crystal zirconia, which exhibits a very high degree of preferred orientation. It appeared that a single grain of zirconium gives rise to quasisingle grain of ZrO_2 even though the grain size is only a few hundred angstroms. The errors in the intensity measurements when monitoring the tetragonal and monoclinic phase are very large because we are oscillating over $3°$ – the highest speed of the detector is $2°/\text{min}$. All I am considering is the ratio of the maximum intensities, which is really only an approximation.

Chapter 28

Space Charge Distribution at Silver Chloride–Aqueous Solution Interface

C. A. Steidel,* H. A. Hoyen, Jr.,† and Che-Yu Li

Department of Materials Science and Engineering
Cornell University, Ithaca, New York

Transport phenomena at the interface between a silver chloride single crystal and aqueous solutions of different silver ion concentrations have been studied. The purpose of this work is to investigate the nature of the space charge distribution at this interface. Single crystals of silver chloride with a controlled doping level and crystallographic orientation are used. The experiments include DC conductance measurements, applied DC potential relaxation studies, and some AC conductance and capacitance measurements. The results of all these experiments are consistent, showing that a significant portion of the space charge distribution exists at the solid side of the interface. The theory of Grimley and Mott on space charge distribution is modified in order to interpret the experimental results and to account for the origin of the space charge distribution.

INTRODUCTION

Extensive theoretical and experimental studies have been carried out on the space charge distribution at ionic crystal–aqueous solution interfaces ([1,2]). The origin of the space charge distribution is an important question which still remains unanswered. Grimley and Mott ([3]) and Grimley ([4]) have proposed that a potential difference exists across the interface in order that the electrochemical potentials of the potential-determining ions are equal in both sides of the interface. A consequence of their argument is that there must be a space charge distribution in the solid side of the interface consisting of point defects, e.g., Frenkel defects in silver halides. There is little direct evidence for the existence of this space charge distribution.

Matejec ([5]) has measured the DC conductance of thin silver bromide single crystals in contact with aqueous solutions of various silver ion concentrations. He found that the conductance of the crystals varied with the silver ion concentration and concluded the applicability of the Grimley–

---

*Present address: Bell Telephone Laboratories, Allentown, Pennsylvania.
†Present address: Kodak Park Research Laboratories, Rochester, New York.

Mott theory. His interpretation of the experimental results, however, ignored the space charge-limited transport theories ([6]) and therefore can be questioned. Davies and Holliday ([7]) and Otterwill and Woodbridge ([8]) have applied the theory of Grimley and Mott to interpret their results of electro-phoresis experiments on silver halide colloidal systems. In their work only the space charge potential in solutions is measured, and the impurity content in solid is unknown. Adsorption experiments on silver halide colloidal sys-tems ([9,10]) of high ionic strength solutions give the excess charge in the solid, but yield no information on the nature and distribution of this charge. An impor-tant aim of this work therefore is to test the validity of the concept of Grimley and Mott by investigating the nature of the space charge distribution in the solid side of the interface. The effects of the impurity level in the solid phase, chemisorbed surface charges, and the changes in the χ potential on the space charge distribution at the interface are not included in the Grimley–Mott analysis. The χ potential is the phase boundary potential as defined by Lange ([11]) and as used by Grimley and Mott. In the next section we will extend the Grimley–Mott analysis to include these effects.

We have made DC conductance, DC applied potential relaxation, and AC conductance and capacitance measurements. These experiments con-firm the validity of the Grimley–Mott theory. However, additional effects mentioned previously must be included in order to explain the origin of the space charge distribution at the silver chloride–aqueous solution interface.

EXTENSION OF THE GRIMLEY AND MOTT ANALYSIS

The space charge potential at the silver chloride–aqueous solution inter-face can be derived from thermodynamic arguments following the concept of Grimley and Mott with the assumptions that the phase boundary potential χ remains constant and that there are no adsorbed surface charges. For the silver chloride–aqueous solution system the silver ion is the most mobile one in the solid and is therefore the potential-determining ion. Let η_{Ag}^S and η_{Ag}^C be the electrochemical potentials of the silver ion in the solution and in the crystal, respectively. Assuming ideal solutions,

$$\eta_{Ag}^S = |e| V_S(0) + \mu_{Ag}^{0,S} + kT \ln C_{Ag}^S \tag{1a}$$

$$\eta_{Ag}^C = |e| V_C(0) + \mu_{Ag}^{0,C} + kT \ln C_{Ag}^C \tag{1b}$$

where $V_C(0)$ is the potential difference between a point just inside the crystal at the interface and a point in the bulk crystal (the sign of the potential is determined relative to the bulk value); $V_S(0)$ is defined similarly for the solu-tion (the sign is determined relative to bulk solution); $\mu_{Ag}^{0,C}$ and $\mu_{Ag}^{0,S}$ are standard chemical potentials; and C_{Ag}^C and C_{Ag}^S are bulk concentrations in mole fractions.

From the equilibrium condition $\eta_{Ag}^S = \eta_{Ag}^C$ the total space charge potention λ in the absence of adsorbed surface charge and for constant potential is

$$|e|\lambda = -kT \ln [C_{Ag}^S/C_{Ag}^S(0)] = |e|V_S(0) - |e|V_C(0) \tag{2}$$

where $C_{Ag}^S(0)$ is the silver ion concentration at the isoelectric point, $\lambda = 0$. It should be noted that at the isoelectric point for this case there is no space charge distribution in either side of the interface.

Following the statistical mechanical method of Grimley and Mott, we have derived an expression evaluating $C_{Ag}^S(0)$ for crystals of various impurity content in the absence of adsorbed surface charge ([2]):

$$\exp[(2U_{Ag} - 2H_{Ag} + 2|e|\chi + E_F)/kT] = \frac{\alpha N_{Ag}^V(\infty)[C_{Ag}^S(0)]^2}{N_{Ag}^i(\infty)[1 - 2pc]} \tag{3}$$

where c is the concentration ratio of ion pairs and water molecules in solution, U_{Ag} is the lattice energy of a silver ion, H_{Ag} is the hydration energy of a silver ion in solution, E_F is the formation energy of a Frenkel pair, $N_{Ag}^V(\infty)$ is the bulk silver ion vacancy concentration, $N_{Ag}^i(\infty)$ is the bulk silver ion interstitial concentration, α is the ratio of the number of interstitial sites to the number of silver ion lattice sites, and p is the number of water molecules in the hydration shell of an ion in solution. From this expression the isoelectric point of the system depends on the impurity level in the crystal.

Let σ_S and σ_C be the total charge per unit interfacial area in the solution and in the crystal, respectively. In the absence of adsorbed surface charge, charge neutrality requires

$$\sigma_S = \sigma_C \tag{4}$$

Using equations (1) and (3), the Poisson–Boltzmann equation, and appropriate boundary conditions, the potential distribution and the charge distribution in the solid and in the solution can be calculated.

In order to investigate the space charge distribution due solely to chemisorbed surface charges σ_{ad}, consider the hypothetical case where $\mu_{Ag}^S = \mu_{Ag}^C$, $\lambda = 0$, and $\chi = 0$. To maintain the charge neutrality condition and the equilibrium condition, a space charge distribution will be produced in both sides of the interface. The equilibrium condition demands that the electrochemical potential of the silver ion be equal in both sides of the interface. If there is a space charge potential $V_S(0)$ in the solution, from Eqs. (1a) and (1b) there must be an equal space charge potential $V_C(0)$ in the solid. For this hypothetical case the charge neutrality condition and the equilibrium condition become

$$\sigma_S + \sigma_C + \sigma_{ad} = 0 \tag{5a}$$

$$V_C(0) = V_S(0) \tag{5b}$$

The isoelectric point for this case corresponds to $\sigma_{ad} = 0$ and no space charge distribution in either side of the interface.

In the presence of adsorbed surface charge the charge neutrality condition and the equilibrium condition used by Grimley and Mott, Eqs. (2) and (4), respectively, will therefore be modified. We note that the χ potential remains constant,

$$\sigma_S + \sigma_C + \sigma_{ad} = 0 \tag{6a}$$

$$|e| V_S(0) - |e| V_C(0) = -kT \ln [C^S_{Ag}/C^S_{Ag}(0)] \tag{6b}$$

It should be noted that the equilibrium conditions, Eqs. (2) and (6b), have the same form whether there is adsorbed surface charge or not. The magnitudes of $V_S(0)$ and $V_C(0)$, however, will differ in these two cases because of the difference in the charge neutrality conditions, Eqs. (4) and (6a). An important consequence of the above consideration is that in the presence of adsorbed surface charge and constant χ potential the space charge potentials $V_S(0)$ and $V_C(0)$ will not be zero at $C^S_{Ag}(0)$, and in fact will not be zero at the same value of silver ion concentration in solution. The definition of isoelectric point therefore becomes ambiguous and it is necessary to define isoelectric point separately for each side of the interface corresponding to values of silver ion concentration in solution where $V_S(0) = 0$ or $V_C(0) = 0$. If only $V_S(0)$ is measured, as in the electrophoresis experiment, the isoelectric point determined for $V_S(0) = 0$ does not ensure that $V_C(0)$ is zero at the same time.

If the phase boundary potential χ varies with silver ion concentration in the solution, another complication is introduced. In this case the equilibrium condition becomes

$$|e|\lambda = |e| V_S(0) - |e| V_C(0) = -kT \ln \frac{C^S_{Ag}}{C^S_{Ag}(0)} - |e|(\chi_C - \chi_0) \tag{7}$$

where χ_C and χ_0 are phase boundary potentials at silver concentrations of C^S_{Ag} and $C^S_{Ag}(0)$, respectively. In the presence of adsorbed surface charge the possibility still exists that $V_S(0)$ and $V_C(0)$ will not be zero at the same value of silver ion concentration in solution.

EXPERIMENTAL PROCEDURE

Single-crystal silver chloride films approximately 0.01 cm thick were prepared from a Harshaw crystal by chemically polishing with 3% KCN solution. From AC conductance measurements the crystals were found to contain nominally 5×10^{17} divalent cation impurities per cm^3. The surfaces were (210) orientation in all experiments and were examined by X-ray diffraction and interference microscopy for perfection.

The schematic diagram of the conductance cell is shown in Fig. 1. The crystals were mounted in the cell using Dow Corning silicone cement such that about 0.3 cm^2 of crystal surface was exposed to solution. On each side of the crystal the cell contained 1 M KNO$_3$ solutions of equal silver ion concentrations saturated with AgCl. The solutions were prepared with Fisher water and the various silver ion concentrations were obtained by adding appropriate amounts of AgNO$_3$ or KCl. The cell was placed in a constant temperature bath controlling to $\pm 0.01°C$. Four electrodes of pure silver wire were placed in the cell for electrical measurements. The silver ion concentration can be represented by P_{Ag}, which is defined as the logarithm of the reciprocal silver ion concentration in moles per liter. The equivalence point of AgCl solution, i.e., where the Ag$^+$ and Cl$^-$ concentrations are equal, is at a P_{Ag} of 4.8 at room temperature.

DC conductance measurements were carried out by measuring the current across electrodes A and A′ using a Sargent MR recorder at various DC applied voltages and over a range of P_{Ag}. The activation energy for DC conductance was obtained using 200 mV applied voltage between 30–40°C.

For DC applied voltage relaxation experiments 100 mV was applied across electrodes A and A′. The potential drop across electrodes B and B′, i.e., essentially the applied potential difference across the crystal, was measured with a Keithley 610 electrometer. After steady state had been obtained the applied voltage was removed and the potential decay measured across electrodes B and B′ was recorded with a Varian G-14A-1 recorder.

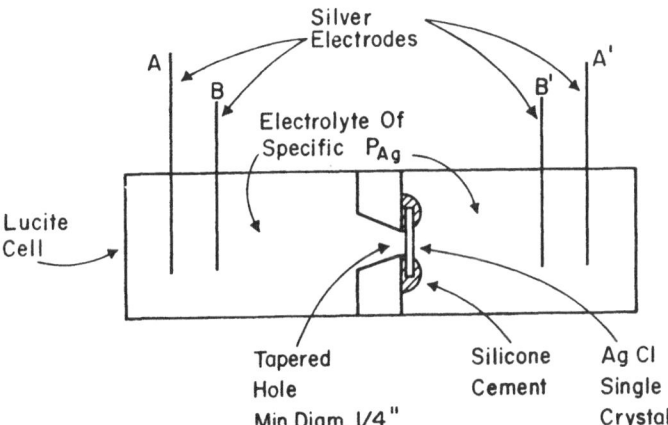

Fig. 1. Schematic diagram of conductance cell used for AC and DC experiments.

In the AC experiments the equivalent circuit of parallel resistor R_M and capacitor C_M for the silver chloride–solution system is measured as a function of frequency, applied AC voltage, and P_{Ag} using a Wayne–Kerr Universal Bridge coupled with a General Radio oscillator 1203-B and tuned amplifier null detector 1232-A The frequency was varied from 50 to 10^4 cps and voltage from 0.025 to 20 V. No voltage dependence was observed up to 0.8 V and the data shown later were taken at 50 and 1592 cps and at 0.3 V.

RESULTS AND DISCUSSION

DC Conductance and Potential Relaxation Measurements

DC conductance data are presented as a function of P_{Ag} and applied voltage in Fig. 2. The current versus P_{Ag} curve is in good qualitative agreement with Matejec's results on silver bromide ([5]). Contrary to Matejec's data, however, the current–voltage relationship at a P_{Ag} of 4.8 is not ohmic, as shown in Fig. 3. This nonohmic behavior suggests the possibility of a space-charge-limited transport process ([6]). The activation energy obtained from an Arrhenius curve of log I versus $1/T$ is plotted in Fig. 4 as a function of P_{Ag}. The activation energies calculated are higher than that for the motion of silver ion vacancies ([12]), which are the predominant current carriers in the AgCl crystal used in this study. This evidence also suggests the existence of a P_{Ag}-dependent space charge potential barrier.

A representative applied potential relaxation curve is shown in Fig. 5. The difference between the applied potential of 100 mV across electrodes A and A' and the steady state voltage $V^a_{BB'}$ across electrodes B and B' corre-

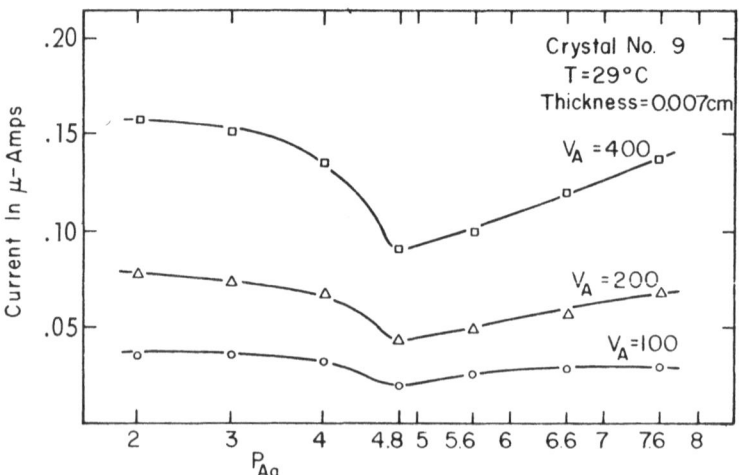

Fig. 2. DC conductance versus P_{Ag} for AgCl at applied voltages of 100, 200, and 400 mV.

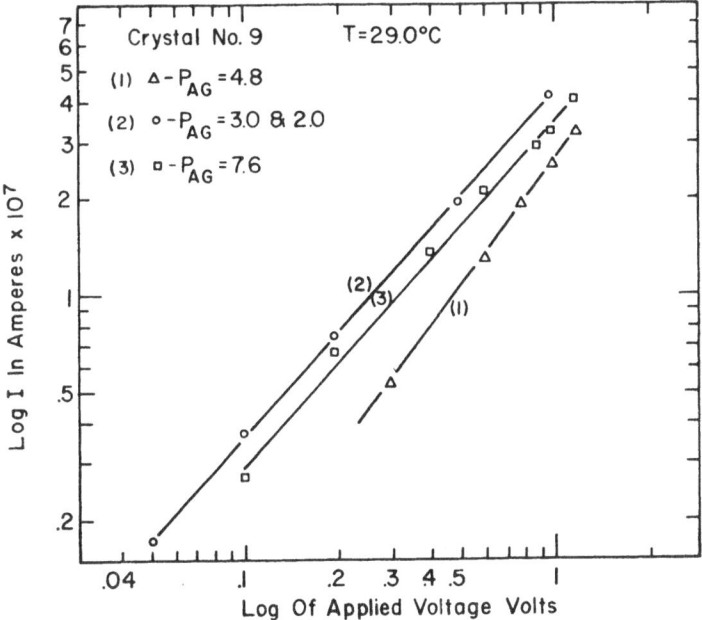

Fig. 3. DC current–voltage characteristics for AgCl crystal.

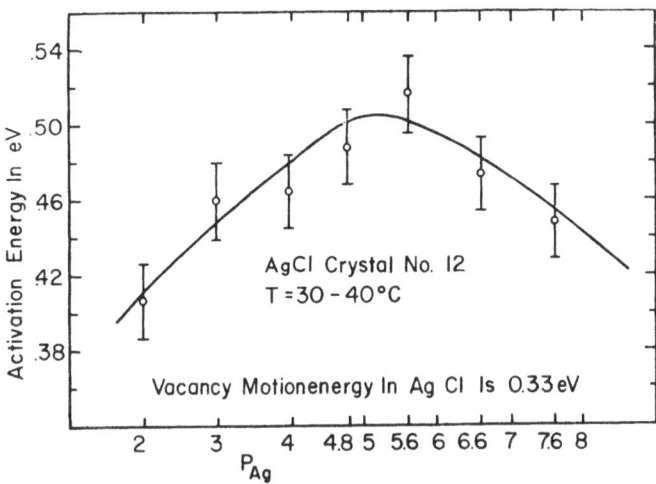

Fig. 4. Activation energy of DC conductance process for AgCl at 200 mV applied potential.

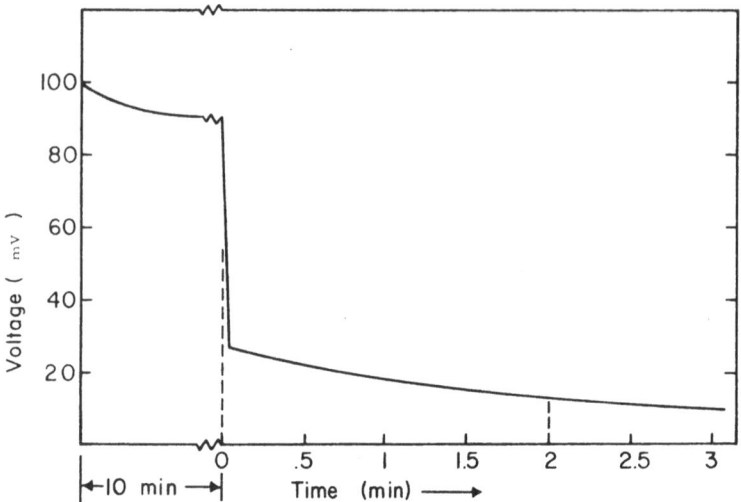

Fig. 5. DC applied relaxation potential versus time.

sponds to the potential barrier at electrodes A and A'. The measured values of $V_{BB'}^a$ at various P_{Ag} are shown in the second column of Table I. From these values it is seen that the potential barrier at electrodes A and A' increases with decreasing silver ion concentrations in the solution. After steady state has been reached, at time $t = 0$ the applied potential is removed. The potential difference between electrodes B and B' decreases immediately to a finite value $V_{BB'}^b$ and then decays slowly to zero. The decay curve is not a simple exponential function of time, suggesting a multiple relaxation process. Two types of models exist describing space-charge-limited current ([13]), the diode theory, applying to thin barrier layers, and the diffusion theory, for thick

TABLE I

Applied Voltage (mV) versus P_{Ag} for Crystal No. 2-45*

P_{Ag}	$V_{BB'}^a$ with appl. pot. steady state	$V_{BB'}^b$ without appl. pot. at $t = 0$	V_B	V_{S1}	V_{S2}
2	94.8	1.8	93	0.9	0.9
3	92.2	4.0	88.2	2.0	2.0
4.8	87.7	29.4	58.3	18.2	11.2
5.6	85.6	19.0	66.6	6.0	13.0
6.6	83.0	9.0	74.0	3.5	5.5
7.6	80.5	6.0	74.5	2.8	3.2

*$T = 28°C$; thickness = 0.01 cm; 100 mV across electrode AA'.

barrier layers. Since the relaxation time is longer than that expected for a thin potential barrier, the latter is adopted in our discussion. The long relaxation process observed is associated with the redistribution of point defects in the space charge region.

In order to apply the diffusion theory to our results, we assume that the applied potential $V_{BB'}^a$ at steady state is the potential difference across the AgCl crystal and that there is a negligible applied potential drop in the solution and across the silver chloride–aqueous solution interface. The $V_{BB'}^a$ can be divided into three portions, V_{S1}, V_{S2}, and and V_B, where V_{S1} and V_{S2} are the applied potentials across the space charge regions and V_B is the remaining applied potential, as shown in Fig. 6. The space charge potential $V_C(0)$ in the solid measured relative to bulk potential is also shown in Fig. 6. In our experiments the silver chloride single crystals have approximately 5×10^{17} divalent cation impurities per cm³, corresponding to 5×10^{17} silver ion vacancies and a negligible concentration of silver ion interstitials. Assuming that the impurities are immobile, the space charge region will consist of an excess or deficiency of silver ion vacancies for a positive or negative $V_C(0)$, respectively.

After the applied voltage has been removed we shall assume that the potential $V_{BB'}^b$ shown in the third column of Table I is equal to the sum of V_{S1} and V_{S2}. This assumption is intuitively reasonable, since the magnitudes of the bulk resistance and capacitance will yield a very short relaxation time for V_B as compared to the relaxation time for V_{S1} and V_{S2}, which involves atomic rearrangement in the space charge region. In the following discussion,

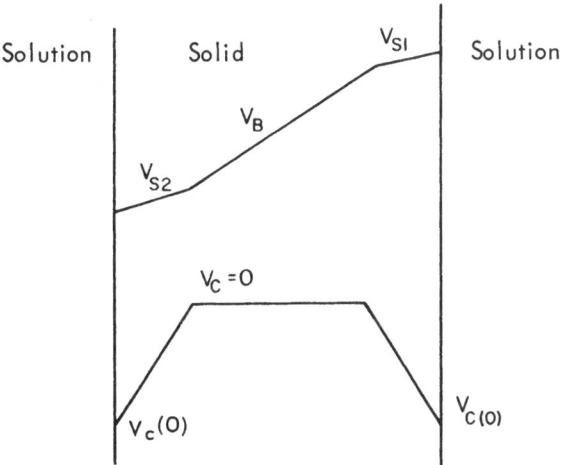

Fig. 6. Applied potential drop $V_{BB'}^a$ and space charge potential $V_C(0)$ change across AgCl crystal.

using the data obtained from the applied potential relaxation measurements and the assumptions for $V^a_{BB'}$ and $V^b_{BB'}$, we shall calculate the DC conductance and activation energy as a function of P_{Ag} and shall explain the nonohmic DC current–voltage relationship.

Following Bardeen ([6]) we can write expressions for the current in the bulk I_B and in the two surface regions I_1 and I_2

$$I_1 = N^V_{Ag}(\infty)|e|u_V\left\{\frac{8\pi N^V_{Ag}(\infty)}{\epsilon}[V_C(0) - V_{S1}]\right\}^{1/2}$$

$$\times \exp\left(\frac{-|e|V_C(0)}{kT}\right)\left[1 - \exp\left(\frac{-|e|V_{S1}}{kT}\right)\right] \qquad (8a)$$

$$I_2 = -N^V_{Ag}(\infty)|e|u_V\left\{\frac{8\pi N^V_{Ag}(\infty)}{\epsilon}[V_C(0) + V_{S2}]\right\}^{1/2}$$

$$\times \exp\left(\frac{-|e|V_C(0)}{kT}\right)\left[1 - \exp\left(\frac{|e|V_{S2}}{kT}\right)\right] \qquad (8b)$$

$$I_3 = N^V_{Ag}(\infty)|e|u_V(V_B/d) \qquad (8c)$$

where ϵ is the permittivity of silver chloride, u_V is the mobility of silver ion vacancies, d is the thickness of the crystal, and potentials are *absolute* values. These equations are applicable for a negative $V_C(0)$, for an applied potential depicted in Fig. 6, and for small values of V_{S1} and V_{S2}. Two additional relationships exist:

$$I_1 = I_2 = I_3 \qquad (9)$$

and

$$V^a_{BB'} = V_{S1} + V_{S2} + V_B \qquad (10)$$

From Eq. (8c) and from measured values of V_B the current versus P_{Ag} curve can be calculated as shown in Fig. 7, curve 2. The values of V_B, which equal the difference of $V^a_{BB'}$ and $V^b_{BB'}$ are shown in the fourth column of Table I for a sample 0.01 cm thick. Curve one in Fig. 7 is reproduced from the lowest curve in Fig. 2 for a crystal 0.007 cm thick. The calculated curve agrees well with that of the measured one except that the latter shows a large current change as a function of P_{Ag}. This is to be expected, because the measured curve was obtained from a thinner crystal, where the surface effects should be more predominant.

The space charge potential in the solid $V_C(0)$ may also be calculated from the applied potential relaxation data. Knowing $V_C(0)$, the activation energy may then be obtained and consequently compared to measured values. From

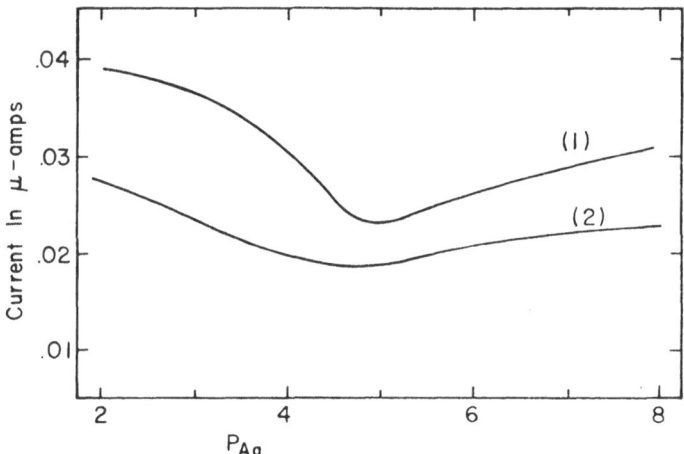

Fig. 7. Current versus P_{Ag} for AgCl. Upper curve: Experimental for 0.007 cm thick crystal. Lower curve: Calculated for 0.01-cm thick crystal relative to 0.007-cm thick crystal.

the values of $V_{BB'}^a$, $V_{BB'}^b$, and V_B and from Eqs. $(8a)$–(10), the applied potentials V_{S1} and V_{S2} in the space charge regions and the space charge potential in the solid, $V_C(0)$, have been calculated-as a function of P_{Ag} by numerical calculations. The result of these calculations are shown in Table I and Fig. 8. The negative values of $V_C(0)$ exhibit a maximum with respect to P_{Ag} near

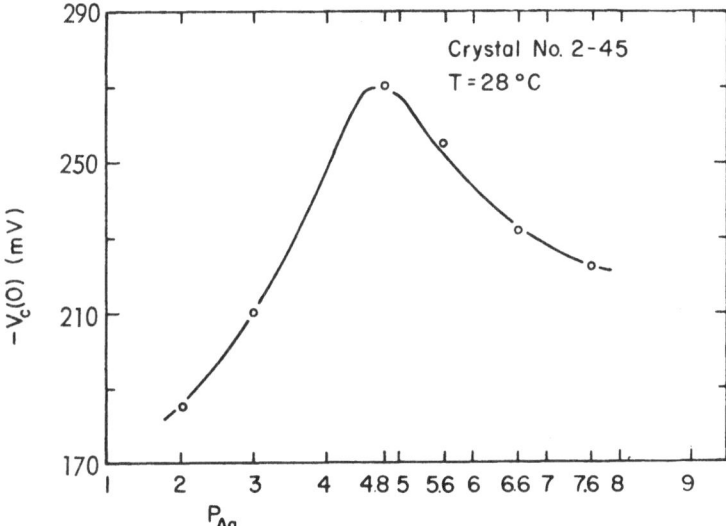

Fig. 8. Barrier height in solid versus P_{Ag} for AgCl.

the equivalence point, corresponding to a maximum in the space charge potential barrier in the solid and to a maximum deficiency of silver ion vacancies. These results are consistent with the minimum in DC conductance observed in current measurements.

From Eqs. (8a) and (8b), it is seen that for small values of V_{S1} and V_{S2} the activation energy for the current is the sum of $V_C(0)$ and an energy of 0.33 eV for the motion of silver ion vacancies[12]. Comparing Figs. 2 and 8, both the experimental activation energy and the calculated activation energy exhibit a maximum near the equivalence point and a greater decrease with P_{Ag} on the silver side of the equivalence point. The calculated activation energy is about 10% greater than the experimental values. Considering the approximations made in (8a) and (8b), however, this agreement is satisfactory.

The current versus applied voltage relationship can be explained from the values of V_{S1} and V_{S2}. In Eqs. (8a) and (8b) the exponential terms involving V_{S1} and V_{S2} contribute to the nonlinear current–voltage relationship. For larger values of V_{S1} and V_{S2} a greater deviation from linearity will be observed, as shown in Fig. 3.

Since our calculations agree well with the experimental data, we may conclude that the assumptions made previously are reasonable and that a space charge distribution is operative in the solid side of the silver chloride–aqueous solution interface. This space charge distribution results from a negative potential $V_C(0)$ in the solid and corresponds to a deficiency in silver ion vacancies, the maximum occurring near the equivalence point. This conclusion will be further supported in the following section on AC conductance and capacitance measurements.

AC Conductance and Capacitance Measurements

The experimental conductance R_M and capacitance C_M as a function of P_{Ag} at a frequency of 50 cps and for a silver chloride crystal 0.01 cm thick are shown in Figs. 9 and 10. Above 1000 cps the conductance and capacitance are dependent on frequency, but are constant with P_{Ag} within experimental accuracy, the value at 1592 cps being 3.1×10^{-6} ohm^{-1} and 5.5×10^{-11} F, respectively. The two space charge regions and the bulk region in the silver chloride crystal can be represented by an equivalent electrical circuit shown in Fig. 11, where R_S and C_S represent the surface space charge region and R_B and C_B the bulk crystal. We assume that the electrode–solution interface, the solution, and the crystal–solution interface do not contribute to the measured R_M and C_M.

The following two relationships exist:

$$\frac{R_M}{1 + \omega^2 R_M^2 C_M^2} = \frac{2R_S}{1 + \omega^2 R_S^2 C_S^2} + \frac{R_B}{1 + \omega^2 R_B^2 C_B^2} \tag{11a}$$

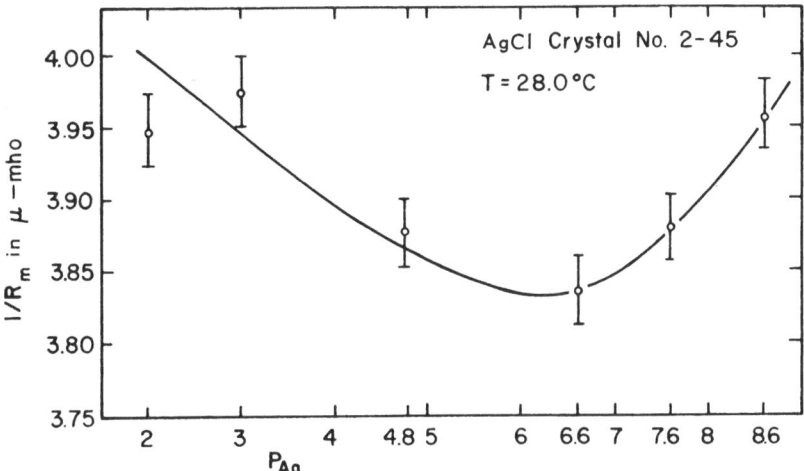

Fig. 9. Experimental conductance versus P_{Ag} for AgCl at 50 cps.

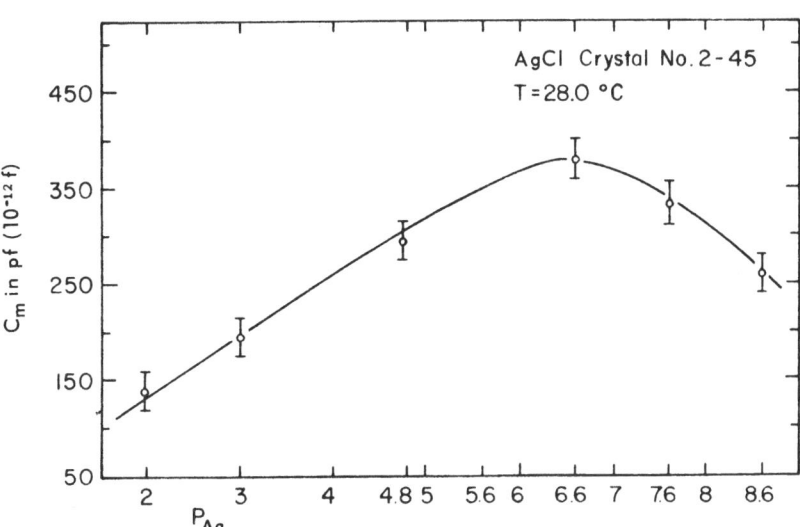

Fig. 10. Experimental capacitance versus P_{Ag} for AgCl at 50 cps.

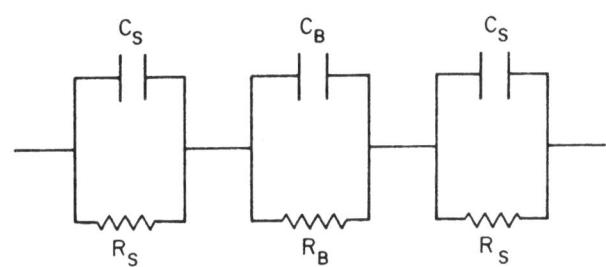

Fig. 11. Equivalent electrical model circuit for charge distribution in solid.

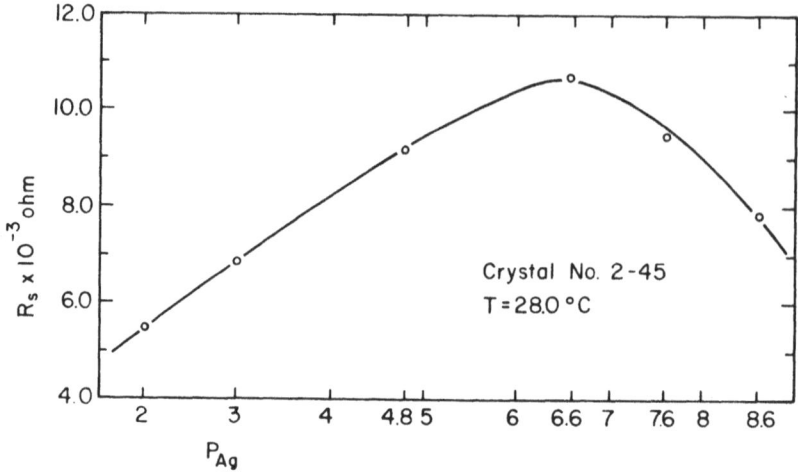

Fig. 12. Calculated surface resistance versus P_{Ag} for AgCl.

$$\frac{R_M^2 C_M}{1 + \omega^2 R_M^2 C_M^2} = \frac{2R_S^2 C_S}{1 + \omega^2 R_S^2 C_S^2} + \frac{R_B^2 C_B}{1 + \omega^2 R_B^2 C_B^2} \qquad (11b)$$

where ω is 2π times the applied frequency. From the values of R_M and C_M at 50 and 1592 cps we may calculate R_S and C_S as a function of P_{Ag} by numerical methods. The results are plotted in Figs. 12 and 13, the values of R_B and C_B being 2.4×10^5 ohms and 5.4×10^{-11} F, respectively. There exist dependences of R_S and C_S on frequency, and this complicated problem will be discussed in future reports. Essentially, R_S and C_S, which represent the space charge region, contain nonuniform elements resulting from poten-

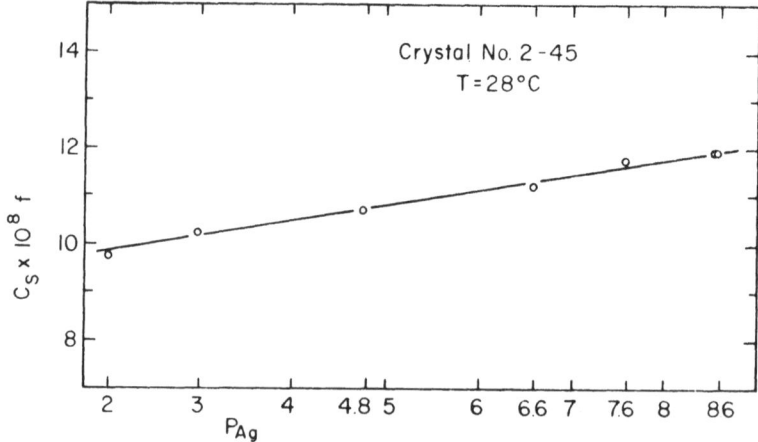

Fig. 13. Calculated surface capacitance versus P_{Ag} for AgCl.

tial and charge distributions. Of interest here is that the calculated surface resistance R_S shows a maximum with P_{Ag} near the equivalence point and a greater change on the silver side of the equivalence point, consistent with the results of our DC studies.

Origin of Space Charge Potential and Isoelectric Point

The isoelectric points of silver iodide and silver bromide colloidal systems determined by electrophoresis experiments, which measure the space charge potential in solution $V_s(0)$, agree well with the isoelectric points found by adsorption experiments, which measure the excess charge in the solid, and $\sigma_{ad} + \sigma_c$ [1]. In these experiments the isoelectric point for AgBr was found to be near the equivalence point, $P_{Ag} = 6.1$. From Matejec's data on the DC conductance experiments, Fig. 14, a finite change in conductance exists near the equivalence point over a range of doping level. If our interpretation of DC conductance is correct, a finite space charge potential $V_c(0)$ must exist in the solid for these AgBr crystals near the equivalence point. We may conclude that for AgBr at the isoelectric point based on $V_s(0) = 0$ there is a finite space charge potential in the solid.

From measurements of the space charge potential in solution the iso-electric point for silver chloride colloidal systems is found to be at $P_{Ag} \simeq 4$ [1]. In our study a finite space charge potential $V_c(0)$ in the solid is found to

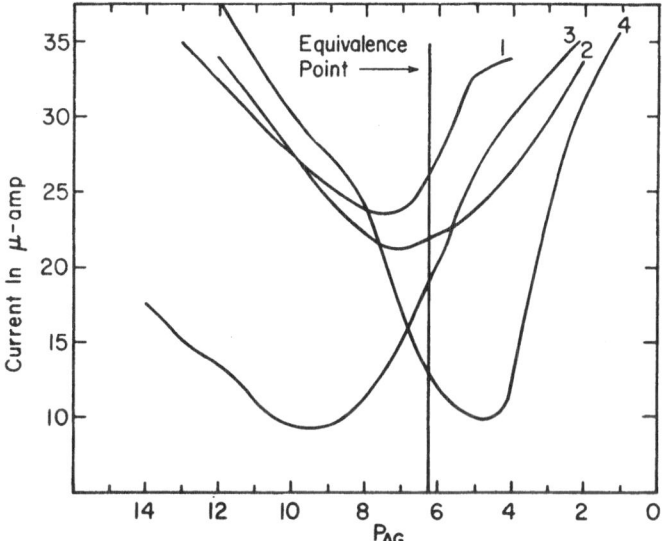

Fig. 14. DC Conductance for AgBr crystals of varied doping levels. Curve 1, normally pure crystal; curve 2, crystal with 0.01 % Ag_2S after bromination; curve 3, crystal with 0.01 % Ag_2S prior to bromination; curve 4, crystal with 0.01 % $CdBr_2$. Data taken from Matejec [5].

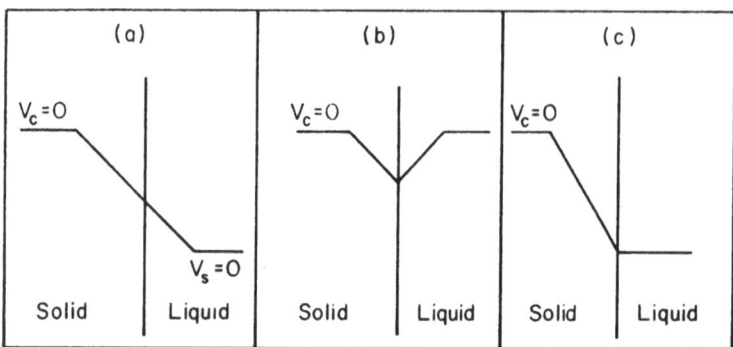

Fig. 15. Space charge potential distribution across solid AgCl–aqueous solution interface. (a) from Grimley–Mott model, (b) from adsorption of negative ions, (c) total effect of (a) and (b).

exist over a range of P_{Ag} from two to eight. We may also conclude that for AgCl a finite space charge potential exists in the solid at $V_s(0) = 0$. Both these conclusions raise a question as to the meaning of the isoelectric point observed from experiments on colloidal systems.

We have stated in our extension of the Grimley and Mott analysis that at the isoelectric point both $V_c(0)$ and $V_s(0)$ have to be zero for the following two cases: first, in the absence of adsorbed surface ions due to chemisorption; second, where the space charge is caused only by adsorbed surface ions with $\mu^C_{Ag} = \mu^S_{Ag}$, $\lambda = 0$ and $\chi = 0$. For the situation described in the above paragraph both these cases do not apply. We may visualize that $V_s(0) = 0$ may arise as follows. In Fig. 15(a) the potential profile shown represents the distribution of space charge potential calculated from Eq. (7). In Fig. 15(b) the profile is due to the chemisorbed surface charge, Eqs. (5a) and (5b). If the space charge potentials in the solution are equal and of opposite sign, combining the potentials in these two figures yields the result in Fig. 15(c), where $V_s(0) = 0$, $V_c(0)$, is finite, and $\sigma_{ad} + \sigma_C = 0$.

In our studies we have assumed an adsorption isotherm and a dependence of the χ potential on P_{Ag}. Using Eqs. (6a) and (7) and the Poisson-Boltzmann equation, we have been able to calculate $V_c(0)$ as a function of P_{Ag} to fit the measured values. Since there is no evidence to support the assumed isotherm and χ potentials, we will not report the results of this calculation.

In conclusion, the results of this study support the concept introduced by Grimley and Mott that at an ionic crystal–aqueous solution interface the equilibrium condition is defined by the electrochemical potentials of potential-determining ions both in the solid and in the solution. The strongest argument for this is the observed space charge distribution in the solid, which is dependent on P_{Ag}. The results of this study also show that the space charge distribution and the isoelectric point at the interface result from this equilibrium condition and from the effects of adsorbed surface ions and χ potential.

ACKNOWLEDGMENTS

This work was supported by the U. S. Atomic Energy Commission Contract No. AT(30–1)3228. The authors wish to thank Dr. A. H. Clark and Mr. B. F. Addis for assistance in the experimental work.

REFERENCES

1. J. T. G. Overbeek, "Electrochemistry of the Double Layer," in: *Colloid Science*, (H. R. Kruyt, ed.), Vol. 1, Elsevier, Amsterdam, 1952.
2. C. A. Steidel, "The Silver Chloride-Aqueous Solution Interface," Ph. D. Thesis, Cornell University (1966).
3. T. B. Grimley and N. F. Mott, *Disc. Faraday Soc.* **1**: 3 (1947).
4. T. B. Grimley, *Proc. Roy. Soc. (London) Ser. A* **201**: 40 (1950).
5. R. Matejec, *Z. Electrochem.* **66**: 326 (1962); *Phot. Sci. and Eng.* **7**: 123 (1963).
6. J. Bardeen, *Bell Systems Tech. J.* **28**: 428 (1949).
7. K. N. Davies and A. K. Holliday *Trans. Faraday Soc.* **48**: 1061 (1952).
8. R. H. Otterwill and R. F. Woodbridge, *J. Colloid Sci.* **19**: 606 (1964).
9. E. L. Mackor, *Rec. Trav. Chim.* **70**: 747 (1951).
10. H. DeBruyn, *Rec. Trav. Chim.* **61**: 12 (1951).
11. E. Lange, *Handb. Exp. Phys.* X11/2, 265 (1933).
12. R. J. Friauf, *J. Phys. Chem.* **66**: 2380 (1962).
13. A. J. Dekker, *Solid State Physics*, Prentice-Hall, Englewood Cliffs, New Jersey, 1957.

DISCUSSION

P. W. M. Jacobs (University of Western Ontario): Could you tell us a little about the frequency dependence of the capacitance? What is the law for the dependency?

Answer (Hoyen, Jr.): The capacitance measured over the frequency range from 40 to 10^4 cps decreased with increasing frequency, approaching a limit at both the high- and low-frequency values. Using an equivalent circuit model as described in the text, the capacitance of the bulk and of the space charge regions for the crystal can be calculated. The dependency is not simple function of the square of the frequency.

H. J. Oel (Max Planck Institute): Maybe I missed the point there, but you are using a kinetic diagram analysis and you are assuming, if I understand you correctly, that you are measuring the projections of the crystal only. I do not see how you can do that because of the contribution from the solution.

Answer: The solution contains a one molar concentration of an inert electrolyte (KNO_3). Therefore essentially all of the applied potential will be across the crystal.

T. J. Gray (Alfred University): I am a little concerned with your measurement insofar as it is obvious that the mechanism you are proposing assumes the presence of a solution capacitative double layer adjacent to the silver fluoride electrode as well as whatever is going on inside, and this can be very severe in the DC case. There isn't any symmetry there. First, why are you not using gauze capacitance electodes adjacent to the two inter-faces and a large electrode close to your silver chloride so that you can actually see what you have got there? You are undoubtedly going to get a double layer, and in addition, it is going to be asymmetric. Second, you simply cannot analyze realistically on a dielectric dispersion curve unless you go over wide limits. Now, if you say you have reached a con-stant level of 1500 cps, which I find absolutely incredible, then I can assure you from all prior experience that for your actual measurements toward zero you are going to have to go down to a hundredth of a cycle to determine your dispersion curve accurately. Our curves are made from approximately 20 cycles or lower up to about one megacycle in order to get full dispersion curves. I am a little concerned when you say you only went from 100 or 200 to 1500 cps.

Answer: First, the space charge contribution in the solution is small compared to the space charge effect in the solid. Second, the size of the electrodes used in our experiments is sufficient as deduced from the data. In addition, the position of the electrodes is not an important factor. Finally, our measurements were obtained over a frequency range from 40 cps to 10^4 cps.

Gray: And you got absolutely no difference all the way through?

Answer: The data of the measured capacitance are independent of P_{Ag} at frequencies greater than 1500 cps, but the data are still dependent on frequency.

Gray: I am not talking about dependence on P_{Ag}. I am talking about the actual dispersion curve for the dielectric loss. Does the curve go up smoothly or does it turn over at very low frequencies? This is the important feature since it is the nature of that curve which allows you to resolve or not to resolve the actual model that you get.

What about at the low frequencies? Does it flatten out and become uniform right at the zero frequency value of the capacitance? If it does not, you do not have a resolvable situation.

Answer: The dependence of the measured capacitance and resistance does indeed yield a typical dispersion curve over the covered frequency range. The capacitance does approach a limit at lower frequencies.

P. W. M. Jacobs (University of Western Ontario): I have a somewhat naive question. What is the actual current, and what is carrying the current? For instance, if you have a silver electrode immersed in an electrolyte and then essentially a silver chloride membrane and then apply a voltage, regardless of how we measure the voltage, let us say we have electrons moving in a particular direction, then in order to have electrons released at this point it seems to me we would have to have some silver ions come in at this level, with electrons coming up at this point. Since this is a silver electrode, we would probably be picking up some silver atoms. How does the current get through here? What is the current mechanism? As I say, perhaps this is a naive question. What is the chemistry of this cell?

Answer: The current-limiting step is the movement of silver ions across the silver chloride crystal via a vacancy mechanism.

Jacobs: But then you are suggesting that there is an electrode potential?

Answer: Yes, but it is small compared to the other potential distributions.

H. J. Oel (Max Planck Institute): Is this why you added potassium nitrate, because this provides the conduction in the electrolyte?

Answer: Yes, this is one of the reasons the solution contained a constant ionic strength of the KNO_3 electrolyte.

Chapter 29

Reaction Kinetics of Wet Chemically Prepared Gadolinium-Iron Garnet

E. A. Giess and R. M. Potemski

IBM Watson Research Center
Yorktown Heights, New York

The reaction kinetics of $Gd_3Fe_5O_{12}$ garnet prepared by a wet-chemical method from alcohol solutions of gadolinium and iron nitrate, thoroughly dried and decomposed, have been investigated at temperatures from 600° to 700°C. Degree of reaction is calculated from magnetic moment measurement at 4.2°K. The data indicate that a complex reaction occurs which does not obey simple reaction equations. The best data fit is obtained from a rate equation of increasing order with increasing temperature. The process has a high activation energy and is complicated by sintering of the extremely small reactant particles and by the formation of a perovskite as a side reaction. At higher temperatures the reaction slows down markedly with time, and it is probable that a diffusion mechanism is operative.

INTRODUCTION

Films of gadolinium-iron garnet, $Gd_3Fe_5O_{12}$, have been prepared by Wade *et al.* ([1]) using a wet-chemical method. The technique involves applying an alcoholic solution of gadolinium-iron nitrate to a substrate as a film which is then dried to a nearly amorphous state and fired for conversion to polycrystalline $Gd_3Fe_5O_{12}$.

In this study the degree of completion of the garnet-forming reaction was calculated from magnetic measurements calibrated with standards. Since precise magnetic analysis of films is difficult, it was decided to study bulk material prepared in a manner analogous to that used for film fabrication. The 600–700°C temperature range for firing this material allowed reaction to proceed in a reasonable period of time.

It is difficult to assign a mechanism to the reaction because the data do not fit any one simple equation or model. Over the temperature range studied the data can be approximated by a reaction equation of increasing order with increasing temperature. The 600–650°C data have been plotted as a pseudo first-order rate process to derive rate constants and an activation energy for the process. The process at higher temperatures, 675° and 700°C, is better represented by an equation resembling higher-order reaction processes. A diffusion mechanism might be operative.

EXPERIMENTAL

A concentrated aqueous solution of iron and gadolinium nitrates corresponding to a 5:3 stoichiometric garnet, $Gd_3Fe_5O_{12}$, was added to ethyl alcohol in a 1:4 volume ratio. Ethyl alcohol improves the viscosity of the solution for application as a film on a substrate and it improves drying of the film. After drying at about 100°C the powder was preheated at 500°C for 16 hr to complete removal of water and decomposition of the nitrate. There was no weight loss upon subsequent firing. Differential thermal analysis showed only an exothermic peak indicating reaction at about 700°C (for a 6°C min^{-1} heating rate), which agrees with the kinetics results. This starting material was brown in color and produced no X-ray pattern, indicating a crystallite size of less than 100 Å. Electron diffraction in reflection resulted in diffuse lines too few in number for any structural interpretation and indicated a crystallite size of from 30 to 50 Å.

The powder was pressed into small disks at a pressure of 1700 psi. In separate experiments it was shown that the disk formation pressure does not affect results. Furnace temperatures were controlled to within ±3°C throughout heat treatment of the disks.

Magnetic measurements were performed at a temperature of 4.2°K with a pendulum magnetometer. Standard samples containing appropriate amounts of Fe_2O_3, Gd_2O_3, and garnet gave the expected magnetization value of 90 emu/g of $Gd_3Fe_5O_{12}$ [2]. The only other possible product is $GdFeO_3$, a perovskite, which, like Fe_2O_3 and Gd_2O_3, is not ferromagnetic. Magnetic saturation is achieved at about 10 kOe and measurements made up to 25 kOe were extrapolated back to zero field to correct for paramagnetic terms.

This magnetic technique was used by Okamura and Shimoizaka [3] to study the reaction kinetics of wet-chemically prepared nickel ferrite and by Economos and Clevenger [4] to study the reaction of dry-mixed nickel oxide and iron oxide powders. Rubinchik et al. [5] detected magnetically the presence of yttrium-iron garnet in the reaction between Fe_2O_3 and Y_2O_3.

RESULTS AND DISCUSSION

The lack of high precision in reaction data makes it difficult to determine exactly the reaction mechanism or mechanisms. For purposes of discussion the 600–650°C results have been plotted in Figs. 1–3 as a first-order reaction. For this relationship the reaction rate

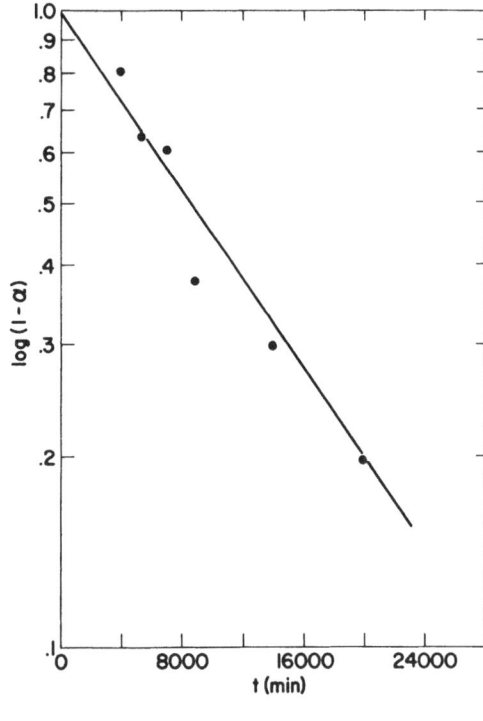

Fig. 1. Log$(1 - \alpha)$ versus time t at 600°C. (Note that ordinate values are $(1 - \alpha)$ plotted on a log scale.)

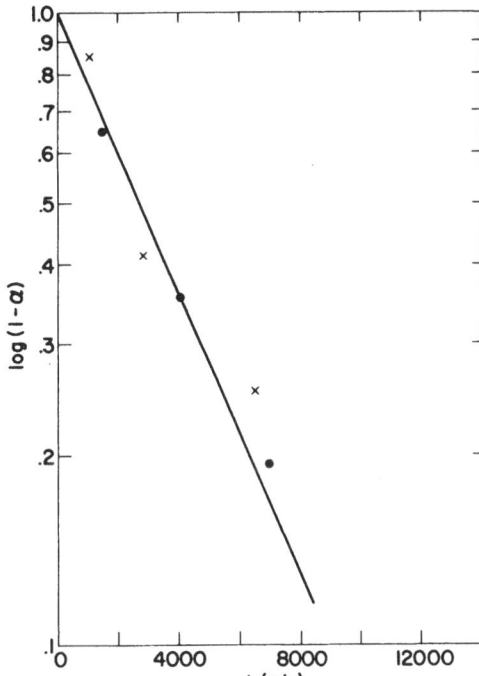

Fig. 2. Log$(1 - \alpha)$ versus time t at 625°C; × represents a second series.

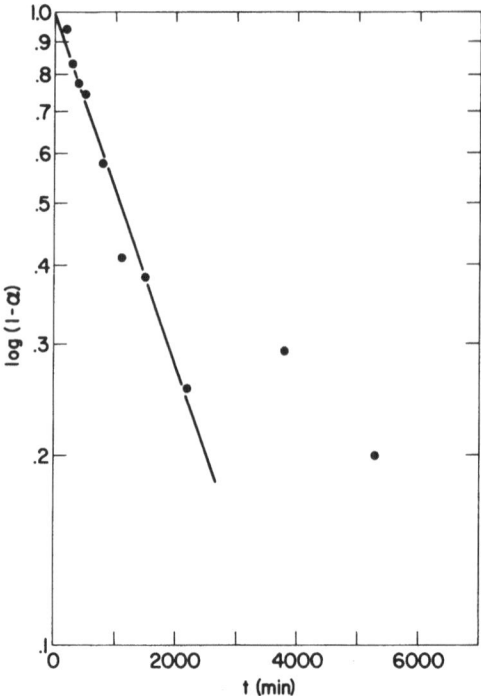

Fig. 3. Log$(1 - \alpha)$ versus time t at 650°C.

$$d\alpha/dt = k(1 - \alpha)^n \tag{1}$$

where α is the fraction of material reacted, t is time, k is a rate constant, and n is the order of reaction. For $n = 1$ (first order) this may be integrated to give

$$\ln(1 - \alpha) = -kt \tag{2}$$

The data yield the rate constants k shown in Table I. The activation energy derived from these rate constants is 69 kcal/mole, which is not unexpected for oxide systems.

TABLE I
Rate Constants for 600°, 625°, and 650°C

Firing temperature (°C)	Rate constant (min⁻¹)
600	8.11×10^{-5}
625	2.59×10^{-4}
650	6.36×10^{-4}

In true first-order reactions the rate is directly proportional to the concentration or volume of the remaining unreacted material, as defined by Eq. (1). Homogeneous first-order gas reactions—namely, decompositions—are well-known and are usually unimolecular. The present system is considerably more complex.

Several solid-state oxide reactions have been found to follow a first-order relationship during some part of the reaction. Murray and White [6], for example, found that the thermal dehydration of fine-grained clay minerals in air is a first-order type of reaction. The decomposition and dissociation of many other solids have also been found to follow first-order kinetics [7].

Perhaps the closest analogy to the present system in the 600–650°C range is the γ- to α-alumina phase transformation studied by Clark and White [8], which also empirically obeyed a first-order relationship and which finally slowed down upon nearing completion. This process also had a high activation energy (79 kcal/mole).

One possible explanation of the reaction mechanism is that of Hume and Colvin [9]. If the linear rate of propagation of the reaction interface is so great that the particle may be regarded as completely reacted as soon as a nucleus develops, first-order kinetics can follow. As long as nucleation of individual particles is independent of neighboring nucleation events, reaction rate will depend on the rate of nucleation, which in turn is proportional to the number of unreacted particles. These conditions will be met best by systems comprised of very small particles such as those found in the present case.

There are, however, other theoretical treatments which lead to first-order-type relationships. Avrami [10] has treated the kinetics of phase changes and derived the equation:

$$\ln (1 - \alpha) = -kt^p \tag{3}$$

where p lies between one and four, depending on whether growth is linear, plate-like, or polyhedral, and whether there is a high or low probability of active nuclei occurrence. In this model a value of $p = 1$ corresponds to linear growth, i.e., one-dimensional, and a high probability of active nuclei occurrence.

Diffusion models can also result in mathematical solutions approximating a first-order relationship. Ham [11] considered diffusion-limited precipitation from a supersaturated solution upon an array of nuclei. This would fit the case where one element in solid solution in a second begins to diffuse to and precipitate on dislocations. The state of aggregation of the present system probably does not resemble this situation.

Dünwald and Wagner [12] derived an equation based on the diffusion of one reactant into particles of a second. The result takes the form of a first-order type equation with a nonzero intercept. The precision of the present results is not sufficient to rule out this possibility.

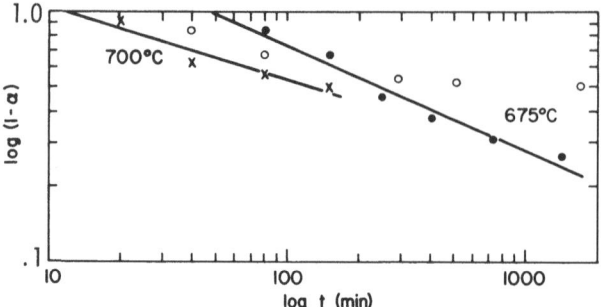

Fig. 4. Log $(1 - \alpha)$ versus log t at 675 and 700°C. Circles represent a second set of samples at 700°C.

Rubinchik et al. ([5]) have determined an activation energy of 20 kcal/mole for reaction between Y_2O_3 and Fe_2O_3 which obeyed a spherical diffusion model. Sphere models treat the case of relatively large particles, i.e., where particle size is significantly greater than the unit cell. It must be recalled that the present system had a starting crystallite size of 30–50 Å, which is equivalent to only three to four unit cell edges, an extremely small size. This could well obscure a diffusion-controlled reaction.

The 675° and 700°C data approximate higher-order reactions according to the least-mean-square straight lines of the log $(1 - \alpha)$ versus log t plot (Fig. 4). Both indicate an initial induction period significantly greater than the heat-up time, which was shown to be less than 5 min in separate heating experiments. The data points for longer time tail off into a final decay period typical of transformation curves in general. At these temperatures the straight lines are of the form

$$(1 - \alpha)^{-m} = k't \qquad (4)$$

The 700°C data approximate a higher-order reaction having a shorter induction period than the 675°C data. The circles in Fig. 4 represent a separate powder fired at 700°C. At much longer times these data asymptotically approach parallelism with the time axis as the reaction slows down.

The final decay period is usually explained in terms of the mutual impingement of regions transforming from separate nuclei which interfere with each other's growth ([13]). This is unlikely here because the decay period begins too soon.

Heavily exposed X-ray patterns of the fired specimens showed both hematite, α-Fe_2O_3, and the perovskite $GdFeO_3$ to be present in small amounts in both high and low temperature samples. Holmquist et al. ([14]) and Rubinchick et al. ([5]) also observed perovskite formation in the reaction between $3Y_2O_3$ and $5Fe_2O_3$ in the same temperature region. The present perovskite had a smaller lattice constant than the literature value and it is therefore believed to be in a defect state. Apparently, the perovskite-structured $GdFeO_3$ forms as an intermediate step or as a product of a side reaction. It is noteworthy that the hematite had a particle size which was large compared with that of the original mixture, so that the reactants must have sintered before reacting completely. This growth of the reactant particles may have been responsible for the observed transition to a slower, diffusion-controlled mechanism.

If a range of particle sizes is considered instead of a monodisperse system in treating the spherical model, a curve of the shape given by the data points in Fig. 4 is obtained. This strongly suggests that a diffusion mechanism may indeed be responsible for the observed behavior.

CONCLUSIONS

An intimate mixture of composition $3Gd_2O_3 \cdot 5Fe_2O_3$ prepared by decomposition of an alcoholic solution of the nitrates did not obey simple reaction equations and showed complex behavior in the 600–700°C temperature range. The data can be represented empirically by a reaction rate equation of increasing order with increasing temperature. The process or processes operating have a fairly high activation energy. The complexity of the system is at least in part related to the extremely small particle size of the reactants.

Sintering of reactant particles and a perovskite side reaction occur. At higher temperatures the latter effects cause the reaction to slow down markedly with time. The possibility of a complex diffusion mechanism cannot be ruled out.

ACKNOWLEDGMENTS

T. J. Mitchell expertly made the magnetization measurements required. R. C. Wnuk helped with X-ray measurements and Miss E. I. Alessandrini did the electron diffraction experiment. F. Holtzberg and A. H. Nethercot contributed several helpful suggestions and conversations.

REFERENCES

1. W. L. Wade, T. Collins, W. W. Malinofsky and W. J. Skudera, *J. Appl. Phys.* **34**: 1219 (1963).
2. F. Bertaut and R. Pauthenet, *Proc. IEE* **B104**: 261 (1957).
3. T. Okamura and J. Shimoizaka, *Sci. Repts. Res. Inst. Tohoku Univ., Ser. A*, **2**: 673 (1950) (in English).
4. G. Economos and T. R. Clevenger, Jr., *J. Am. Ceram. Soc.* **43**: 48 (1960).

5. Ya. S. Rubinchik, M. M. Pavlyuchenko, I. A. Tsybul'ko, and V. G. Leitsina, Russ. *J. Inorg. Chem.* **10**: 907 (1965).

6. P. Murray and J. White, *Trans. Brit. Ceram. Soc.* **48**: 187 (1949).

7. P. W. M. Jacobs and F. C. Tompkins, *Chemistry of the Solid State*, Butterworth and Co., London, 1955.

8. P. W. Clark and J. White, *Trans. Brit. Ceram. Soc.* **49**: 305 (1950).

9. J. Hume and J. Colvin, *Phil. Mag.* **8**: 589 (1929).

10. M. Avrami, *J. Chem. Phys.* **8**: 212 (1940).

11. F. S. Ham, *J. Phys. Chem. Solids* **6**: 335 (1958).

12. H. Dünwald and C. Wagner, *Z. Phys. Chem.* (*Leipzig*) **B24**: 53 (1934).

13. J. W. Christian, *The Theory of Transformation in Metals and Alloys*, Pergamon Press, New York, 1965.

14. W. R. Holmquist, C. F. Kooi, and R. W. Moss, *J. Am. Ceram. Soc.* **44**: 194 (1961).

DISCUSSION

Sutarno (*Department of Energy and Mines, Ottawa*): From Fig. 4 you said that the reaction never goes to completion. It seems to me that you probably get some segregation during your drying. The reason I say this is that we did almost the same kind of work with barium and iron oxides. It segregates very badly and we also found some other ionic process and at the same time disproportionation. Did you check for this? Was there any volatilization of constituents?

Answer (*E. A. Giess*): Disproportionation or segregation would be possible in this system. I don't think we had any volatile constituent left in our starting material after we preheated it at 500°C. We checked pellets for weight loss by firing at high temperatures and there was none. Either no volatile constituent was there or it didn't come out in the firing.

R. J. Bratton (*Westinghouse Research Laboratories*): I have experienced similar problems of segregation during drying of nitrates, and it is necessary to agitate and continue to heat or somehow keep the thing mixed up. An easy way of checking to determine the segregation after a certain amount of reaction is simply to remix and refire it and you'll find, I believe, additional reaction. At least, this is what we found.

Answer: Actually, we were really concerned about what the films were going to do and we found that the films followed pretty much the same trend as the 600–650°C data. In fact, one of the principal things we were seeking was to find out how long we had to fire these films, and using these rate data we extrapolated to 99% reaction and came up with a figure of 5 days at 650°C. Using the same extrapolation for 700° we came up with 3 hr. We tried firing some films for 3 or 4 days and observed them optically, and they were not completely reacted; after 5 days they were. After firing films at 700°C for just 1 rather than 3 hr they were extensively reacted, so we really could not fix the reaction time that closely. In other words, I do not think we were as badly off (with the films) as it appears from the 675° and 700°C data for the dried bulk powders. These results might reflect on the possibility of segregation or disproportionation occurring in bulk material too.

V. D. Fréchette (*Alfred University*): George Economos made a suggestion years ago that these solutions be sprayed into a burner and the volatiles burned off. Has anybody done that?

Answer: I think so—there are some technical ferrites made by spray-drying. We tried to flame-spray various garnet compositions from alcohol–nitrate solutions but it does not work very well. This was in connection with the making of films on glass substrates.

Chapter 30

Kinetics of Crystallization in TiO_2-Opacified Porcelain Enamels

R. A. Eppler

Ceramics Group, Glidden-Durkee Div., SCM Corp.
Baltimore, Md.

Three reactions occur when a TiO_2-opacified porcelain enamel frit is fired: crystallization of anatase, crystalllization of rutile, and inversion of anatase to rutile. The concepts of nucleation and growth theory have been applied to these three reactions to develop an overall kinetic law of transformation which predicts the concentrations of anatase and of rutile as function of the firing conditions. This theoretical development has been found to agree with the results of quantitative X-ray studies.

INTRODUCTION

TiO_2-opacified porcelain enamels are alkali borosilicate glasses containing TiO_2 which have been fused onto metal substrates for decorative purposes and for corrosion resistance.

The opacifier in these enamels, which imparts the properties of reflectance and color, is TiO_2 in the form of either anatase or rutile. The enamels are made by melting together a combination of appropriate ingredients including the TiO_2 and then quenching in water. The shattered glass is then milled with water and small amounts of suspending media and electrolytes and is then applied to a ground-coated metal substrate. The coated substrate is then dried and placed into a preheated furnace and fired at a temperature in the range from 700° to 900°C. During the firing process the frit particles flow together to form an impervious glass coating. In addition, owing to a greatly reduced solubility, much of the titania in the frit crystallizes, primarily, although not exclusively, as anatase. Subsequently some of the anatase inverts to rutile.

This chapter is concerned with an attempt to predict the concentrations of anatase and of rutile as a function of the parameters of time and temperature of firing. Using a typical enamel, 4 in. × 6 in. panels were prepared and fired over a wide range of times and temperatures ([1]). The composition

TABLE I		TABLE II	
Composition of Enamel		**Mill Formula**	
SiO_2	39.6	Frit	1000
TiO_2	21.09	Bald clay	40
P_2O_5	1.31	Bentonite	$2\frac{1}{2}$
B_2O_3	16.2	$NaAlO_2$	$2\frac{1}{2}$
F	6.4	K_2CO_3	$2\frac{1}{2}$
Na_2O	9.4	Gum tragacanth	$\frac{3}{10}$
Li_2O	0.9	H_2O	450
K_2O	5.1		

of the glass is shown in Table I and the mill formula in Table II. The firing temperatures ranged from 660° to 940°C and the times from 1 to 64 min. The concentrations of anatase and of rutile in these panels were then determined by quantitative X-ray techniques ([2]).

Some typical data are shown in Fig. 1. Complete data can be found in ([1]). Here we have plotted the concentrations of anatase and of rutile as a function of time for a given temperature. Of course, similar plots can be constructed at other temperatures. From a consideration of these curves it is possible to discern the essential aspects of the crystallization phenomena

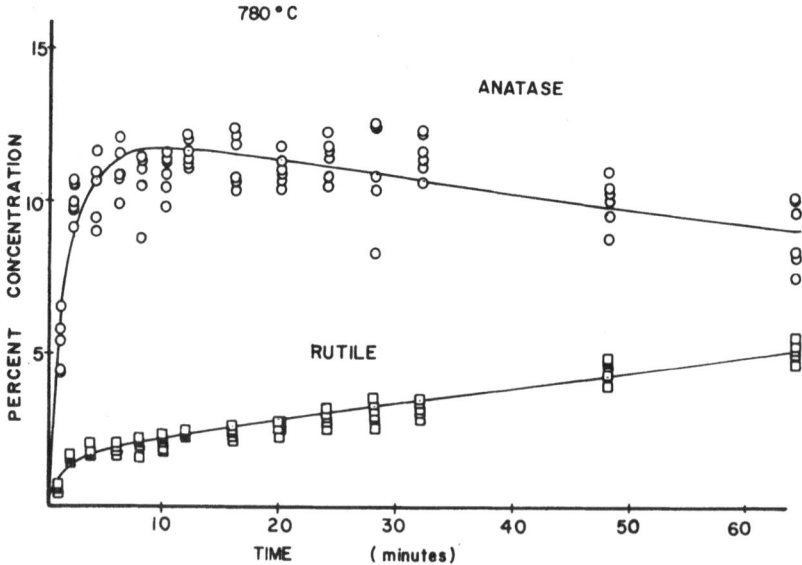

Fig. 1. Concentrations of anatase and rutile in enamel fired at 780°C. Points are experimental data; curves are calculated from Eqs. (17) and (18).

[3]. In the first place, since anatase is a metastable phase at these temperatures, it must appear by a crystallization reaction if at all. Secondly, from the behavior of the curves at longer times we see that we must also consider the inversion of anatase to rutile. Finally, if the inversion reaction were the only source of rutile, then the rutile curve would be S-shaped, with a maximum rate coinciding with the maximum in anatase concentration. We can easily see that this is not the case. The rate of rutile formation decreases monotonically with time. Therefore one must conclude that some of the rutile is formed by direct crystallization. These observations lead to the following model for the firing of TiO_2-opacified enamels [3]:

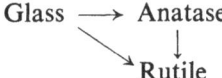

The model states that there are three reactions to be considered in the treatment of the reaction kinetics of the firing of a TiO_2-opacified porcelain enamel: crystallization of anatase, crystallization of rutile, and inversion of anatase to rutile. Our task is to find kinetic expressions for each of these reactions and then combine them to derive overall equations for the anatase and rutile concentrations. Our previous work [3] had indicated that a simple first-order kinetic expression is an excellent description for the inversion reaction:

$$-dA/dt = k_1 A \qquad (1)$$

For the two crystallization reactions, however, this simple treatment is not valid. In particular, a higher order of reaction is required. It was therefore decided to examine the theories of nucleation and growth to see if a suitable equation could be derived therefrom.

THEORY

Since nucleation and growth theory is phenomenological in nature, the physical reasoning involved in the development of the theory will be emphasized here, and the mathematical details will be discussed elsewhere [5]. The theory begins with the postulate that in the unfired glass there are a number N_0 of potential sites for nucleation, or embryos as they are sometimes called, per unit volume. This postulate was first made by Avrami [4] around 1940 in treating a different problem. As will be seen, the mathematical formalism follows Avrami in many respects, although the postulates do not.

Let us write an equation for the change in N, the number of sites/unit volume, with time of firing at a constant temperature:

$$dN = -nN\,dt - [N/(V_1 - V)]\,dV \qquad (2)$$

The first term on the right-hand side allows for the rate at which sites become nuclei and the second term allows for ingestion of embryos by the growing crystals. The terms $V_1 - V$ represent the volume of material which is still crystallizable; it arises from a concept of the enamel as being a volume of crystallizable material imbedded in a completely inert matrix or carrier.

We then introduce the dummy variable $\tau = \int_0^t n\,dt$ and integrate to obtain

$$N = N_0 e^{-\tau}(V_1 - V)/V_1 \qquad (3)$$

Now, define a quantity V_{ex} to be the volume of crystals which would be present if no ingestion of nuclei occurred. Then by definition:

$$V_{ex} = \int_0^\tau V(\tau, Z)N(Z)\,dZ \qquad (4)$$

where $V(\tau, Z)$ is the volume at time τ of a crystal which began to grow at Z, or:

$$V(\tau, Z) = \sigma V_g \int_Z^\tau (G/n)\,d\tau \qquad (5)$$

G in this equation is the volumetric growth rate (cm^3/sec-cm^3 of glass).

The critical assumption in this and all other similar treatments is the choice of what to do about the factors G and n in Eq. (5). In our case both of these are concentration- and time-dependent quantities. However, in vitreous media diffusion is often the rate-controlling step in all processes. If this is so, then in our case it would be reasonable to assume that both have the same concentration-time dependence, so that their ratio is a constant. With this assumption we can substitute Eqs. (3) and (5) into Eq. (4), integrate, and obtain

$$V_{ex} = \sigma N_0 V_g \gamma [e^{-\tau} + \tau - 1] \qquad (6)$$

where γ is the ratio between the nucleation and growth rates.

One final assumption, that of a random distribution of nuclei, permits us to write:

$$dV/dV_{ex} = (V_1 - V)/V_1 \qquad (7)$$

Combining Eqs. (6) and (7), integrating, passing to the limit of large τ, and then differentiating leads us to an overall kinetic law

$$dL/dt \approx (\sigma N_0 V_g / V_1) GL \tag{8}$$

where L is the weight per cent of crystallizable (but still vitreous) material.

This equation states that the rate of disappearance of crystallizable material L is proportional to the growth rate and to the concentration of crystallizable material. The latter term arises in Eq. (7) from the fact that as the crystals grow, they ingest nuclei.

Now we will assume that the growth rate is diffusion controlled, as is customary in treating vitreous systems [6]. This assumption was first examined by Nernst [7], who proposed that

$$G = (DA\rho_g / V_g \delta\rho)(L - S) \tag{9}$$

Substituting this in Eq. (8) gives

$$dL/dt = (-\sigma N_0 DA\rho_g / V_1 \delta\rho)(L - S)L \tag{10}$$

which states that the rate of crystallization is proportional to the excess concentration times the total concentration of crystallizable material. The first term can be considered to be the expression of a dynamic equilibrium between the rate at which atoms arrive at the crystal ($\sim L$) and the rate at which they leave ($\sim S$).

The Nernst theory makes the questionable assumption of a constant geometry for diffusion. A more accurate treatment, due to Nielsen [8], yields the relationship:

$$G = \frac{4\pi r_1 D\rho_g}{V_1 V_g \rho}(L - S)\left(1 - \frac{L - S}{L_0 - S}\right)^{1/3} \tag{11}$$

The essential difference here from Eq. (9) is the addition of the final term. Substituting this equation into Eq. (8) we get

$$\frac{dL}{dt} = \frac{-\sigma N_0 4\pi r_1 D\rho_g}{V_1^2 \rho}(L - S)L\left(1 - \frac{L - S}{L_0 - S}\right)^{1/3} \tag{12}$$

APPLICATION TO TiO$_2$-OPACIFIED ENAMELS

The final step is to apply Eq. (10) or Eq. (12) to the model for TiO$_2$-opacified porcelain enamels in order to obtain an equation which can be compared with experiment. Both cases will be considered. Equation (10) leads to an explicit relation which is easy to use, while Eq. (12) is a more realistic theoretical development leading to a less useful final result.

First we will consider the Nernst theory as represented by Eq. (10).

Using such an equation for each of the crystallization reactions and Eq. (1) for the inversion, one can write

$$\frac{-dL}{dt} = \left(\frac{\sigma_A N_{0A} D A_A \rho_g}{V_1 \delta_A \rho_A} + \frac{\sigma_R N_{0R} D A_R \rho_g}{V_1 \delta_R \rho_R}\right)(L - S)L \tag{13}$$

$$\frac{dA}{dt} = \left(\frac{\sigma_A N_{0A} D A_A \rho_g}{V_1 \delta_A \rho_A}\right)(L - S)L - k_1 A \tag{14}$$

$$\frac{dR}{dt} = \left(\frac{\sigma_R N_{0R} D A_R \rho_g}{V_1 \delta_R \rho_R}\right)(L - S)L + k_1 A \tag{15}$$

Equation (13) is directly integrable, yielding

$$L = \frac{S}{1 + [(S/L_0) - 1]e^{-k_2 t}} \tag{16}$$

where k_2/S is equal to the constant in Eq. (13).

The integration of Eq. (14) is described in detail elsewhere ([5]). Briefly, Eq. (16) is substituted for L in Eq. (14). Then an integrating factor $\theta = e^{-k_2 t}$ is introduced. Finally, the assumption is made that $k_2 \gg k_1$. This assumption is equivalent to stating that crystallization is faster than inversion, which is experimentally verified. The result is

$$A = L_0 k_s e^{-k_1 t}\left[1 - \frac{k_3}{1 + (k_3 - 1)e^{-k_2 t}}\right] \tag{17}$$

whence, by stoichiometry,

$$R = L_0(1 - k_s e^{-k_1 t})\left[1 - \frac{k_3}{1 + (k_3 - 1)e^{-k_2 t}}\right] \tag{18}$$

Since some of the constants have been redefined in order to save space, it is well to indicate their significance: k_1 is the inversion rate constant; k_s the percentage of crystallization which is to anatase; $k_3 L_0 = S$, the solubility of TiO_2 in the matrix; therefore k_3 is the solubility ratio or fraction of uncrystallizable TiO_2; and k_2 is a type of melting rate constant; i.e., the rate at which atoms leave the crystal surface. Therefore $(k_2 k_s/k_3 L_0)$ is the crystallization rate of anatase and $[k_2(1 - k_s)/k_3 L_0]$ is the crystallization rate of rutile.

The correspondence between this theory and experiment is illustrated in Fig. 1, where the two curves shown were calculated by fitting Eqs. (17) and (18) to the anatase and rutile data, respectively. A computerized grid search technique was used in the fitting procedure ([1]). Similar curves can be drawn at the other temperatures studied ([1]). The rate constants and standard deviations for anatase and for rutile at all temperatures studied are given in

TABLE III

Rate Constants and Standard Deviations of the Experimental Data from the Nernst Formulation

Temp. (°C)	k_1	k_2	k_3	k_5	S_A	S_R
940	0.6	9.0	0.52	0.965	1.480	1.190
860	0.11	2.2	0.43	0.94	0.963	1.072
820	0.028	0.64	0.35	0.91	1.091	1.111
800	0.013	0.35	0.305	0.90	1.091	0.530
780	0.005	0.28	0.33	0.885	1.059	0.288
740	0.0006	0.28	0.38	0.865	0.650	0.175
700	0.0003	0.15	0.455	0.87	0.567	0.226
660	0.0009	0.12	0.67	0.92	0.775	0.157

Table III. The differences in standard deviations appear to be due primarily to variable scatter in the data rather than to any systematic discrepancies between theory and experiment. The increase in S_R at higher temperatures is the result of the much larger values of R at these temperatures. We thus conclude that even this very simple theory leads to results which agree very well with experiment.

Let us now turn to the Nielsen formation as represented by Eq. (12). Using such an equation for each of the two crystallization steps and Eq. (1) for the inversion, one can write

$$\frac{-dL}{dt} = \left(\frac{\sigma_A N_{0A} D 4\pi r_{1A}\rho_g}{V_1^2 \rho_A} + \frac{\sigma_e N_{0R} D 4\pi r_{1R}\rho_g}{V_1^2 \rho_R} \right)(L - S)L\left(1 - \frac{L - S}{L_0 - S} \right)^{1/3} \tag{19}$$

$$\frac{dA}{dt} = \left(\frac{\sigma_A N_{0A} D 4\pi r_{1A}\rho_g}{V_1^2 \rho_A} \right)(L - S)L\left(1 - \frac{L - S}{L_0 - S} \right)^{1/3} - k_1 A \tag{20}$$

$$\frac{dR}{dt} = \left(\frac{\sigma_R N_{0R} D 4\pi r_{1R}\rho_g}{V_1^2 \rho_R} \right)(L - S)L\left(1 - \frac{L - S}{L_0 - S} \right)^{1/3} + k_1 A \tag{21}$$

Equation (19) is solved by a change of variables to $\alpha = 1 - [(L - S)/(L_0 - S)]$. Making this substitution and rearranging, we obtain

$$\frac{(L_0 - S)}{S} k_2' t = \int_0^\alpha \frac{d\alpha}{\alpha^{1/3}(1 - \alpha)\{[L_0/(L_0 - S)] - \alpha\}} \tag{22}$$

where k_2'/S is the constant in Eq. (19). Equation (22) is integrated by partial fractions ([5]) to give a function I_N which is a function solely of α and S:

$$k_2' t = I_N(\alpha, S) \tag{23}$$

The function I_N is described explicitly elsewhere [5]. Suffice it to say that it involves a complex function only suitable for computer manipulation.

Equation (20) is then divided by Eq. (19) and the terms in L converted to terms in α to give

$$\frac{dA}{d\alpha} = k_s'(L_0 - S) - \frac{k_1 SA}{k_2'(L_0 - S)(1 - \alpha)\{[L_0/(L_0 - S)] - \alpha\}^{1/3}} \quad (24)$$

Where k_s' is the ratio of the constant in Eq. (20) to that in Eq. (19). This equation is solved with the integrating factor method [5] and the above-mentioned assumption that $k_1 \ll k_2'$ to give

$$A = (L_0 - S)\alpha k_s' e^{-k_1 t} \quad (25)$$

and, by stoichiometry

$$R = (L_0 - S)\alpha(1 - k_s' e^{-k_1 t}) \quad (26)$$

where the term α is given by Eq. (23).

Comparison of these equations with Eqs. (17) and (18) reveals that the constant k_s' is identical in function with k_s of the Nernst development. Since the k_1 are the same, and by definition $(1 - k_3)L_0 = (L_0 - S)$, it is seen that the only difference between the two formulations lies in the specification of the kinetic constants k_2 or k_2'. Furthermore, when Eqs. (25) and (26) are compared with experiment the results are equivalent to the results with Eqs. (17) and (18). Therefore the rather questionable assumption in the Nernst treatment of a constant surface area perpendicular to diffusion does not introduce errors large enough to be detected in our experimental work. Thus for practical purposes one can use the simple Nernst relationship.

CONCLUSION

The concepts of nucleation and growth theory have been applied to develop a set of equations which predict the concentrations of anatase and of rutile as a function of the time and temperature of firing over the whole of the useful firing range—from 660° to 940°C and from 1 to 64 min.

REFERENCES

1. R. A. Eppler, paper presented before the American Ceramic Society, April 1967 in New York; see *Bull. Am. Ceram. Soc.* **46**(4): 361 (1967), *J. Am. Ceram. Soc.*, to be published.
2. H. P. Klug and L. E. Alexander, *X-Ray Diffraction Procedures for Polycrystalline and Amorphous Materials*, John Wiley and Sons, New York, 1954.

3. R. A. Eppler and W. A. McLeran, Jr., *J. Am. Ceram. Soc.* **50**(3): 152–56 (1967); see also *Bull. Am. Ceram. Soc.* **45**(4): 387 (1966).
4. M. Avrami, *J. Chem. Phys.* **7**: 1103–12 (1939); **8**: 212–24 (1940); **9**: 177–84 (1941).
5. R. A. Eppler, *J. Am. Ceram. Soc.*, to be published.
6. D. Turnbull and M. H. Cohen, in: *Modern Aspects of the Vitreous State* (J. D. Mackenzie, ed.), Butterworth and Co., 1960, Volume 1, pp. 38–62.
7. W. Nernst, *Z. Physik. Chem. (Liepzig)* **47**: 52–55 (1904).
8. A. E. Nielsen, *Kinetics of Precipitation*, Pergamon Press, Oxford, 1964.

DISCUSSION

V. D. Fréchette (Alfred University): The titania-opacified frits are not usually melted to a homogeneous glass, are they? In addition, clay and other mill additions are present.

Answer: This particular set of data applies to a commercial frit that I pulled off the commercial lines. However, you must remember that after being melted to a clear glass and then quenched it is then milled in water, and it is amazing how much inhomogeniety you can statistically average out by a milling procedure. For example, I have seen glazes that had whole pieces of unmelted batch in them. Yet they are sold and the customer is perfectly happy, because by the time he mills them and refires them, there is no noticeable difference.

C. H. Greene (Alfred University): Would there be a differentiation between anatase and rutile as far as the models are concerned? What I am getting at is that from a program in Alfred some years ago on recrystallizing glass fired at higher temperature it seems the surface defects remain in the glass at the firing temperature for crystallization control. On this basis we hot pressed the ground frit and we were able to get about double the concentration of crystals, confirming significant surface nucleation. I wondered how this is explained.

Answer: The only thing I can say is that I have looked at electron micrographs, including transmission micrographs, quite often, and there is no question whatsoever but that there is volume nucleation for the most part. This kind of firing cycle is short enough that you can still find where the particle boundaries were. Therefore you can tell just by looking whether or not the crystal could possibly have been formed at the surface. The answer is that the vast majority of them are not. Therefore the overall kinetic behavior treated here will not be greatly affected by any surface effects.

John H. Hensler (University of Melbourne): I would like to compliment Dr. Eppler on being one of the few people who has put in results with some statistical analysis, rather than theory alone.

Answer: I should admit that I am really forced to this statistical treatment. The particle size distribution in these materials is rather broad, and therefore we are forced into using X-ray methods and other methods that take in a large number of particles so that we can statistically average out the particle size problem; I have not yet figured out how to allow for this in a systematic way.

Chapter 31

Effect of Pressure on High-Temperature Dissolution

M. A. Stett and R. M. Fulrath

Inorganic Materials Research Division, Lawrence Radiation Laboratory and Department of Mineral Technology, College of Engineering, University of California, Berkeley, California

The effect of pressure on high-temperature dissolution has been investigated. The solubility of NiO in $Na_2O \cdot 2SiO_2$ glass was found to decrease with applied pressure, indicating that the entry rate of atoms from a solute into a solvent is decreased by increased pressure. The saturation concentration as measured with the electron beam microprobe is an accurate measurement of the liquidus concentration for a given temperature.

INTRODUCTION

Physical properties such as the compressive strength of clays and zirconia have been enhanced by the application of pressure during a chemical reaction ([1-5]). Chemical reactions can occur at lower temperatures under an applied pressure than under vacuum. Carruthers and Wheat ([6]) found mullite formation at 650°C in a china clay hot pressed during the dehydroxylation reaction with 30,000 psi. The process of dissolution can also be affected by an applied pressure.

The dissolution process consists of two steps, the entry of atoms of solute into the solvent and a subsequent diffusion of these atoms through the solvent. Either or both of these steps can be rate controlling. These steps and their characteristics have been discussed previously ([7]). When the migration of the solute atoms in the solvent is limited by the interfacial reaction the expression for the concentration at any point contains error functions and a constant of proportionality ([8]):

$$C = C_s\left[\operatorname{erfc}\left(\frac{X}{2\sqrt{Dt}}\right) - \exp(hX + h^2 Dt)\operatorname{erfc}\left(\frac{X}{2\sqrt{Dt}} + h\sqrt{Dt}\right)\right] \quad (1)$$

where $h = \alpha/D$ and α is the reaction coefficient defined as

$$\alpha(C_s - C_i) = -D \, \partial C/\partial X \qquad (2)$$

and C is the concentration at any point, C_s the saturation coefficient, C_i the interface concentration, X is distance, D the diffusion coefficient, and t time. It is assumed that the reaction is proportional to the difference in chemical activity between the two phases. Simpson and Carter [9] used these equations to study the diffusion of oxygen in calcia-stabilized zirconia, taking into account the surface oxygen exchange.

The self-diffusion coefficient of lead has been shown by Hudson and Hoffman [10] to decrease with increased pressure. The other step in dissolution, the entry, will be shown here to be inhibited by an applied hydrostatic pressure.

The type of data obtained can be used to plot the liquidus of a phase diagram as a function of pressure. Schwerdtfeger [11] has shown that the saturation concentration as measured by the electron microprobe technique used in this study is an accurate measurement of a liquidus. He measured a portion of the liquidus in the Na_2O–SiO_2 system and found agreement to within 1 % or better. The results of his work can be seen in Fig. 1. The dotted line shows the accepted liquidus of Kracek and the solid line shows Schwerdtfeger's determination. In the present work the NiO–$Na_2O \cdot 2SiO_2$ glass system was investigated and the effect of pressure on the entry rate demonstrated.

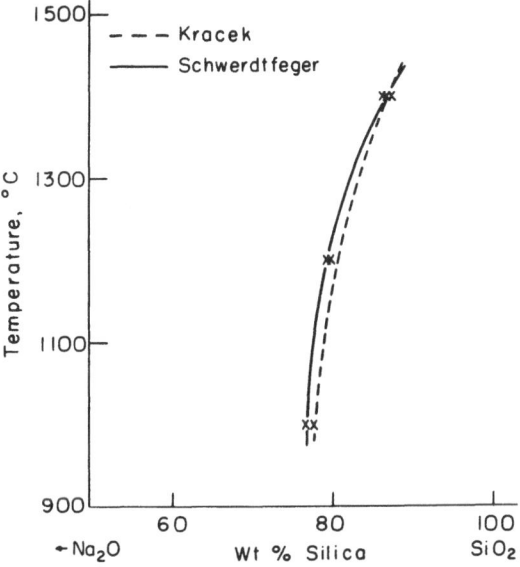

Fig. 1. A portion of the liquidus curve of silica in the system $Na_2O \cdot SiO_2$. From Schwerdtfeger [11].

EXPERIMENTAL PROCEDURE

Materials

Powdered sodium disilicate glass and oxidized nickel spheres were used as the starting materials. The glass was made in the laboratory from silica flour and reagent grade sodium carbonate by melting and refining in platinum crucibles in an inclined, rotating electric furnace at 1300°C. The glass was cast and then crushed and dry ground to −325 mesh and had a density of 2.57.

Commercially obtained nickel microspheres were oxidized at 800°C for 1 hr to provide a spherical source of NiO. The majority of the particles were about 100 μ in diameter.

Hot Pressing

The powdered glass was mixed with about 10% oxidized nickel spheres. Precise control of the composition was not necessary, since microprobe examination was performed on isolated spheres within the matrix. The glass-oxidized nickel-sphere mixture was hot pressed at 900°C in a graphite die for 2 hr at pressures ranging from 0 to 1000 psi. A description of the hot-pressing arrangement has been given in a previous paper ([7]).

The hot-pressed samples were sectioned and mounted in a clear casting resin. All samples were polished with a sequence of silicon carbide papers (240, 400, and 600 grit) and then finished on a series of diamond paste laps (6, 2, and $\frac{1}{2}$ μ diamond). Carbon was vapor-deposited on the finished samples to provide a conductive surface suitable for electron microprobe analysis.

Electron Microprobe

A Materials Analysis Company electron beam microprobe was used for the analysis of nickel. Characteristic radiation emitted by the nickel in the specimen was resolved by a properly positioned lithium fluoride crystal and the intensity was measured with a proportional detector. A motor-driven gear mechanism moved the sample in a step-wise fashion relative to the electron beam. Integrated counts were taken at various intervals along the radius of an oxidized nickel sphere and into the glass sufficient to give a smooth curve for the concentration as a function of distance. The electron beam diameter was 1 μ and the depth of penetration was a maximum of 3 μ. These values are small compared with the $\sim$100 μ oxidized nickel particle size. The electron beam microprobe was equipped to take photographs of the X-ray image for any element from an oscilloscope. In order to minimize the

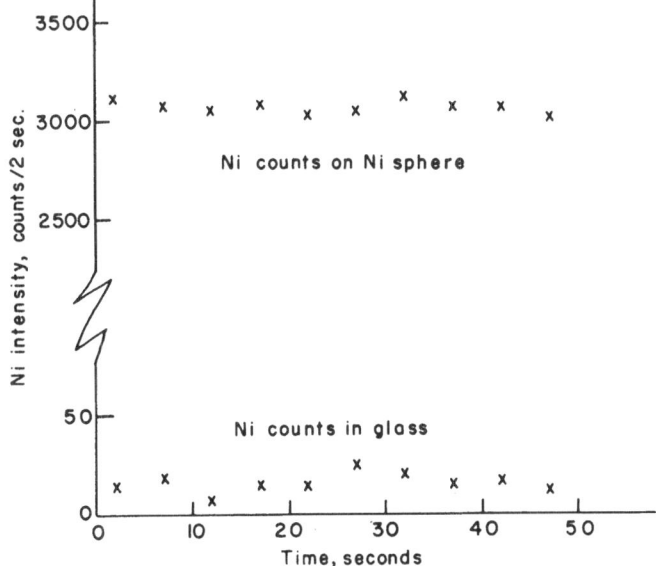

Fig. 2. Effect of surface damage on count rate.

instability of the alkali glass as a result of the electron beam, short counting times (10 sec), a low sample current (0.02 μA), low voltage (15 kV), and a relatively heavy carbon coating were used. The stability obtained can be seen in Fig. 2.

All data obtained from the proportional counters were corrected for background, absorption, counter dead time and nonlinearity, fluorescence, and atomic number. In this case the counter dead time, fluorescence, and atomic number corrections are negligible and the background can be accounted for by a simple subtraction. The absorption, however, requires more careful consideration. Detailed procedure and mass absorption coefficients were taken from Smith [12].

RESULTS AND DISCUSSION

The composites pressed at 0, 250, and 1000 psi were examined in the microprobe and photographs were taken of the oscilloscope image of Ni around a sphere. The background concentration of nickel can be seen around an untreated oxidized sphere in Fig. 3. This background concentration was measured with the counter to be 0.2 wt. % and a correction was made for this in each of the values given. Figure 4 shows the nickel concentration around

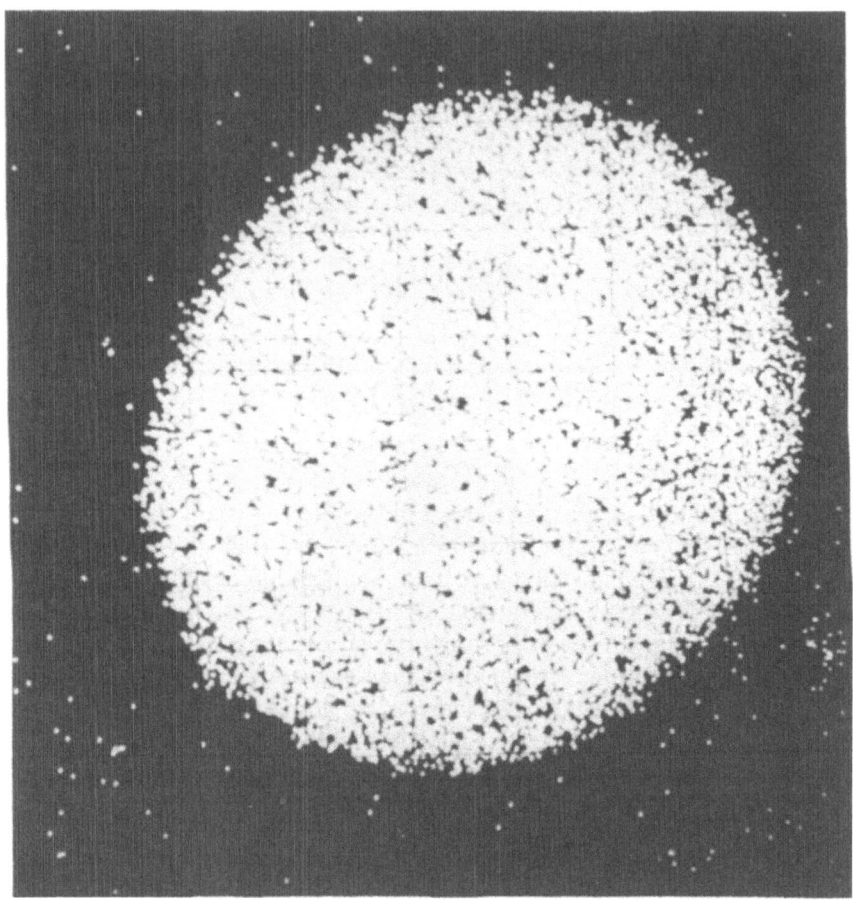

Fig. 3. Oscilloscope photograph showing background Ni concentration about an untreated sphere.

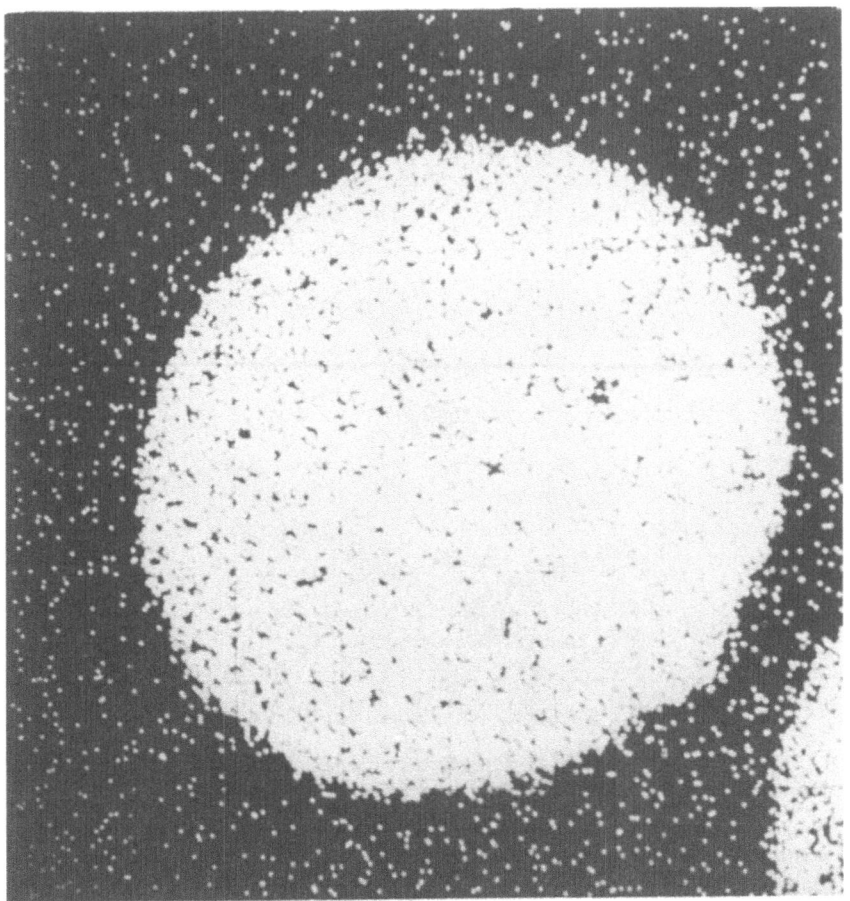

Fig. 4. Oscilloscope photograph showing Ni concentration about a sphere in a composite formed at 900°C, zero psi, and 2 hr.

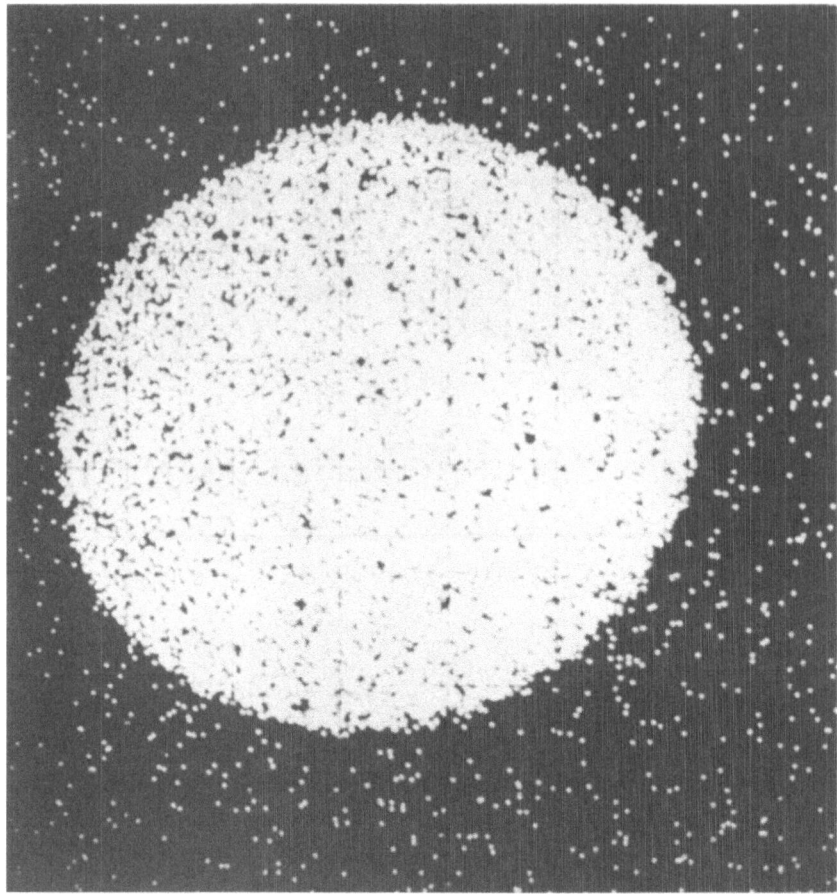

Fig. 5. Oscilloscope photograph showing Ni concentration about a sphere in a composite formed at 900°C, 250 psi, and 2 hr.

an oxidized nickel sphere in a composite that was treated at 900°C and zero pressure for 2 hr in vacuum. Measured concentrations here were 5.1, 4.7, 5.0, and 4.7 wt.%.

The Ni concentration around spheres in composites formed at 900°C for 2 hr in vacuum and at 250 psi and 1000 psi can be seen in Figs. 5 and 6, respectively. Measured concentrations were 2.6 wt.% for the 250-psi sample and 0.4 wt.% for the 1000-psi sample.

A composite was fabricated from sodium disilicate glass and nickel oxide particles to ensure that the oxidized nickel spheres were actually acting as nickel oxide spheres. The composite was fabricated at 900°C for 2 hr and at zero pressure. The measured concentration in the matrix was

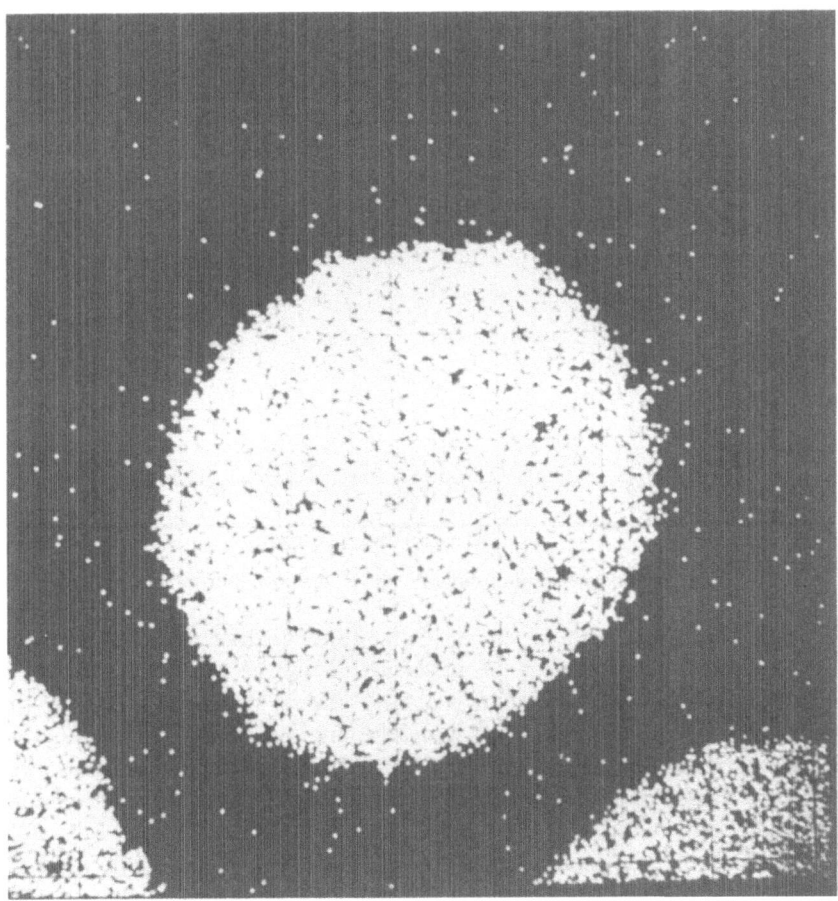

Fig. 6. Oscilloscope photograph showing Ni concentration about a sphere in a composite formed at 900°C, 1000 psi, and 2 hr.

found to be 4.7 wt. %, the exact agreement indicating that the oxidized nickel spheres did in fact act as nickel oxide spheres in the composites. A concentration profile across the particle interface in the glass–nickel oxide composite can be seen in Fig. 7. The concentration gradient went from nickel oxide to glass in only 8 μ. Differences in polishing characteristics between the nickel oxide and the glass contributed almost all of this.

Another sample was made in order to verify this polishing difference and to determine the method of material transfer. Oxidized nickel spheres and sodium disilicate particles were sintered at 750°C for 2 hr under zero pressure. The concentration gradient in this sample went from nickel oxide to glass in 7 μ, verifying the polishing difference. A similar effect due to polish-

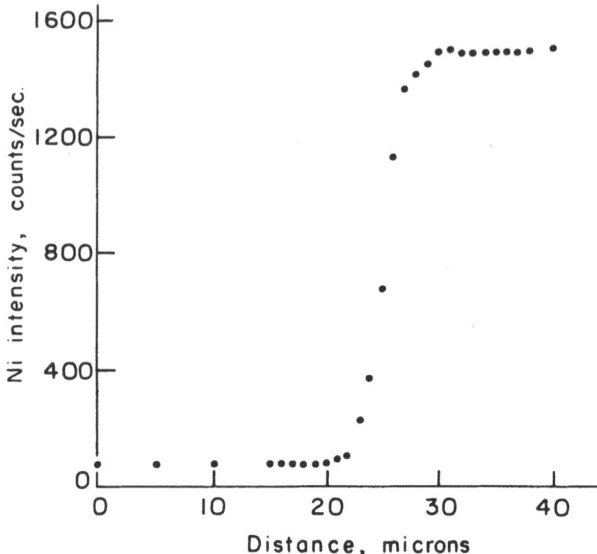

Fig. 7. Concentration profile of 900°C, zero psi, 2-hr sample.

ing differences was seen in the Al_2O_3–sodium disilicate glass system ([7]). The measured concentration of nickel oxide in the glass was 0.3 wt.%, indicating that surface diffusion provided almost no contribution to the mechanism of material transport from the oxidized nickel sphere into the glass matrix and that almost all of the transport away from the sphere occurs by bulk diffusion once the glass particles soften enough to provide a minimum fluidity.

As was seen earlier, the saturation concentration is an accurate measure of the concentration of the liquidus at a given temperature ([11]). With a relatively long length of time at temperature the saturation concentration will be reached and will be, in effect, the measured concentration in the matrix. A plot of these compositions as a function of pressure would therefore yield the liquidus as a function of pressure at a given temperature. This function can be seen in the nonequilibrium liquidus plotted in Figs. 8 and 9, and can be seen to be the correct shape as would be predicted from free energy considerations.

The free energy of a two-component system can be expressed as

$$dF = V\,dP - S\,dT + \mu_1\,dn_1 + \mu_2\,d(1 - n_1) \tag{3}$$

where F is the Gibbs free energy, V the volume, P the pressure, S the entropy, T the temperature, μ the chemical potential, and n the mole fraction. Since the temperature is constant,

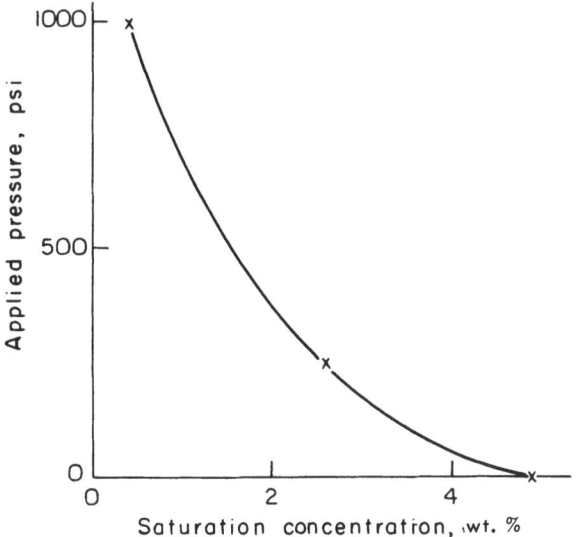

Fig. 8. Effect of pressure on saturation concentration.

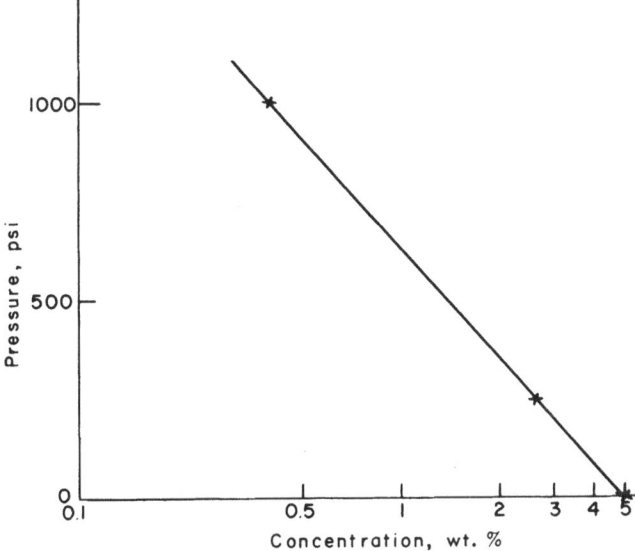

Fig. 9. Effect of pressure on concentration.

$$\frac{dF}{dn_1} = \frac{V\,dP}{dn_1} + (\mu_1 - \mu_2). \tag{4}$$

Differentiating this,

$$d^2F/dn_1^2 = V\,d^2P/dn_1^2 \tag{5}$$

Therefore the slope of a *P–n* diagram changes in a manner directly proportional to the change of slope of an *F–n* diagram. For a system which, like this, tends toward compound formation, the shape of curve shown in Fig. 8 would be expected.

In all cases the concentration became constant deep in the matrix, indicating that the entry step was affected by the pressure. Had the diffusion step alone been affected by the pressure, there would have been a concentration gradient in the matrix.

Another interesting feature can be seen in the fact that the solubility decreases with pressure. This decrease would illustrate that in the hot pressing of two-phase systems reaction between the two phases becomes less important as greater pressures are applied. Under high enough pressures two-phase systems actually are distinctly two-phase and the problem of a third, interfacial phase is minimized.

SUMMARY

The rate at which solute atoms enter the solvent has been shown to be dependent upon pressure. In the $NiO–Na_2O \cdot 2SiO_2$ system as the pressure increases, the solubility of the NiO in the glass decreases and the system is truly two-phase. This would indicate that in the hot-pressing of two-phase systems once a high enough pressure is reached the problem of interference from a third, interfacial phase is minimized.

ACKNOWLEDGMENTS

Acknowledgment is extended to G. Georgakopoulos and G. Dahl for experimental assistance. This work was done under the auspices of the U. S. Atomic Energy Commission.

REFERENCES

1. A. C. D. Chaklader and M. N. Shetty, "Ceramic–Metal Composites by Reactive Hot Pressing," *Trans. Met. Soc. AIME* **233**(7): 1440–43 (1965).
2. A. C. D. Chaklader and L. G. McKenzie, "Reactive Hot Pressing of Clays," *Bull. Am. Ceram. Soc.* **43**(12): 892–93 (1964).

3. A. C. D. Chaklader and V. T. Baker, "Reactive Hot Pressing: Fabrication and Densification of Non-Stabilized ZrO₂," *Bull. Am. Ceram. Soc.* **44**(3): 258–59 (1965).

4. P. E. D. Morgan and E. Scala, "The Formation of Fully Dense Oxides by Pressure Calcintering of Hydroxides," Department of Materials Science and Engineering, Cornell University, Ithaca, New York.

5. A. C. D. Chaklader, "Reactive Hot Pressing: A New Ceramic Process," *Nature* **206**: 392–93 (1965).

6. T. G. Carruthers and T. A. Wheat, "Hot Pressing of Kaolin and of Mixtures of Alumina and Silica," Proc. Roy. Soc., #3, Fabrication Science, September 1965.

7. Mark A. Stett and R. M. Fulrath, "Chemical Reactivity in an Al_2O_3–Glass Composite," Lawrence Radiation Laboratory Report UCRL–16892 Rev., October 1966.

8. J. Crank, *The Mathematics of Diffusion*, Oxford Univ. Press, Oxford, 1956.

9. L. A. Simpson and R. E. Carter, "Oxygen Exchange and Diffusion in Calcia-Stabilized Zirconia," *J. Am. Ceram. Soc.* **49**(3): 139–144 (1966).

10. J. Hudson and R. Hoffman, "The Effect of Hydrostatic Pressure on Self-Diffusion in Lead," *Trans. AIME* **221**(4): 761–68 (1961).

11. Klaus Schwerdtfeger, "Dissolution of Solid Oxides in Oxide Melts. The Rate of Dissolution of Solid Silica in Na_2O–SiO_2 and K_2O–SiO_2 Melts," *J. Phys. Chem.* **70**(7): 2131–37 (1966).

12. J. V. Smith, "X-ray Emission Microanalysis of Rock-Forming Minerals, I. Experimental Techniques," *J. Geol.*, **73** (6): 830–864 (1965).

DISCUSSION

R. S. Gordon (University of Utah): We have attempted to hot press tungsten fibers in fused silica without any success with respect to bonding. I was wondering why you were successful.

Answer (R. M. Fulrath): In the system tungsten–fused silica there apparently is a lack of a chemical reaction between the two phases. Thus by comparing the values for the linear thermal expansion coefficient of the two:

	Temp. range	α
Tungsten	RT–700°C	5.6×10^{-6}
Fused silica	0–1000°C	0.5×10^{-6}
Soda borosilicate	RT–450°C	7.7×10^{-6}

we can see that upon cooling, the tungsten will shrink away from the fused silica and form a system composed essentially of fused silica and cylindrical pores. In our work glasses were usually ternary compositions that contained Na_2O and therefore would be expected to chemically bond to tungsten. There are numerous literature references to the sodium tungstate system in the manufacture of tungsten–glass seals. In cases where our thermal expansions were reversed, i.e., the glass expansion lower than that of tungsten, the chemical reaction will form the necessary stress-bearing interface required for strengthening.

Chapter 32

The Initial Stages of Reaction Between Quartz and Calcium Carbonate*

M. R. Montierth[†], R. S. Gordon[‡], and I. B. Cutler[§]

Department of Ceramic Engineering
University of Utah

A study was made of the low-temperature (700–850°C) reaction between calcium carbonate and quartz using a thermogravimetric technique. The samples prepared for reaction were pressed disks composed of large quartz particles (10%) in a matrix of fine calcium carbonate particles (90%). The reaction occurred in two steps: (1) a rapid initial reaction, markedly influenced by water vapor present either in the atmosphere or adsorbed on the surfaces of the calcium carbonate particles, and (2) a final stage in which the rate was controlled by diffusion through a protective product layer around the quartz particles. On the basis of optical microscope, electron microprobe, and X-ray lattice parameter studies, the mechanism of the initial reaction was postulated as a rapid diffusion of calcium into the quartz structure promoted by the presence of water vapor. As the water vapor was removed during the reaction, a dense protective layer of product formed and coated the quartz particles, thereby accounting for the observed parabolic kinetics in the final stages of reaction.

INTRODUCTION

Commercial interest in the solid-state reaction between calcium carbonate (or lime) and quartz has spurred a considerable amount of research in this system. The preliminary work by Dykerhoff ([1]), Weyer ([2]), Nagai ([3,4]), and Hild and Tromel ([5]) all indicated that calcium orthosilicate (Ca_2SiO_4) was the first reaction product to appear, regardless of the temperature or the initial mole ratio of the reactants. Agreement as to the order of appearance of subsequent phases, such as $Ca_3Si_2O_7$ and $CaSiO_3$, was lacking in these early studies. In order to resolve the apparent conflicts, Jander and Hoff-

*Portion of a thesis submitted by Max Montierth in partial fulfillment of the requirements for the Ph.D. degree, University of Utah.
[†]Presently Staff Scientist, Corning Glass Works.
[‡]Assistant Professor, Ceramic Engineering, University of Utah.
[§]Professor, Ceramic Engineering, University of Utah.

man ([6]) undertook a comprehensive investigation in which the effects of temperature, initial mole ratio of reactants, atmosphere, and purity were systematically studied. Jander and Hoffman confirmed the conclusion of the previous work that Ca_2SiO_4 was always the first product to form in a dry atmosphere. The $Ca_3Si_2O_7$ was the second product, and later reacted with SiO_2 to form $CaSiO_3$ as the final product. Using X-ray and chemical analyses, Jander and Hoffman determined the relative amounts of these three phases as a function of time at temperatures of 1000° and 1200°C. Jander ([7]) theorized a parabolic reaction rate for the $CaO–SiO_2$ reaction, but did not give any experimental data to confirm the theory. A final result of the study was that the rate, but not the course, of reaction was enhanced markedly in the presence of water vapor. In studies of the reaction at lower temperatures Kakitani and Fukisaka ([8]) found that after 272 hr at 700°C the only product identifiable by X-ray was Ca_2SiO_4. Other workers ([7,9,10]) also detected the presence of reaction at temperatures as low as 700°C.

In a recent study by Verduch ([11]) of the reaction between calcium carbonate and cristobalite at temperatures between 500–700°C the solid solution of calcium ion in cristobalite was suggested as an important part of the mechanism in the initial stages of reaction. Verduch reported the sequence of events for the reaction between calcium carbonate and excess cristobalite as follows: (1) an initial calcium ion migration into the cristobalite structure prior to the appearance of Ca_2SiO_4; (2) nucleation and growth of Ca_2SiO_4 following attainment of a critical calcium concentration in the solid solution layer; (3) exsolution of the calcium from solution in cristobalite during the growth of the Ca_2SiO_4 layer, along with depletion of the calcium carbonate; (4) continued exsolution of calcium from solid solution following depletion of the calcium carbonate, resulting in the formation of the $Ca_3Si_2O_7$ and $CaSiO_3$ phases; and (5) continued exsolution of calcium from the solid solution and the final growth of $CaSiO_3$ at the expense of the $Ca_3Si_2O_7$ and Ca_2SiO_4 products.

Verduch analyzed the "d" spacing of the primary diffraction line of cristobalite to determine the extent of calcium solubility. Since the procedure was not the usual lattice parameter determination, his results may be questionable. It is noted that Verduch confirmed the work of Jander and Hoffman with respect to the order in appearance of products.

As is evident from the foregoing, most of the work has been concerned with product identification and the general overall progress of the reaction. No detailed kinetic study has been made with the aim of deducing actual reaction mechanisms. This situation is due to a number of reasons. The reaction rates are slow, especially at lower temperatures. Conventional product identification by X-ray and chemical analyses has limited accuracy. X-ray analysis requires the presence of well-defined crystals in sufficient amounts, a requirement difficult to meet, especially in low-temperature studies, where reaction products are often crystals of extremely small size or amorphous.

The purpose of this work was to make a kinetic study of the $CaCO_3$–SiO_2 reaction utilizing a thermogravimetric technique particularly suitable for reaction at temperatures between 700° and 850°C. By conducting the reaction in a CO_2 atmosphere and at temperatures below that of the decomposition of calcium carbonate, it is possible to attribute the measured weight loss to the actual reaction between calcium carbonate and quartz in the formation of the first product according to the reaction $2CaCO_3 + SiO_2 \rightarrow Ca_2SiO_4 + 2CO_2$. The results of this study are now presented.

EXPERIMENTAL

The samples prepared for the reaction were thin-pressed disks composed of high-purity powders of calcium carbonate and quartz. The disks (~ 500 mg) were prepared by mixing 10 vol.% coarse quartz particles in a matrix of very fine calcium carbonate particles. This procedure was used in order to approach the ideal condition, reported by Komatsu ([12]), in which the large reactant particles (quartz) are completely surrounded and in continuous contact with the fine matrix particles (calcium carbonate). The materials used, their reported purity, and particle sizes are listed in Table I. The coarse particle size was controlled by selecting the desired size after screening through a standard screen series. The average large particle sizes used were 35, 40, 48, and 57 μ, respectively. The powders were mixed in approximately 50-g batches, pressed dry at 50,000 psi in a $\frac{5}{8}$-in. steel die, and dried in a 100°C drying oven prior to placement in the reaction furnace.

The thermogravimetric studies were conducted in both dry and wet CO_2 atmospheres in a standard Kanthal-wound furnace. All weight changes during reaction were calculated from measured extensions of fused silica springs.

TABLE I

Sample Materials

Coarse particle (10% by volume)	Fine matrix particle (90% by volume)
Type 204 GE	Baker Reagent Grade $CaCO_3$ 99.96%, 7–10 μ
Ground Quartz	GE I μ-$CaCO_3$ (vaterite) 99.98%, 2–3 μ
99.97%	GE II $CaCO_3$ 99.99%, 17–20 μ

A Gaertner Model 912 cathetometer was used to record the spring extensions. The precision with this instrument was ± 0.001 cm, yielding an error
in measurement of ± 0.1 mg, or approximately $\pm 0.5\%$ for the total observed
weight loss of a typical sample. Temperature control in the furnace was
within $\pm 0.3\%$ for the 700–850°C range. For the dry CO_2 studies the CO_2
stream was dried by passing it through concentrated H_2SO_4 and then Drierite
($CaSO_4$), resulting in a water partial pressure of approximately 10^{-5} atm.
The partial pressure of water in the wet CO_2 was approximately 23 mm Hg.
The gas stream was passed through approximately 12 in. of Fiberfrax packed
in the bottom of the furnace tube to ensure gas preheating and to eliminate
eddy currents. Temperature measurement was accomplished using a chromel–
alumel thermocouple placed within 1 cm of the sample pan.

Unreacted and reacted samples were prepared for microscopic examination in order to study the product layer formation on the surface of the quartz
particles. These samples were also used in the electron microprobe studies.
Due to the fragile nature of the pressed samples, successful micrographic
examination of product layers on quartz particles was accomplished only
after employing an epoxy vacuum impregnation technique. Samples were
placed in a bell jar and subjected to a vacuum of approximately 100 μ Hg
for more than 15 min. When the sample disks were covered with Buehler
#20–8130 AB plastic Epoxide the vent was opened, allowing the bell jar to
return to atmospheric pressure and forcing the Epoxide material into the
porous samples. After curing the samples were ground and polished with
successively finer polishing papers and diamond pastes. The diamond paste
was necessary in order to ensure an even surface cut across the hard quartz
particles and soft $CaCO_3$ matrix. A final polish was accomplished using a
0.3 μ alumina on a fabric wheel. The photomicrographs were obtained
using a Vickers Fifty-five microscope.

RESULTS

Weight Loss Results—Dry CO_2 Atmosphere

Typical weight-loss isotherms for samples of 57-μ quartz in a 90 vol.$\%$
matrix of 17–20-μ calcium carbonate can be seen in Fig. 1. All of the weight
loss data are reported on the basis of normalized 500-mg initial sample
weights. The large initial portion of the reaction evident in these curves was
observed in all cases. The amount of the initial reaction was found to be
extremely dependent on the experimental conditions and varied from approximately 20 to 80$\%$ of the total amount of reaction. The same data are shown
in Fig. 2 in parabolic plots of the square of the weight loss versus time. In
the latter stages of reaction a parabolic dependence is observed, indicating the

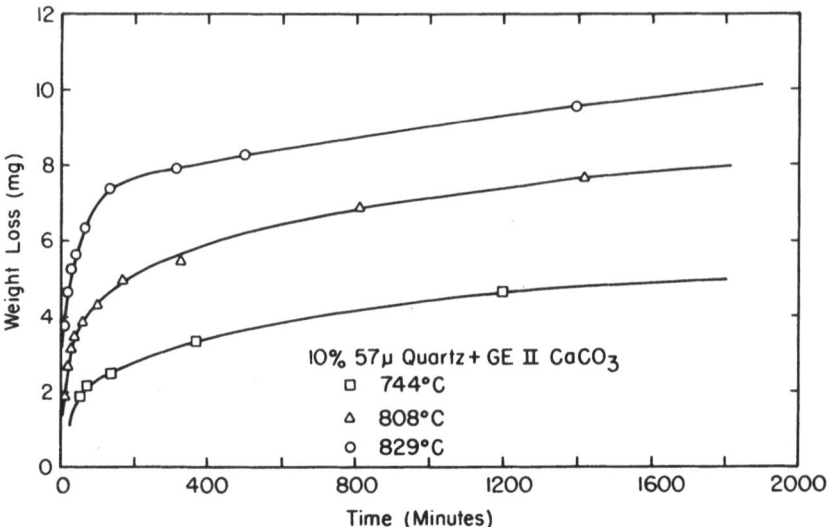

Fig. 1. Isothermal weight loss for 57-μ quartz + GE II CaCO$_3$ reacted in dry CO$_2$.

possibility of a diffusion-controlled reaction. The significant initial nonlinear portion of these isotherms is readily apparent.

The rate constants obtained from the slopes of the linear portion of the parabolic plots are plotted against the reciprocal temperature in Fig. 3 for reactions involving various calcium carbonate reactant materials. Data for the

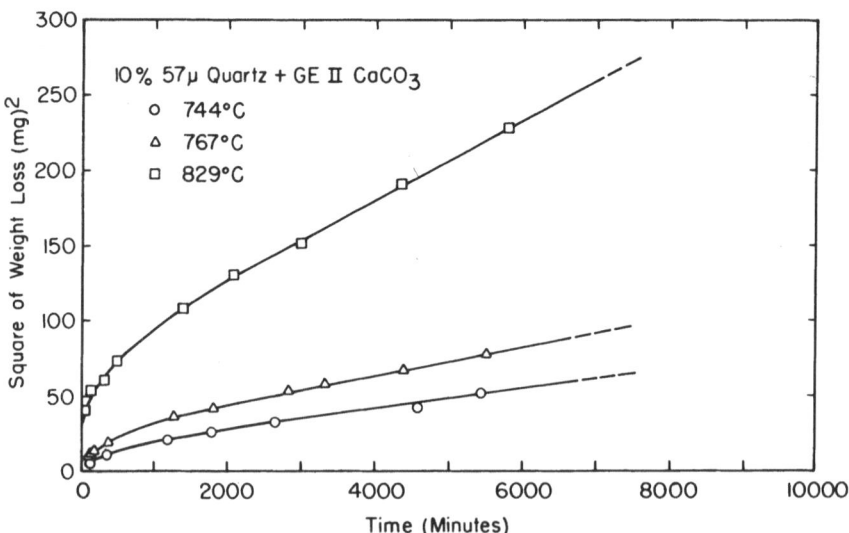

Fig. 2. Parabolic rate plot for 57-μ quartz + GE II CaCO$_3$ reacted in dry CO$_2$.

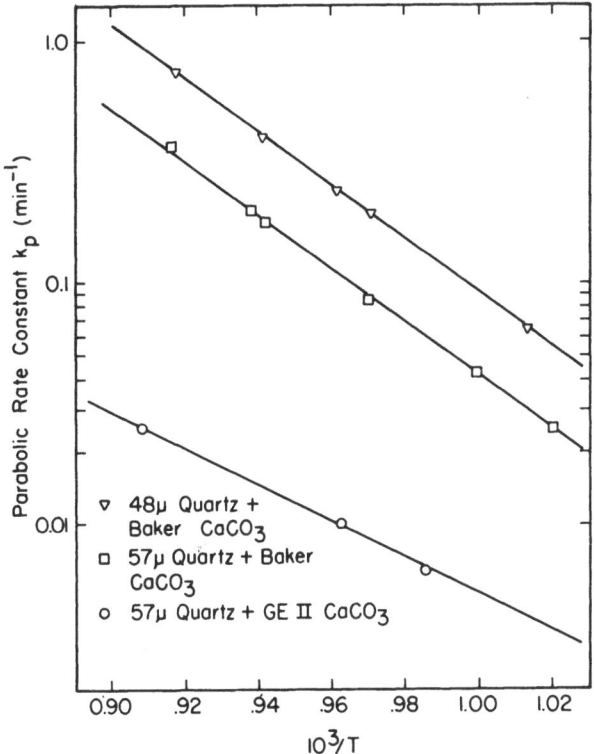

Fig. 3. Temperature dependence of parabolic rate constant for different calcium carbonate materials.

2–3-μ calcium carbonate (vaterite) are not shown due to a lack of reproducibility. However, the rate data obtained using this reactant resulted in an approximate activation energy of 100 kcal/mole, much higher than those obtained in reactions with the other two calcium carbonate reactants (51, 78 kcal). The results for the quartz particle size shown in Figs. 1–3 are representative of the particle size range studied. Two quartz particle sizes are shown for the Baker calcium carbonate matrix samples. The quartz particle size difference accounts for only approximately 30% of the displacement between the two curves. The cause of the remaining portion of the displacement was undetermined. It could possibly be due to the fact that the 48-μ quartz particles were small enough such that conditions of ideal reactant contact were not attained. The slopes of the curves for the two quartz particle sizes were identical.

The activation energies obtained for the materials studied are given in Table II. The 51 kcal/mole activation energy for the 57- and 48-μ quartz

TABLE II

Dependence of Activation Energy on Matrix Material

Matrix material	Observed activation energy (kcal/mole)
GE II CaCO$_3$ (calcite)	51
Baker CaCO$_3$ (calcite)	78
GE I (vaterite)	~100

particles in a 90 vol.% matrix of 17–20-μ calcite agreed well with the 55 kcal/mole value reported by Lindner ([13]) for the diffusion of calcium through well-crystalized Ca$_2$SiO$_4$. It is difficult to understand the range of activation energies. They may be due to the presence of reaction layers of variable composition.

In addition to differences in the activation energy for the parabolic (diffusion-controlled) portion of the reaction, the extent of the nonparabolic initial portion of the reaction was very dependent on the nature of the calcium carbonate matrix material. This effect is readily apparent in the data presented in Fig. 4, which is a plot of the weight loss data for samples consisting of 57-μ quartz particles in various matrix materials. Although a portion of the curves in the initial few minutes of reaction could be a result of differences in the number of reaction sites, as determined by the number of SiO$_2$–CaCO$_3$ point contacts, the relative displacement of the curves was found to be far greater than these differences would indicate.

No evidence of calcite decomposition was found in the temperature

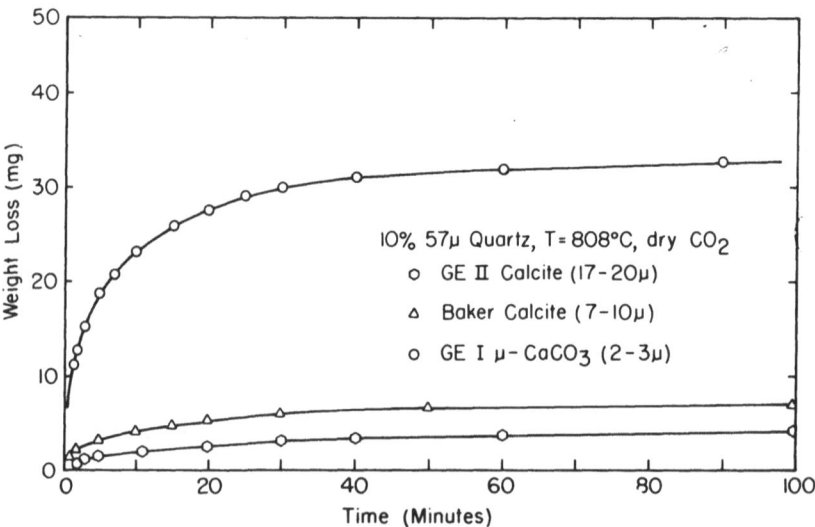

Fig. 4. Effect of calcium carbonate matrix material on the initial reaction in dry CO$_2$.

range studied for any of the samples. The possibility of calcium carbonate decomposition without reaction with quartz was checked primarily by X-ray diffraction analysis and evidence of weight gain upon sudden furnace temperature decreases following the completion of a weight loss experiment. A sudden decrease in furnace temperature would lead to recarbonation and weight gain if decomposition had occurred. The X-ray diffraction traces failed to show any evidence of CaO or Ca(OH)$_2$. The latter check was found to be the most sensitive. Reaction at a temperature deliberately high enough to independently decompose calcium carbonate gave large weight gains with temperature reductions into the stability range of calcium carbonate.

An HF acid test was made of the calcium carbonate matrix after mechanically separating it from the reacted quartz particles to determine if the initial portion of the reaction was caused by a rapid silicon transport into the calcium carbonate bulk and subsequent reaction. This analysis showed negligible silicon in the matrix for all the samples. This indicated that differences in the amount of the initial reaction for the various calcium carbonate matrix materials were not due to reaction away from the periphery of the quartz particles.

In all cases the extremely fine-grained products and their probable amorphous structure made X-ray identification impossible. Past investigations utilizing product analysis by X-ray techniques had been conducted at much higher temperatures $(T > 1000°C)$, and usually for longer times, permitting the formation of well-defined crystalline products.

The weight loss data did not fit a Jander plot, nor did they fit a changing reaction order model such as the linear-to-parabolic model suggested by Wagner and Grunewald ([14]). It was not possible to fit the results to a combination of parabolic rate and higher-order rate equations.

Reaction in the Presence of Water Vapor

After Jander ([6]) reported that water vapor accelerated the rate of reaction it was thought initially that water vapor might in some way be responsible for the extensive initial reaction. When water vapor ($P_{H_2O} \approx 23$ mm Hg) was introduced into the CO_2 gas stream the initial amount and rate of reaction in samples composed of 2–3-μ calcium carbonate (vaterite) and quartz were observed to increase. The weight loss curves in Fig. 5 illustrate this effect.

The pre-dried sample was lowered into the 550°C region of the furnace for 1 hr before lowering into the hot reaction zone of the furnace. The weight loss of the sample while in the 550°C zone was 1.5 mg, approximately 30 % of the difference between the curves for the pre-dry sample and the normal sample (24-hr drying at 100°C). The weight loss during the drying period indicated the presence of adsorbed water, and the large displacement

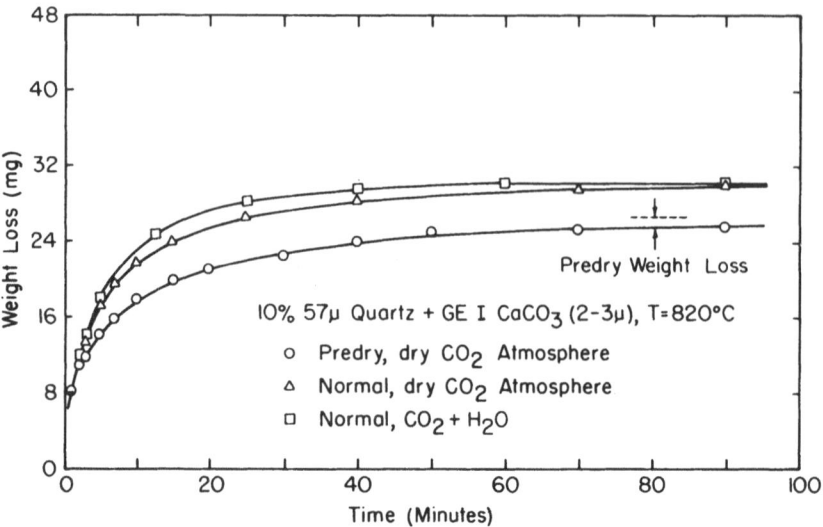

Fig. 5. Effect of water vapor on the initial reaction.

between these curves indicated the large effect of small amounts of water upon the reaction in the initial stages. The sample reacted in CO_2 containing water vapor reached a maximum amount of reaction within approximately 1 hr and showed no further reaction in an additional 48 hr. This behavior in water vapor is remarkably different than that observed in the samples reacted in dry CO_2 in that no parabolic rate law was observed in the later stages of reaction. This behavior was observed for all samples with the GE I $CaCO_3$ matrix, which were reacted in the presence of water vapor, and is represented by the weight loss isotherms in Fig. 6 for samples of 57-μ quartz reacted in a matrix of GE I $CaCO_3$. For each temperature the reaction was observed to cease within approximately 1 hr.

The same general behavior was observed for samples with a Baker calcium carbonate matrix. In this case the presence of water vapor greatly accelerated the reaction for the initial few minutes. Following the first 10 min the reaction rates in dry CO_2 and wet CO_2 atmospheres were approximately equivalent, but were displaced by an amount equal to the initial difference in the reaction for the two conditions. The reaction in wet CO_2 for these samples was not observed to terminate as the GE I $CaCO_3$ had. The initial reaction with this material was enhanced more by water vapor than was the case with the GE I $CaCO_3$ samples. It is believed that less water was initially present in this material in an adsorbed condition due to a larger particle size; therefore a more pronounced effect was produced when water vapor was added.

For the samples with a GE II $CaCO_3$ matrix (17–30-μ calcite) water vapor increased the reaction rate for the first 20 min, but the rate beyond

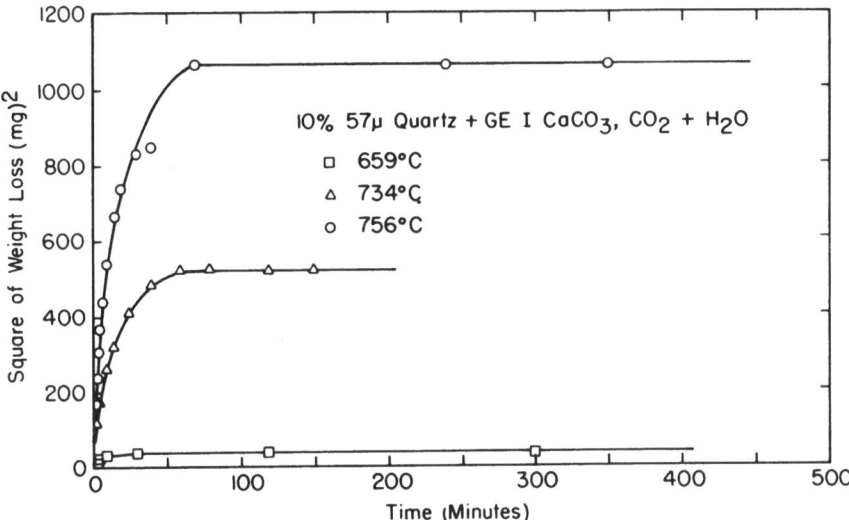

Fig. 6. Effect of temperature on the initial reaction in water vapor for GE I CaCO₃ samples.

that decreased, such that the reaction for the sample in dry CO_2 initially lagged that for the sample in wet CO_2, then surpassed it at about 70 min. Evident here was the retardation of the final rate by the presence of water vapor. The presence of the parabolic region in these samples reacted in the presence of water vapor was not definite.

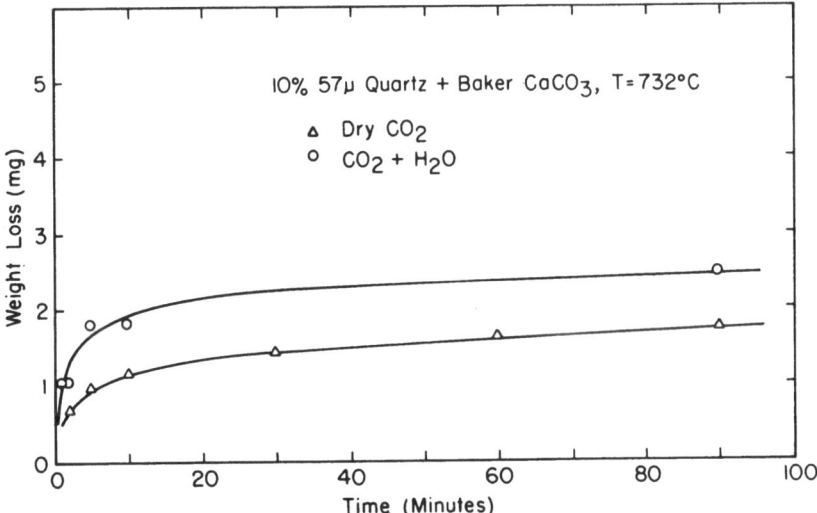

Fig. 7. Effect of water vapor on the initial reaction for Baker calcium carbonate samples.

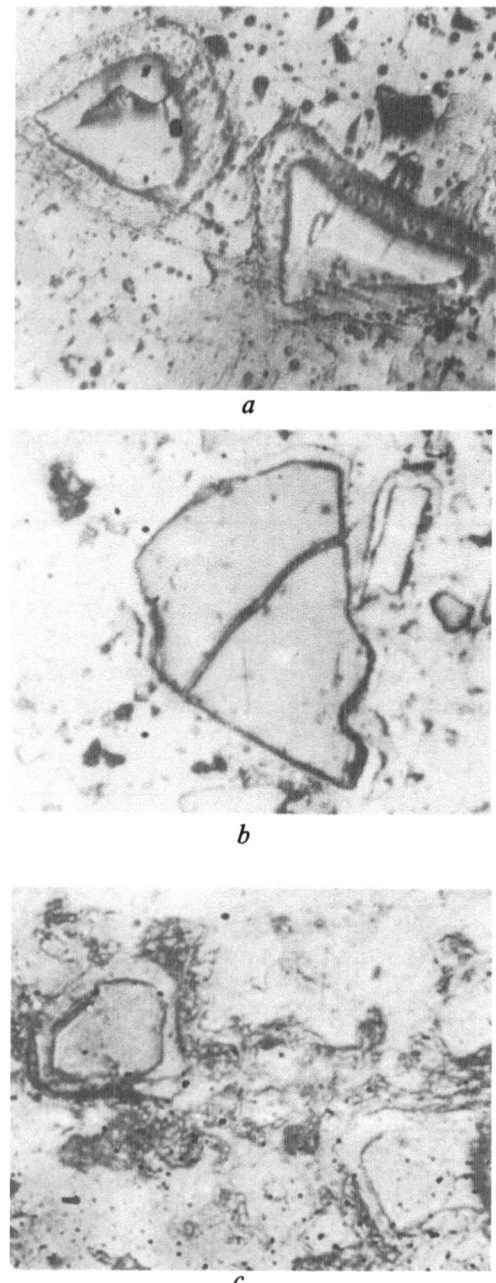

Fig. 8. Optical micrographs illustrating the effect of water vapor. (a) 57-μ quartz + GE I $CaCO_3$, 46% reaction, dry CO_2 atmosphere, 410 ×. (b) 40-μ quartz + Ge I $CaCO_3$, 45% reaction, CO_2 + H_2O atmosphere, 600 ×. (c) 40-μ quartz + GE I $CaCO_3$, 46% reaction, reheated in dry CO_2, 1140 ×.

Optical Micrograph Evidence

A typical product layer on a quartz particle after being reacted with calcium carbonate under dry CO_2 conditions at 842°C can be seen in Fig. 8(a). The sample consisted of 57-μ quartz particles in a matrix of Ge I $CaCO_3$. The weight loss data indicated a 46% reaction. Assuming Ca_2SiO_4· as the reaction product, layer thickness measurements indicated a 58% reaction. The product layer for these conditions was observed to be well defined. The amount of reaction calculated from weight loss data agreed well with values calculated from product layer thicknesses on the basis of 20 different measurements. This agreement was found for all three calcium carbonate matrix materials and confirmed the HF acid test in that reaction was restricted to the quartz particle surfaces.

When similar samples were reacted at 756°C in CO_2 atmospheres containing water vapor the micrographs [see Fig. 8(b)] showed no clearly defined product layer on the quartz particles. However, the weight loss of the sample of Fig. 8(b) indicated 45% reaction to Ca_2SiO_4. Under optical observation the quartz particles in these samples were found to exhibit a blue-black color. The color intensity and weight loss were far greater for samples employing a vaterite matrix than for samples with the Baker or GE II calcite matrices.

When a second portion of the sample shown in Figure 8(b) was re-reacted at 830°C in dry CO_2 the blue-black color decreased in intensity. In addition, a reaction product layer formed during the reheating, although the weight loss, and, therefore further reaction of the sample, during the second firing was negligible.

Electron Microprobe Results

The electron microprobe results for samples reacted in dry and wet CO_2 are shown in Figs. 9 and 10, respectively. The samples were comprised of 57-μ quartz particles in a matrix of GE I $CaCO_3$ (2–3μ vaterite) and are representative of samples shown in Fig. 8(a) and (b). The silicon and calcium traces shown in these figures represent concentration profiles for the respective elements for a beam traverse across the quartz particle, the product layer, and into the calcium carbonate bulk material.

The microprobe analysis for the sample reacted in dry CO_2 indicated:

1. No silicon present in the bulk calcium carbonate.
2. A constant silicon concentration across the product layer until the quartz–product interface was approached. At this point the silicon concentration increased while still in the product layer, perhaps indicating the presence of silicate phases rich in SiO_2 near the unreacted quartz interface. The silicon concentration in the constant composition region of the product

layer corresponded to approximately 11.4 ± 2 wt.% Si. The theoretical silicon concentration in dicalcium silicate is 16 wt.%.

3. No calcium concentration in the quartz.

4. An unexplained calcium decrease within the product layer.

The microprobe analysis for the sample reacted in CO_2 containing water vapor indicated:

1. A silicon concentration in the large particles equal to that for a quartz

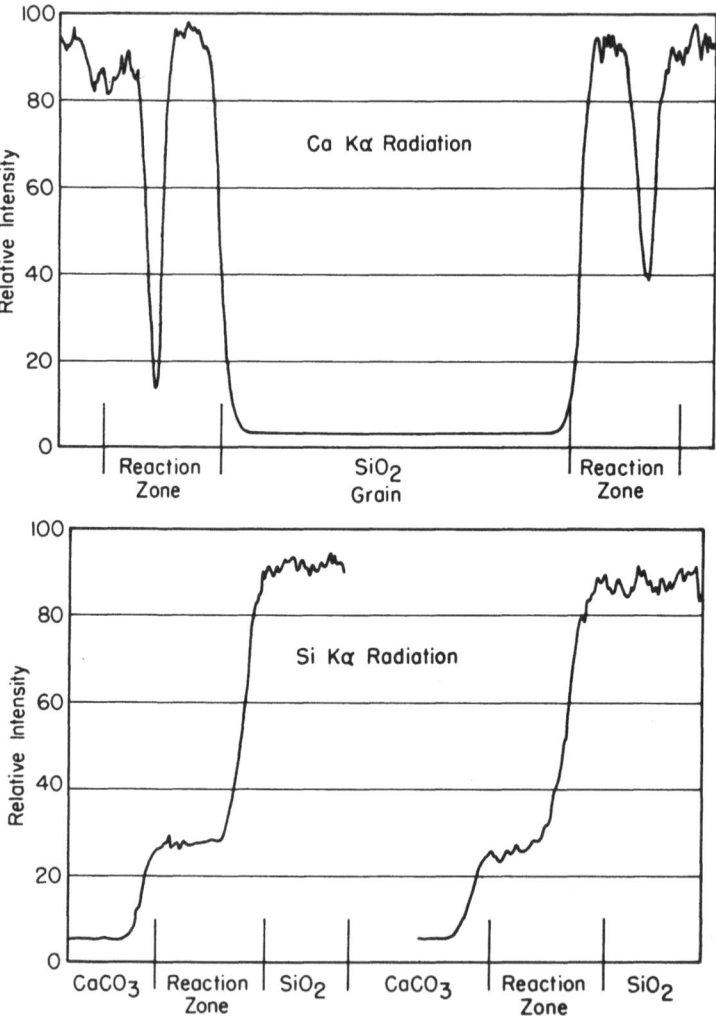

Figs. 9. Electron microprobe analysis of sample reacted in dry CO_2.

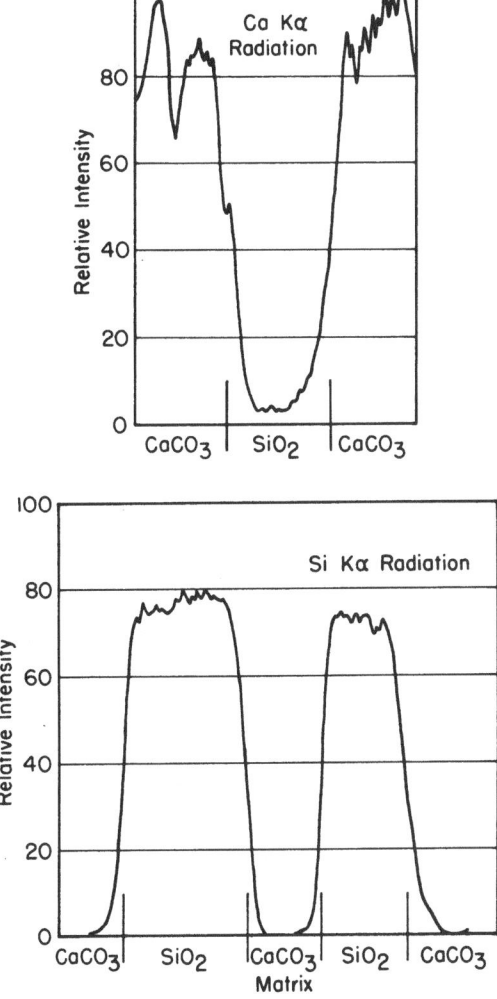

Fig. 10. Electron microprobe analysis of sample reacted in wet CO_2.

reference, indicating the particles to be the quartz phase even after reaction. The silicon plateau across the quartz particles was flat, suggesting a single-phase structure throughout each of the particles.

2. No observable reaction layers.

3. No silicon in the calcium carbonate matrix.

4. A definite calcium concentration in the quartz particles. Concentration checks on eight separate quartz particles indicated the calcium concen-

tration to be as high as 12.5 wt. % just inside the quartz particle surface. This calcium concentration dropped rapidly with distance into the quartz particles, until it reached a minimum concentration at a distance corresponding to an approximate 10–15-μ penetration.

Microprobe image patterns (Figs. 11 and 12) were taken on an ARL

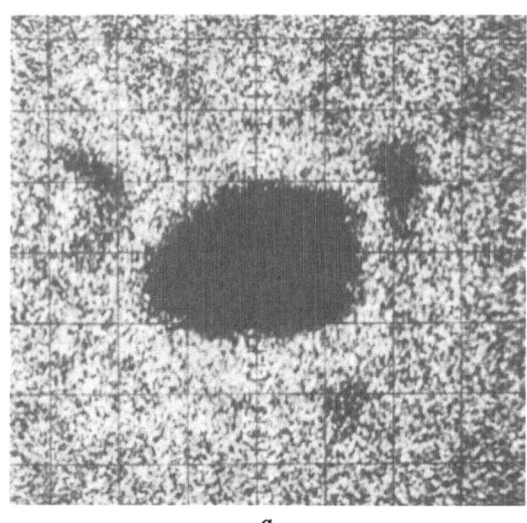

a

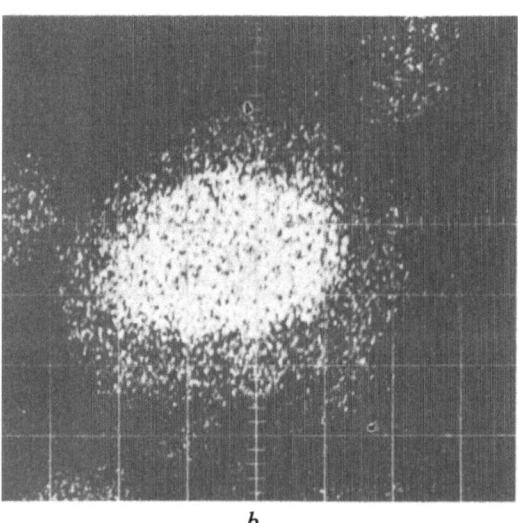

b

Fig. 11. Microprobe image patterns, 57-μ quartz + GE I CaCO$_3$, $T = 842°C$, 2$\frac{1}{2}$ hr, dry CO$_2$. (*a*) Ca $K\alpha$. (*b*) Si $K\alpha$.

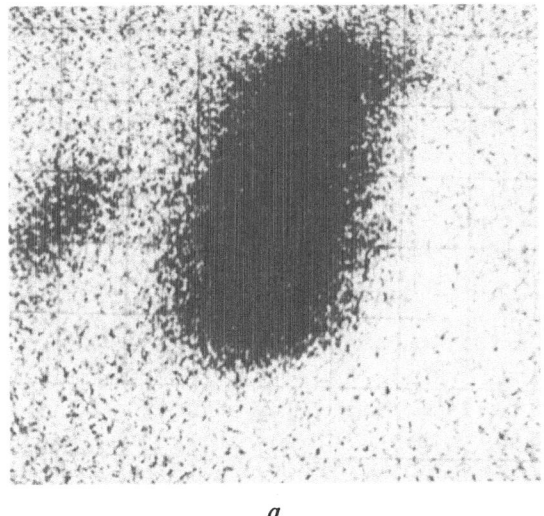

a

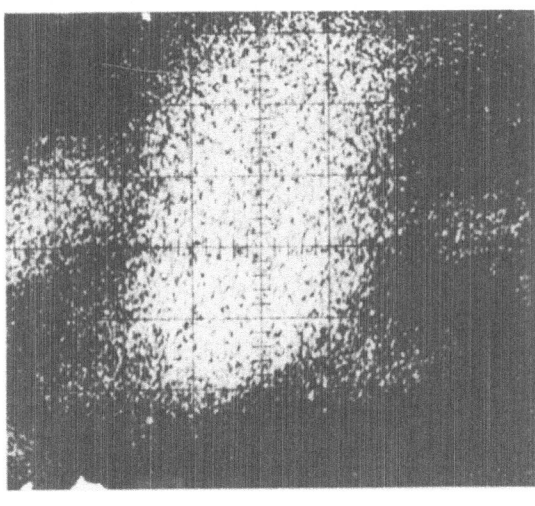

b

Fig. 12. Microprobe image patterns, 40-μ quartz + GE I
CaCO$_3$, $T = 756°C$, 3 hr, wet CO$_2$. (*a*) Ca $K\alpha$. (*b*) Si $K\alpha$.

microprobe for the quartz-GE I CaCO$_3$ samples reacted in dry and wet
CO$_2$ atmospheres. These were obtained from the same specimens from which
the samples for the microprobe traces in Figs. 9 and 10 were taken. The
reaction temperature and time for the samples reacted in dry and wet CO$_2$
were 842°C for 2$\frac{1}{2}$ hr and 756°C for 3 hr respectively.

The reaction product layer can be distinguished easily in both the calcium and silicon $K\alpha$ patterns (Fig. 11) of the quartz particle reacted in dry CO_2. No calcium can be detected in the unreacted portion of the quartz particle. However, for the quartz particle reacted in wet CO_2 (Fig. 12) the area rich in silicon appears significantly larger than the area of calcium exclusion, strongly indicating the presence of calcium within the quartz particle. In addition, no distinct product layer can be observed in this particle even though significant reaction has occurred.

Lattice Parameter Studies

Taylor ([15]) made a lattice parameter study of quartz before and after reaction with $CaCO_3$ in dry and wet CO_2 atmospheres. Quartz particles were reacted in GE I and Baker $CaCO_3$ matrices at temperatures between 800° and 805°C for a period of 1 hr. Following reaction the quartz was separated from the $CaCO_3$ matrix by first crushing the sample and then leaching it with HCl. The remaining quartz residue was then washed in distilled water, dried, and crushed to pass a -400 mesh. X-ray lattice parameter studies were made on this material on a Norelco X-ray machine using a Norelco Precision Symmetrical Back-Reflection Focusing Camera. The lattice parameters "a" and "c" of the quartz unit cell (hexagonal) were calculated by Cohen's method using the film spacings for five quartz X-ray diffraction lines. The calculation, conducted on a Univac 1108 digital computer, took into account all of the necessary corrections.

The results of Taylor's study are presented in Table III. The best experimental results were obtained with samples C-1 and E-1. Both of these samples (GE I $CaCO_3$) exhibited a lattice parameter shift, consistent with the proposed calcium solid solution. The increased lattice parameters were observed for all the vaterite (GE I $CaCO_3$) samples reacted in both dry and wet CO_2 atmospheres. When sample E-1 was reannealed in dry CO_2 at 750°C for 24 hr the lattice parameters decreased in magnitude, indicating perhaps the formation of a product layer and loss of calcium from the quartz structure.

The thermogravimetric data for the vaterite matrix samples (Fig. 5) suggested that large amounts of adsorbed water were present. Consequently, the similar lattice parameter shifts observed for the quartz particles reacted under wet and dry conditions are consistent with the large extent of the initial reaction present under both conditions.

The shift in the lattice parameters for the quartz particles reacted with the Baker $CaCO_3$ was much more pronounced when the reaction was conducted in the presence of water vapor. This result is consistent with the observed effect of water vapor on the extent of the initial reaction in these samples (see Fig. 7).

TABLE III

Lattice Parameters for Reacted Quartz*

Sample description			Lattice parameter (Å)	
			a	c
Unreacted standard			4.905	5.366
GE I $CaCO_3$ (vaterite) matrix samples				
B-1	$CO_2 + H_2O$	800°C; 60 min	4.914	5.399
C-1	$CO_2 + H_2O$	803°C; 60 min	4.920	5.408†
D-1	CO_2—Dry	803°C; 60 min	4.905	5.378
E-1	CO_2—Dry	801°C; 60 min	4.917	5.404†
E-1	Reannealed 750°C max. 24 hr		4.910	5.392†
Baker's $CaCO_3$ (calcite) matrix				
O	$CO_2 + H_2O$	801°C; 60 min	4.912	5.392
T	CO_2—Dry	805°C; 60 min	4.905	5.384

*After Taylor ([15]).
†Sharpest patterns produced with these samples.

It is emphasized that the data in Table III do indicate a trend of increased lattice distortion and possibly calcium solution with increased water contamination. It is noted that all the samples investigated by Taylor were reacted for only 60 min, which is near or slightly past the end of the initial portion of the reaction in most samples. In contrast, Verduch in earlier work reacted the samples for 24 hr before conducting X-ray analyses. As a result, a quantitative comparison of the two studies cannot be made. The low values of calcium solid solution reported by Verduch as compared to Taylor's study may actually be due to the longer reaction times. Taylor's experiments performed at the peak of the initial reaction may have caught the calcium solid solution at its maximum. This conclusion is supported by the fact that the lattice parameters decreased in reheating experiments (sample E).

DISCUSSION

The microprobe results on the GE I $CaCO_3$ matrix samples reacted in dry and wet CO_2 clearly suggested a different reaction occurring under the two different conditions. For the dry CO_2 atmospheres a distinct product layer of essentially constant chemical composition was observed. When the same type of sample was reacted in the presence of water vapor no product layer was observed, but a finite amount of calcium was found within the quartz particles. This observation was checked by making 10-sec counts of the calcium characteristic peak at various positions in the quartz particle for eight separate particles and comparing these counts with those on a pure quartz standard. This examination showed a minimum calcium concentration of 0.5 wt. % at the center of the quartz particles and a maximum of 12.5 wt. % just inside the particle surface.

This calcium "solution" in quartz is similar to that reported by Verduch ([11]) in his investigation of the reaction between cristobalite and calcium carbonate. Verduch suggested that a solid solution of calcium in cristobalite was the first stage of the reaction at 500–700°C. Upon reaching a critical value of approximately 0.08 % CaO the orthosilicate phase was nucleated from this solid solution. Verduch interpreted the sequence of the reaction for the SiO_2–$CaCO_3$ system as that shown in Fig. 13. However, the maximum amount of calcium in solid solution was determined by Verduch to be approximately 0.08 %, much smaller than that observed in this study. Verduch in his study either failed to consider the effect of water vapor or did not mention the effect in his paper. In addition, Verduch reacted his samples for 24 hr, which according to the results of the present study, is sufficiently long for a product layer to form. His longer reaction times may actually be responsible for the small amounts of solid solution reported.

The results of this investigation indicate the formation of a product other than the Ca_2SiO_3 phase in the initial stages of the reaction, possibly due to the presence of water adsorbed on the surface of the reactants. It is thought that the presence of water vapor promoted the formation of a metastable solid solution of calcium in quartz, as indicated by the electron microprobe results, the X-ray lattice parameter studies, and the optical micrographic evidence. Upon elimination of the bulk of the water the reaction proceeded to the formation of what is thought to be the Ca_2SiO_4 phase.

For the fine calcium carbonate matrix relatively large amounts of adsorbed water probably caused excessive solid solution. Following the elimination of water vapor the Ca_2SiO_4 layer nucleated and formed. The excessive initial reaction in this material resulted in the formation of very

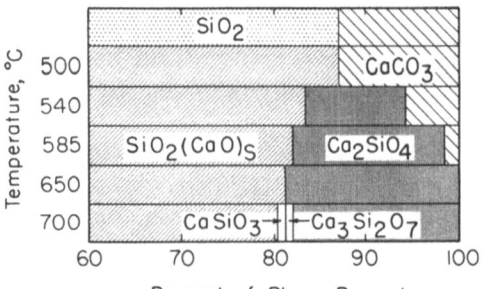

Fig. 13. Sequence of $CaCO_3$–SiO_2 reaction. After Verduch ([11]).

thick layers on the quartz particles following the elimination of the water. Any further reaction could take place only by diffusion of calcium through the reaction layer, resulting in very little additional weight loss at subsequent periods of time. The resultant weight loss curves consisted of a large initial weight loss followed by a parabolic weight loss. For the Baker and the GE II calcium carbonates the smaller surface areas of matrix material contributed small amounts of adsorbed water, thereby leading to smaller amounts of the initial reaction. The subsequent reaction for these samples followed the expected parabolic kinetics more closely.

CONCLUSIONS

1. The solid-state reaction between calcium carbonate and quartz occurs in two distinct steps, as determined by a standard thermogravimetric technique. The initial portion is believed to be a solid solution reaction of calcium into the quartz. The final stage, obeying parabolic reaction kinetics, is believed to be a diffusion-controlled process in a well-defined product layer.

2. The initial reaction is enhanced markedly in the presence of water vapor present either in the reaction atmosphere or as adsorbed water on the surface of the calcium carbonate reactant particles. This initial reaction is more extensive with the fine GE I $CaCO_3$ than with either of the coarser Baker or GE II $CaCO_3$ materials and is probably due to a larger amount of adsorbed water on the finer material.

3. The proposed solid solution of calcium into quartz, which is enhanced by water vapor, is supported by the electron microprobe, X-ray diffraction, and microstructure studies.

4. The parabolic portion of the reaction is probably controlled by the diffusion of calcium ions through a product layer which is difficult to define. The variations in activation energy are difficult to understand and may be due to a product layer which is variable in composition. The product layer is believed to be Ca_2SiO_4 in the GE II $CaCO_3$ samples. For the other two calcium carbonate reactants the product layer is believed to be primarily Ca_2SiO_4, but it may contain in addition a variable-composition amorphous phase or the crystalline phases $Ca_3Si_2O_7$ and $CaSiO_3$ richer in silica than the orthosilicate. There may be a correlation between the amount of the initial reaction and the activation energy for diffusion in the final stage.

ACKNOWLEDGMENTS

The authors gratefully acknowledge the financial support of the National Science Foundation through NSF Grant GK 215. One of the authors (M.R. M.) was financially supported as an NDEA Title IV Fellow.

REFERENCES

1. W. Dykerhoff, "Uber Bildung and Eigenschaften der Kalziumsilicate," *Zement* **14**: 3–6, 21–24 (1925).
2. I. Weyer, "Uber den Verlauf der Reaktion im Festen Zustande Zwischen Kalk und Kieselsaure," *Z. Anorg. Allg. Chem.* **209**: 409 (1932).
3. S. Nagai, "Hydrothermale Synthesen von Kalziumsilikaten," *Z. Anorg. Allg. Chem.* **106**: 177 (1932).
4. S. Nagai, "Hydrothermale Synthesen von Kalziumsilikaten, II," *Z. Anorg. Allg. Chem.* **207**: 321 (1932).
5. K. Hild and G. Tromel, "Die Reaktion von Kalziumoxyd und Kieselsaure im Festen Zustand," *Z. Anorg. Allg. Chem.* **215**: 333 (1933).
6. W. Jander and E. Hoffman, "Reactionen im Festen Zustande bei Hoheren Temperaturen," *Z. Anorg. Allg. Chem.* **218**: 211 (1934).
7. W. Jander, "Reaktionen ion Festen Zustande bie Hoheren Temperature," *Z. Anorg. Allg. Chem.* **163**: 1–30 (1927).
8. S. Kakitani and M. Fujisaka, "Solid Phase Reaction between Calcium Carbonate and Silica," *Yogoyo Kyokai Shi* **66**: 133 (1958); abstracted in *Ceramic Abstracts* (6) 173b (1959).
9. J. W. Cobb, "The Synthesis of a Glaze, Glass or Other Complex Silicate: The Interaction of Lime and Silica," *J. Soc. Chem. Ind.* **29**: 69 (1910).
10. W. E. S. Turner, W. Miskill, and G. H. Whiting, "The Reaction between Calcium Carbonate and Silica," *J. Soc. Glass Tech.* **16**: 61, 94–110 (1932).
11. A. G. Verduch, "Initial Stages of the Reaction between Cristobalite and Calcium Carbonate," *Bull. Soc. Esp. Ceram.* **3**(6): 594–602 (1964).
12. W. Komatsu, "The Kinetic Equation of the Solid State Reaction; the Effect of Particle Size and Mixing Ratio on the Reaction Rate in a Mixed Powder System," in: *Reactivity of Solids*, (R. Schwab, ed.), Am. Elsevier, New York, 1965, pp. 182–191.
13. R. Lindner, "Studies on Solid State Reactions with Radiotracers," *J. Chem. Phys.* **23**: 410 (1955).
14. C. Wagner and K. Grunewald, "Theory of the Tarnishing Process, III," *Z. Physik. Chem.* **B40**: 455–75 (1938).
15. D. Taylor, "Calcium Solid Solution in Quartz in the Presence of Water Vapor," unpublished B. S. thesis, University of Utah, 1967.

DISCUSSION

S. J. Hulbert (Clemson University): How did you make the correction for the initial heating period?

Answer: We attempted some sample heating corrections in log–log plots similar to those used in sintering studies. We found that the data at long periods of time were not significantly affected by these corrections. They amounted to a sample heating time on the order of 1–2 min for $\frac{1}{2}$-g samples. In the kinetic analysis, in which activation energies were being computed from the temperature dependence of parabolic rate constants, we were concerned with weight loss data at time periods in excess of a thousand minutes. The sample heating correction was negligible.

In the initial stages of the reaction, even with the corrections, we found no correlation with the Jander or Bronstein equations, or any other related model. We discovered that the effect of water vapor dominated the initial reaction and that a sample heating time of several minutes had little effect on the interpretation of the data.

Hulbert: How much variation was there in sample mass?

Answer: Very little. It was on the order of $\pm 15\%$. We encountered many of the same problems that you mentioned in your work. As a result, we attempted to press samples under uniform conditions of sample mass and overall dimension.

R. J. Bratton (*Westinghouse Research Laboratories*): In the wet CO_2 experiments did you find any evidence for the hydration of CaO?

Answer: No. In fact, we examined the reacted samples by X-ray diffraction in an attempt to detect the presence of CaO or $Ca(OH)_2$. No such phases were detected. In addition to X-ray analysis, temperature reductions were made following several experiments. If CaO were present, it would recarbonate with a resulting weight gain. No recarbonation effects were observed.

Bratton: The hydration of CaO is believed to occur by a solution-precipitation reaction ("through solution mechanism"). If this hydration process occurred, could the observed enhancement in the initial reaction be a result of better coverage of the silica grains by calcium ions?

Answer: It might be possible. However, the microprobe, X-ray lattice parameter, and micrographic results suggest that there is actually a difference in reaction mechanism in the dry and wet CO_2 atmospheres. These results strongly suggest a solid solution of calcium in quartz which is promoted by water vapor. We don't think these results can be explained in terms of more intimate reactant contacts.

Bratton: Did you observe fissuring in the reaction layer for the wet CO_2 experiments?

Answer: No, we didn't observe any fissuring in the $CaCO_3$ matrix samples. While we were investigating samples composed of large calcite particles in a fine-particle matrix of quartz we performed multiple temperature experiments. In these experiments the sample, after reacting at a given temperature, was removed from the furnace, the temperature was changed, and the sample was relowered into the furnace. For this type of specimen geometry the diffusion of calcium ions was in the reverse direction. Instead of diffusing into a quartz particle, they diffused out into the quartz matrix. It is probable that the product layer formed in this case would not be nearly as strong as one formed by diffusing calcium into a quartz particle. The temperature fluctuations during the cooling and reheating process probably were sufficient to fracture the product layer. As a consequence, the reaction progressed through a nonparabolic region prior to reestablishing a coherent reaction layer leading to diffusion-controlled kinetics.

H. J. Oel (*Max Planck Institute*): It has generally been observed that water vapor enhances reactions in which SiO_2 is present in one form or another and in which other oxides are involved. I am especially thinking of the corrosion of refractory materials by molten glass and slags. It seems there exists some interaction between SiO_2 and H_2O to cause this.

Answer: We have no conclusive evidence or model suggesting the mechanism by which water vapor enhances the reaction. It is known, however, that water can diffuse into silica glass and form hydroxyl groups in the lattice with the rupture of $\equiv$Si–O–S$\equiv$ bridges.[*] Perhaps water diffuses into quartz and disrupts the structure, allowing calcium to diffuse more easily into the quartz lattice.

Donald L. Branson (*University of Missouri*): I notice you conducted experiments in a CO_2 atmosphere. Was this a pure CO_2 atmosphere?

---

[*]T. Drury, G. J. Roberts, and J. P. Roberts, "Diffusion of H_2O in Silica Glass," in: *Advances in Glass Technology*, Plenum Press, New York, 1962.

Answer: In the dry CO_2 experiments the atmosphere was pure CO_2 with the possible exception of small amounts of water which were expelled from the reactants during reaction. In the wet CO_2 experiments the partial pressure of water was ~ 23 mm Hg.

Branson: I also noticed that you prepared these specimens by pressing them at 50,000 psi. Had they been pressed at 25,000 psi, would there be any difference in the kinetics of the reaction? In addition, would the reaction rate change if the powders were only loosely mixed, and would you be able to repeat the results?

Answer: We attempted mixing some loose powders and we believe that it is necessary to press them beyond a certain pressure. Perhaps 50,000 psi was excessive. We discovered early in the study that the kinetics were independent of sample pressing pressure only when it exceeded 15,000–20,000 psi. Any pressing pressure above 25,000 psi probably would be sufficient. At lower pressures the reaction kinetics were difficult to reproduce probably because of reactant contact problems.

   The main point we are attempting to illustrate, apart from the kinetic data, is the fact that considerable evidence exists for a water-vapor-enhanced solid solution of calcium in quartz.

Chapter 33

Crystallization of Barium Aluminum Borate Glasses— Rhodium as a Specific Catalyst

C. H. Greene

Glass Department, State University of New York College of Ceramics Alfred University, Alfred, New York

R. L. Wahlers

International Resistance Division T.R.W. Philadelphia, Pennsylvania

Experimental work is described on the nucleation and crystallization of various $BaO-B_2O_3-Al_2O_3$ glasses both with and without metallic nucleating agents. Previous studies of the crystallization behavior of glasses are reviewed. Experiments are reported which indicate that small additions of metallic rhodium catalyzed internal crystallization in glasses of the barium aluminum borate system. Other noble metals were studied, but none was as effective as rhodium.

INTRODUCTION

The crystallization of glass has been studied for many years. Crystallization, or devitrification, as it is frequently called, has generally been regarded as an evil to be avoided since it often ruins the mechanical and optical properties of articles made from glass. However, in 1957 Stookey [1] of the Corning Glass Works showed that by making suitable additions to a wide range of glass compositions and heat treating the resulting glass according to a suitable time and temperature schedule, materials containing a large proportion of crystals and having desirable mechanical and electrical properties could be obtained. Products harder than steel, lighter than aluminum, and much stronger than ordinary glass have been produced. The crystalline "glass ceramics" obtained by this process have much better resistance to high temperatures than the glasses from which they are produced and in some cases have remarkably low coefficients of thermal expansion.

In spite of much study, the mechanism of the nucleation of glass and its subsequent crystallization is not completely understood. It is our purpose here to review some general treatments of the process and to describe some observations on a specific case of nucleated homogeneous crystallization in which the catalyst which initiates the process appears to be quite specific.

Preliminary studies indicated that small additions of metallic rhodium to certain glasses in the $BaO-B_2O_3-Al_2O_3$ system catalyzed internal crystallization. The problem selected was a study of the nucleation and crystallization of these glasses both with and without metallic nucleating agents. It was hoped that this study might provide new information on the mechanism of glass-ceramic formation.

THE GENERAL PROCESS OF GLASS CRYSTALLIZATION

One of the first investigators to study glass and its crystallization behavior in detail was Tammann ([2]). Tammann's definition of glass as a supercooled liquid showed considerable insight into the nature of the materials which he was studying, for glasses, like supercooled liquids, are metastable with respect to one or more crystalline phases, and the X-ray diffraction patterns of glasses and liquids are similar. Another similarity is that both liquids and glasses flow when they are stressed.

Tammann showed that the crystallization of glasses can be separated into two steps, nucleation and growth. He showed that in general the rates of these steps versus temperature behave as in Fig. 1. In the high-temperature metastable zone crystals cannot be nucleated, but crystals, once formed, can grow in this region. The maximum in the rate of growth usually occurs at a higher temperature than the maximum in the nucleation rate. The existence of this high-temperature metastable zone is believed to be due to the excess solubility of very small crystals. More recently Yee and Andrews have published the same type of diagram for the crystallization of enamels.

The nucleation step in the crystallization process is usually divided into two categories. If the crystallization starts on a foreign particle or a foreign

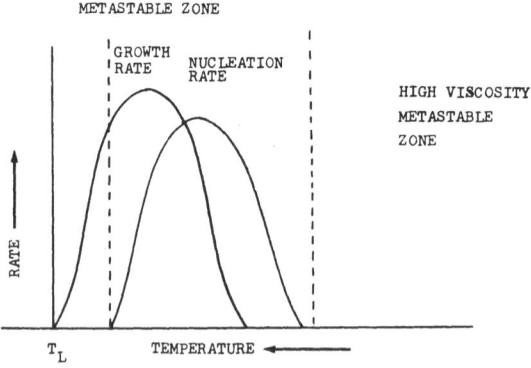

Fig. 1. Rates of nucleation and crystallization as a function of temperature.

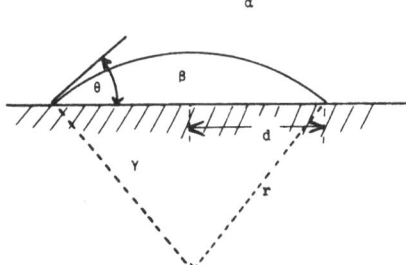

Fig. 2. Wetting of substrate by liquid.

surface (such as a container wall), the nucleation is termed heterogeneous. Crystallization from a free surface is included in this category. On the other hand, if crystallization starts in the interior of a homogeneous phase without the aid of foreign particles or surfaces, the nucleation is called homogeneous.

Volmer and Weber (^{3}) and Becker and Döring (^{4}), by theoretical treatment of the energetics and kinetics of nucleus formation derive the equation

$$I = Zns^*(\sigma \, \Delta G_v/2\pi i^* kT)^{\alpha_c/2} \exp(16\pi\sigma^3/3 \, (\Delta G_v)^2 kT)$$

for the nucleation rate I. Here Z is the collision frequency of molecules with the surface of a nucleus per unit area, n is the number of molecules per unit volume, s^* is the surface area of a critical size nucleus, σ is the interfacial free energy between nucleus and liquid, ΔG_v is the free energy of formation of the crystalline material from liquid (taken positive), i^* is the number of molecules in a nucleus of critical size, and α_c is an accommodation constant (the fraction of the molecules which "stick" when they hit a nucleus).

Volmer and Flood (^{5}) have found this equation to agree with experimental findings on the condensation of water, alcohols, and ethyl acetate from the vapor.

For homogeneous nucleation in condensed systems diffusion to the phase boundary becomes an important factor and Δf^*, the energy of activation for this diffusion, must be added to ΔF^*, the free energy of formation of a critical size nucleus, $\Delta F^* = 16\pi\sigma^3/3(\Delta G_v)^2$. A detailed treatment of this problem has been given by Turnbull and Fisher (^{6}).

For heterogeneous nucleation the simple surface energy equation for a simple spherical nucleus must be modified to take account of the interfacial energy between nucleus and catalytic substrate (see Fig. 2). This has been done by Turnbull and Vonnegut (^{7}), with experimental verification by Turnbull (^{8}). The volume energy of the transformation from liquid to crystal is also modified by the properties of the catalytic surfaces. Turnbull and Vonnegut (^{7}) incorporated this factor in their theory of heterogeneous nucleation, considering principally the decrease in free energy of the transformation due to disregistry of crystal nucleus with catalytic crystal surfaces.

After a nucleus has been formed the rate of crystal growth becomes the important factor determining the character of glass ceramics. Papers by Hillig ([9]) and Hillig and Turnbull ([10]) may be cited. Heat flow and diffusion gradients in the liquid phase enter into the problem and may lead to the dendritic crystallization considered by Chalmers ([11]) or the spherulitic crystalization considered by Keith and Padden ([12]).

In a comprehensive treatment of the nucleation and crystallization of glass, it would be necessary to consider spinodal separation into two phases by a diffusion-like process without initial formation of surfaces, and consequently without the surface energy barrier encountered in classical nucleation theory ([14]). However, this type of nucleation does not appear to enter into the crystallization we have observed in the $BaO-B_2O_3-Al_2O_3$ system.

GENERAL PROCEDURE

Glasses were made from reagent grade $BaCO_3$, H_3BO_3, and Al_2O_3. Noble metals to catalyze crystallization were added in solution, chlorides of Pt, Pd, Au, and Ir and nitrates of Ag and Rh being used. Batches were dried before melting to eliminate much of the combined water. This diminished the loss of B_2O_3 during melting. This loss was controlled by weighing the glass produced from each batch as well as by analysis of selected compositions. It varied from less than 0.1% from melts low in B_2O_3 and high in Al_2O_3 to as much as 3% from melts high in B_2O_3 and low in Al_2O_3. The concordance between weight loss and analysis was satisfactory, as shown by Table I.

Melts were made in an electric furnace with a programmed rate of heating. Figure 3 is a typical temperature program. Platinum or platinum-80–rhodium-20 crucibles were used, except for compositions where the effects of traces of noble metals were being investigated. Here dense 99.5% Al_2O_3, crucibles were used. Unfortunately, alumina dissolved in the fluid borate melts, adding from 2 to 8% of the weight of glass produced. In spite of its fluidity, the resulting glass was not completely homogeneous. It was, however,

TABLE I

Glass Composition

From weight loss			By chemical analysis		
%BaO	%Al$_2$O$_3$	%B$_2$O	%BaO	%Al$_2$O$_3$	%B$_2$O$_3$
47.1	1.0	51.0	47.1	1.0	51.0
52.0	5.0	43.0	52.1	4.9	43.0
56.5	15.5	28.0	56.6	15.7	27.7
42.0	20.0	38.0	41.9	20.1	38.0

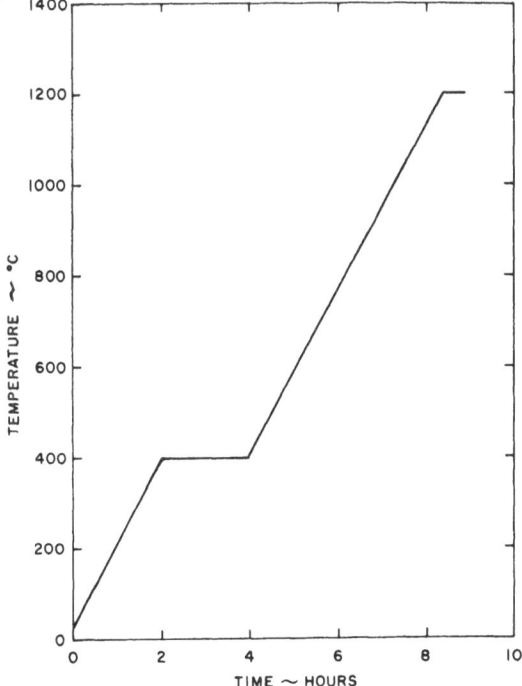

Fig. 3. Typical melting schedule.

entirely free from traces of platinum and of rhodium, which was found to be a potent catalyst for the crystallization of some compositions. Indeed, the amount of rhodium dissolved from commercial platinum crucibles was found to have a detectable effect on the crystallization of compositions in the glass ceramic region.

Phase equilibria in a triangular region shown in Fig. 4 were studied by the classical "quench" technique described by Schairer ([14]).

RESULTS

Forty four compositions were crystallized to obtain the diagram shown in Fig. 4. Detailed results of this phase equilibrium study have been described elsewhere ([15]). Three new compounds and two new polymorphic forms of $BaO \cdot B_2O_3$ and $BaO \cdot 2B_2O_3$ were discovered. One of the new compounds was a binary phase with a composition close to $2BaO \cdot 5B_2O_3$. The other two were ternary compounds, one probably being $BaO \cdot Al_2O_3 \cdot B_2O_3$.

All of these compounds crystallized from the surface of the glass. One,

β-BaO2·B$_2$O$_3$, always developed in spherulites at the surface, so that its refractive indices could not be determined.

Glasses in the region shown in Fig. 5 crystallized throughout the volume of the sample at temperatures much below the liquidus temperatures for these compositions. Two metastable phases appeared to separate at first. One was a binary compound, possibly a form of BaO·2B$_2$O$_3$, while the other contained Al$_2$O$_3$ and appeared to take Al$_2$O$_3$ or B$_2$O$_3$ into solid solution.

A trace of metallic rhodium was essential for the crystallization of these compounds throughout the volume of the sample. Samples melted in alumina crucibles did not show internal crystallization, but crystallized from the surface with separation of the same phases. Even the trace of rhodium picked up by melting in commerical platinum was sufficient to cause a small amount of internal crystallization. When more rhodium was added to the composition this internal crystallization increased, although it was necessary to keep the

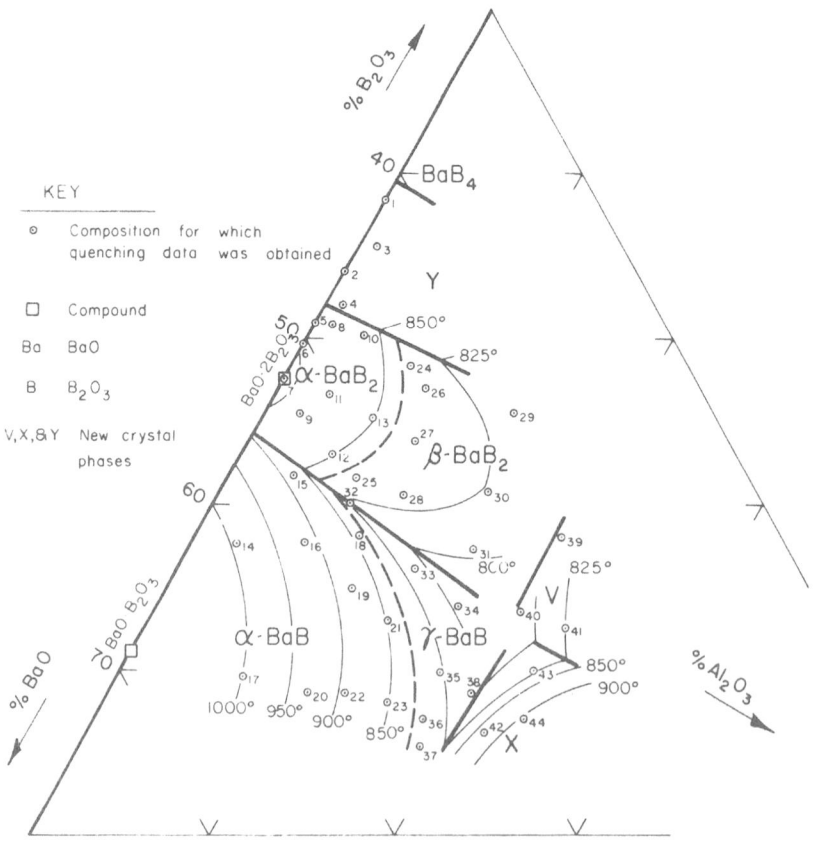

Fig. 4. Phase equilibrium diagram.

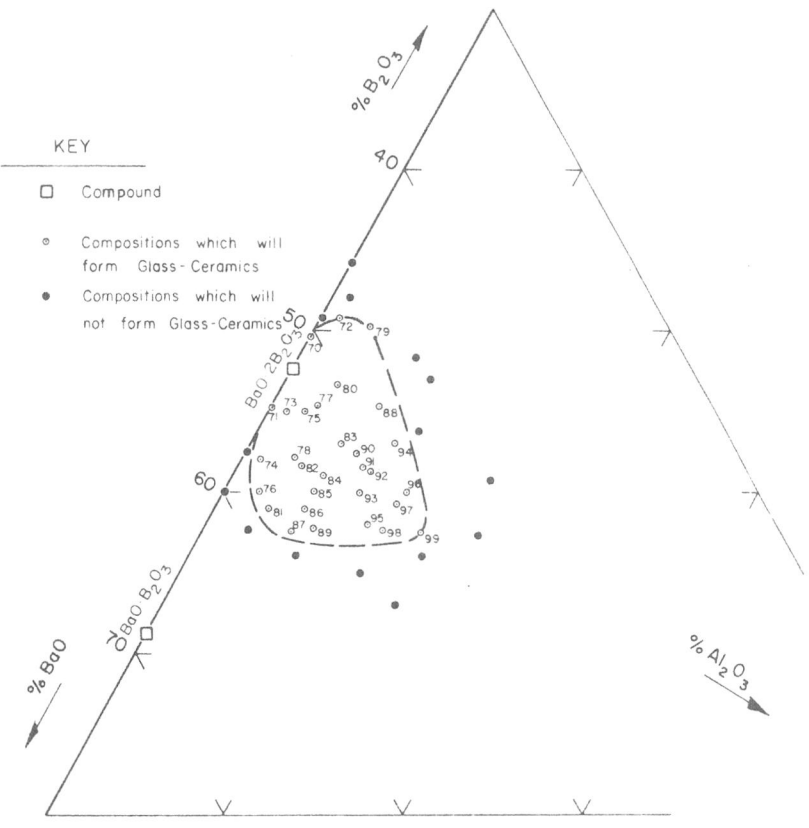

Fig. 5. Glass-ceramic region.

rhodium in the metallic state and to avoid high concentrations of dissolved water.

Prolonged melting in the oxidizing atmosphere of an electric furnace converted rhodium in these glasses to an ionized state characterized by yellow-brown color and lack of "catalytic power" to initiate bulk crystallization. On the other hand, very rapid melting to prevent the oxidation of starch which was added to reduce rhodium to the metallic state caused the retention of high levels of dissolved water in these glasses, and this also retarded the bulk crystallization desired for the production of "glass ceramic" bodies.

The initial phases separating in this "glass ceramic" region were meta-stable, and changed, after a time which decreased as the temperature increased from 650°C to 800°C, to more stable phases. In some compositions low in Al_2O_3 these were the $BaO \cdot B_2O_3$, $BaO \cdot 2B_2O_3$, and $2BaO \cdot 5B_2O_3(?)$ phases stable at the liquidus surface. These crystals grew inward from the surfaces of the glass specimens and gradually replaced the metastable crystals.

In other compositions in the "glass ceramic" region containing more Al_2O_3, $> 4\%$, the original metastable crystals were replaced by a new phase characterized by X-ray diffraction and stable at temperatures below 785 (± 15)°C. Above this temperature, 800°C, it is converted to the phases stable at the liquidus surface, but below 770°C they changed slowly into the new phase.

Electron micrographs and optical microscopy of these glasses did not show any separation of two liquid phases.

Spherulites of the metastable phases were observed to grow throughout the interior of samples in the glass ceramic region which had been nucleated with metallic rhodium. Tiny black specks were observed at the centers of some of the spherulites, but using electron diffraction techniques we did not succeed in proving that these were metallic rhodium.

A comparative study was made of noble metals other than rhodium as catalysts for this bulk "glass ceramic" type of crystallization. None of those tried, which included platinum, palladium, gold, silver, and iridium, were as effective as rhodium in causing uniform crystallization throughout the volume of the glass. This is perhaps associated with the fact that rhodium, with a lattice constant of 3.796 Å, has a smaller unit cell than any of the other metals tried, which have constants ranging from 3.831 for iridium to 4.086 for silver. Ease of oxidation to an ionic form may also enter as a factor in the poor catalytic activity of the other metals.

ACKNOWLEDGMENTS

We are most grateful to the Texas Instrument Company, which provided a graduate fellowship for one of us (R.L.W.) during work on this problem.

REFERENCES

1. S. D. Stookey, "Catalyzed Crystallization of Glass in Theory and Practice," in: *Glasstech. Ber.*, **32K**, V. Internationaler Glaskongress (fifth International Congress on Glass), Verlag der Deutschen Glastechnischen Gesellschaft, Frankfurt am Main, 1959, pp. V/1–8. S. D. Stookey (Corning Glass Works), "Method of Making Ceramics and Product Thereof," U. S. Pat. 2, 920,971, January 12, 1960.
2. G. Tammann, *Der Glaszustand* (*The Glassy State*), Leopold Voss, Leipzig, 1933. G. Tammann, *Kristallisieren und Schmelzen* (*Crystallization and Fusion*), J. A. Barth, Leipzig, 1903.
3. M. Volmer and A. Weber, "Nucleus Formation in Supersaturated Systems," *Z. Physik. Chem.* **119**: 277–301 (1926).
4. R. Becker and W. Doring, "The Kinetic Treatment of Nucleus Formation in Supersaturated Vapors," *Ann. Physik* **24**: 719–52 (1935).

5. M. Volmer and H. Flood, "Formation of Droplets in Vapors," *Z. Physik. Chem.* **A170**: 273–85 (1934).
6. D. Turnbull and J. C. Fisher, "Rate of Nucleation in Condensed Systems," *J. Chem. Phys.* **17**: 71–73 (1949).
7. D. Turnbull and B. Vonnegut, "Nucleation Catalysis," *Ind. Eng. Chem.* **44**: 1292–8 (1952).
8. D. Turnbull, "Kinetics of solidification of Supercooled Liquid-Mercury Droplets," *J. Chem. Phys.* **20**: 411–24 (1952).
9. W. B. Hillig, "The Kinetics of Freezing in Ice in the Direction Perpendicular to the Basal Plane," in: *Growth and Perfection of Crystals* (R. H. Doremus, B. W. Roberts, and D. Turnbull, eds.), John Wiley and Sons, New York, 1958, p. 356.
10. W. B. Hillig and D. Turnbull, "Theory of Crystal Growth in Undercooled Pure Liquids," *J. Chem. Phys.* **24**: 914 (1956).
11. B. Chalmers, *Principles of Solidification*, John Wiley and Sons, New York, 1964.
12. H. D. Keith and F. J. Padden, "A Phenomenological Theory of Spherulitic Crystallization," *J. Applied Phys.* **34**(8): 2409 (1963).
13. J. W. Cahn and J. E. Hilliard. *J. Chem. Phys.* **31**: 688 (1959).
14. J. F. Schairer, "Phase Equilibria with Particular Reference to Silicate Systems", in: *Physicochemical Measurements at High Temperatures* (J. O. Bockris *et al.*, eds.), Butterworth and Co., London, 1959, Chapter 5.
15. R. L. Wahlers, "Nucleation and Crystallization of Glasses in a Portion of the BaO–B_2O_3–Al_2O_3 System," Ph.D. Thesis, Alfred University, 1967.

DISCUSSION

H. J. Oel (Max Planck Institute, Würzburg): I am a little surprised that the rhodium oxide does not affect the system at all as the nucleating agent.

Answer: It seems to be totally ineffective. If you have the glass oxidized sufficiently so that the rhodium is in the yellow ionic form, the crystallization in this low-temperature region starts at the surface and grows in from the surface. The rhodium material in the glass seems to be totally ineffective. It seems to take a small particle of the metal, but particles of platinum, palladium, and gold are ineffective. I am not sure about the silver— we may have had that as dissolved silver ions. We need to do a little more work along that line.

J. A. Pask (University of California, Berkeley): What was the solubility of rhodium oxide in the glass? Did you have any amount beyond the solubility limit, and if so, did precipitation occur?

Answer: We did not precipitate the oxide of rhodium. When the rhodium was present as particles it was present as the metal under reducing conditions. We could dissolve as much as 0.1% metallic rhodium under oxidizing conditions, developing color in the glass. We studied the absorption of this color. It follows Beers' law fairly well and you can estimate the amount of rhodium from the amount of color. We did not, however, investigate how much you could put in.

We thought the most interesting thing about this was the specificity of the rhodium as a catalyst for this particular form of crystallization. I would like to find some face-centered metal with a smaller lattice constant. It happens that rhodium is at the lower end of the series of lattice constants for the series of noble metals which we put in the glass.

Author Index

Subject Index